DANA'S MANUAL OF

MINERALOGY

18TH EDITION

CORNELIUS S. HURLBUT, JR.

Professor of Mineralogy
Harvard University

JOHN WILEY & SONS, INC.

NEW YORK · LONDON · SYDNEY · TORONTO

Library of Congress Catalogue Card Number: 72-114411

ISBN 0-471-42225-8

Printed in the United States of America

10 9 8 7 6 5

PREFACE

This book, as with the preceding editions of *Dana's Manual of Mineralogy*, is designed for the beginning course in mineralogy. Since the most recent edition, advances in the science have indicated the need for inclusion in the book of new techniques and data. Adding force to these needs has been the changing character of the course itself. For example, curricular pressures at many institutions have made it impossible to offer more than one undergraduate mineralogy course. Consequently, some topics formerly considered in advanced courses are now presented in this revision for inclusion in the beginning course. The most prominent example is discussion of the elementary aspect of optical mineralogy.

The general organization of the book remains the same, with the major sections being crystallography, followed by physical, chemical, and descriptive mineralogy, in that order. In the crystallography section, I have included a short account of space groups so as to prepare students for later encounters with space group notation so common in modern mineralogical references and descriptions. The optical properties material comes in the physical mineralogy section of the book. While this treatment is far from complete, I have found with my students that it is adequate for the routine determination of optical properties by the immersion method. It is also sufficient for a basic understanding of the optical data given in mineral descriptions. The chemical mineralogy section is marked by the return from earlier editions of chemical and blow pipe tests for the elements found in the common minerals. The main thrust of this section, however, still has to do with crystal chemistry.

In the descriptive mineralogy chapter there remain descriptions of some 200 minerals. With the descriptions the following crystallographic data have been added: interfacial angles for minerals commonly in good crystals; space group data; unit cell dimensions; axial ratios for nonisometric minerals; and the d spacings of the five strongest lines of the X-ray powder photograph. Included with physical properties of the mineral descriptions are indices of refraction and other pertinent optical properties. Since the additional information more fully characterizes the individual minerals, the book is now more valuable as a reference source. This was one of my chief aims in preparing this revision. Further to document this aim, the index of refraction is added to the data given in the mineral index.

I have been fortunate, in the preparation of the manuscript for this edition, to have had the consultative assistance of several of my colleagues at Harvard; to Clifford Frondel, Charles W. Burnham, Cornelis Klein, and Donald C. Noble, I gratefully acknowledge their advice and criticism. To Dr. Joel Arem for the X-ray photographs and for his help in reducing to a few pages the material on space groups, I am also indebted.

Cornelius S. Hurlbut, Jr.

CONTENTS

DANA'S MANUAL OF MINERALOGY

INTRODUCTION

Although the emergence of mineralogy as a science is relatively recent, the practice of mineralogical arts is as old as human civilization. Tomb paintings in the Nile Valley executed nearly 5000 years ago show busy artificers weighing malachite and precious metals, smelting mineral ores, and contriving delicate gems of lapis and emerald. Minerals and products derived from minerals have figured largely in the growth of our present technologic culture, from the prized flints of Stone Age man to the uranium ores of the present-day atomic scientist. Mineral substances and products are indispensable to the welfare, health, and standard of living of modern man and are the most valued and jealously guarded of the natural resources of a nation.

In view of the age-old dependence of man on minerals for his weapons, his comforts, his adornments, and often for his pressing needs, it is surprising that many persons have only a vague idea about the nature of minerals and are unaware of the existence of a systematic science concerning them. Yet anyone who has climbed a mountain, walked on a sea beach, or worked in a garden has seen minerals in their natural occurrence. The rocks of the mountain, the sand on the beach, and the soil in the garden are completely or in large part made up of minerals. Even more familiar in everyday experience are the products made from minerals, for all articles of commerce that are inorganic, if not minerals themselves, are mineral in origin. All the common materials used in a modern building such as steel, cement, brick, glass, and plaster had their origin in minerals.

In general, we may think of minerals as the materials of which the rocks of the earth's crust are made, and, as such, minerals constitute the most important physical and tangible link with the earth's history. Because one of the goals of mineralogy is the elucidation of the physical, chemical, and historical aspects of the earth's crust, the term *mineral* and the study of mineralogy are limited to materials of *natural occurrence*. Thus, steel, cement, plaster, and glass are not regarded as minerals themselves, since they have been processed by man. A synthetic ruby, although it is chemically, structurally, and physically identical with naturally occurring ruby, is thus not a mineral. A further limitation imposed on minerals is that they be of *inorganic* origin. Thus coal, oil, amber, and the bones of animals are excluded even though they occur naturally in the earth's crust. Also the pearl of the oyster and the shell itself, although chemically and structurally identical with the minerals aragonite and calcite, are not classed as minerals.

Perhaps the most important and significant limitation placed on the definition of a mineral is that it must be a *chemical element or compound*. This restriction arises from the consistent picture of the structure of a crystalline solid as an indefinitely extended framework of atoms, ions, or groups of atoms arranged in regular geometric patterns. Such a solid must of necessity obey the laws of definite and multiple proportions and be as a whole electrically neutral; hence it must have a composition expressible by a chemical formula. Thus all mechanical mixtures, even if quite uniform and homogeneous, are eliminated.

Now that we have determined what shall be included and what excluded, we may frame a definition of a mineral as *a naturally occurring chemical element or compound formed as a product of inorganic processes*. This definition sets logical limits for the sphere of activity of the mineralogist and permits the construction of a consistent classification of minerals.

The science of mineralogy is an integrated field of study intimately related to geology on the one hand and to physics and chemistry on the other. The mineralogist may in the field map rock formations, mineral deposits, and structural features of the earth's crust, and then subject the specimens he collects to scrutiny in the laboratory, using the techniques of the chemist and physicist. It is neither necessary nor desirable to view mineralogy as compartmented into definite divisions, for the chemical, physical, and crystallographic properties are closely related and interdependent. However, to make for easier treatment in this book, the following arbitrary divisions are used: *crystallography*, *physical mineralogy*, *chemical mineralogy*, and *descriptive mineralogy*.

Although modern mineralogists share in the scientific disciplines of the physicist and the chemist and use physical and chemical techniques to construct a truer and more exact picture of the intrinsic nature of crystalline minerals, they have never forgotten that they are natural scientists whose primary purpose is the search for keys to the problems of earth history. Nor have mineralogists, in their concern with atoms and space lattices, become insensitive to the imaginative appeal of the world of ordered beauty beneath their feet.

The history of mineralogy shows that techniques and philosophy have been repeatedly and profoundly changed by the introduction of new tools and new concepts. Mineralogy today is a living growth, and we may expect similar drastic revisions of viewpoint and method in the future.

CRYSTALLOGRAPHY

INTRODUCTION

Minerals, with few exceptions, possess the internal ordered arrangement that is characteristic of the solid state. When conditions are favorable, they may be bounded by smooth plane surfaces and may assume regular geometric forms known as *crystals*. Today, most crystallographers use the term crystal in reference to any solid with ordered internal structure, regardless of whether it possesses external faces. Since bounding faces are mostly an accident of growth and their destruction in no way changes the fundamental properties of a crystal, this usage is reasonable. Thus we may frame a broader definition of a crystal as a *homogeneous solid possessing long-range three-dimensional internal order*. The study of these solid bodies and the laws that govern their growth, external shape, and internal structure is called *crystallography*. Although crystallography was developed as a branch of mineralogy, today it has become a separate science that deals not only with minerals but with all crystalline matter.

This chapter presents the essential facts and principles of crystallography most useful in elementary mineralogy. The discussion will be concerned primarily with the external geometry, or *morphology*, of well-formed crystals with only a brief introduction to the internal structure.

The general term *crystalline* is used in this book to denote the ordered arrangement of atoms in the structure; whereas the term *crystal*, without a modifier, is used in its traditional sense of a regular geometric solid bounded by smooth plane surfaces. *Crystal* is also used in its broader sense with modifiers indicating perfection of development. Thus, a crystalline solid with well-formed faces is *euhedral*; if it has imperfectly developed faces, it is *subhedral*; and without faces, *anhedral*.

Certain crystalline substances are in such fine subdivisions that the crystalline nature can be determined only with the aid of the microscope. These are designated as *microcrystalline*. Crystalline aggregates so finely divided that the individuals cannot be resolved with the microscope yet give a diffraction pattern with x-rays are *cryptocrystalline*. Although most substances, both natural and synthetic, are crystalline, a few lack any ordered internal structure. These are said to be *amorphous*. The naturally occurring amorphous substances are designated as *mineraloids*.

Crystallization. Crystals are formed from solutions, melts, and vapors. The atoms in these disordered states have a random array but with changing temperature, pressure, and concentration they may join in an ordered arrangement characteristic of the crystalline state.

As an example of crystallization from solution consider sodium chloride dissolved in water. If the water is allowed to evaporate, the solution contains more and more salt per unit volume. Ultimately, the point is reached when the remaining water can no longer hold all the salt in solution, and solid salt begins to precipitate. If the evaporation of the water is very slow, the ions of sodium and chlorine will group themselves together to form a crystal or a few crystals with characteristic shapes. If evaporation is rapid, many centers of crystallization will be set up and the resulting crystals will be small.

Crystals also form from solution by lowering the temperature or pressure. Hot water will dissolve slightly more salt, for instance, than cold water; and, if a hot solution is allowed to cool, a point will be reached where the solution becomes supersaturated and salt will crystallize. Again, the higher the pressure, the more salt water can hold in solution. Thus, with lowering the pressure of a saturated solution, supersaturation will result and crystals will form. Therefore, in general, crystals may form from a solution by evaporation of the solvent, by lowering the temperature, or by decreasing the pressure.

A crystal is formed from a melt in much the same way as from a solution. The most familiar example of crystallization from fusion is the formation of ice crystals when water freezes. Although it is not ordinarily considered in this way, water is fused ice. When the temperature is lowered sufficiently, the particles of water which were free to move in any direction now become fixed and arrange themselves in a definite order to build up a solid, crystalline mass. The formation of igneous rocks from molten magmas, though more complicated, is similar to the freezing of water. In the magma there are many elements in a dissociated state. As the magma cools the various ions are attracted to one another to form crystal nuclei of the different minerals. Crystallization proceeds with the addition of more ions in the same ratios to form the mineral particles of the resulting rock.

Although crystallization from a vapor is less common than from a solution or a melt, the underlying principles are much the same. As the vapor is cooled, the dissociated atoms or molecules are brought closer together, eventually locking themselves into a crystalline solid. The most familiar example of this mode of crystallization is the formation of snowflakes from air laden with water vapor.

INTERNAL STRUCTURE OF CRYSTALS

The forces that bind atoms together in crystals cause the atoms to assume specific geometrical arrangements called "motifs." The shape of such a unit depends on the number and types of atoms involved. However, when a crystal forms from these units, additional forces demand positioning of the motifs at specific distances and directions from one another. The result is an almost infinite repetition of the motif in three dimensions. The repeat mechanisms have been likened to those found in wallpaper, in which, by repetition, a motif, or fundamental pattern unit, builds up a two-dimensional pattern of indefinite extent.

In crystals the repeat distances along each of the three axes of coordinate space may be all the same, or all different. Also, the directions along which repetition occurs may be orthogonal (all at right angles to one another) or non-orthogonal. In any case, if the motif is replaced by a single point, the result is a three-dimensional periodic array of points, which define a *lattice*. Lattices are frameworks on which crystals are built, and the *unit cell* is the smallest parallelepiped outlined by a lattice. It has been proved that only fourteen types of space lattices are possible. Other arrangements do not obey the restrictions that (a) the environment about all lattice points must be identical, and (b) the unit cells outlined must fill all space, leaving no "holes." In 1842 Frankenheim described fifteen possible arrangements, but six years later Bravais pointed out that two of these were identical. Thus the fourteen space lattices are known as the *Bravais lattices* (Fig. 1). The simplest units of the lattices, illustrated in Fig. 1, differ from each other in the lengths of the edges and the angles (α, β, and γ) between edges. Some, with lattice points only at the corners, are called *primitive*, each containing one unit of pattern. Others with additional points at face centers or along body diagonals are *multiple cells* containing 2, 3, or 4 units of pattern.

The restrictions placed on space lattices impose additional restrictions on the motifs that populate the lattice points. The motifs themselves must outline units that fill all space. For example, 5-fold arrangements cannot exist in crystals for they will not pack together without leaving spaces between. It has been shown that

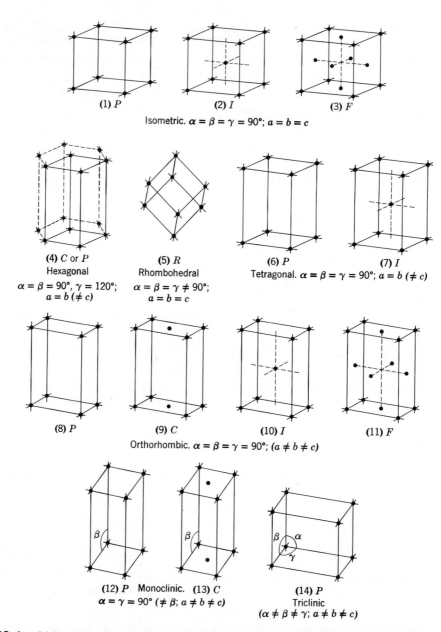

FIG. 1. **14 Bravais space lattices.** Each lattice type has its own constraints regarding length of edges a, b, and c and angles between edges, α, β, and γ. In the above notations the nonequivalence of angles or edges that usually exist but are not mandatory are set off by parentheses.

there are a limited number of ways of arranging objects, such as atoms of a motif, about a point (in our example, a lattice point). Only 32 such ways exist; these are known as the 32 *point groups*.

When the 32 possible ways of arranging objects about a point are combined with the 14 Bravais lattices, the result is 230 possible types of arrangements of objects in space, the so-called *space groups*. When the objects are atoms, such an arrangement constitutes a crystal structure. The motifs repeated at lattice nodes may be atoms, molecules, or ionic clusters and may vary extensively with atom type and bonding scheme. Thus while the number of arrangement types is limited (230), the number of possible crystal structures is much larger.

In some minerals, such as the native elements, the atoms are uncharged, but more commonly they carry electrical charges and are called *ions*. Positively charged ions are *cations*; negatively charged ions are *anions*. Most minerals are made of ions or clusters of ions held together by the electrical forces that arise between oppositely charged bodies.

The unit cell can never be as small as the individual atom, since the relations of the atoms to each other and the forces binding them together are important factors in determining the properties of the crystal. The number of atoms in a unit cell is generally some whole number or a multiple of the number shown by the simplest chemical formula. Thus, in quartz the structural unit has been shown to be $3(SiO_2)$, in halite it has been shown to be $4(NaCl)$. Any smaller subdivisions would not have the properties of the minerals.

PROPERTIES OF CRYSTALS

Outward Form.   Since crystals are formed by the repetition in three dimensions of a unit of structure, the limiting surfaces depend in part on the shape of the unit. They also depend on the environment in which the crystal grows. Environment as used here includes all the external influences such as temperature, pressure, nature of solution, and direction of movement of the solution.

If a cubic unit cell is repeated in three dimensions to build up a crystal having n units along each edge, the crystal will be a cube containing n^3 units. With the same orderly repeat mechanism, different shapes may result. If, for example, the stacking yields n units along edges in two directions but $n/5$ units in the third direction, the crystal will be tabular; and stacking of n units in one direction and $n/5$ in the other directions yields an elongated crystal (Fig. 2a). By the systematic omission of units because of environmental conditions, a variety of external forms results as the outward expression of the same internal structure. This is illustrated by the cube, the octahedron (Fig. 2b), and the dodecahedron (Fig. 2c).

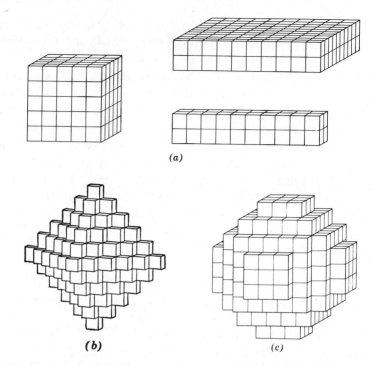

(a)

(b)   (c)

FIG. 2. **Different shapes from stacking of small cubic units. (a) Cube and malformed cubes. (b) Octahedron. (c) Dodecahedron.**

The octahedral and dodecahedral forms are common on many crystals, but the building units are so small that the resulting faces are smooth plane surfaces.

With a given internal structure, there are a limited number of planes that serve to bound a crystal, and only a comparatively few are common. In considering the distribution of faces on a crystal, we are concerned only with the arrangement of the structural units which can be represented diagrammatically by lattice points or nodes. The position of crystal faces is determined by those directions through the structure having a high density of these nodes. The frequency with which a given face is observed on a crystal is roughly proportional to the number of nodes lying in it; the larger the number, the more common the face. Consider Fig. 3 which represents one layer of nodes in a cubic crystal lattice. The nodes are equally spaced from one another and have a rectilinear arrangement. It will be noted that there are several possible lines through this network that include a greater or lesser number of nodes. These lines represent the trace on this section of possible crystal planes; and it is found that of these possible planes those that include the largest number of lattice points, those cutting along *AB* and *AC*, are most common. The above rule, known as the *law of Bravais*, is generally confirmed

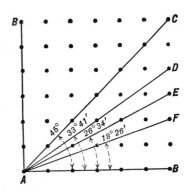

FIG. 3. Layer of nodes in a crystal lattice.

by observations. Although there are exceptions to the law, as pointed out by Donnay and Harker in 1937, it is usually possible to choose the lattice in such a way that the rule holds true.

Since the internal structure of any crystalline substance is constant and since the crystal faces have a definite relationship to that structure, it follows that the faces must also have a definite relationship to each other. This fact was observed in 1669 by Nicolaus Steno who pointed out that the angles between corresponding faces on crystals of quartz were always the same. This observation is generalized today as Steno's law of the constancy of interfacial angles which states: *the angles between equivalent faces of crystals of the same substance, measured at the same temperature, are constant.* For this reason crystal morphology is frequently a valuable tool in mineral identification. A mineral may be found in crystals of widely varying shapes and sizes, but the angles between pairs of corresponding faces are always the same.

Because crystals possess regular orderly internal structure, different planes and directions within them have different atomic environments. Consider Fig. 4, a photograph of a model illustrating the packing of atoms in sodium chloride, the mineral halite. It will be seen that any plane parallel to the front face of the cubic model contains alternating atoms of sodium and chlorine along directions parallel to the edges, and alternating strings of sodium and chlorine atoms parallel to the diagonals. On the other hand, planes cutting the corner from the cube, as shown in Fig. 4, contain either all sodium atoms or all chlorine atoms, forming rather widely spaced sheets. Planes through the crystal cutting edges from the cube and inclined 45° to the cube faces contain rather widely separated strings of alternating sodium and chlorine atoms parallel to the cube edges.

These different atomic arrangements along different crystal planes or directions give rise to *vectorial properties.* Since the magnitude of the property is dependent on direction, it differs with changing crystallographic direction. Some of the vectorial properties of crystals are hardness, conductivity for heat and electricity,

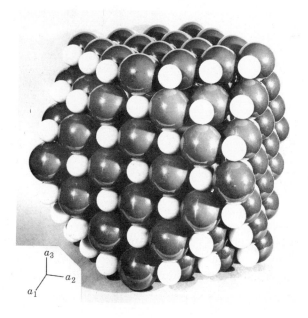

FIG. 4. Halite, NaCl, packing model, cubo-octahedron. Na white, Cl dark. Note that the {001} planes consist of sheets having equal numbers of Na and Cl ions, whereas the {111} planes consist of alternating sheets of Na ions and sheets of Cl ions. Na and Cl are both in 6 coordination in a face-centered lattice. This structure is also found in galena PbS, MgO and many other AX compounds.

thermal expansion, speed of light, growth rate, solution rate, diffraction of x-rays, and many others.

Of these properties, some vary continuously with direction within the crystal; that is, any direction chosen at random can be said to have a characteristic magnitude for the property being considered. Such properties could be represented graphically by a smoothly rounded solid whose surface is everywhere at a distance from the center proportional to the magnitude of the property. Hardness, electrical and heat conductivity, thermal expansion, and the speed of light in the crystal are all examples of such *continuous vectorial properties*.

The *hardness* of some crystals varies so greatly with crystallographic direction that the difference may be detected by simple scratch tests. Thus, kyanite, a mineral that characteristically forms elongate bladed crystals, may be scratched with an ordinary pocketknife in a direction parallel to the elongation, but it cannot be scratched by the knife perpendicular to the elongation. Some directions in the diamond crystal are much harder than others. When diamond dust is used for cutting or grinding, a certain percentage of the grains always presents the hardest surface and, hence, is capable of cutting along planes in the crystal of lesser hardness. If a perfect sphere cut from a crystal is placed in a cylinder with abrasive and tumbled for a long time, the softer portions of the crystal wear away most rapidly. The nonspherical solid resulting serves as a hardness model for the substance being tested.

The directional character of *electrical conductivity* is of great importance in the manufacture of silicon and germanium diodes, tiny bits of silicon and germanium crystals used to rectify alternating current. In order to obtain the optimum rectifying effect, the small bit of semimetal must be oriented crystallographically, since the conduction of electricity through such crystals varies greatly with orientation.

Ball bearings of synthetic ruby sound very attractive, as the great hardness of ruby would cut down wear and give long life to the bearing. However, when heated, ruby expands vectorially, and ballbearing ruby balls would rapidly become nonspherical with the rise of temperature from friction during operation. However, since the *thermal expansion* figure of ruby is an ellipsoid of revolution with a circular cross section, roller bearings are practical. Most minerals have unequal coefficients of thermal expansion in different directions, leading to poor resistance to thermal shock and easy cracking with heating or cooling. Quartz glass, which has a much less regular internal structure than quartz itself, is more resistant to thermal shock than the mineral.

The *velocity of light* in all transparent crystals except those which are isotropic (see page 144) varies continuously with crystallographic direction. Of all the vectorial properties of crystals, it is most easily determined quantitatively and is expressed as the index of refraction, the reciprocal of the velocity.

Discontinuous vectorial properties, on the other hand, pertain only to certain definite planes or directions within the crystal and may not be represented by a smoothly rounded solid. There are no intermediate values of such properties connected with intermediate crystallographic directions. An example of such a property is *rate of growth*. The rate of growth of a plane in a crystal is intimately connected with the density of lattice points in the plane. We saw that a plane such as *AB* in Fig. 3 has a much greater density of points than plane *AD*, *AE*, or *AF*. Calculations of the energy involved indicate that the energy of particles in a plane such as *AB*, in which there is a high density of points, is less than the energy of particles in less densely populated planes such as *AF*. Hence, the plane *AB* will be the most stable, since in the process of crystallization the configuration of lowest energy is that of maximum stability. Planes *AF*, *AD*, *AE*, etc., will however be faster growing than *AB* since fewer particles need to be added per unit area. In the growth of a crystal from a nucleus, the early forms that appear will be those of relatively high energy and rapid growth. Continued addition of material to these planes will build them out while the less rapidly growing planes lag behind. Thus, the edges and corners of a cube may be built out by addition of material to the planes cutting off the corners and edges, while little material is added to the cube faces. As growth progresses, the rapidly growing faces disappear, literally growing themselves out of existence, building the slower-growing, more stable forms in the process. After this stage is complete, growth is much slower, as addition is now entirely to the slower-growing, lowest-energy form. Thus, crystals themselves, if taken at various stages of their development, serve as models of the rate of growth for the compound being studied.

The rate of solution of a crystal in a chemical solvent that attacks it is likewise a discontinuous vectorial process, and solution of a crystal or of any fragment of a single crystal may yield a more or less definite solution polyhedron. A clearer illustration of the vectorial nature of the rate of solution is afforded by etch pits. If a crystal is briefly treated with a chemical solvent that attacks it, the faces are etched or pitted. The shape of these pits is regular and depends on the structure of the crystal, the face being attacked, and the nature of the solvent. Valuable information about the internal geometry of arrangement of crystals may be obtained from a study of such etch pits.

Cleavage may be thought of as a discontinuous vectorial property and, like crystal form, reflects the internal structure, since cleavage always takes place along those planes across which there exist the weakest electrical forces. Those planes are generally the most widely spaced and the most densely populated.

CRYSTAL SYMMETRY

Point Group Symmetry. On close examination most crystals will show a regularity or symmetry of arrangement of similar faces and angles, and by performing certain operations on them the like faces and angles can be brought into coincidence. There are three types of symmetry operations that characterize point groups and that can be detected by morphology. They are (1) reflection across a plane, (2) rotation about an axis, and (3) inversion through a center.

Symmetry Plane. A symmetry plane is an imaginary plane that divides a crystal into halves, each of which, in a perfectly developed crystal, is the mirror image of the other. The shaded portion of Fig. 5 illustrates the nature and position

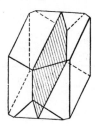

FIG. 5. Symmetry plane.

of a plane of symmetry. For each face, edge, or point on one side of the plane there is a corresponding face, edge, or point in a similar position on the other side of the plane.

Symmetry Axis of Rotation. A symmetry axis is an imaginary line through a crystal about which the crystal may be rotated and repeat itself in appearance two or more times during a complete rotation. In Fig. 6, the line CC' is an axis of symmetry for, when rotated about it, the crystal will have, after a rotation of 180°, the same appearance as at first; or, in other words, similar planes, edges, and solid angles will appear in the places of the corresponding planes, edges, and solid angles of the original position. Since the crystal is repeated twice during a complete rotation, the axis is one of 2-fold symmetry. In addition, crystals may have rotation axes of 3-fold (trigonal), 4-fold (tetragonal), and 6-fold (hexagonal) symmetry. Assuming the concept of crystal lattices, no symmetry axes other than 1-, 2-, 3-, 4-, and 6-fold can exist.

Symmetry Center. A crystal is said to have a center of symmetry if an imaginary line can be passed from any point on its surface through its center and a similar point is found on the line at an equal distance beyond the center. This operation is known as *inversion*. The crystal in Fig. 7 thus has a center of symmetry, for the point A is repeated at A' on the line passing from A through the center, C; and $AC = A'C$. Similar and parallel faces on opposite sides of the crystal indicate a symmetry center.

Symmetry Axis of Rotary Inversion. This composite symmetry element combines a rotation about an axis with inversion through the center. Both operations must be completed before the new position is obtained. If the only symmetry possessed by a crystal is a center, the symmetry notation is usually given as a 1-fold axis of rotary inversion. There are also 2-, 3-, 4-, and 6-fold axes of rotary inversion. Let us consider the mechanism of an axis of rotary inversion. In the operation of a 4-fold rotation axis, four identical points will be located—one each 90° of rotation—all at the top or all at the bottom of the

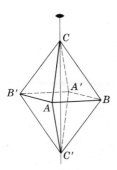

FIG. 6. Symmetry axis.

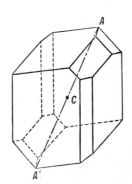

FIG. 7. Symmetry center.

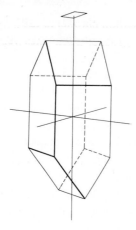

FIG. 8.   Four-fold axis of rotary inversion.

crystal. In the operation of a 4-fold axis of rotary inversion, four identical points will also be located, but two will be at the top and two at the bottom of the crystal. The operation of such an axis involves four rotations of 90°, each followed by an inversion. Thus, if the first point is at the top of the crystal, the second is at the bottom, the third at the top, and the fourth at the bottom as illustrated in Fig. 8.

Space Group Symmetry. Space group theory was developed, as a kind of mathematical exercise, long before anyone realized that real crystals obeyed its restrictions. In fact, the theory was derived independently and at about the same time (1880–1890) by Schoenflies (in Germany), Federov (in Russia), and Barlow (in England). It was soon realized that allowing point groups and space lattices to interact created two new elements of symmetry. These arose as a consequence of allowing the objects within a motif, possessing only inherent point symmetry, to be repeated by the repetition scheme characteristic of lattices, that is, translation.

What happens when a symmetry plane is combined with a lattice translation? Think of two objects oppositely disposed (reflected) across such a plane and allow them to be displaced from one another, parallel to the plane, one-half a lattice translation. Such "zig-zag" symmetries of reflection-plus-translation are unique to space group symmetries. Point groups, not allowing for translation, are devoid of them. They are known as *glide planes* (Fig. 9) and are described in terms

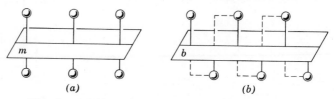

(a)   *(b)*

FIG. 9.   (a) Symmetry plane, *m*. (b) Glide plane *b*.

of the *orientation* of the plane and the *direction* of translation parallel to the plane.

Now consider the rotation axes typical of point groups. Such axes repeat objects in a "ring," "depositing" copies of the original at fixed angular intervals as every $90°$ (4-fold), every 60 degrees (6-fold), etc. But all the "copies" and the original object lie in a plane perpendicular to the axis. If these objects are now allowed to translate parallel to the axis, the result is a "spiral" form of repetition and the axis is known as a *screw axis* (Fig. 10). As with any screw motion, the axis is right- or left-handed, and the spiral has a specific pitch. The axis is described in terms of the rotational aspect of the symmetry and the unit of translation parallel to the axis. Screw axes are found only in space groups.

It is important to the development of space group theory to recognize that: *space group and point group operations are identical, except for the presence of translations in space groups.* A mirror plane plus a translation produces a glide plane; a rotation axis, plus translation parallel to the axis, results in a screw axis.

The earliest geometrical theories of crystallography, based on crystal morphology produced point group theory. The sublattice translations of glide planes and screw axes were detected only through their x-ray effects.

Summary. A crystal lattice is a repetitive array of points outlining a fundamental unit of space, the unit cell. In real crystals at the points (lattice nodes) are "hung" the motifs of atoms, ions, or molecules, which have point symmetry. The combined arrangement, now capable of translation due to glide planes and screw axes, is called a space group.

Symmetry Notation of Point Groups. The symmetry axis, plane, axis of rotary inversion, and center are known as the *elements of symmetry*. To express them in describing the symmetry of a crystal several schemes of notation have been proposed. The easiest for the beginning student to understand is as follows: A rotation axis is indicated by A_n, where n is 2, 3, 4, or 6; a plane by P; and a center by C. Using these notations, the symmetry of a crystal (Fig. 11) with a

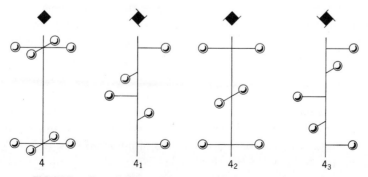

FIG. 10.   Four-fold rotation axis and 4-fold screw axis.

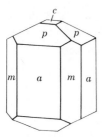

FIG. 11. Idocrase crystal.

symmetry center, one 4-fold rotation axis, four 2-fold rotation axes, and five planes of symmetry would be written: C, $1A_4$, $4A_2$, $5P$. In this way it is possible to express all but two possible symmetry conditions. One of these is the case of no symmetry; the other, illustrated by Fig. 8, is an axis of 4-fold rotary inversion. It is represented by the symbol $\mathcal{P}_4$.

Of the many proposed symmetry notations, the *Hermann-Mauguin symbols* have been most widely accepted by crystallographers, and thus their use is essentially universal. By means of them, one can express not only the outward symmetry (point group) but also the internal symmetry (space group). Using the Hermann-Mauguin symbols, the symmetry elements are indicated as follows: Axes of rotation are denoted by numbers 1, 2, 3, 4, 6; and axes of rotary inversion by numbers with lines above, as $\bar{1}, \bar{3}, \bar{4}, \bar{6}$ (read: bar one, bar three, etc.). Symmetry planes are indicated by m (mirror plane). An axis of symmetry with a symmetry plane normal to it is given as a number over m; for example, $2/m$, $4/m$. Thus, the symmetry of the crystal illustrated in Fig. 11 and given above as C, $1A_4$, $4A_2$, $5P$ can be expressed as $4/m2/m2/m$.

At first glance it may appear that the Hermann-Mauguin symbols do not express the symmetry completely. However, it should be pointed out that symmetry elements operate on one another as well as on crystal faces. Consider the symbol $4/m2/m2/m$. The 4-fold axis is vertical, at right angles to a horizontal symmetry plane. The 2-fold axes lie in a horizontal plane at 45° to each other with symmetry planes normal to them. Therefore, operating with the 4-fold axis generates two additional 2-fold axes and two additional symmetry planes.

The Hermann-Mauguin symbols used here are complete, but some crystallographers use more abbreviated symbols. For example, $4/m2/m2/m$ can be expressed as $4/mmm$; for 2-fold axes at the intersections of the horizontal plane and vertical planes are implied. The symbol $2/m2/m2/m$ may be shortened to mmm. The shorter symbol indicates three mutually perpendicular symmetry planes which implies 2-fold symmetry axes at their lines of intersection.

Symmetry Notation of Space Groups. Since space groups are generated by combining point groups with space lattices, their symbols must indicate the lattice type. This is done by beginning the symbol with a letter as: *P* (primitive), *I* (body centered), *F* (face centered).

Otherwise, space group symbols are exactly like point group symbols, but with some indication of translation components which may be present. It is important to note that the symmetry elements of some space groups do not have additional translation components. For example, the point group $2/m2/m2/m$ has mirror planes perpendicular to the 2-fold axes. The space group $P2/m2/m2/m$ also exists, and the corresponding symmetry elements in the two groups are identical. But, since translations *are* allowed in space groups, objects related by mirror planes in $2/m2/m2/m$ may be displaced in directions parallel to the planes. Thus, we might have $P2/b2/a2/m$ (sometimes seen shortened to *Pbam*), where objects originally reflected across mirror planes perpendicular to the *a*- and *b*- axes are now displaced in the *b*- and *a*- axis directions.[1] This is clarified in Fig. 9.

A critical point is that angular relationships are unchanged in going from point groups to space groups. Rotation axes, in spite of the possibility of translations parallel to the axis, are still 2-, 3-, 4-, and 6-fold. Thus, it is a trivial task to determine the point group "isogonal" with a given space group—merely strip off the translation components. Thus, in our example $P2/b2/a2/m$ is isogonal with $2/m2/m2/m$, when the *a*- and *b*- glide planes are replaced by mirror planes (which reflect without translating). It is also obvious that any given point group can have several isogonal space groups; for example isogonal with $2/m2/m2/m$ are:

$$P2/n2/n2/n \qquad P2/c2/c2/m \qquad P2/b2/c2/m$$
$$P2/b2/a2/n \qquad P2/n2/n2/a \qquad P2/b2/c2/n$$

The screw axes, 6_1, 6_2, 6_3, 6_4, and 6_5 are all isogonal with a 6-fold rotation axis—they all rotate objects by 60 degrees. They also respectively translate objects a distance of $\frac{1}{6}$, $\frac{1}{3}$, $\frac{1}{2}$, $\frac{2}{3}$, and $\frac{5}{6}$ the length of the axial repeat of the lattice. Remove the subscript from the screw axis symbol to obtain the isogonal point group symmetry element.

Symmetry Classes or Point Groups. It has been shown (page 7) that, with the restrictions placed by space lattices, there are only 32 ways of arranging objects about a point. This point group symmetry is the same as exhibited by crystal morphology and can be expressed in terms of planes and axes of rotation and inversion. The 32 combinations of these symmetry elements are also called the *crystal classes*. Since point groups and crystal classes possess the same symmetry they are often used interchangeably. On page 18 the crystal classes are listed with their symmetry notations; the fifteen most important to the mineralogist are in boldface type.

[1] See under Crystal Notation page 19 for explanation of *a*, *b*, and *c*.

The Thirty-Two Crystal Classes

Crystal System	Crystal Class	Hermann-Mauguin Symbols	Symmetry
Isometric *Cubic*	**Hexoctahedral**	$4/m\bar{3}2/m$	$C, 3A_4, 4A_3, 6A_2, 9P$
	Gyroidal	432	$3A_4, 4A_3, 6A_2$
	Hextetrahedral	$\bar{4}3m$	$3A_2, 4A_3, 6P$
	Diploidal	$2/m\bar{3}$	$C, 3A_2, 4A_3, 3P$
	Tetartoidal	23	$3A_2, 4A_3$
Hexagonal Hexagonal division	**Dihexagonal-dipyramidal**	$6/m2/m2/m$	$C, 1A_6, 6A_2, 7P$
	Hexagonal-trapezohedral	622	$1A_6, 6A_2$
	Dihexagonal-pyramidal	$6mm$	$1A_6, 6P$
	Ditrigonal-dipyramidal	$\bar{6}m2$	$1A_3, 3A_2, 4P$
	Hexagonal-dipyramidal	$6/m$	$C, 1A_6, 1P$
	Hexagonal-pyramidal	6	$1A_6$
	Trigonal-dipyramidal	$\bar{6}$	$1A_3, 1P$
Hexagonal Rhombohedral division	**Hexagonal-scalenohedral**	$\bar{3}2/m$	$C, 1A_3, 3A_2, 3P$
	Trigonal-trapezohedral	32	$1A_3, 3A_2$
	Ditrigonal-pyramidal	$3m$	$1A_3, 3P$
	Rhombohedral	$\bar{3}$	$C, 1A_3$
	Trigonal-pyramidal	3	$1A_3$
Tetragonal	**Ditetragonal-dipyramidal**	$4/m2/m2/m$	$C, 1A_4, 4A_2, 5P$
	Tetragonal-trapezohedral	422	$1A_4, 4A_2$
	Ditetragonal-pyramidal	$4mm$	$1A_4, 4P$
	Tetragonal-scalenohedral	$\bar{4}2m$	$3A_2, 2P$
	Tetragonal-dipyramidal	$4/m$	$C, 1A_4, 1P$
	Tetragonal-pyramidal	4	$1A_4$
	Tetragonal-disphenoidal	$\bar{4}$	$1 \bar{A}_4$
Orthorhombic	**Rhombic-dipyramidal**	$2/m2/m2/m$	$C, 3A_2, 3P$
	Rhombic-disphenoidal	222	$3A_2$
	Rhombic-pyramidal	$mm2$	$1A_2, 2P$
Monoclinic	**Prismatic**	$2/m$	$C, 1A_2, 1P$
	Sphenoidal	2	$1A_2$
	Domatic	m	$1P$
Triclinic	**Pinacoidal**	$\bar{1}$	C
	Pedial	1	No symmetry

(handwritten annotation: "Common symbol")

Many workers have suggested names for the crystal classes, but those most generally accepted and used here were proposed by Groth in 1895. They are derived from the name of the general form in each crystal class (see page 24). In the mineral descriptions of this book the class is designated by the Hermann-Mauguin symbol and the name omitted.

CRYSTAL NOTATION

Crystallographic Axes. It is found convenient in describing crystals to assume, after the methods of analytic geometry, certain lines passing through the crystal as axes of reference. These imaginary lines, the *crystallographic axes*, are taken parallel to the intersection edges of major crystal faces. In addition, the positions of the crystallographic axes are more or less fixed by the symmetry of the crystals, for in most crystals they are symmetry axes or normals to symmetry planes. For some crystals there may be more than one choice of crystallographic axes when the selection is made by morphology alone. Ideally they should be parallel to and their lengths proportional to the edges of the unit cell.

All crystals, with the exception of those belonging to the hexagonal system (see page 56), are referred to three crystallographic axes designated as a, b, and c. In the general case (triclinic system), all the axes are of different lengths and at oblique angles to one another; but for simplicity in describing their orientation consider those illustrated in Fig. 12. Here three axes of different lengths are mutually perpendicular and when placed in the conventional orientation are as follows: The a axis is horizontal and in a front-back position; the b axis is

Summary of Space Group Operations

Glide Plane	Symbol	Translation Component	Comments
Axial	a	$a/2$	
Axial	b	$b/2$	
Axial	c	$c/2$	
Diagonal	n	$a/2 + b/2$, $b/2 + c/2$ or $c/2 + a/2$	
Diamond	d	$a/4 + b/4$, $b/4 + c/4$ or $a/4 + c/4$	

Screw Axis	Symbol	Translation Component	
6-fold	6_1	1/6	Only in hexagonal
	6_2	1/3	space groups
	6_3	1/2	
	6_4	2/3	
	6_5	5/6	
3-fold	3_1	1/3	Only in hexagonal or
	3_2	2/3	rhombohedral space groups
4-fold	4_1	1/4	Only in tetragonal
	4_2	1/2	or isometric
	4_3	3/4	space groups
2-fold	2_1	1/2	

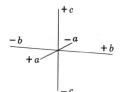

FIG. 12.   Orthorhombic crystal axes.

horizontal and in a right-left position; the *c* axis is vertical. The ends of each axis are designated plus or minus; the front end of *a*, the right-hand end of *b*, and the upper end of *c* are positive; the opposite ends are negative.

Crystal Systems.   Certain of the thirty-two crystal classes have symmetry characteristics in common with others which permit their assignment to six larger groups called *crystal systems.* The crystal systems are listed below with the crystallographic axes and the characterizing symmetry for each.

Isometric System.   Crystals in the isometric system have four 3-fold symmetry axes and are referred to three mutually perpendicular axes of equal lengths.

Hexagonal System.   Crystals in the hexagonal system have either a single 3-fold or 6-fold symmetry axis. They are referred to four crystallographic axes; three equal horizontal axes intersect at angles of 120°, the fourth is of different length and perpendicular to the plane of the other three.

Tetragonal System.   A unique 4-fold symmetry axis characterizes crystals of the tetragonal system. Crystals are referred to three mutually perpendicular axes; the two horizontal axes are of equal length, but the vertical axis is either shorter or longer than the other two.

Orthorhombic System.   Crystals in the orthorhombic system have three 2-fold symmetry elements, that is, symmetry planes or 2-fold symmetry axes. They are referred to three mutually perpendicular axes, all of different length.

Monoclinic System.   Crystals of the monoclinic system are characterized by a single 2-fold symmetry axis or a single symmetry plane or a combination of a 2-fold axis and a symmetry plane. The crystals are referred to three unequal axes, two of which are inclined to each other at an oblique angle and the third is perpendicular to the plane of the other two.

Triclinic System.   Crystals in the triclinic system have a 1-fold symmetry axis as their only symmetry. This may be a simple rotary axis or an axis of rotary inversion. The crystals are referred to three unequal axes all intersecting at oblique angles.

Axial Ratio.   In all the crystal systems, with the exception of the isometric, there are crystallographic axes differing in length. If it were possible to isolate a unit cell and measure the dimensions along the edges, which are parallel to the crystallographic axes, we would be able to write ratios between the edge lengths. The x-ray crystallographer cannot isolate the cell, but he can measure accurately the cell dimensions in angstrom units, (Å).[2] Thus for the orthorhombic mineral sulfur the cell dimensions are given as: $a = 10.45$ Å, $b = 12.84$ Å, and $c = 24.37$ Å (Fig. 13). Setting the b value equal to 1, we can write the ratios $a:b:c = 0.814:1:1.898$. The ratios thus express the relative, not the absolute, lengths of the cell edges that correspond to the crystallographic axes.

We saw, on page 8, that the number of crystal planes appearing as faces is limited and that the angular relation between them is dependent on the unit cell, that is, the length of its edges and the angles between the edges. Long before x-ray investigation made it possible to determine the absolute dimensions of the unit cell, this relation of crystal morphology to internal structure was recognized, and axial ratios were calculated. By measuring the interfacial angles on the crystal and making certain calculations, it is possible to arrive at axial ratios that express the relative lengths of the crystallographic axes. It is interesting to see how closely the new axial ratios calculated from the dimensions of the unit cell compare with the older ratios derived from morphological measurements. For example, the ratios for sulfur are: $a:b:c = 0.8131:1:1.9034$ from morphological calculations; $a:b:c = 0.8138:1:1.8980$ from unit cell measurements.

Parameters. Crystal faces are defined by indicating their intercepts on the crystallographic axes. Thus, in describing a crystal face it is necessary to determine whether it is parallel to two axes and intersects the third, or is parallel to one axis and intersects the other two, or intersects all three. In addition, one must determine at what relative distance the face intersects the different axes. We have seen in the discussion of axial ratios that the ratios express the relative lengths of the axes. For sulfur: $a = 0.8$, $b = 1$, and $c = 1.9$. For a crystal face that intersects the crystallographic axes at these relative distances (taken as unit distances), the intercepts are given as: one on a, one on b, and one on c, or $1a$, $1b$, $1c$ (see Fig. 14). These are the parameters of the face. Once the units along the axes are established,

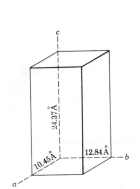

FIG. 13. Unit cell of sulfur.

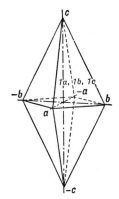

FIG. 14. Dipyramid of sulfur.

[2] An angstrom unit is 0.00000001 centimeter, i.e., 10^{-8} centimeter.

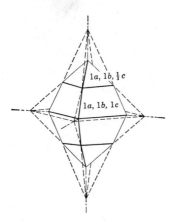

FIG. 15.   Sulfur crystal; two dipyramids.

other faces would be found to intersect the axes at a small multiple of the units—as 2, 3, 4; or some fractional part as $\frac{1}{2}$, $\frac{1}{3}$, $\frac{1}{4}$.

A face that cuts the two horizontal axes at distances that are proportional to their unit lengths and cuts the vertical axis at a distance twice its relative unit length will have for parameters $1a$, $1b$, $2c$. It is to be emphasized that these parameters are strictly relative values and do not indicate any actual cutting lengths. To illustrate this further, consider Fig. 15, which represents a possible sulfur crystal. The forms present upon it are two dipyramids of different slope, but each intersects all three of the crystal axes when properly extended.

The problem exists here, as in many crystals, of choosing which dipyramid to select as the unit form, that is, which one shall be considered to cut all three of the crystallographic axes at unit lengths. Lacking other evidence, it is customary to select as the unit the form with the largest faces. This, in Fig. 15, is the lower dipyramid, and the parameters of the face of this form which cuts the positive ends of the three axes are $1a$, $1b$, and $1c$. The upper dipyramid would cut the two horizontal axes, as shown by the dotted lines, also at distances which, although greater than in the lower dipyramid, are still proportional to the unit lengths. It cuts the vertical axis, however, at a distance which, when considered in respect to its intersections with the horizontal axes, is proportional to one-third of the unit length of c. The parameters of a face of this form would therefore be $1a$, $1b$, $\frac{1}{3}c$. From this it will be seen that the parameters $1a$, $1b$ do not in the two examples represent the same actual cutting distances but express only relative values. The parameters of a face do not in any way determine its size, for a face may be moved parallel to itself for any distance without changing the relative values of its intersections with the crystallographic axes.

Indices.   Various methods of notation have been devised to express the intercepts of crystal faces upon the crystal axes. But the most universally employed is the system of indices proposed by W. H. Miller.

The Miller indices of a face consist of a series of whole numbers that have been derived from the parameters by their inversion and, if necessary, the subsequent clearing of fractions. The indices of a face are always given so that the three numbers (four in the hexagonal system) refer to the a, b, and c axes respectively, and therefore the letters that indicate the different axes are omitted. Like the parameters, the indices express a ratio, but for the sake of brevity the ratio sign is also omitted. The faces of the dipyramids illustrated in Fig. 15 with parameters $1a$, $1b$, $1c$ and $1a$, $1b$, $\frac{1}{3}c$ have respectively (111) (read one, one, one) and (113) for indices. The face, Fig. 16, which has $1a$, $1b$, αc for parameters and on inversion $\frac{1}{1}$, $\frac{1}{1}$, $\frac{1}{\alpha}$ would have (110) for indices. If the parameters are $1a$, $1b$, $2c$, inversion yields $\frac{1}{1}$, $\frac{1}{1}$, $\frac{1}{2}$, and, on clearing of fractions, (221).

It is sometimes convenient when the exact intercepts are unknown to use a general symbol (hkl) for the Miller indices; here h, k, and l each represents a simple whole number. In this symbol h, k, and l are, respectively, the reciprocals of rational but undefined intercepts along the a, b, and c axes. The symbol (hkl) would indicate that a face cuts all three of the crystallographic axes. If a face is parallel to one of the crystallographic axes and intersects the other two, the general symbols would be written as $(0kl)$, $(h0l)$, and $(hk0)$. A face parallel to two of the axes is considered to intersect the third at unity, and the indices would, therefore, be: (100), (010), and (001).

In the above discussion only those faces that intercept the positive ends of the crystallographic axes have been considered. To indicate that a face intercepts the negative end of an axis, a line is placed over the appropriate letter or number, as shown in Fig. 17.

Early in the study of crystals it was discovered that for given faces the indices would always be expressed by simple whole numbers or zero. This is known as the *law of rational indices*.

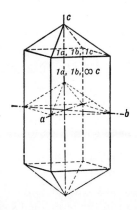

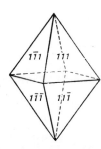

FIG. 16.   Prism and dipyramid.   FIG. 17.   Dipyramid.

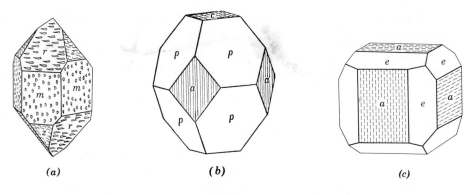

FIG. 18.   Different appearance of different forms. (*a*) Quartz. (*b*) Apophyllite. (*c*) Pyrite.

Form. In its most familiar meaning the term *form* is used to indicate general outward appearance. In crystallography, external shape is denoted by the word *habit*, whereas *form* is used in a special and restricted sense. Thus, a form consists of a group of crystal faces all of which have the same relation to the elements of symmetry and display the same chemical and physical properties because all are underlain by like atoms in the same geometrical arrangement. Although faces of a form may be of different sizes and shapes because of malformation of the crystal, the similarity is frequently evidenced by natural striations, etchings, or growths as shown in Fig. 18. On some crystals the differences between faces of different forms can be seen only after etching with acid.

In Figs. 18a and 18b there are three forms, each of which has a different physical appearance from the others, and in Fig. 18c there are two forms.

In our discussion of Miller indices we saw that a crystal face may be designated by a symbol enclosed in parentheses as (hkl), (010), or (111). Miller indices also may be used as form symbols and are then enclosed in braces as $\{hkl\}$, $\{010\}$, etc. Thus in Fig. 17, (111) refers to a specific face; whereas $\{111\}$ embraces all eight faces. In choosing a form symbol it is desirable to select, if possible, the face symbol with positive digits; $\{111\}$ rather than $\{1\bar{1}1\}$, $\{010\}$ rather than $\{0\bar{1}0\}$.

In each crystal class there is a form, the faces of which intersect all of the crystallographic axes at different lengths; this is the *general form*, $\{hkl\}$. All other forms that may be present are *special forms*. In the orthorhombic, monoclinic, and triclinic crystal systems $\{111\}$ is a general form, for the unit length along each of the axes is different. In the crystal systems of higher symmetry in which the unit distances along two or more of the crystallographic axes are the same, a general form must intersect the like axes at different multiples of the unit length. Thus $\{121\}$ is a general form in the tetragonal system but a special form in the isometric system, and $\{123\}$ is a general form in the isometric system.

FIG. 19. Pedion.

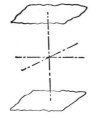

FIG. 20. Pinacoid.

In Fig. 17 is illustrated a single crystal form known as a dipyramid. In the symmetry class to which it belongs the three crystal axes are axes of 2-fold symmetry, and the axial planes are planes of symmetry. With this symmetry, if the presence of face (111) is assumed, there must be the seven other faces, for they all have the same relations to the elements of symmetry. These eight faces constitute a form, and, since they inclose space, it is called a *closed* form. Also see Figs. 34–43. The forms illustrated in Figs. 24–33 do not inclose space and, therefore, are called *open forms*.

A crystal usually displays several forms in combination with one another but may have only one, provided that it is a closed form. Since any combination of forms must inclose space, a minimum of two open forms is necessary. The two may exist by themselves or be in combination with closed forms or other open forms. Altogether there are forty-eight different types of crystal forms which can be distinguished by the angular relations of their faces. Thirty-two of these are represented by the general forms of the thirty-two crystal classes; ten are special closed forms of the isometric system; and six are special open forms (prisms) of the hexagonal and tetragonal systems. The various forms will be discussed under the symmetry class or classes in which they are found. Each form of the isometric system has a special name, but the same general names are used for the forms present in other systems; they are as follows:

Pedion. A single face comprising a form (Fig. 19).

Pinacoid. A form made up of two parallel faces (Fig. 20).

Dome. Two nonparallel faces symmetrical with respect to a symmetry plane (Fig. 21).

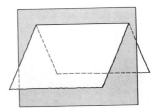

FIG. 21. Dome.

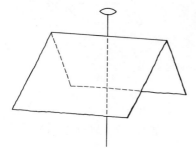

FIG. 22.   Sphenoid.

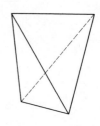

FIG. 23.   Disphenoid.

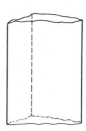

FIG. 24.   Trigonal prism.

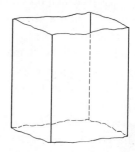

FIG. 25.   Tetragonal prism.

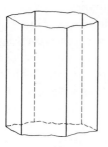

FIG. 26.   Hexagonal prism.

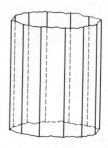

FIG. 27.   Dihexagonal prism.

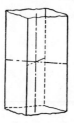

FIG. 28.   Vertical orthorhombic prism.

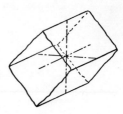

FIG. 29.   Horizontal orthorhombic prism.

Sphenoid. Two nonparallel faces symmetrical with respect to a 2-fold or 4-fold symmetry axis (Fig. 22).

Disphenoid. A four-faced form in which two faces of the upper sphenoid alternate with two of the lower sphenoid (Fig. 23).

Prism. A form composed of 3, 4, 6, 8, or 12 faces, all of which are parallel to the same axis. Except for certain prisms in the monoclinic system, the axis is one of the principal crystallographic axes (Figs. 24–29).

Pyramid. A form composed of 3, 4, 6, 8, or 12 nonparallel faces that meet at a point (Figs. 30–33).

Scalenohedron. 8-faced (tetragonal, Fig. 34) or 12-faced (hexagonal, Fig. 35) closed forms with the faces grouped in symmetrical pairs. For the 8-faced forms there are two pairs of faces above and two pairs below in alternating positions. For the 12-faced forms there are three pairs of faces above and three pairs below

FIG. 30. Trigonal pyramid.

FIG. 31. Tetragonal pyramid.

FIG. 32. Hexagonal pyramid.

FIG. 33. Dihexagonal pyramid.

FIG. 34. Tetragonal scalenohedron.

FIG. 35. Hexagonal scalenohedron.

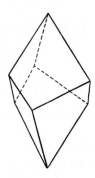

FIG. 36.   Trigonal
trapezohedron.

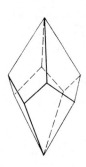

FIG. 37.   Tetragonal
trapezohedron.

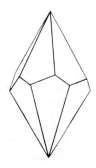

FIG. 38.   Hexagonal
trapezohedron.

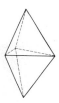

FIG. 39.   Trigonal
dipyramid.

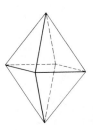

FIG. 40.   Tetragonal
dipyramid.

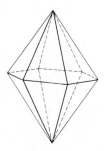

FIG. 41.   Hexagonal
dipyramid.

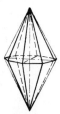

FIG. 42.   Ditetragonal dipyramid.

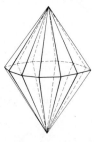

FIG. 43.   Dihexagonal dipyramid.

in alternating positions. In perfectly developed crystals, each face is a scalene triangle.

Trapezohedron. 6-, 8-, or 12-faced forms with 3, 4, or 6 faces above offset from 3, 4, or 6 faces below (Figs. 36–38). In addition there is an isometric trapezohedron, a 24-faced form. In well-developed trapezohedrons each face is a trapezium.

Dipyramid. 6-, 8-, 12-, 16-, or 24-faced closed forms (Figs. 39–43). The dipyramids can be considered as formed from pyramids by reflection across a horizontal symmetry plane.

Rhombohedron. A closed form composed of six faces, the intersection edges of which are not at right angles. Rhombohedrons are found only on crystals of the rhombohedral division of the hexagonal system.

In crystal drawings it is found convenient to indicate the faces of a form by the same letter. The choice of which letter shall be assigned to a given form rests largely with the person who first describes the crystal. However, there are certain simple forms that, owing to convention, usually receive the same letter. Thus the three pinacoids that cut the *a*, *b*, and *c* axes are lettered respectively *a*, *b*, and *c*, the letter *m* is usually given to {110}, and *p* to {111} (Fig. 44).

Zones. One of the conspicuous features on many crystals is the arrangement of a group of faces in such a manner that their intersection edges are mutually parallel. Considered collectively, these faces form a *zone*. A line through the center of the crystal that parallels the lines of face intersections is called the *zone axis*. In Fig. 45 the faces *m'*, *a*, *m*, and *b* are in one zone, and *b*, *r*, *c*, and *r'* in another. The lines given as [001] and [100] are the zone axes.

A zone is indicated by a symbol similar to that for Miller indices of faces, the generalized expression of which is [*uvw*]. Any two nonparallel faces determine a zone. Assume one wishes to determine the zone axis of two such faces, (*hkl*) and (*pqr*). The method usually used is to write twice the symbol of one face and directly beneath, twice the symbol of the other face. The first and last digits of

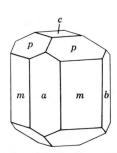

FIG. 44. Conventional lettering on crystal drawings.

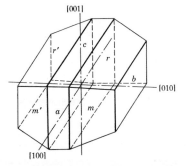

FIG. 45. Crystal zones and zone axes.

each line are disregarded and the remaining numbers, joined by sloping arrows, multiplied. In each set the product of 2 is subtracted from the product of 1 as:

$$
\begin{array}{c|c|c|c|c|c}
h & k & l & h & k & l \\
\end{array}
$$

$$
\begin{array}{c|c|c|c|c|c}
p & q & r & p & q & r \\
\end{array}
$$

kr-lq, lp-hr, hq-kp

$$
\begin{array}{c|c|c|c|c|c}
1 & 1 & 0 & 1 & 1 & 0 \\
\end{array}
$$

$$
\begin{array}{c|c|c|c|c|c}
0 & 1 & 0 & 0 & 1 & 0 \\
\end{array}
$$

0-0, 0-0, 1-0 or [001]

As a specific example assume m, Fig. 45, is (hkl) with index (110) and b is (010), the zone axis is thus [001] as shown above. It should be noted that the zone symbols are inclosed in brackets, as $[uvw]$, to distinguish them from face and form symbols.

Crystal Habit. The term *habit* is used to denote the general shapes of crystals as cubic, octahedral, prismatic. Because habit is controlled by the environment in which crystals grow, it may vary with locality. At one place it may be equant and at others tabular or fibrous. Only rarely do crystals present an ideal geometrical shape. But even in malformed crystals evidence of the symmetry is present in the physical appearance of the faces and in the symmetrical arrangement of the interfacial angles. In Figs. 46 to 48 are given various crystal forms, first ideally developed and then malformed.

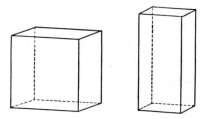

FIG. 46. Cube and malformed cube.

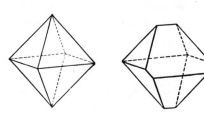

FIG. 47. Octahedron and malformed octahedron.

FIG. 48. Dodecahedron and malformed dodecahedron.

MEASUREMENT OF CRYSTAL ANGLES

Since, as stated by Steno's Law the angles between equivalent faces of the same substance are constant, they characterize a crystal and should be measured carefully. Angles are measured with goniometers and for accurate work, a *reflecting goniometer* is used. This is an instrument upon which the crystal is mounted so as to rotate about a zone axis and to reflect a collimated beam of light from its faces through a telescope to the eye. The angle through which a crystal must be turned in order to throw successive beams of light from two adjacent faces into the telescope determines the angle between the faces.

It will be seen from Fig. 49 that the angle thus determined is the *internal* angle. These internal angles, supplements of the external interfacial angles, are the ones given in crystallographic data.

A simpler instrument used for approximate work and with larger crystals is known as a *contact goniometer* (Fig. 50). In using the contact goniometer, it is imperative that the plane determined by the two arms of the goniometer be exactly at right angles to the edge between the measured faces. It must also be remembered that it is the internal angle that is recorded. Thus in Fig. 50, the angle should be read as 40°, not as 140°.

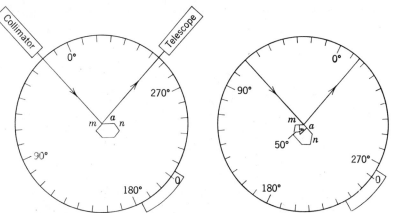

FIG. 49.   Angular measurement by reflecting goniometer.

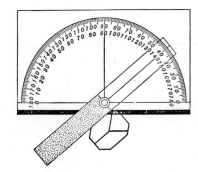

FIG. 50.   Contact goniometer.

CRYSTAL PROJECTIONS

Introduction.   A crystal projection is a means of representing the three-dimensional crystal on a two-dimensional plane surface. Different projections are used for different purposes, but each is made according to some definite rule so that the projection bears a known and reproducible relationship to the crystal. The crystal drawings in this book are known as *clinographic projections* and are a type of perspective drawing which yields a portraitlike picture of the crystal in two dimensions. This is the best means of conveying the appearance of a crystal and generally serves much better for the purpose than a photograph.

Spherical Projection.   Since the actual size and shape of the different faces on a crystal are chiefly the result of accidents of growth, we wish to minimize this aspect of crystals in projecting, but at the same time emphasize the angular relation of the faces. We can do this using the *spherical projection*. Although the spherical projection is itself three dimensional, an understanding of it is necessary, for it serves as the basis for other crystal projections. We can envisage the construction of such a projection in the following way. Imagine a hollow model of a crystal containing a bright point source of light. Now let us place this model within a large hollow sphere of translucent material in such a way that the light source is at the center of the sphere. If now we make a pinhole in each of the faces so that the ray of light emerging from the hole is perpendicular to the face, these

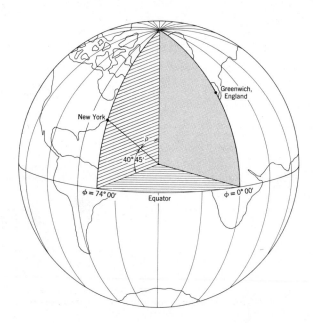

FIG. 51.   Latitude and longitude of New York City.

rays of light will fall on the inner surface of the sphere and make bright spots. The whole situation resembles a planetarium in which the crystal model with its internal light source and pinholes is the projector and the translucent sphere is the dome. If we now mark on the sphere the position of each spot of light, we may remove the model and have a permanent record of its faces. Each of the crystal faces is represented on the sphere by a point called the face *pole*. This is the spherical projection.

The position of each pole and thus its angular relationship to other poles can be fixed by *angular coordinates* on the sphere. This is done in a manner similar to the location of points on the earth's surface by means of longitude and latitude. For example, the angular coordinates $74°\,00'$ west longitude, $40°\,45'$ north latitude, locate a point in New York City. This means that the angle, measured at the center of the earth, between the plane of the equator and a line from the center of the earth through that point in New York is exactly $40°\,45'$; and the angle between the Greenwich meridian and the meridian passing through the point in New York, measured west in the plane of the equator, is exactly $74°\,00'$. These relations are shown in Fig. 51.

A similar system may be used to locate the poles of faces on the spherical projection of a crystal. There is one major difference between locating points on a spherical projection and locating points on the earth's surface. On the earth, the latitude is measured in degrees north or south from the equator, whereas the angle used in the spherical projection is the colatitude, or *polar angle*, which is measured in degrees from the north pole. The north pole of a crystal projection is thus at a colatitude of $0°$, the equator at $90°$. The colatitude of New York City is $49°\,15'$. This angle is designated in crystallography by the Greek letter ρ (rho).

The "crystal longitude" of the pole of a face on the spherical projection is measured, just as are longitudes on earth, in degrees up to $180°$, clockwise and counterclockwise from a starting meridian analogous to the Greenwich meridian of geography. To locate this starting meridian, the crystal is oriented in conventional manner with the (010) face to the right of the crystal. The meridian passing through the pole of this face is taken as zero. Thus, to determine the crystal longitude of any crystal face, a meridian is passed through the pole of that face, and the angle between it and the zero meridian is measured in the plane of the equator. This angle is designated by the Greek letter ϕ (phi).

If any plane is passed through a sphere, it will intersect the surface of the sphere in a circle. The circles of maximum diameter are those formed by planes passing through the center and having a diameter equal to the diameter of the sphere. These are called *great circles*. All other circles formed by passing planes through the sphere are *small circles*. The meridians on the earth are great circles, as is the equator, whereas the parallels of latitude are small circles.

The spherical projection of a crystal brings out interesting zonal relationships,

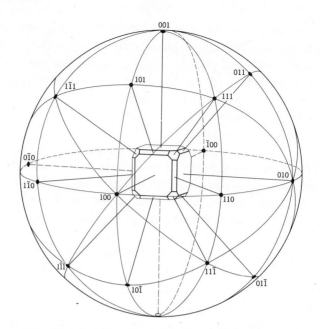

FIG. 52. Spherical projection of isometric forms.

for the poles of all the faces in a zone lie along a great circle of the projection. In Fig. 52 (spherical projection) faces (001), (101), (100), (10$\bar{1}$), and (00$\bar{1}$) lie in a zone with the zone axis [010]. Since the great circle along which the poles of these faces lie passes through the north and south poles of the projection, it is called a *vertical great circle*. The zone axis is always perpendicular to the plane containing the face poles and thus all vertical circles have horizontal zone axes.

Stereographic Projection. It is important in reducing the spherical projection of a crystal to a plane surface that the angular relation of the faces be preserved in such a way as to reveal the true symmetry. This can best be done with the stereographic projection.

The stereographic projection is a representation in a plane of half of the spherical projection, usually the northern hemisphere. The plane of the projection is the equatorial plane of the sphere, and the *primitive* circle (the circle outlining the projection) is the equator itself. If one were to view the poles of crystal faces located in the northern hemisphere of the spherical projection with the eye at the south pole, the intersection of the lines of sight with the equatorial plane would be the corresponding poles on the stereographic projection. We can thus construct a stereographic projection by drawing lines from the south pole to the face poles in the northern hemisphere. The corresponding poles on the stereographic projection are located where these lines intersect the equatorial plane. The relationship of the two projections is brought out in Fig. 53.

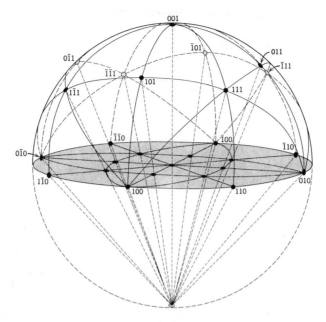

FIG. 53. Relation of spherical and stereographic projections. (After E. E. Wahlstrom, *Optical Crystallography*, John Wiley and Sons, New York, 1951.)

Since in practice one plots poles directly on the stereographic projection, it is necessary to determine stereographic distances in relation to angles of the spherical projection. Figure 54 shows a vertical section through the spherical projection of a crystal in the plane of the "zero meridian," that is, the plane containing the pole of (010). The ϕ angle of any face which lies in this section is $0°$ if to the right of the center, and $180°$ if to the left of center. N and S are respectively the north and south poles of the sphere of projection, O is the center of the projected crystal. Consider the face (011). OD is the perpendicular to the face (011), and D is the pole of this face on the spherical projection. The line from the south pole, SD, intersects the trace of the plane of the equator, FG, in the point D', the stereographic pole of (011). The angle NOD will be recognized as the angle ρ (rho). In order to plot D' directly on the stereographic projection, it is necessary to determine the distance OD' in terms of angle ρ. Since $\triangle SOD$ is isosceles, $\angle ODS = \angle OSD$. $\angle ODS + \angle OSD = \angle NOD = \rho$. Therefore, $\angle OSD = \rho/2$. $OS = r$, the radius of the primitive of the projection.

$$\tan \rho/2 = OD'/r, \text{ or } OD' = r \tan \rho/2.$$

To sum up, in order to find the stereographically projected distance from the center of the projection of the pole of any face, find the natural tangent of one-

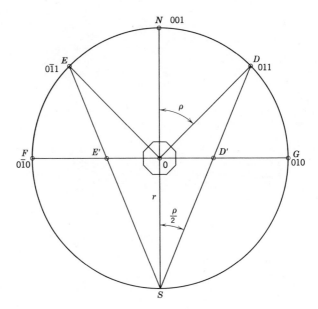

FIG. 54. Section through sphere of projection showing relation of spherical to stereographic poles.

half of ρ of that face and multiply by the radius of the projection. The distance so obtained will be in whatever units are used to measure the radius of the primitive of the projection.

In addition to determining the distance a pole should lie from the center of the projection, it is also necessary to determine its "longitude" or ϕ (phi) angle. Since the angle is measured in the plane of the equator, which is also the plane of the stereographic projection, it may be laid off directly on the primitive by means of a circular protractor. It is first necessary to fix the "zero meridian" by making a point on the primitive circle to represent the pole of (010). A straight line drawn through this point and the center of the projection is the zero meridian. With the protractor edge along this line and center point at the center of the projection, the ϕ angle can be marked off. On a construction line from the center of the projection through this point lie all possible face poles having the specified ϕ angle. Positive ϕ angles are laid off clockwise from (010); negative ϕ angles are laid off counterclockwise, as shown in Fig. 55.

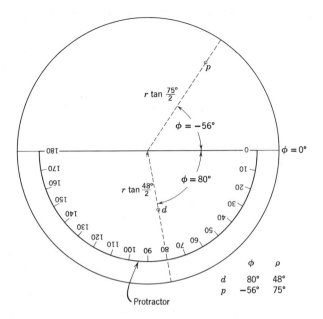

FIG. 55. Stereographic projection of crystal faces.

In order to plot the pole of the face having this given ϕ value, it is necessary to find the natural tangent of one-half of ρ, to multiply by the radius of the projection, and to lay off the resulting distance along the ϕ line. Although any projection radius may be chosen, one of 10 centimeters is usually used. This is large enough to give accuracy, but not be unwieldy, and at the same time simplifies the calculation. With a 10-centimeter radius, it is only necessary to look up the natural tangent, move the decimal point one place to the right, and plot the result as centimeters from the center of the projection.

When the poles of crystal faces are plotted stereographically as explained above, their symmetry of arrangement should be apparent. (See Fig. 56.) We have seen (page 34) that a great circle in the spherical projection is the locus of poles of faces lying in a crystal zone. When projected stereographically, vertical great circles become diameters of the projection; all other great circles project as circular arcs which subtend a diameter. The limiting case of such great circles is the primitive of the projection which is itself a great circle common to both the spherical and stereographic projections. The poles of vertical crystal faces lie on the primitive and thus are projected without angular distortion.

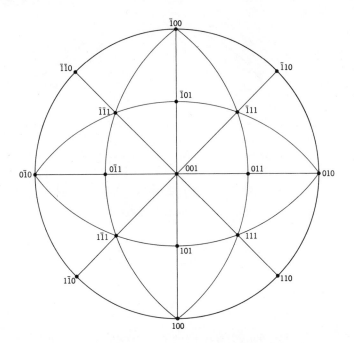

FIG. 56. Stereographic projection of isometric crystal faces. (After E. E. Wahlstrom, *Optical Crystallography*, John Wiley and Sons, New York, 1951.)

Stereographic Net. Both the measurement and plotting of angles on the stereographic projection are greatly facilitated by means of the stereographic net.[3] (Fig. 57.) Both great and small circles are drawn on the net at intervals of 1° or 2°. In Fig. 57 the intervals are 2°. A projection made on tracing paper may be placed over the net and the angles read directly. The ϕ angles are determined where a straight line from the center of the projection through the face pole intersects the primitive. To determine the ρ angle, the projection must be rotated about the center until the face pole lies on one of the vertical great circles. The angle can then be read directly from the net.

If the ϕ and ρ angles are known, the stereogram can be constructed by reversing the process. First, one should make a mark on the primitive at $\phi = 0°$ so that it is always possible to return the projection to its initial setting. The pole of a face is located as follows: (1) Mark a point on the primitive at the ϕ angle. (2) Rotate the projection until this point is at the end of a vertical great circle. (3) Mark off the ρ angle along this line at the proper distance from the center. This is the face pole.

[3] The stereographic net is also called the Wulff net, named after G. V. Wulff, Russian crystallographer (1863–1925).

See inside back cover for stereographic net of 10 centimeter radius.

Instead of measuring ϕ and ρ angles the elementary student usually measures interfacial angles which can be plotted easily with the help of the stereographic net. As an example consider Fig. 58, a crystal drawing of the orthorhombic mineral, anglesite. The starting point, as in all projections, is (010), face b. (Fig. 59.) The pole of this face should be placed on the primitive at $0°$. The interfacial angles $b \wedge n = 32\frac{1}{2}°$, and $b \wedge m = 52°$ can be measured and plotted as ϕ angles on the primitive. Face c is (001); it makes an angle of $90°$ with b and its pole should be placed at the center of the projection. Face o is in zone with c and b and thus has a ϕ angle of $0°$. Its ρ angle, $c \wedge o = 52°$, can be measured directly and plotted along the vertical great circle. Face d lies in a vertical zone at $90°$ to the zone c, o, b. It thus has $\phi = 90°$ and $c \wedge d = 39\frac{1}{2}°$ which can be plotted along the vertical great circle of the net. The pole of face y cannot be plotted directly, but the angles $b \wedge y = 50°$ and $c \wedge y = 57°$ can be measured. To locate this

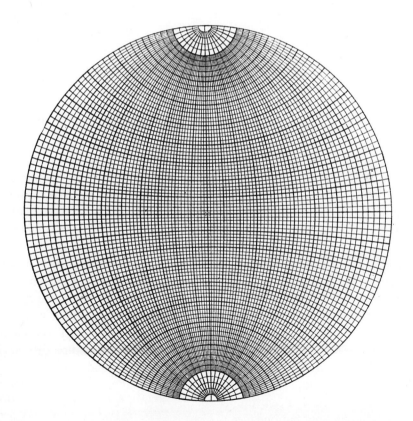

FIG. 57. Stereographic (Wulff) net. Radius equals 5 centimeters.

pole, the projection is rotated $90°$ so that b lies along the radii of the small circles of the net, and a tracing of the $50°$ circle is made. This small circle is the locus of all poles $50°$ from b. The projection is again rotated until this tracing of the circle intersects a vertical great circle at $57°$ $(c \wedge y)$. This is the pole of y.

To check this position, measure the angle $m \wedge y = 38°$. Now with the pole m placed along the radii of the small circles, trace the $38°$ small circle. It should also intersect at the common point. Having located pole y, its ϕ $(32\frac{1}{2}°)$ and ρ $(57°)$ angles can be read directly from the stereographic net.

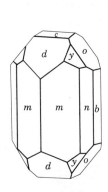

FIG. 58. Anglesite.

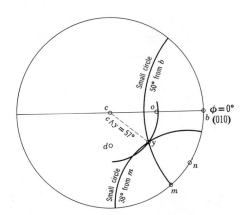

FIG. 59. Location of poles using stereographic net.

THE THIRTY-TWO CRYSTAL CLASSES

In the following section the thirty-two classes, listed on page 18, are described under the crystal systems in which they are grouped. The symmetry of each class is given in terms of planes and axes and Hermann-Mauguin symbol, but it is also shown by means of stereograms, giving projections of all the faces of the general forms. These are the forms from which the classes derive their names. In the stereograms it is necessary to show faces in the southern hemisphere as well as in the northern hemisphere in order to give completely the symmetry of the class. This is done by superimposing stereographic projections of the two hemispheres with the poles in the northern hemisphere represented by solid points and those in the southern hemisphere by circles. Thus, if two poles lie directly one above the other on the sphere, they will be represented by a solid point surrounded by a circle. A vertical face is represented by a single point on the primitive, for, although such a pole would appear on projections of both top and bottom of the crystal, it would represent but one face.

Figure 60a is a drawing of a crystal with a horizontal symmetry plane.[4] The stereogram of this crystal, Fig. 60b, consequently, has for all faces solid points circled to indicate corresponding faces at the top and bottom of the crystal. Figure 61a is a drawing of a crystal lacking a horizontal symmetry plane. Its stereogram, Fig. 61b, has twelve solid points as poles of faces in the northern hemisphere and twelve circles as poles of faces in the southern hemisphere.

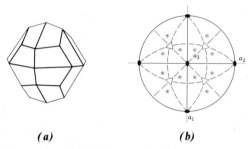

(a)   **(b)**

FIG. 60. Crystal with symmetry 2/m3̄. The stereogram shows a symmetry plane at right angles to each of the 2-fold rotation axes, and four 3-fold axes of rotary inversion. The solid primitive circle denotes a horizontal symmetry plane which indicates faces at the bottom of the crystal directly below those at the top.

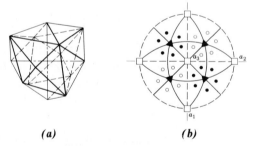

(a)   **(b)**

FIG. 61. Crystal with symmetry 4̄3m. Stereogram shows three 4-fold axes of rotary inversion, six symmetry planes, and four 3-fold rotation axes. The broken primitive circle indicates lack of a horizontal symmetry plane and faces at the top of the crystal do not lie above those at the bottom.

[4] A solid primitive circle of the projection indicates a horizontal plane of symmetry; a broken primitive circle indicates the lack of a horizontal symmetry plane. A solid line drawn as a great circle (vertical or otherwise) indicates a plane of symmetry. Broken straight lines indicate the positions of crystal axes or symmetry axes; crystal axes are labeled a, b, or c, whereas symmetry axes are so indicated by symbols at the ends of these lines. Symmetry axes of rotation are denoted with solid symbols as follows: ◆2-fold axis, ▲ 3-fold axis, ■ 4-fold axis, ● 6-fold axis. Axes of rotary inversion are denoted with open symbols as: △ 3-fold, □ 4-fold, ○ 6-fold.

ISOMETRIC SYSTEM

Crystallographic Axes. The crystal forms of all the classes of the isometric system are referred to three axes of equal length that make right angles with each other. Since the axes are identical, they are interchangeable, and all are designated by the letter a. When properly oriented, one axis, a_1, is horizontal and oriented front to back, a_2 is horizontal and right to left, and a_3 is vertical (see Fig. 62).

Form Symbols. Although the symbol of any face of a crystal form might be used as the form symbol, it is conventional, when possible, to use one in which

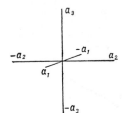

FIG. 62.   Isometric crystal axes.

h, k, and l are all positive. In forms that have two or more faces with h, k, and l positive the rule followed is to take the form symbol with $h < k < l$. For example, the form with a face symbol (123) also has faces with symbols (132), (213), (231), (312), and (321). Following the rule, {123} would be taken as the form symbol, for here $h < k < l$.

In giving the angular coordinates of a form, it is customary to give those for only one face; the others can be determined knowing the symmetry. The face for which these coordinates are given is the one with the least ϕ and ρ values. This is the face of the form in which $h < k < l$.

Hexoctahedral Class—$4/m\bar{3}2/m$

Symmetry—C, $3A_4$, $4A_3$, $6A_2$, $9P$. There is a center of symmetry. The three crystallographic axes are axes of 4-fold symmetry (see Fig. 63a). There are also four diagonal axes of 3-fold symmetry. These axes emerge in the middle of each of the octants formed by the intersection of the crystallographic axes (Fig. 63b). Further, there are six diagonal axes of 2-fold symmetry, each of which bisects one of the angles between two of the crystallographic axes, as illustrated in Fig. 63c.

This class has nine planes of symmetry: three of them are known as the axial planes, since each includes two crystallographic axes (Fig. 64a); and six are called diagonal planes, since each bisects the angle between two of the axial planes (Fig. 64b).

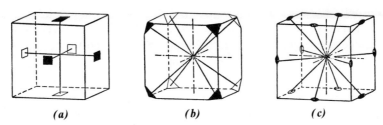

FIG. 63.   Symmetry axes, hexoctahedral class.

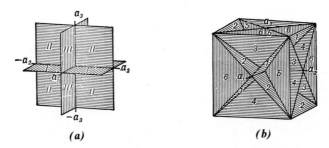

FIG. 64.   Symmetry planes, hexoctahedral class.

This symmetry, the highest possible in crystals, defines the hexoctadedral class of the isometric system. Every crystal form and every combination of forms that belong to this class must show its complete symmetry. It is important to remember that in this class the three crystallographic axes are axes of 4-fold symmetry. Thus one can easily locate the crystallographic axes and properly orient the crystal.

The hexoctadedron, the general form from which the class derives its name, is shown in Fig. 65 with a stereogram showing the symmetry by giving the face poles of the hexoctahedron. Once the student understands the use of the stereogram, it becomes apparent how elegantly the symmetry can be conveyed by its use, for in one diagram is shown the same symmetry as in the five diagrams of Figs. 63 and 64.

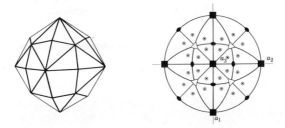

FIG. 65.   Hexoctahedron and stereogram.

Forms. 1. *Cube* or *Hexahedron* {001}. The cube is composed of six square faces that make 90° angles with each other.[5] Each face intersects one of the crystallographic axes and is parallel to the other two. (Fig. 66.)

2. *Octahedron* {111}. The octahedron is composed of eight equilateral triangular faces, each of which intersects all three of the crystallographic axes equally. Figure 67 represents a simple octahedron, and Fig. 68 shows combinations of a cube and an octahedron. When in combination the octahedron can be recognized by its eight similar faces, each of which is equally inclined to the three crystallographic axes. It is to be noted that the faces of an octahedron truncate symmetrically the corners of a cube.

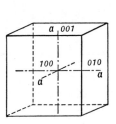

FIG. 66.   Cube.

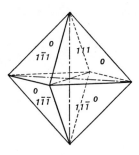

FIG. 67.   Octahedron.

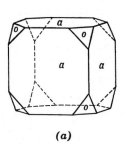

(a)

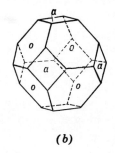

(b)

FIG. 68.   Cube and octahedron.

[5] In the description of forms on the following pages the geometrically perfect model of the unmodified form is considered in each case. It should be kept in mind that in nature this ideal is rarely obtained, and that crystals not only are frequently malformed but also are usually bounded by a combination of forms.

3. *Dodecahedron* or *Rhombic Dodecahedron* {011}. The dodecahedron is composed of twelve rhomb-shaped faces. Each face intersects two of the crystallographic axes equally and is parallel to the third. Figure 69 shows a simple dodecahedron; Fig. 70 shows a combination of dodecahedron and cube; Fig. 71, combinations of dodecahedron and octahedron; and Fig. 72, a combination of cube, octahedron, and dodecahedron. Note that the faces of a dodecahedron truncate the edges of both the cube and the octahedron. The dodecahedron is sometimes called the rhombic dodecahedron to distinguish it from the pentagonal dodecahedron and the regular geometrical dodecahedron.

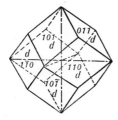

FIG. 69. Dodecahedron.

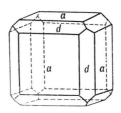

FIG. 70.   Cube and dodecahedron.

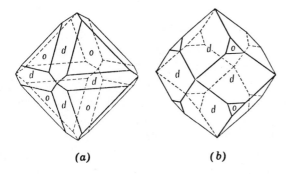

(a)   (b)

FIG. 71.   Octahedron and dodecahedron.

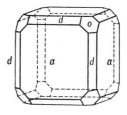

FIG. 72.   Cube, octahedron, and dodecahedron.

4. *Tetrahexahedron* {0kl}.   The tetrahexahedron is composed of twenty-four isosceles triangular faces, each of which intersects one axis at unity and the second at some multiple and is parallel to the third. There are a number of tetrahexahedrons which differ from each other in respect to the inclination of their faces. The commonest has the parameter relations $\infty a_1, 2a_2, 1a_3$, the symbol of which would be {012}. The indices of other forms are {013}, {014}, {023}, etc., or, in general, {0kl}. It is helpful to note that the tetrahexahedron, as its name indicates,

is like a cube the faces of which have been replaced by four others. Figure 73 shows a simple tetrahexahedron, and Fig. 74 shows a cube with its edges beveled by the faces of a tetrahexahedron.

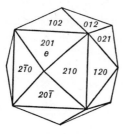

FIG. 73. Tetrahexahedron.

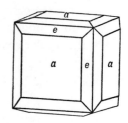

FIG. 74. Cube and tetrahexahedron.

5. *Trapezohedron* or *Tetragonal Trisoctahedron* {*hhl*}. The trapezohedron is composed of 24 trapezium-shaped faces, each of which intersects one of the crystallographic axes at unity and the other two at equal multiples. There are various trapezohedrons with their faces having different angles of inclination, but the most common is {112}, Fig. 75. This form is sometimes called a tetragonal trisoctahedron to indicate that each of its faces has four edges and to distinguish it from another 24-faced form, the trigonal trisoctahedron.

Figure 76 shows the common trapezohedron *n* {112} truncating the edges of the dodecahedron. Both forms by themselves and in combination are common on the mineral garnet.

6. *Trisoctahedron* or *Trigonal Tristoctahedron* {*hll*}. The trisoctahedron is composed of twenty-four isosceles triangular faces, each of which intersects two of the crystallographic axes at unity and the third axis at some multiple. There are various trisoctahedrons, the faces of which have different inclinations, but most common is {122} (Fig. 77). The trisoctahedron, like the trapezohedron, is a form that may be conceived as an octahedron, each face of which has been replaced by three others. It is frequently spoken of as the trigonal trisoctahedron, indicating that its faces have three edges and so differ from those of the trapezohedron. Figure 78 shows a combination of an octahedron and trisoctahedron.

7. *Hexoctahedron* {*hkl*}. The hexoctahedron is composed of forty-eight triangular faces, each of which intersects all three crystallographic axes at different lengths. There are several hexoctahedrons, which have varying ratios of axial intercepts. A common hexoctahedron has for its parameter relations $6a_1, 3a_2, 2a_3$, with indices {123}. Other hexoctahedrons have indices {124}, {135}, etc., or,

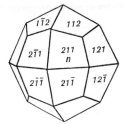

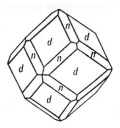

FIG. 75.   Trapezohedron.   **FIG. 76.   Dodecahedron and trapezohedron.**

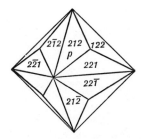

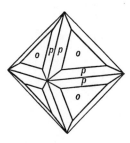

FIG. 77.   Trisoctahedron.   **FIG. 78.   Octahedron and trisoctahedron.**

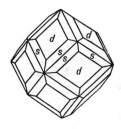

FIG. 79.   Hexoctahedron.   **FIG. 80.   Dodecahedron and hexoctahedron.**

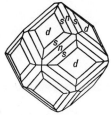

FIG. 81.   Dodecahedron, trapezohedron, and hexoctahedron.

in general, $\{hkl\}$. Figure 79 shows a simple hexoctahedron; Fig. 80, a combination of dodecahedron and hexoctahedron; and Fig. 81, a combination of dodecahedron, trapezohedron, and hexoctahedron.

Determination of Indices of Isometric Forms.   In determining the forms present on any crystal of the hexoctahedral class, it is first necessary to locate the crystallographic axes (axes of 4-fold symmetry). Once the crystal has been oriented by these axes, the faces of the cube, dodecahedron, and octahedron are easily recognized, since they intersect respectively one, two, and three axes at unit distances. The indices can be quickly obtained for faces of other forms which truncate symmetrically the edges between known faces. The algebraic sums of the h, k, and l indices of two faces give the indices of the face symmetrically truncating the edge between them. Thus, in Fig. 82 the algebraic sum of the two dodecahedron faces (101) and (011) is (112), or the indices of a face of a trapezohedron.

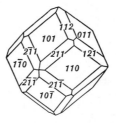

FIG. 82.   Dodecahedron and trapezohedron.

Occurrence of Isometric Forms of the Hexoctahedral Class.   The cube, octahedron, and dodecahedron are the most common isometric forms. The trapezohedron is also frequently observed as the only form on a few minerals. The other forms, the tetrahexahedron, trisoctahedron, and hexoctahedron, are rare and are ordinarily observed only as small truncations in combinations.

A large group of minerals crystallize in the hexoctahedral class. Some of the most common are:

analcime	galena	leucite
cerargyrite	garnet	silver
copper	gold	spinel group
cuprite	halite	sylvite
fluorite	lazurite	uraninite

Gyroidal Class—432

Symmetry.—$3A_4, 4A_3, 6A_2$. The gyroidal class has all the symmetry axes of the hexoctahedral class but lacks symmetry planes and a center (Fig. 83).

Forms. The *gyroid* $\{hkl\}$ right, $\{khl\}$ left. These forms have twenty-four faces each and are *enantiomorphic*. That is, they have the relation to one another as a right hand and a left hand; the reflection of one form yields the other. All the forms of the hexoctahedral class with the exception of the hexoctahedron can be present in the gyroidal class.

For many years cuprite was considered gyroidal, but more recent studies have shown that it is probably hexoctahedral. With the elimination of cuprite, no known mineral crystallizes in the gyroidal class.

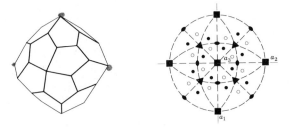

FIG. 83. Gyroid and stereogram.

Hextetrahedral Class—$\bar{4}3m$

Symmetry—$3A_2, 4A_3, 6P$. The three crystallographic axes are axes of apparent 2-fold symmetry; actually they are 4-fold axes of rotary inversion. The four diagonal axes are axes of 3-fold symmetry (Fig. 84a). There are six diagonal planes of symmetry (Fig. 84b), the same planes shown in Fig. 64b for the hexoctahedral class. The general form, the hextetrahedron, is illustrated in Fig. 85.

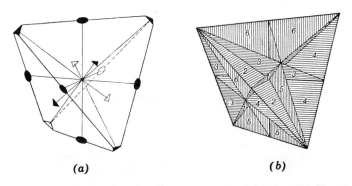

(a)   (b)

FIG. 84. Symmetry of hextetrahedral class. (a) Axes. (b) Planes.

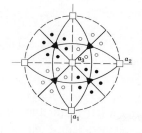

FIG. 85. Hextetrahedron and stereogram.

Forms. 1. *Tetrahedron* {111} *positive*, {1Ī1} *negative*. The tetrahedron is composed of four equilateral triangular faces, each of which intersects all the crystallographic axes at equal lengths. It can be considered as derived from the octahedron of the hexoctahedral class by the omission of the alternate faces and the extension of the others, as shown in Fig. 86. This form, shown also in Fig. 87a, is known as the positive tetrahedron {111}. If the other four faces of the octahedron had been extended, the tetrahedron resulting would have had a different orientation, as shown in Fig. 87b. This is the negative tetrahedron, {1Ī1}. The positive and negative tetrahedrons are geometrically identical. The existence of both must be recognized for they may occur together as shown in Fig. 87c. If the positive

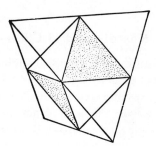

FIG. 86. Relation of octahedron and tetrahedrons.

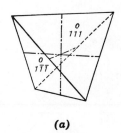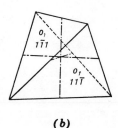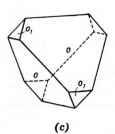

(a)   (b)   (c)

FIG. 87. Tetrahedrons. (a) Positive. (b) Negative. (c) Combination (+) and (−).

and negative tetrahedron are equally developed on the same crystal, the combination could not be distinguished from an octahedron unless, as often happens, the faces of the two forms showed different lusters, etchings, or striations that would serve to differentiate them.

2. *Tristetrahedron* {*hhl*} *positive*, {*hh̄l*} *negative*. These forms have twelve faces which correspond to one-half of the faces of a trapezohedron (Fig. 88) taken in alternating groups of three above and three below. The positive form may be made negative by a rotation of 90° about the vertical axis.

3. *Deltoid Dodecahedron* {*hll*} *positive*, {*h̄ll*} *negative*. This is a twelve-faced form in which the faces correspond to one-half of those of the trisoctahedron taken in alternate groups of three above and three below (Fig. 89).

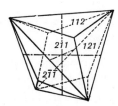

FIG. 88.   Tristetrahedron.

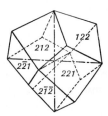

FIG. 89.   Deltoid dodecahedron.

4. *Hextetrahedron* {*hkl*} *positive*, {*hk̄l*} *negative*. The hextetrahedron (Fig. 90) has twenty-four faces which correspond to one-half of the faces of the hexoctahedron taken in groups of six above and six below.

The cube, dodecahedron, and tetrahexahedron are also present on crystals of the hextetrahedral class. Figure 91 shows combinations of cube and tetrahedron. Figure 92 shows the combination of tetrahedron and dodecahedron and Fig. 93 represents a combination of cube, dodecahedron, and tetrahedron. Figure 94 shows a combination of tetrahedron and tristetrahedron.

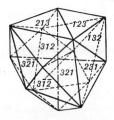

FIG. 90.   Positive hextetrahedron.

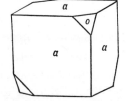 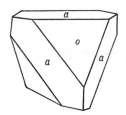

FIG. 91.   Cube and tetrahedron.

Tetrahedrite and related tennantite are the only common minerals that ordinarily show distinct hextetrahedral forms. Sphalerite occasionally exhibits them, but commonly its crystals are complex and malformed.

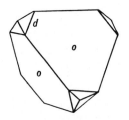

FIG. 92. Tetrahedron and dodecahedron.

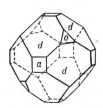

FIG. 93. Dodecahedron, cube and tetrahedron.

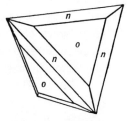

FIG. 94. Tetrahedron and tristetrahedron.

Diploidal Class—$2/m\bar{3}$

Symmetry—$C, 3A_2, 4A_3, 3P$. The three crystallographic axes are axes of 2-fold symmetry; the four diagonal axes, each of which emerges in the middle of an octant, are axes of 3-fold symmetry; the three axial planes are planes of symmetry (Figs. 95a and 95b). Figure 96 illustrates a positive diploid and its stereogram.

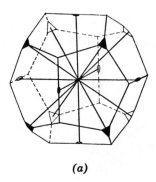

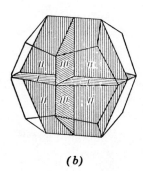

(a)   *(b)*

FIG. 95. Symmetry of diploid class. (*a*) Axis. (*b*) Planes.

Forms. 1. *Pyritohedron or Pentagonal Dodecahedron* {h0l} *positive*, {0kl} *negative.* This form consists of twelve pentagonal-shaped faces, each of which intersects one crystallographic axis at unity, intersects the second axis at some multiple of unity, and is parallel to the third. A rotation of 90° about a crystallographic axis brings the positive pyritohedron into the negative position. There are a number of pyritohedrons which differ from each other in respect to the

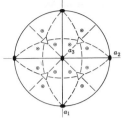

FIG. 96. Diploid and stereogram.

inclination of their faces. The most common positive pyritohedron has indices {102} (Fig. 97a). Figure 97b shows the corresponding negative pyritohedron.

2. *Diploid {khl} positive, {hkl} negative.* The diploid is a rare form composed of twenty-four faces which correspond to one-half of the faces of a hexocta-hedron. The diploid may be pictured as having two faces built up on each face of the pyritohedron. As in the case of the pyritohedron, a rotation of $90°$ about one of the crystallographic axes brings the positive diploid into the negative position.

In addition to the pyritohedron and the diploid, there may be present the cube, dodecahedron, octahedron, trapezohedron, and trisoctahedron. On some crystals these forms may appear alone and so perfectly developed that they cannot be distinguished from the forms of the hexoctahedral class. This is often true of octahedrons of pyrite. Usually, however, they will show by the presence of stria-tion lines or etching figures that they conform to the symmetry of the diploidal class. This is shown in Fig. 98, which represents a cube of pyrite with characteristic striations showing the lower symmetry. Figures 99, 100, and 101 show combina-tions of pyritohedron with forms of the hexoctahedral class. Figure 102 represents a combination of cube and diploid {124}.

The chief mineral of the diploidal class is pyrite; other rarer minerals of this class are skutterudite, chloanthite, gersdorffite, and sperrylite.

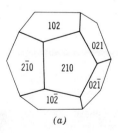
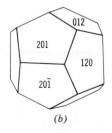

FIG. 97. Pyritohedron. (*a*) Positive. (*b*) Negative.

FIG. 98. Striated cube.

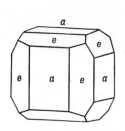

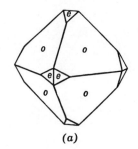

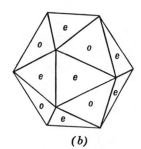

(a) *(b)*

FIG. 99.   Cube and pyritohedron.   **FIG. 100.   Pyritohedron and octahedron.**

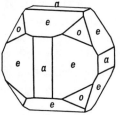

**FIG. 101.   Pyritohedron,
cube, and octahedron.**

FIG. 102.   Diploid and cube.

Tetartoidal Class—23

Symmetry—$3A_2, 4A_3$.   The three crystallographic axes are axes of 2-fold symmetry, and the four diagonal axes are axes of 3-fold symmetry. The symmetry axes are the same as those of the diploid class (Fig. 95a), but there are no symmetry planes and no center. Figure 103 is a drawing of the positive right tetartoid and its stereogram.

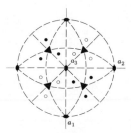

FIG. 103.   Tetartoid and stereogram.

Forms. There are four separate forms of the tetartoid. These are: positive right $\{hkl\}$, positive left $\{khl\}$, negative right $\{k\bar{h}l\}$, negative left $\{h\bar{k}l\}$. They comprise two enantiomorphic pairs, positive right and left, and negative right and left. Other forms that may be present are the cube, dodecahedron, pyritohedron, tetrahedron, and deltoid dodecahedron.

Cobaltite and ullmanite, NiSbS, are the most common mineral representatives crystallizing in the tetartoidal class.

Characteristics of Isometric Crystals

Four 3-fold symmetry axes are common to all isometric crystals. Symmetrically developed crystals are equidimensional in the three directions of the crystallographic axes. The crystals commonly show faces that are squares, equilateral triangles or these figures with truncated corners. All forms are *closed forms*. Thus, crystals are characterized by the large number of similar faces; the smallest number of any form of the hexoctahedral class is six. The same indices are used for forms in the different classes, and therefore, in referring to forms by indices, it is necessary to give the class.

Some important *interfacial angles* of the isometric system which may aid in the recognition of the commoner forms are as follows:

$$\text{Cube (100)} \wedge \text{cube (010)} = 90°\,00'$$
$$\text{Octahedron (111)} \wedge \text{octahedron } (\bar{1}11) = 70°\,32'$$
$$\text{Dodecahedron (011)} \wedge \text{dodecahedron (101)} = 60°\,00'$$
$$\text{Cube (100)} \wedge \text{octahedron (111)} = 54°\,44'$$
$$\text{Cube (100)} \wedge \text{dodecahedron (110)} = 45°\,00'$$
$$\text{Octahedron (111)} \wedge \text{dodecahedron (110)} = 35°\,16'$$

HEXAGONAL SYSTEM

The twelve crystal classes in the hexagonal system are divided into two groups—the *hexagonal division* and the *rhombohedral division*. The crystal classes in the hexagonal division have a 6-fold symmetry axis of either rotation or rotary inversion, whereas the crystal classes in the rhombohedral division have a 3-fold axis of either rotation or rotary inversion. However, even on well-formed crystals, it may be difficult to determine the crystal class from morphology alone. Rhombohedral crystals may appear hexagonal because of the presence of dominant hexagonal forms; and hexagonal crystals may appear rhombohedral, for a 6-fold axis of rotary inversion ($\bar{6}$) is equivalent to a 3-fold axis with a plane at right angles ($3/m$).

Crystallographic Axes. The forms of both hexagonal and rhombohedral divisions are referred to four crystallographic axes as proposed by Bravais. Three of these, designated a_1, a_2, a_3, lie in the horizontal plane and are of equal length with angles of $120°$ between the positive ends; the fourth axis, c, is vertical. When properly oriented, one horizontal crystallographic axis, a_2, is left to right, and the other two make $30°$ angles on either side of a line perpendicular to it (Fig. 104). The positive end of a_1 is to the front and left, the positive end of a_2 is to the right, and the positive end of a_3 is to the back and left. Figure 104b shows

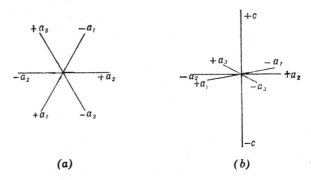

(a)   (b)

FIG. 104. Hexagonal crystal axes.

the four axes in clinographic projection. In stating the indices for any face of a hexagonal crystal, four numbers (the Bravais symbol) must be given. The numbers expressing the reciprocals of the intercepts of a face on the axes are given in the order a_1, a_2, a_3, c. The general Bravais form symbol is $\{hkil\}$ with $h > k$. The third digit of the index is the sum of the first two times -1; or, stated another way, $h + k + i = 0$.

In the Hermann-Mauguin symbols the first number refers to the principal axis of symmetry coincident with c. The second and third symbols refer to the axial and intermediate symmetry elements respectively.

HEXAGONAL DIVISION

Dihexagonal-Dipyramidal Class—$6/m\ 2/m\ 2/m$

Symmetry—C, $1A_6$, $6A_2$, $7P$. The vertical axis is an axis of 6-fold symmetry. There are six horizontal axes of 2-fold symmetry, three of them coincident with the crystallographic axes and the other three lying midway between them (Fig. 105a). There are six vertical planes of symmetry perpendicular to the 2-fold axes and a horizontal plane of symmetry (Fig. 105b). Figure 106 shows a dihexagonal dipyramid and its stereogram.

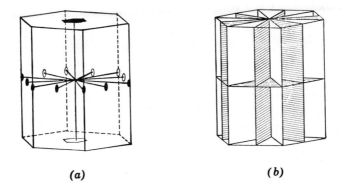

(a)   (b)

FIG. 105.   Symmetry of dihexagonal-dipyramidal class. (a) Axes. (b) Planes.

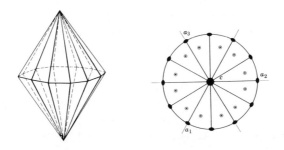

FIG. 106.   Dihexagonal dipyramid and stereogram.

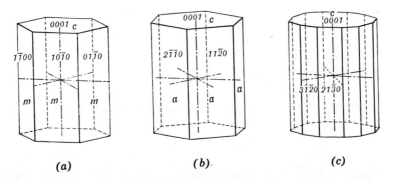

(a)   (b)   (c)

FIG. 107.   Hexagonal prisms. (a) First-order, (b) Second-order. (c) Dihexagonal.

Forms. 1. *Basal Pinacoid* {0001} is composed of two horizontal faces. It is shown in combination with different prisms in Fig. 107.

2. *Prism of First-Order* {10$\bar{1}$0} consists of six vertical faces, each of which intersects two of the horizontal crystallographic axes equally and is parallel to the third (Fig. 107a).

3. *Prism of Second-Order* {11$\bar{2}$0} has six vertical faces, each of which intersects two of the horizontal axes equally and the intermediate horizontal axis at one-half this distance. (Fig. 107b.) The prisms of the first- and second-order are geometrically identical forms, the distinction between them is only in orientation.

4. *Dihexagonal Prism* {hk$\bar{i}$0} has twelve vertical faces, each of which intersects all three of the horizontal crystallographic axes at different lengths. There are various dihexagonal prisms, depending upon their different relations to the horizontal axes. A common dihexagonal prism with indices {21$\bar{3}$0} is shown in Fig. 107c.

5. *Dipyramid of First-Order* {h0$\bar{h}$l} consists of twelve isosceles triangular faces, each of which intersects two horizontal crystallographic axes equally, is parallel to the third, and intersects the vertical axis. Various dipyramids of the first-order are possible, depending upon the inclination of the faces to the c axis. The unit form has the indices {10$\bar{1}$1} (Fig. 108a).

6. *Dipyramid of Second-Order* {hh$\overline{2h}$l} is composed of twelve isosceles triangular faces. Each face intersects two of the horizontal axes equally and the third (the intermediate horizontal axis) at one-half this distance; it also intersects the vertical axis. Various dipyramids of the second-order are possible, depending upon the inclination of the faces to c. A common form (Fig. 108b) has the indices {11$\bar{2}$2}. The relations between the dipyramids of the first- and second-order are the same as between the corresponding prisms.

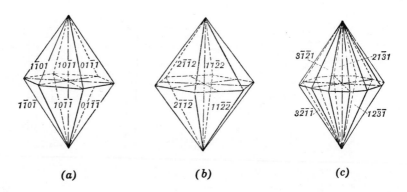

FIG. 108. Hexagonal dipyramids. (*a*) First-order. (*b*) Second-order. (*c*) Dihexagonal.

If only one dipyramid is present on a crystal, it is usually set as of the first-order. If dipyramids of both orders are present, the dominant one, in the absence of other evidence, is considered the first-order. If a dipyramid is combined with prismatic forms, the orientation of the crystal is usually determined by the dipyramid.

7. *Dihexagonal Dipyramid* $\{hk\bar{i}l\}$ is composed of twenty-four triangular faces. Each face is a scalene triangle which intersects all three of the horizontal axes differently and also intersects the vertical axis. A common form, $\{21\bar{3}1\}$, is shown in Fig. 108c.

Combinations of the forms of this class are shown in Fig. 109.

Beryl affords the best example of a mineral representative in this class. Other minerals are molybdenite, pyrrhotite, and niccolite.

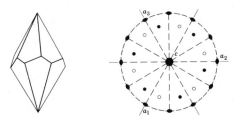

FIG. 109.   Combinations of hexagonal forms.

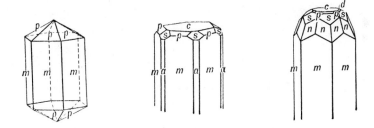

FIG. 110.   Hexagonal trapezohedron and stereogram.

Hexagonal-Trapezohedral Class—622

Symmetry—$1A_6, 6A_2$. The symmetry axes are the same as those in the dihexagonal-dipyramidal class (Fig. 105), but there are no symmetry planes nor symmetry center.

Forms. The *hexagonal trapezohedrons* {*hkil*} *right* and {*ihkl*} *left* are enantio-morphic forms, each with six trapezium-shaped faces (see Fig. 110). Other forms that may be present are the pinacoid, first- and second-order hexagonal prisms and dipyramids and dihexagonal prisms.

High quartz and kalsilite, $KAlSiO_4$, are the only mineral representatives in this class.

Dihexagonal-Pyramidal Class—6*mm*

Symmetry—$1A_6, 6P$. There is a vertical axis of 6-fold symmetry and six vertical symmetry planes intersecting in this axis. Figure 111 shows a dihexagonal pyramid and its stereogram.

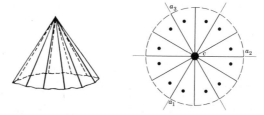

FIG. 111.   Dihexagonal pyramid and stereogram.

Forms. The forms of the dihexagonal-pyramidal class are similar to those of the dihexagonal-dipyramidal class, but, inasmuch as a horizontal plane of symmetry is lacking, different forms appear at the top and bottom of the crystal. The *dihexagonal pyramid* is thus two forms: {*hkīl*} upper and {*hkīl̄*} lower. The hexagonal-pyramidal forms are: {*h0h̄l*} upper and {*h0h̄l̄*} lower; and {*hh2̄hl*} upper and {*hh2̄hl̄*} lower. The pinacoid cannot exist here, but instead there are two pedions ·{0001} and {0001̄}. The first- and second-order hexagonal prisms and the dihexagonal prism may be present.

Wurtzite, greenockite, and zincite are the commonest mineral representatives in this class. Figure 112 represents a zincite crystal with a hexagonal prism termi-nated above by a hexagonal pyramid and below by a pedion.

FIG. 112.   Zincite.

Ditrigonal-Dipyramidal Class—$\bar{6}m2$

Symmetry—$1A_3, 3A_2, 4P$.   The vertical axis is a 6-fold axis of inversion which is equivalent to a 3-fold axis of rotation with a horizontal symmetry plane. Three symmetry planes intersect in the vertical axis, and three horizontal 2-fold axes lie in the vertical symmetry planes. A ditrigonal dipyramid and its stereogram are shown in Fig. 113.

Forms.   The ditrigonal dipyramid $\{hkil\}$ is a twelve-faced form with six faces at the top of the crystal and six at the bottom. Additional forms that may be present are: pinacoid, trigonal prisms (see page 26), second-order hexagonal prism, ditrigonal prisms, trigonal dipyramids, and second-order hexagonal dipyramids.

Benitoite is the only mineral that has been described as definitely crystallizing in this class.

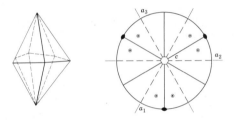

FIG. 113.   Ditrigonal dipyramid and stereogram.

Hexagonal-Dipyramidal Class—$6/m$

Symmetry—$C, 1A_6, 1P$.   There is a vertical axis of 6-fold symmetry, a horizontal plane of symmetry, and a symmetry center. Figure 114 illustrates a hexagonal dipyramid and its stereogram.

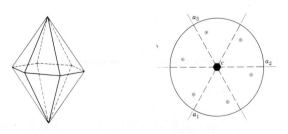

FIG. 114.   Hexagonal dipyramid and stereogram.

Forms. The general forms of this class are the dipyramids, {*hkīl*} *positive*, {*hkīl̄*} *negative*. These forms consist of twelve faces, six above and six below, which correspond in position to one-half the faces of a dihexagonal dipyramid. Other forms that may be present are: pinacoid and hexagonal prisms {*hki*0}.

The hexagonal-dipyramidal class has as its chief mineral representatives the minerals of the apatite group. The dipyramid revealing the symmetry of the class is rarely seen but is illustrated as face μ, Fig. 115.

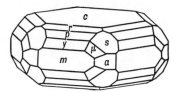

FIG. 115. Apatite crystal.

Hexagonal-Pyramidal Class—6

Symmetry—$1A_6$. A vertical 6-fold rotation axis is the only symmetry of this class. A hexagonal pyramid and its stereogram are shown in Fig. 116.

Forms. There are four hexagonal pyramids {*hkil*}, two at the top and two at the bottom of the crystal, each of which corresponds to six faces of the dihexagonal dipyramid. Other forms that may be present are pedions and hexagonal prisms {*hki*0}.

The form development of crystals is rarely sufficient to enable one without other evidence to place unequivocally a crystal in this class. The mineral nepheline is the chief mineral representative.

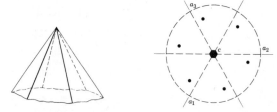

FIG. 116. Hexagonal pyramid and stereogram.

Trigonal-Dipyramidal Class—$\bar{6}$

Symmetry—$1A_3$, $1P$. The vertical axis is a 6-fold axis of rotary inversion ($\bar{6}$) which is equivalent to a 3-fold axis of rotation with a symmetry plane at right angles to it ($3/m$). Figure 117 shows a trigonal dipyramid and its stereogram.

Forms. There are four trigonal dipyramids $\{hkil\}$, each with six faces corresponding to six faces of the dihexagonal dipyramid. Other forms that may be present are the pinacoid and trigonal prisms $\{hki0\}$. The symmetry does not permit hexagonal prisms, but instead there are two trigonal prisms. For example, the first-order hexagonal prism $\{10\bar{1}0\}$ becomes the two trigonal prisms $\{10\bar{1}0\}$ and $\{01\bar{1}0\}$.

There is no example of an authenticated mineral or other crystalline substance belonging to this crystal class.

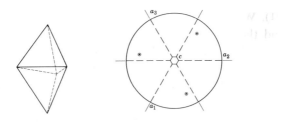

FIG. 117. Trigonal dipyramid and stereogram.

RHOMBOHEDRAL DIVISION

The forms of the crystal classes of the rhombohedral division of the hexagonal system are referred to the same crystallographic axes as the forms in the classes of the hexagonal division. However, some workers use three axes parallel to the three culminating edges of the unit rhombohedron to describe the forms. All the classes are characterized by a 3-fold axis of either rotation or rotary inversion. The crystals usually show a lower symmetry than those of the hexagonal division, and are generally recognized by a 3-fold distribution of faces at the ends of the principal axis. In the section on descriptive mineralogy the crystal system of minerals crystallizing in the rhombohedral division is designated as *Hexagonal–R*.

Hexagonal-Scalenohedral Class—$\bar{3}2/m$

Symmetry—$C, 1A_3, 3A_2, 3P$. The vertical crystallographic axis is one of 3-fold symmetry, and the three horizontal crystallographic axes are axes of 2-fold symmetry (see Fig. 118a). There are three vertical planes of symmetry bisecting the angles between the horizontal axes (see Fig. 118b). Figure 119 illustrates a hexagonal scalenohedron and its stereogram.

Forms. 1. *Rhombohedron* $\{h0\bar{h}l\}$ *positive,* $\{0h\bar{h}l\}$ *negative.* The rhombohedron is a form consisting of six rhomb-shaped faces, which correspond in

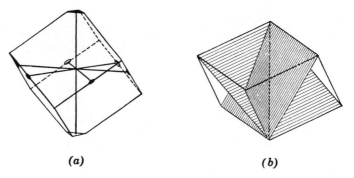

(a)   (b)

FIG. 118.   Symmetry of hexagonal scalenohedral class. (a) Axes. (b) Planes.

their positions to the alternate faces of a hexagonal dipyramid of the first order. The relation of these two forms to each other is shown in Fig. 120. The rhombohedron may also be thought of as a cube deformed in the direction of one of the axes of 3-fold symmetry. The deformation may appear either as an elongation along the symmetry axis producing an acute solid angle, or compression along the symmetry axis producing an obtuse solid angle. Depending on the angle, the rhombohedron is known as acute or obtuse.

Depending on the orientation the rhombohedron may be positive or negative (Fig. 121). When properly oriented, the positive rhombohedron has one of its faces, and the negative rhombohedron one of its edges, toward the observer. There are various rhombohedrons that differ from each other in the inclination of their faces to the c axis. The index symbol of the unit positive rhombohedron is $\{10\bar{1}1\}$ and of the unit negative rhombohedron $\{01\bar{1}1\}$.

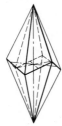 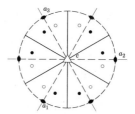

FIG. 119.   Hexagonal scalenohedron and stereogram.

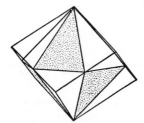

FIG. 120.   Relation between first-order hexagonal dipyramid and rhombohedron.

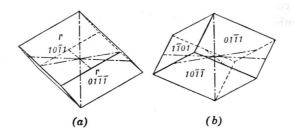

FIG. 121. Rhombohedrons. (*a*) Positive. (*b*) Negative.

The rhombohedron is such an important form in the hexagonal system that it need not appear externally on a crystal to determine the orientation. The orientation of calcite is determined by the rhombohedral cleavage, and the orientation of corundum is determined by rhombohedral parting (see page 98). Thus, in calcite the only external rhombohedral form may be negative, and in corundum the rhombohedral parting invariably necessitates orienting the crystal so that the prism is of the second-order. However, if but one rhombohedron is present on a crystal, it is oriented, in the absence of other evidence, in the positive position.

2. *Scalenohedron* $\{hk\bar{\imath}l\}$ *positive*, $\{kh\bar{\imath}l\}$ *negative*. This form consists of twelve scalene triangular faces corresponding in position to alternate *pairs* of faces of a dihexagonal dipyramid. (Fig. 122.) The scalenohedron is differentiated from the dipyramid by the zigzag appearance of the middle edges. It is in the positive position when the angle between the upper and lower faces points down toward the observer (Fig. 123a) and in the negative position when the angle points up (Fig. 123b).

There are many different scalenohedrons; most frequently seen is $\{21\bar{3}1\}$, a common form on calcite (Fig. 123a).

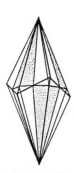

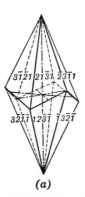

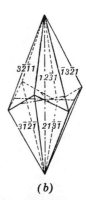

FIG. 122. Relation of dihexagonal dipyramid and scalenohedron.

FIG. 123. Hexagonal scalenohedrons. (*a*) Positive. (*b*) Negative.

The rhombohedron and scalenohedron of the hexagonal-scalenohedral class may combine with forms found in classes of higher hexagonal symmetry. Thus they may be in combination with the first- and second-order hexagonal prisms, dihexagonal prisms, second-order dipyramid, and basal pinacoid (Fig. 124). (See also Figs. 125 and 126.)

Several common minerals crystallize in this class. Chief among them is calcite and the other members of the calcite group. Other minerals are corundum, hematite, brucite, soda niter, arsenic, millerite, antimony, and bismuth.

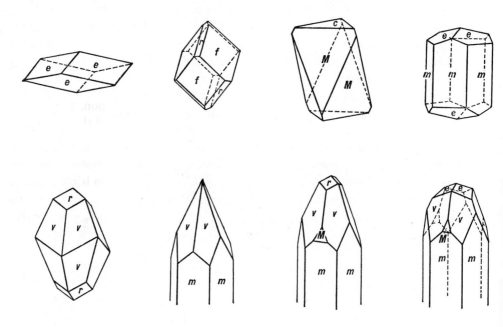

FIG. 124.   Calcite crystals. Forms: $e\{01\bar{1}2\}$ and $f\{02\bar{2}1\}$, negative rhombohedrons. $r\{10\bar{1}0\}$ and $M\{40\bar{4}1\}$ positive rhombohedrons. $m\{10\bar{1}0\}$, first-order prism. $c\{0001\}$, basal pinacoid. $v\{21\bar{3}1\}$, scalenohedron.

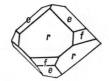

FIG. 125.   Chabazite.

FIG. 126.   Corundum.

Trigonal-Trapezohedral Class—32

Symmetry—$1A_3$, $3A_2$. The vertical crystallographic axis is an axis of 3-fold symmetry, and the three horizontal crystallographic axes are axes of 2-fold symmetry. The symmetry axes are the same as in the scalenohedral class, but planes of symmetry are lacking. Figure 127 shows a trigonal trapezohedron and its stereogram.

Forms. There are four trigonal trapezohedrons, each made up of six trapezium-shaped faces. These faces correspond in position to one-quarter of the faces of a dihexagonal dipyramid and thus have similar symbols, as follows: positive right $\{hk\bar{i}l\}$, positive left $\{i\bar{k}\bar{h}l\}$, negative left $\{kh\bar{i}l\}$, negative right $\{\bar{k}i\bar{h}l\}$. These forms can be grouped into two enantiomorphic pairs each with a *right* and *left* form. Other forms that may be present are: pinacoid, first-order hexagonal prism, ditrigonal prisms, and rhombohedrons. The second-order hexagonal prism is not present, but instead there are two trigonal prisms $\{11\bar{2}0\}$ and $\{2\bar{1}\bar{1}0\}$. Likewise the second-order hexagonal dipyramid becomes two trigonal dipyramids $\{hh\bar{2}hl\}$ and $\{2h\bar{h}\bar{h}l\}$.

Low-temperature quartz is the most common mineral crystallizing in this class, but only rarely can faces of the trigonal trapezohedron be observed. When this form is present, the crystals can be distinguished as right-handed or left-handed (Fig. 128), depending on whether, with a prism face fronting the observer, the trigonal trapezohedral faces, x, truncate the edges between prism and the top rhombohedron faces at the right or at the left. The faces marked s are trigonal dipyramids.

Cinnabar and the rare mineral berlinite, $AlPO_4$, also crystallize in the trigonal trapezohedral class.

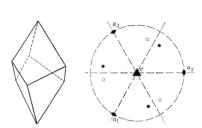

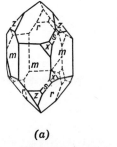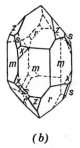

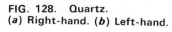

(a)   (b)

FIG. 127. Trigonal trapezohedron and stereogram.

FIG. 128. Quartz.
(a) Right-hand. (b) Left-hand.

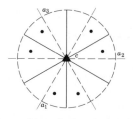

FIG. 129. Ditrigonal pyramid and stereogram.

Ditrigonal-Pyramidal Class—3m

Symmetry—$1A_3$, $3P$. The vertical axis is a 3-fold symmetry axis, and there are three symmetry planes that intersect in this axis. A ditrigonal pyramid and its stereogram are shown in Fig. 129.

Forms. The forms are similar to those of the hexagonal scalenohedral class but with only half the number of faces. Because of the lack of 2-fold symmetry axes, the faces at the top of the crystals belong to different forms from those at the bottom. There are four ditrigonal pyramids, each corresponding to half the faces of either the positive or negative scalenohedron with indices: $\{hk\bar{i}l\}$, $\{kh\bar{i}l\}$, $\{hk\bar{i}\bar{l}\}$, $\{kh\bar{i}\bar{l}\}$. Other forms that may be present are pedions, second-order hexagonal prism and pyramids, trigonal pyramids, trigonal prisms, and ditrigonal prisms. There are four trigonal pyramids; two of these are at the top of the crystal, one corresponding to the top three faces of the positive rhombohedron, the other corresponding to the top three faces of the negative rhombohedron. Their respective indices are $\{h0\bar{h}l\}$ and $\{0h\bar{h}l\}$. The two trigonal pyramids at the bottom of the crystal correspond to the bottom faces of the rhombohedron with indices $\{0h\bar{h}\bar{l}\}$ and $\{h0\bar{h}\bar{l}\}$. The trigonal prisms $\{10\bar{1}0\}$ and $\{01\bar{1}0\}$ each corresponds to three faces of the first-order hexagonal prism. There are two ditrigonal prisms, $\{hk\bar{i}0\}$ and $\{kh\bar{i}0\}$, each corresponding to half the faces of a dihexagonal prism.

Tourmaline (Fig. 130) is the most common mineral crystallizing in this class, but in addition there are pyrargyrite, proustite, and alunite.

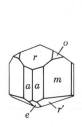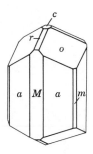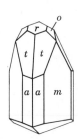

FIG. 130. Tourmaline crystals: Forms: $r = \{10\bar{1}1\}$, $o\{02\bar{2}1\}$, $t\{21\bar{3}1\}$, $c\{0001\}$, $a\{11\bar{2}0\}$, $M\{10\bar{1}0\}$, $m\{01\bar{1}0\}$, $e\{10\bar{1}2\}$, $r'\{01\bar{1}\bar{1}\}$.

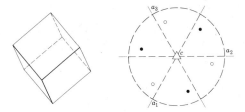

FIG. 131. Rhombohedron and stereogram.

Rhombohedral Class—$\bar{3}$

Symmetry—C, $1A_3$. There is one 3-fold axis of inversion. This is equivalent to a 3-fold rotation axis and a center. Figure 131 illustrates a rhombohedron and its stereogram.

Forms. The rhombohedron, which is here the general form $\{hkil\}$, is described on page 63 as a special form of the hexagonal-scalenohedral class. If it were to appear alone on a crystal, it would have the morphological symmetry of that class ($\bar{3}2/m$). It is only in combination with other forms that its true symmetry becomes apparent. In addition to the rhombohedron there may be the pinacoid and hexagonal prisms, $\{hki0\}$.

Dolomite is the most common mineral crystallizing in this class; other representatives are ilmenite, willemite, and phenacite.

Trigonal-Pyramidal Class—3

Symmetry—$1A_3$. One 3-fold rotation axis is the only symmetry. Figure 132 shows a trigonal pyramid and its stereogram.

Forms. The trigonal pyramid, $\{hkil\}$, is the general form. In combination with the pedion it appears to have the symmetry of the ditrigonal-pyramidal class ($3m$) with three vertical symmetry planes (Fig. 129). It is only when certain trigonal pyramids are in combination with one another that the true symmetry is revealed. Pedions and trigonal prisms may be present.

Possibly the mineral gratonite, $Pb_9As_4S_{15}$ belongs in this class; there are no other mineral representatives.

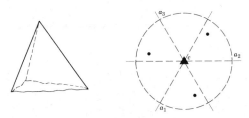

FIG. 132. Trigonal pyramid and stereogram.

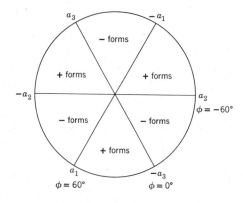

FIG. 133.   Distribution of rhombohedral forms.

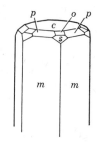

FIG. 134.   Beryl.

Axial Ratios in the Hexagonal System

With the exception of the hexagonal system, crystals are oriented so that (010) is to the right with $\phi = 0°\,00'$. In the hexagonal system the negative end of the a_3 axis is taken as $\phi = 0°$. According to this convention ϕ of second-order forms is $0°$, whereas ϕ of first-order forms is $30°$. This apparent inconsistency has definite advantages in working with rhombohedral crystals; it gives positive forms $+\phi$ values and negative forms $-\phi$ values. (See Fig. 133.)

To determine an axial ratio of a crystal, the forms must first be indexed and their ϕ and ρ angles determined. For many crystals ϕ and ρ can be measured directly as interfacial angles. For other crystals it may be necessary to project the measurable interfacial angles and determine ϕ and ρ from the projection. (See p. 40.)

Consider the drawing of the beryl crystal, Fig. 134. It is oriented so that m and p are first order forms and therefore $\phi = 30°$. o and s are second-order forms with $\phi = 0°$. The angles $c \wedge p$, $c \wedge o$, $c \wedge s$ are respectively the ρ angles of p, o, and s. If c were not present the ρ angle of p could be determined as the complement of $m \wedge p$. If only faces of form p were present, an interfacial angle could be measured over the top of the crystal between p and p' (p' a face at the back of the crystal $180°$ removed from p). ρp equals half of this measured angle.

An axial ratio in the hexagonal system expresses the length of c in terms of a as unity or $a:c = 1:?$. For ease of calculations the $-a_3$ axis is taken as unity since it is at the position of $\phi = 0°$. The reciprocal of the intercept on this axis is $i = -(h + k)$.

The formula for determining c from the angles of the general form is:

$$c = \frac{l \tan \rho_{hk\bar{i}l} \cos \phi}{h + k}.$$

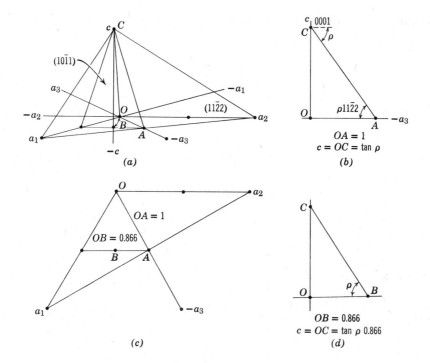

FIG. 135. First- and second-order hexagonal forms.

If a second-order form is used, $\phi = 0°$, and $\cos \phi = 1$; hence $\cos \phi$ disappears from the equation, leaving:

$$c = \frac{l \tan \rho_{hh\bar{2}hl}}{h + k}$$

A face of the second-order form $\{11\bar{2}2\}$ intersects $-a_3$ and c at unit distances but a_1 and a_2 at twice unity (Fig. 135a). Thus for this form $l = 2, h + k = 2$ and $c = \tan \rho$. For the form $\{11\bar{2}1\}$, $c = \tan \rho/2$.

For a first-order form: $\phi = 30°$, $\cos \phi = 0.8660$ and the equation becomes:

$$c = \frac{l \tan \rho_{h0\bar{h}l} \times 0.8660}{h + k}$$

TETRAGONAL SYSTEM

Crystallographic Axes. The forms of the tetragonal system are referred to three crystallographic axes that make right angles with each other. The two horizontal axes, a, are equal in length and interchangeable, but the vertical axis, c,

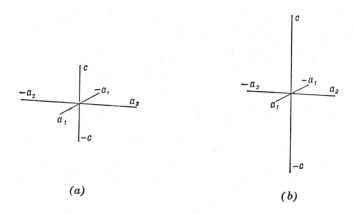

FIG. 136. Tetragonal crystal axes.

is of a different length. Figure 136a represents the crystallographic axes for the tetragonal mineral zircon with c less than a. Figure 136b represents the crystallographic axes of the mineral octahedrite with c greater than a. The length of the horizontal axes is taken as unity, and the relative length of the vertical axis is expressed in terms of the horizontal. The axial ratio must be determined for each tetragonal crystal by measuring the interfacial angles and making the proper calculations (see page 80). For zircon, the length of the vertical axis is expressed as $c = 0.901$, for octahedrite as $c = 1.777$. The proper orientation of the crystallographic axes and the method of their notation are shown in Fig. 136. When the general form symbols are used, $h < k$.

In the Hermann Mauguin symbols of the tetragonal system, the first part of the symbol refers to the principal axis of symmetry. The second and third parts refer to the axial and diagonal symmetry elements respectively.

Ditetragonal-Dipyramidal Class—4/m 2/m 2/m

Symmetry—C, $1A_4$, $4A_2$, $5P$. The vertical crystallographic axis is an axis of 4-fold symmetry. There are four horizontal axes of 2-fold symmetry, two of which are coincident with the crystallographic axes, and the others at $45°$ to them. There are five planes of symmetry, one horizontal and four vertical. One of the horizontal axes of symmetry lies in each of the vertical symmetry planes. The position of the axes and planes of symmetry is shown in Fig. 137. Figure 138 illustrates a ditetragonal dipyramid and its stereogram.

Forms. 1. *Basal Pinacoid* {001} is a form composed of two horizontal faces. It is shown in combination with different prisms in Fig. 139.

2. *Prism of First-Order* {110}} consists of four rectangular vertical faces, each of which intersects the two horizontal crystallographic axes equally (Fig. 139a).

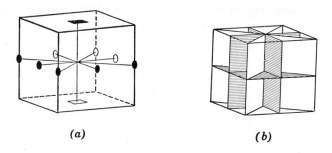

(a)   (b)

FIG. 137. Symmetry of ditetragonal dipyramidal class. (*a*) Axes. (*b*) Planes.

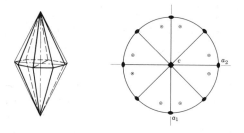

FIG. 138. Ditetragonal dipyramid and stereogram.

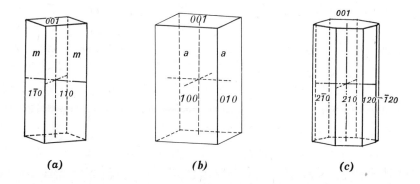

(a)   (b)   (c)

FIG. 139. Tetragonal prisms. (*a*) First-order. (*b*) Second-order. (*c*) Ditetragonal.

3. *Prism of Second-Order* {010} consists of four rectangular vertical faces, each of which intersects one horizontal crystallographic axis and is parallel to the other two axes. The form is represented in Fig. 139. The prisms of the first- and second-orders are identical forms, except for their orientation. They can be converted into each other by a revolution of 45° about the vertical axis. Since both may occur together upon the same crystal, it is necessary to recognize the two forms.

4. *Ditetragonal Prism* {hk0} consists of eight rectangular vertical faces, each of which intersects the two horizontal crystallographic axes unequally. There are various ditetragonal prisms, depending upon their differing relations to the horizontal axes. A common form, represented in Fig. 139c, has indices {120}.

5. *Dipyramid of First-Order* {hhl} has eight isosceles triangular faces, each of which intersects all three crystallographic axes, with equal intercepts upon the two horizontal axes. There are various dipyramids of the first-order, depending upon the inclination of their faces to *c*. The unit dipyramid {111} (Fig. 140a) which intersects all the axes at their unit lengths, is most common. Indices of other dipyramids of the first-order are {221}, {331}, {112}, {113}, etc., or, in general, {hhl}.

6. *Dipyramid of Second-Order* {0kl} is composed of eight isosceles triangular faces, each of which intersects one horizontal axis and the vertical axis and is parallel to the second horizontal axis. There are various dipyramids of the second-order, with different intersections upon the vertical axis. The most common is the unit dipyramid {011} (Fig. 140b). Other dipyramids of the second-order have indices {021}, {031}, {012}, {013}, or, in general {0kl}.

The relationship between the dipyramids of the first- and second-order is similar to that between the prisms of the first- and second-order.

As a general rule in the absence of other evidence, if one dipyramid is present, it is set as first-order. If two dipyramids of different orders are present, the

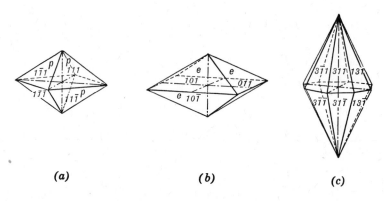

(a)   (b)   (c)

FIG. 140.   Tetragonal dipyramids. (a) First-order. (b) Second-order. (c) Ditetragonal.

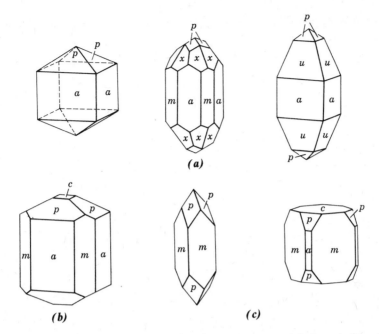

FIG. 141. Tetragonal crystals. (*a*) Zircon. (*b*) Idocrase. (*c*) Apophylite. Forms: *p*{011} and *u*{012}, second-order dipyramids. *c*{001}, basal pinacoid. *a*{010}, second-order prism. *m*{110}, first-order prism. *x*{211}, ditetragonal dipyramid.

dominant one is usually set as first-order. In the orientation of a crystal, the prisms are subordinate to the dipyramids. Thus, an important prism may be relegated to second-order by the presence of a small dipyramid.

7. *Ditetragonal Dipyramid* {*hkl*} is composed of sixteen triangular faces, each of which intersects all three of the crystallographic axes, cutting the two horizontal axes at different lengths. There are various ditetragonal dipyramids, depending upon the different intersections on the crystallographic axes. One of the most common is the dipyramid {131} shown in Fig. 140c.

Several common minerals crystallize in the ditetragonal-dipyramidal class. Major representatives are rutile, anatase, cassiterite, apophyllite zircon, and idocrase.

Tetragonal Combinations. Characteristic combinations of ditetragonal-dipyramidal forms as found on crystals of different minerals are represented in Fig. 141.

Tetragonal-Trapezohedral Class—422

Symmetry—$1A_2, 4A_2$. The vertical axis is one of 4-fold symmetry, and there are four 2-fold axes at right angles to it. The symmetry axes are the same as those in the ditetragonal-dipyramidal class (Fig. 137), but symmetry planes and center are lacking.

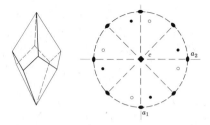

FIG. 142. Tetragonal trapezohedron and stereogram.

Forms. The tetragonal trapezohedron has eight faces, corresponding to half the faces of the ditetragonal dipyramid. There are two enantiomorphic forms, right $\{hkl\}$ (Fig. 142) and left $\{h\bar{k}l\}$. Other forms that may be present are: pinacoid, first- and second-order tetragonal prisms, ditetragonal prism, and first- and second-order tetragonal dipyramids.

Phosgenite, $Pb_2CO_3Cl_2$, is the only mineral representative in this class.

Ditetragonal-Pyramidal Class—4*mm*

Symmetry—$1A_4$, $4P$. The vertical axis is a 4-fold symmetry axis, and four planes of symmetry intersect in this axis. Figure 143 illustrates a ditetragonal pyramid and its stereogram.

Forms. The lack of a horizontal symmetry plane gives rise to different forms at the top and bottom of crystals of this class. There are pedions $\{001\}$ and $\{00\bar{1}\}$. The first-order $\{hhl\}$ and second-order $\{h0l\}$ tetragonal pyramids have corresponding lower forms, $\{hh\bar{l}\}$ and $\{h0\bar{l}\}$. The ditetragonal pyramid $\{hkl\}$ is an upper form, whereas $\{hk\bar{l}\}$ is the lower form. The first- and second-order tetragonal prisms as well as ditetragonal prism may be present.

The rather rare mineral diaboleite, $Pb_2Cu(OH)_4Cl_2$, is the only mineral representative in this crystal class.

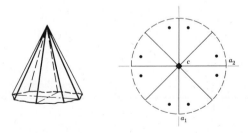

FIG. 143. Ditetragonal pyramid and stereogram.

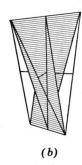

(a)   (b)

FIG. 144.   Symmetry of tetragonal scalenohedral class. (a) Axes. (b) Planes.

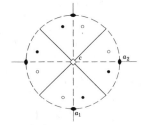

FIG. 145.   Tetragonal scalenohedron and stereogram.

Tetragonal-Scalenohedral Class—$\bar{4}2m$

Symmetry—$3A_2, 2P$.   The vertical crystallographic axis is a 4-fold axis of rotary inversion that appears morphologically as a 2-fold rotation axis. The *a* crystallographic axes are axes of 2-fold symmetry.   At 45 degrees to these axes are two vertical symmetry planes intersecting in the vertical axis. (See Fig. 144.) Figure 145 illustrates a tetragonal scalenohedron and its stereogram.

Forms.   1. *Disphenoid* $\{hhl\}$ *positive*, $\{h\bar{h}l\}$ *negative* are the only important forms in this class. They consist of four isosceles triangular faces which intersect all three of the crystallographic axes, with equal intercepts on the two horizontal axes. The faces correspond in their position to the alternating faces of the tetragonal dipyramid of the first-order. There may be different disphenoids, depending upon their varying intersections with the vertical axis. Two different disphenoids are shown in Figs. 146a and 146b. There may also be a combination of a positive and a negative disphenoid as represented in Fig. 146c.

The disphenoid differs from the tetrahedron in the fact that its vertical crystallographic axis is not the same length as the horizontal axes. The only common mineral in the tetragonal-scalenohedral class is chalcopyrite, crystals of which

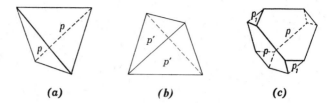

FIG. 146. Tetragonal disphenoids. (*a*) Positive. (*b*) Negative. (*c*) Positive and negative.

ordinarily show only the disphenoid {112}. This disphenoid closely resembles a tetrahedron, and it requires accurate measurements to prove its tetragonal character.

2. *Tetragonal Scalenohedron {hkl}.* This form, Fig. 145, if it were to occur by itself, is bounded by eight similar scalene triangles, which correspond in their position to alternate *pairs* of faces of the ditetragonal dipyramid. It is a rare form and observed only in combination with others. Other forms that may be present are: pinacoid, first- and second-order tetragonal prisms, ditetragonal prisms, and second-order tetragonal dipyramids.

Chalcopyrite and stannite are the only common minerals that crystallize in this class.

Tetragonal-Dipyramidal Class—4/m

Symmetry—*C*, 1*A₄*, 1*P*. There is a vertical 4-fold symmetry axis with a symmetry plane at right angles. Figure 147 illustrates a tetragonal dipyramid and its stereogram.

Forms. The tetragonal dipyramid, {*hkl*}, is an eight-faced form having four upper faces directly above four lower faces. This form by itself appears to have higher symmetry, and it must be in combination with other forms to reveal the absence of vertical symmetry planes. The basal pinacoid and tetragonal prisms {*hk0*} may be present. The tetragonal prism {*hk0*} is equivalent to four alternate faces of the ditetragonal prism and is present in those classes which have no vertical symmetry planes or 2-fold horizontal symmetry axes.

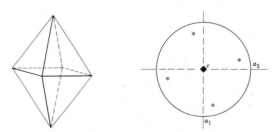

FIG. 147. Tetragonal dipyramid and stereogram.

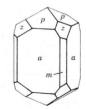

FIG. 148. Scapolite.

Mineral representatives in this class are: fergusonite, scheelite, powellite, and scapolite. Figure 148 illustrates a crystal of scapolite showing the true symmetry of this class.

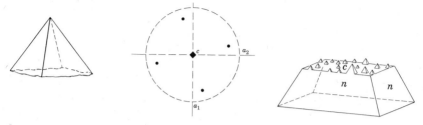

FIG. 149.   Tetragonal pyramid and stereogram.   FIG. 150.   Wulfenite.

Tetragonal-Pyramidal Class—4

Symmetry—$1A_4$.   The vertical axis is one of 4-fold symmetry. There are no symmetry planes or center. The tetragonal pyramid and its stereogram are shown in Fig. 149.

Forms.   The tetragonal pyramid is a four-faced form. The upper form $\{hkl\}$ is different from the lower form $\{hk\bar{l}\}$, and each has a right- and left-hand variation. There are thus two enantiomorphic pairs of tetragonal pyramids. Pedions and tetragonal prisms may also be present.

As in some other classes, the true symmetry is not shown morphologically unless the general form is in combination with other forms. Figure 150, is a drawing of wulfenite. Other mineral representatives are unknown.

Tetragonal-Disphenoidal Class—$\bar{4}$

Symmetry—$1\,P_4$.   The vertical axis is a 4-fold axis of rotary inversion. There is no other symmetry. Figure 151 illustrates a tetragonal disphenoid and its stereogram.

Forms.   The tetragonal disphenoid $\{hkl\}$ is a closed form composed of four wedge-shaped faces. In the absence of other modifying faces, the form appears

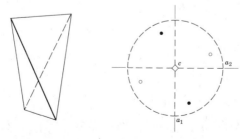

FIG. 151.   Tetragonal disphenoid and stereogram.

to have two vertical symmetry planes giving it the symmetry $\bar{4}2m$. The true symmetry is shown only in combination with other forms. The pinacoid and tetragonal prisms may be present.

The only mineral representative in this class is the rare mineral, cahnite, $CaB(OH)_4AsO_4$.

Tetragonal Axial Ratios

The axial ratio of a tetragonal crystal is expressed as $a:c$, with the length of the two equal a axes taken as unity. It is calculated from ϕ and ρ angles derived from interfacial angles, by the general formula:

$$c = \left(\frac{l}{k}\right) \tan \rho \cos \phi \quad \text{where} \quad k \text{ and } l \text{ are Miller indices.}$$

Tetragonal crystals are so oriented that (010), perpendicular to a_2, has $\phi = 0°$. Thus, for second-order forms, $\{0kl\}$, $\cos \phi = 1$ and the formula becomes: $c = (l/k) \tan \rho$. For the trigonometry involved, consider the calculation of c from angular measurements of face (021), (Fig. 152a, b). Tan $\rho_{021} = CO/OA$. $CO = 2c$, $AO = a = 1$. Thus, $c = \tan \rho_{021}/2$. For the unit form $\{011\}$, $c = \tan \rho$.

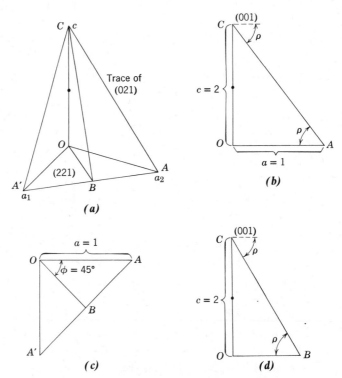

FIG. 152. Tetragonal crystal, angular relationships.

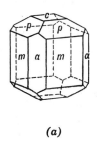

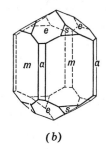

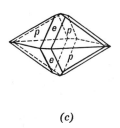

(a)   (b)   (c)

FIG. 153. Tetragonal crystals. Forms: $a\{010\}$, $c\{001\}$, $m\{110\}$, $e\{011\}$, $p\{111\}$, $s\{221\}$.

For first-order tetragonal forms, $\phi = 45°$, $\cos \phi = 0.7071$ and

$$c = \left(\frac{l}{k}\right) \tan \rho \cdot 0.7071.$$

Consider the calculation of c using the angular measurements of face (221), (Fig. 152a, c, d). In triangle AOB, $OB = \cos 45°$. In triangle COB, $OC = 2c$, $\tan \rho_{221} = OC/OB$. Thus $c = (\tan \rho \cos 45°)/2$.

When (001), face c Fig. 153a, is present, ρ of face p can be measured directly as $c \wedge p$. If (001) is not present (Fig. 153b), ρ of face e can be determined as $90° - a(010) \wedge e(011)$ and ρ of face $s = 90° - m(110) \wedge s(111)$. If the pyramidal form and a prism do not lie in a horizontal zone (Fig. 153c), measure the interfacial angle over the top of the crystal as $p \wedge p'$ (where p and p' are faces of the same form differing in ϕ by 180°). ρ of p is one-half this interfacial angle.

ORTHORHOMBIC SYSTEM

Crystallographic Axes. The forms of the crystal classes in the orthorhombic system are referred to three crystallographic axes of unequal length that make angles of 90° with each other. The relative lengths of the axes, or the axial ratios, must be determined for each orthorhombic mineral. In orienting an orthorhombic crystal, the convention is to set the crystal so that $c < a < b$. In the past, however, this convention has not necessarily been observed, and it is customary to conform to the orientation given in the literature. One finds, therefore, that any one of the three axes may have been chosen as c. The longer of the other two is then taken as b and the shorter as a.

In the past, the decision as to which of the three axes should be chosen as the vertical axis rested largely upon the crystal habit of the mineral. If its crystals commonly showed an elongation in one direction, this direction was usually chosen as the c axis (see Fig. 161). If, on the other hand, the crystals showed a

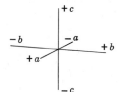

FIG. 154.   Orthorhombic crystal axes.

prominent pinacoid and therefore were tabular, this pinacoid was usually taken as {001} with c normal to it (see Fig. 162). Cleavage also aided in orienting orthorhombic crystals. If, as in topaz, there was one pinacoidal cleavage, it was taken as {001}. If, as in barite, there were two equivalent cleavage directions, they were set vertical and their intersection edges determined c. After the orientation has been determined, the length of the axis chosen as b is taken as unity, and the relative lengths of a and c are given in terms of it. Figure 154 represents the crystallographic axes for the orthorhombic mineral sillimanite with axial ratios $a:b:c = 0.98:1:0.75$.

In the Hermann Mauguin notation for the orthorhombic system, the symbols refer to the symmetry elements in the order a, b, c. For example, in the class $mm2$, the a and b axis lie in vertical symmetry planes and c is an axis of 2-fold symmetry.

Rhombic-Dipyramidal Class—2/m 2/m 2/m

Symmetry—$C, 3A_2, 3P$.   The three crystallographic axes are axes of 2-fold symmetry and perpendicular to each of them is a plane of symmetry (Fig. 155). A rhombic dipyramid and its stereogram is shown in Fig. 156.

Forms.   There are three types of forms in the rhombic-dipyramidal class—pinacoids, prisms, and dipyramids.

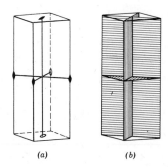

(a)   (b)

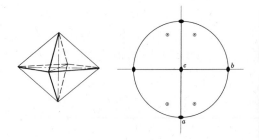

FIG. 155.   Symmetry of rhombic dipyramidal class. (a) Axes. (b) Planes.

FIG. 156.   Rhombic dipyramid and stereogram.

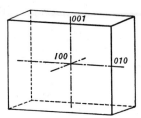

FIG. 157. Orthorhombic pinacoids.

1. *Pinacoids.* There are three pinacoids, the faces of which intersect one crystallographic axis and parallel the other two. (See Fig. 157.) They are:

Front or *a* pinacoid, {100}. Intersects *a*; parallels *b* and *c*.
Side or *b* pinacoid, {010}. Intersects *b*; parallels *a* and *c*.
Basal or *c* pinacoid, {001}. Intersects *c*; parallels *a* and *b*.

2. *Prisms.* Orthorhombic prisms consist of four faces which are parallel to one axis and intersect the other two.

First-order prism, {0*kl*}. Faces parallel *a* (1st axis); intersect *b* and *c*.
Second-order prism, {*h0l*}. Faces parallel *b* (2nd axis); intersect *a* and *c*.
Third-order prism, {*hk*0}. Faces parallel *c* (3rd axis); intersect *a* and *b*.

There are various prisms of each order with different axial intercepts. The unit forms are illustrated in Fig. 158. Since all prisms intersect two axes and parallel the third, one prism will be transformed into another by a different choice of axes.

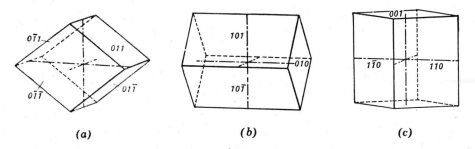

(a)   (b)   (c)

FIG. 158. Orthorhombic prisms, unit forms. (a) First-order. (b) Second-order. (c) Third-order.

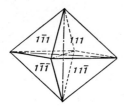

FIG. 159. Orthorhombic dipyramid.

3. *Dipyramid {hkl}.*   An orthorhombic dipyramid has eight triangular faces, each of which intersects all three of the crystallographic axes. It is the general form from which the orthorhombic-dipyramidal class receives its name. Figure 159 represents the unit dipyramid {111}.

Combinations.   Practically all orthorhombic crystals consist of combinations of two or more forms. Characteristic combinations of the various forms are given in Figs. 160–162.

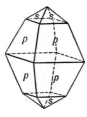

FIG. 160. Sulfur crystals. Forms: *p*{111}, *S*{113}, first-order dipyramids. *n*{011}, first-order prism. *c*{001}, basal pinacoid.

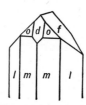

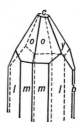

FIG. 161. Topaz crystals. Forms: *c*{001}, *b*{010}, pinacoids. *f*{011}, *y*{021}, *d*{101}, *m*{110}, *l*{120}, prisms. *o*{111}, dipyramid.

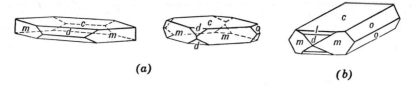

FIG. 162. (*a*) Barite. (*b*) Celestite. Forms: *c*{001}, *o*{011}, *d*{101}, *m*{210}, *l*{104}.

There are many mineral representatives in this class. Among the more common are the following:

andalusite	columbite	marcasite
anthophyllite	cordierite	olivine
aragonite (group)	danburite	sillimanite
barite (group)	enstatite	stibnite
brookite	goethite	sulfur
chrysoberyl	lawsonite	topaz

Rhombic-Dishphenoidal Class—222

Symmetry—$3A_2$. There are three axes of 2-fold symmetry coincident with the crystallographic axes. There are no planes and no center of symmetry. Figure 163 illustrates a rhombic disphenoid and its stereogram.

Forms. The rhombic disphenoid is composed of four faces, two in the upper hemisphere and two in the lower. It resembles the tetragonal disphenoid, but each face is a scalene triangle; whereas in the tetragonal disphenoid each face is an isosceles triangle. There are two disphenoids. The right $\{hkl\}$ and left $\{h\bar{k}l\}$ are enantiomorphic forms.

The three pinacoids and the three prisms may be present in this class.

Although there are several representative minerals crystallizing in this class, they are all comparatively rare. The most common are epsomite and olivenite.

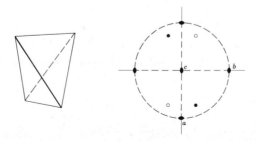

FIG. 163. Rhombic disphenoid and stereogram.

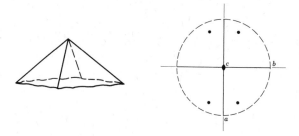

FIG. 164.   Rhombic pyramid and stereogram.

Rhombic-Pyramidal Class—*mm2*

Symmetry—$1A_2$, $2P$.   The *c* crystallographic axis is an axis of 2-fold symmetry. Two planes of symmetry at right angles to each other intersect in this axis. A rhombic pyramid and its stereogram are shown in Fig. 164.

Forms.   Because of the absence of a horizontal symmetry plane, the forms at the top of the crystal are different from those at the bottom. The rhombic dipyramid thus becomes two rhombic pyramids, {*hkl*} at the top and {*hkl̄*} at the bottom. Likewise the first-order and second-order prisms do not exist. In the place of each there are two domes (two-faced forms). {*0kl*} and {*0kl̄*} are the first-order domes, and {*h0l*} and {*h0l̄*} are second-order domes. In addition to these forms there are also pedions, {001} and {00ī}, and third-order prisms.

Only a few minerals crystallize in this class; the most common representatives are hemimorphite, Fig. 165, and bertrandite.

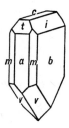

FIG. 165.   Hemimorphite.

Orthorhombic Axial Ratios

To express the relative lengths of the axes in the orthorhombic system there are two ratios, $a:b$ and $b:c$, where the length of *b* is taken as unity. They are given as $a:b:c = -:1:-$, and can be calculated using ϕ and ρ angles.

As an example, consider the face of a general form *ABC*, Fig. 166. Assume this to be face (132) of aragonite with $\phi = 28° \ 11'$, $\rho = 50° \ 48'$. *OP* is the face normal, and *OD* is normal to *AB*. Therefore, angle *BOD* is ϕ and angle *COP* is ρ. Reducing

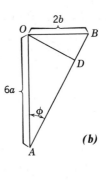

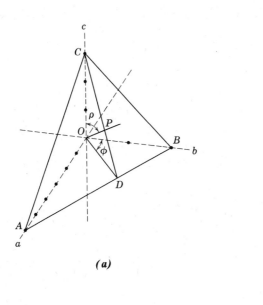

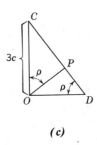

(a)

(b)

(c)

FIG. 166.   Intercepts of face (132).

the indices (132) to intercepts we find $AO = 6a$, $OC = 3c$, and $OB = 2b = 2$, since $b = 1$. Hence, we may find a (see Fig. 166b) by

$$\cot \phi = \frac{6a}{2b} \quad \text{or} \quad a = \frac{b \cot \phi}{3}$$

$$b = 1 \quad a = \frac{1 \times \cot 28° \, 11'}{3} \quad a = 0.6221$$

In triangle COD (Fig. 166c), $\tan \rho = 3c/OD$. In triangle BOD, $\cos \phi = OD/2$. Setting both of these expressions equal to OD,

$$OD = \frac{3c}{\tan \rho} = 2 \cos \phi \quad \text{or} \quad c = \frac{2 \tan \rho \cos \phi}{3}$$

Substituting ϕ and ρ values of face (132)

$$c = \frac{2 \tan 50° \, 48' \cos 28° \, 11'}{3} = 0.7205$$

Using Miller indices h, k, and l rather than intercepts, a and c relative to b are obtained using the formulas: $a = h/k \cot \phi$ and $c = l/k \tan \rho \cos \phi$. Crystallographic calculations generally involve the variables: (1) axial ratios, (2) indices,

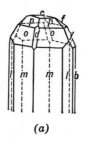 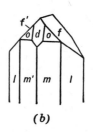

FIG. 167.   Topaz crystals.

and (3) angles ϕ and ρ. When two of these variables are known, the third may be calculated using the above formulas.

The ϕ and ρ angles may be obtained from interfacial angles. For example (Fig. 167a), the angle between b (010) and another face, as m in the [001] zone is the ϕ angle of m. Other faces, as p and o, in the same horizontal zone with m, also have $b \wedge m$ as their ϕ angle. When c (001) is present (Fig. 167a) the ρ angles of f and o are, respectively, $c \wedge f$ and $c \wedge o$. When (010) and (001) are not present (Fig. 167b), ϕ and ρ angles must be calculated from interfacial angles. For example, $\phi m = 90° - (m \wedge m')/2$; and $\rho f = (f \wedge f')/2$.

MONOCLINIC SYSTEM

Crystallographic Axes. Monoclinic crystals are referred to three axes of unequal lengths. The only restrictions in the angular relations are that $a \wedge b$ (γ) and $c \wedge b$ $(\alpha) = 90°$. For most crystals the angle between $+a$ and $+c$ is greater than 90° but in rare instances it too may equal 90°. In such cases the monoclinic symmetry is not apparent from the morphology. When properly oriented, the b axis is horizontal and placed in a left-to-right position; the a axis is inclined downward toward the front; and c is vertical. The crystal constants of monoclinic crystals include the axial ratios expressed as $a:b:c$ with b taken as unity, and the angle β. Calculations of axial ratios that in the orthogonal systems are simple and straightforward, become involved and tedious in the inclined systems. For the formulas used in their calculations, one is referred to more advanced books on crystallography. Figure 168 represents the crystallographic axes of the monoclinic mineral orthoclase, the axial constants of which are expressed as $a:b:c = 0.658:1:0.553$; $\beta = 116° 01'$.

In monoclinic crystals the positions of the b axis and of the plane of the a and c axes are fixed by the symmetry (see Fig. 169). The directions that serve as the a and c axes, however, are matters of choice and depend upon crystal habit and cleavage.

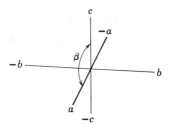
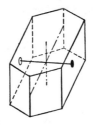
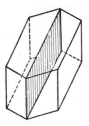

FIG. 168. Monoclinic crystal axes. **FIG. 169.** Symmetry of prismatic class.

If crystals show an elongated development (prismatic habit) parallel to a direction in the a-c plane, that direction often serves as the c axis. Further, if there is a prominent sloping plane or planes, such as planes c or r in the drawings of Fig. 176, the a axis may be taken as parallel to these. It is quite possible that there may be two, or even more, choices that are equally good, but in the description of a new mineral it is conventional to orient the crystals so that $c < a$.

Cleavage is also an important factor in orienting a monoclinic crystal. If there is a good pinacoidal cleavage parallel to the b axis, as in orthoclase, it is usually taken as the basal cleavage. If there are two equivalent cleavage directions, as in the amphiboles and pyroxenes, they are usually taken to be vertical prismatic cleavages.

Prismatic Class—2/m

Symmetry—$C, 1A_2, 1P$. The b axis is an axis of 2-fold symmetry, and the a-c plane is a symmetry plane (Fig. 169). The stereogram, Fig. 170, shows the symmetry of the prism (fourth-order). Since the a axis slopes down and to the front, it does not lie in the equatorial plane, and the positive end intersects the sphere of projection in the southern hemisphere.

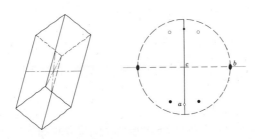

FIG. 170. Monoclinic prism and stereogram.

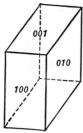

FIG. 171.   Monoclinic pinacoids. Front {100}, side {010}, base {001}.

Forms.   There are but two types of forms in the prismatic class of the mono-clinic system—pinacoids and prisms.

1. *Pinacoids* (see Fig. 171).

> Front or *a* pinacoid, {100}. Intersects *a*, parallels *b* and *c*.
> Side or *b* pinacoid, {010}. Intersects *b*, parallels *a* and *c*.
> Basal or *c* pinacoid, {001}. Intersects *c*, parallels *a* and *b*.
> Second-order pinacoids: {$h0l$} positive, {$\bar{h}0l$} negative.

Because the opposite ends of the *a* axis are not interchangeable, there is no second-order prism. Instead there are pinacoids; {$h0l$} between {100} and {001} and {$\bar{h}0l$} between {$\bar{1}00$} and {001}. It should be emphasized that these two forms are independent of each other, and the presence of one does not necessitate the presence of the other. (See Figs. 172 and 173.)

2. *Prisms.*   In the monoclinic system the four-faced prism is the general form and also the special forms {$0kl$} and {$hk0$}.

First-order prism {$0kl$}.   Faces intersect the *b* and *c* axis and parallel the *a* axis (see Fig. 174).

Third-order or vertical prism {$hk0$}.   Faces intersect the *a* and *b* axis and parallel the *c* axis.

Fourth-order prism {hkl} *positive*, {$\bar{h}kl$} *negative*.   The faces of these two independent forms intersect all three crystallographic axes. If the faces of the prism at the upper end of the crystal intersect the positive end of the *a* axis, the prism is positive; if they intersect the negative end of *a*, the prism is negative (see Fig. 175).

The only monoclinic form that is fixed is the side pinacoid. The other forms may vary with the choice of the *a* and *c* axes. For instance, the front pinacoid, basal pinacoid, and second-order pinacoid may be converted into each other by a rotation about the *b* axis. In the same manner the three prisms can be changed from one position to another.

Characteristic combinations of the forms described above are given in Fig. 176.

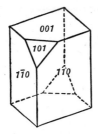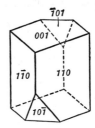

FIG. 172.   Second-order pinacoids. {101} positive, {1̄01} negative with third-order prism {110}, and basal pinacoid {001}.

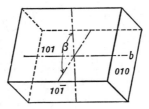

FIG. 173.   Positive and negative second-order pinacoids with side pinacoid.

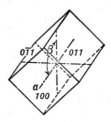

FIG. 174.   First-order prism {011} and front pinacoid {100}.

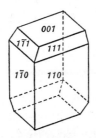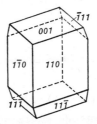

FIG. 175.   Fourth-order prisms. {111} positive, {1̄11} negative with third-order prism {110}, and basal pinacoid {001}.

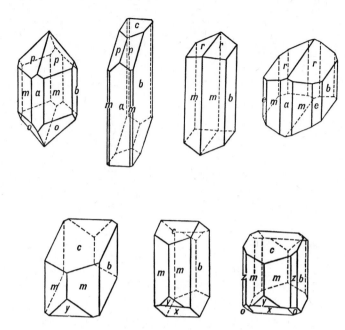

FIG. 176. Monoclinic crystals. Forms: $a\{100\}$, $b\{010\}$, $c\{001\}$, $m\{110\}$, $p\{111\}$, $o\{\bar{2}21\}$, $r\{011\}$, $e\{120\}$, $x\{\bar{1}01\}$, $y\{\bar{2}01\}$, $z\{130\}$.

Many minerals crystallize in the monoclinic, prismatic class; some of the most common are:

azurite	gypsum	orthoclase
borax	heulandite	realgar
calaverite	kaolinite	sphene
chlorite	malachite	spodumene
colemanite	monazite	talc
datolite	muscovite (and	tremolite (and
diopside (and	other micas)	other amphiboles)
other pyroxenes)	orpiment	wolframite
epidote		

Sphenoidal Class—2

Symmetry—$1A_2$. The b crystallographic axis is an axis of 2-fold symmetry. Figure 177 shows a sphenoid and its stereogram.

Forms. With the absence of the a-c symmetry plane, the b axis is polar, and different forms are present at opposite ends. The $\{010\}$ pinacoid of class $2/m$ becomes two *pedions*, $\{010\}$ and $\{0\bar{1}0\}$. Likewise the prisms, first-, third-, and fourth-order, each degenerates into a pair of enantiomorphic sphenoids. A

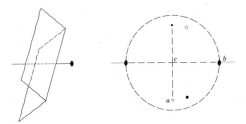

FIG. 177. Monoclinic sphenoid and stereogram.

sphenoid is a two-faced form with symmetry about an axis, *b*, in contrast to a *dome* with symmetry across a plane. The general form is thus the sphenoid $\{hkl\}$ with its enantiomorphic equivalent $\{h\bar{k}l\}$. The other special prismatic forms of class $2/m$ become sphenoids $\{0kl\}$, $\{0\bar{k}l\}$, and $\{hk0\}$, $\{h\bar{k}0\}$. The pinacoids $\{100\}$, $\{001\}$, $\{h0l\}$ and $\{\bar{h}0l\}$ may be present.

Mineral representatives in the sphenoidal class are rare, but chief among them are the members of the halotrichite isostructural group, of which pickeringite, $MgAl_2(SO_4)_4 \cdot 22H_2O$, is the most common member.

Domatic Class—*m*

Symmetry—1*P*. There is one vertical symmetry plane (010) that includes the *a* and *c* crystallographic axes. A monoclinic dome and its stereogram are shown in Fig. 178.

Forms. The *dome* is a two-faced form symmetrical across a plane of symmetry in contrast to a sphenoid (Fig. 177) which is symmetrical about a 2-fold symmetry axis. The general forms $\{hkl\}$ and $\{\bar{h}kl\}$ each correspond to two of the faces of the fourth-order prism of the prismatic class. The special prisms $\{0kl\}$ and $\{hk0\}$ of class $2/m$ also become domes, $\{0kl\}$, $\{0k\bar{l}\}$, and $\{hk0\}$, $\{\bar{h}k0\}$. The form $\{010\}$ is a pinacoid, but all faces lying across the symmetry plane as $\{100\}$, $\{\bar{1}00\}$, $\{001\}$, $\{00\bar{1}\}$, and $\{h0l\}$, $\{\bar{h}0l\}$ are pedions.

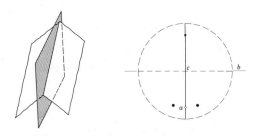

FIG. 178. Monoclinic dome and stereogram.

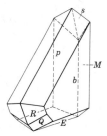

FIG. 179. Hilgardite.

The rare minerals hilgardite, $Ca_8B_{18}O_{33}Cl_4 \cdot 4H_2O$ (Fig. 179), and clinohedrite, $H_2CaZnSiO_5$, crystallize in this class.

TRICLINIC SYSTEM

Crystallographic Axes. In the triclinic system the crystal forms are referred to three crystallographic axes of unequal lengths that make oblique angles with each other (Fig. 180). The three rules to follow in orienting a triclinic crystal and thus in determining the position of the crystallographic axes are: (1) The most pronounced zone should be vertical. The axis of this zone then becomes c. (2) {001} should slope forward and to the right. (3) Two forms in the vertical zone should be selected: one as {100} the other as {010}. The directions of the a and b axes are determined respectively by the intersections of {010} and {100} with {001}. The b axis should be longer than the a axis. In reporting on the crystallography of a new triclinic mineral or one that has not been recorded in the literature, the convention should be followed that $c < a < b$. The relative lengths of the three axes and the angles between them can be established only with difficulty and must be calculated for each mineral from appropriate measurements. The angles between the positive ends of b and c, c and a, and a and b are designated respectively as α, β, and γ (see Fig. 180). For example, the crystal constants of the triclinic mineral axinite are as follows $a:b:c = 0.779:1:0.978$; $\alpha = 88°\ 4'$, $\beta = 81°\ 36'$, $\gamma = 77°\ 42'$.

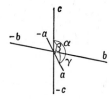

FIG. 180. Triclinic crystal axes.

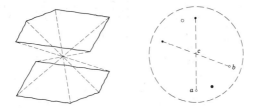

FIG. 181. Triclinic pinacoid and stereogram.

Pinacoidal Class—$\bar{1}$

Symmetry—C. The symmetry consists of a 1-fold axis of rotary inversion which is equivalent to a symmetry center. Figure 181 illustrates a triclinic pinacoid and its stereogram.

Forms. All the forms are pinacoids and thus consist of two similar and parallel faces. Once a crystal is oriented, the Miller indices of a crystal face establish its position. However names are given to the various pinacoids which, in a general way, designate their relation to the crystallographic axes. In addition to the front, side, and basal pinacoids, there are first-, second-, third-, and fourth-order pinacoids.

1. *Front, Side, and Basal Pinacoids.* Each of these pinacoids intersects one crystallographic axis and is parallel to the other two. The front or *a* pinacoid, {100}, intersects the *a* axis and is parallel to the other two; the side or *b* pinacoid, {010}, intersects the *b* axis; the basal or *c* pinacoid, {001}, intersects the *c* axis (Fig. 182).

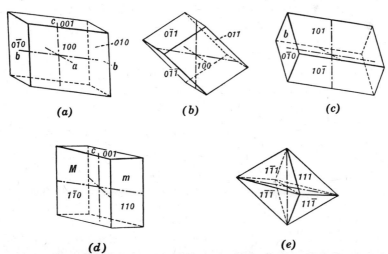

FIG. 182. Triclinic pinacoids. (*a*) Front {100}, side {010}, and base {001}. (*b*) First-order {011} positive, {0$\bar{1}$1} negative. (*c*) Second-order {101} positive, {10$\bar{1}$} negative. (*d*) Third-order {110} positive, {1$\bar{1}$0} negative. (*e*) Fourth-order (four different forms).

2. *First-, Second-, and Third-Order Pinacoids.* For each of these designations there is a positive and negative form whose faces are parallel to one axis and intersect the other two as follows:

First-order: parallel to a; $(0kl)$ positive, $(0\bar{k}l)$ negative.
Second-order: parallel to b; $(h0l)$ positive, $(\bar{h}0l)$ negative.
Third-order: parallel to c; $(hk0)$ positive, $(h\bar{k}0)$ negative.

3. *Fourth-Order Pinacoids* $\{hkl\}$ *Positive Right,* $\{h\bar{k}l\}$ *Positive Left,* $\{\bar{h}kl\}$ *Negative Right,* $\{\bar{h}\bar{k}l\}$ *Negative Left.* Each of these two-faced forms can exist independently of the others.

Various first-, second-, third-, and fourth-order pinacoids may be present depending on the axial intercepts.

Among the minerals that crystallize in the pinacoidal class are:

amblygonite	polyhalite
chalcanthite	rhodonite
microcline	turquoise
pectolite	ulexite
plagioclase feldspars	wollastonite

Of the above mentioned minerals only rhodonite and chalcanthite are at all common in well-formed crystals (see Fig. 183).

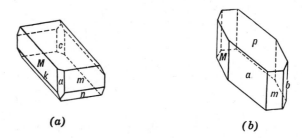

FIG. 183.   Triclinic crystals. (a) Rhodonite. (b) Chalcanthite.

Pedial Class—1

Symmetry.   There is merely a 1-fold rotation axis, which is equivalent to no symmetry. Figure 184 illustrates a triclinic pedion and its stereogram.

Forms.   The general form $\{hkl\}$ as well as all other forms are pedions and thus each face stands by itself. Each pinacoidal form of class $\bar{1}$ becomes two pedions.

Axinite is the only common mineral that crystallizes in the pedial class.

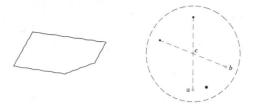

FIG. 184. Triclinic pedion and stereogram.

TWIN CRYSTALS

When two or more crystals of the same mineral have grown together so that certain directions of the lattices are parallel whereas other directions are in reverse position, they are called *twin crystals*. The parts of a twin crystal are related to each other in the following ways: (1) The relation may be as if one part were derived from the other by reflection over a common plane, the *twin plane*. (2) One part of the twin may appear to have been derived from the other by a revolution about a crystal direction common to both, the *twin axis*. Although there are some exceptions, the angular revolution is usually 180°. (3) The two individuals may be symmetrical about a point, a *twin center*. Twinning is defined by a *twin law*, which states whether there is a center, an axis, or a plane of twinning, and gives the crystallographic orientation for the axis or plane.

The surface on which two individuals are united is known as the *composition surface*. If this surface is a plane, it is called the *composition plane*. The composition plane is commonly, but not invariably, the twin plane. However, if the twin law can be defined only by a twin plane, the twin plane is always parallel to a possible crystal face but *never parallel to a plane of symmetry*. The twin axis is a zone axis or a direction perpendicular to a possible crystal face; but it can *never be an axis of even symmetry* (2-, 4-, 6-fold) if the rotation involved is 180°. In some crystals a 90° rotation about a 2-fold axis can be considered a twin operation.

Twin crystals are usually designated as either *contact twins* or *penetration twins*. Contact twins have a definite composition surface separating the two individuals, and the twin law is defined by a twin plane (Fig. 189). Penetration twins are made up of interpenetrating individuals having an irregular composition surface, and the twin law is usually defined by a twin axis. (See Fig. 190.)

Repeated or multiple twins are made up of three or more parts twinned according to the same law. If all the successive composition surfaces are parallel, the resulting group is a *polysynthetic twin* (Figs. 185 and 186). If successive composition planes are not parallel, a *cyclic twin* results (Fig. 187). When a large number of individuals in a polysynthetic twin are closely spaced, crystal faces or cleavages crossing the composition planes show striations owing to the reversed positions of adjacent individuals.

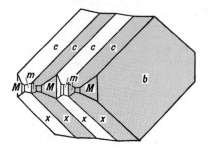

FIG. 185. Albite twinned on {010}.

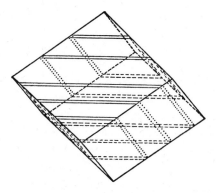

FIG. 186. Calcite twinned on negative rhombohedron {$\bar{1}012$}.

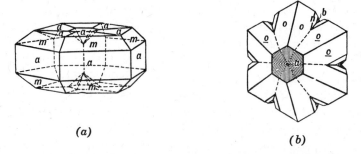

(a)

(b)

FIG. 187. Cyclic twinning. (a) Rutile. (b) Chrysoberyl.

Twinning in the lower symmetry groups generally produces a resulting aggregate symmetry higher than that of each individual because the twin plane is an added symmetry plane.

Common Twin Laws. *Isometric System.* In the hexoctahedral class of the isometric system the twin axis, with a few rare exceptions, is a 3-fold symmetry

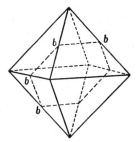

FIG. 188. Octahedron.
Twin plane b-b ($1\overline{1}1$).

FIG. 189. Twinned octa-
hedron (spinel twin).

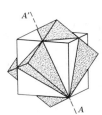

FIG. 190. Fluorite
penetration twin.

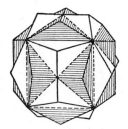

FIG. 191. Pyrite iron
cross.

axis, and the twin plane is thus parallel to a face of the octahedron. Figure 188 shows an octahedron with plane bb a possible twin plane, and Fig. 189 shows an octahedron twinned according to this law, forming a contact twin. This type of twin is especially common in gem spinel and hence is called a *spinel twin*. Figure 190 shows two cubes forming a penetration twin with the 3-fold symmetry axis, A-A', the twin axis.

In the diploidal class, two pyritohedrons may form a penetration twin (Fig. 191) with the twin axis normal to $\{011\}$. A 90° rotation about the 2-fold axis would produce the same result. This twin is known as the *iron cross*.

Hexagonal System. In the hexagonal division of this system, twins are rare and unimportant, but in the rhombohedral division twins are common. The rhombohedral carbonates, especially calcite, serve as excellent illustrations of three twin laws. The twin plane may be $\{0001\}$, with c the twin axis (Fig. 192a), or it may be the positive rhombohedron $\{10\overline{1}1\}$. But twinning on the negative rhombohedron $\{01\overline{1}2\}$, is most common and may yield contact twins (Fig. 192a) or polysynthetic twins as the result of pressure (Fig. 186). The ease of twinning according to this law can be demonstrated by the artificial twinning of a cleavage fragment of Iceland spar by the pressure of a knife blade (Fig. 192).

In the trigonal trapezohedral class, quartz shows several types of twinning. Figure 193a illustrates the *Brazil law* with the twin plane perpendicular to an *a*

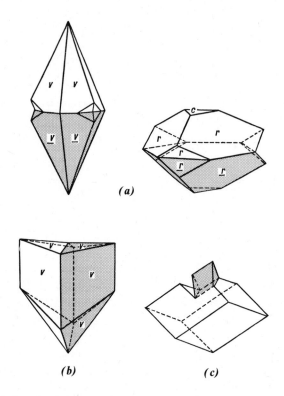

FIG. 192. Twinning in calcite. (*a*) Twinned on {0001}. (*b*) Twinned on {01$\bar{1}$2}. (*c*) Artificial twinning on {01$\bar{1}$2}.

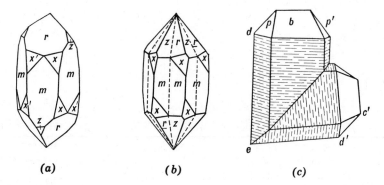

FIG. 193. Twinning in quartz. (*a*) Brazil twin. (*b*) Dauphine twin. (*c*) Japanese twin.

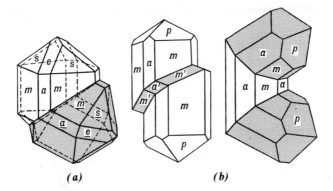

FIG. 194. Cassiterite and rutile twinned on {011}. (*a*) Cassiterite. (*b*) Rutile.

axis. Here, right- and left-hand individuals have formed a penetration twin. Figure 193b shows a *Dauphiné twin*, a penetration twin with *c* the twin axis. Such twins are composed either of two right- or two left-hand individuals. Figure 193c illustrates the *Japanese law* with the twin plane {11$\bar{2}$2}. The reentrant angles usually present on twinned crystals do not show on either Brazil or Dauphiné twins.

Tetragonal System. The most common type of twin in the tetragonal system has {011} as the twin plane. Crystals of cassiterite and rutile, twinned according to this law are shown in Fig. 194.

Orthorhombic System. In the orthorhombic system the twin plane is most commonly parallel to a prism face. The contact twin of aragonite (Fig. 195a), the cyclic twin of the same minerals, in cross section (Fig. 195b) and the cyclic twin of cerussite (Fig. 196) are all twinned on {110}. The pseudohexagonal appearance of Figs. 195b and 196 results from the fact that (110) $\wedge$ (1$\bar{1}$0) of nearly 60°.

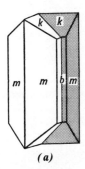

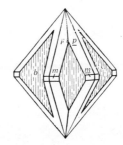

(*a*)   (*b*)

FIG. 195. Twinning in aragonite.   FIG. 196. Cerussite cyclic twin.

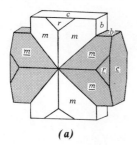 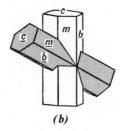 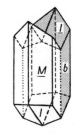

(a) *(b)*

FIG. 197. Staurolite twins. (*a*) {032} Twin plane. (*b*) {232} Twin plane.

FIG. 198. Gypsum. Twin plane {100}.

In the orthorhombic mineral staurolite two types of penetration twins are common. In one with {031} as twin plane a right angle cross results (Fig. 197), in the other with twin plane {232}, a 60° cross is formed.

Monoclinic System. In the monoclinic system twinning on {100} and {001} is most common. Figure 198 of gypsum illustrates twinning with {100} the twin plane, and Fig. 199b shows a *Manebach twin* of orthoclase in which {001} is the twin plane. Orthoclase also forms penetration twins according to the *Carlsbad law*, in which the *c* crystallographic axis is the twin axis, and the individuals are united on an irregular surface roughly parallel to {010} (Fig. 199a). The *Baveno twin* also found in orthoclase has {021} as the twin plane (Fig. 199c).

Triclinic System. The feldspars best illustrate the twinning of the triclinic system. They are almost universally twinned according to the *albite law*, with {010} the twin plane, as shown in Figure 199d. Another important type of twinning in triclinic feldspar is according to the *pericline law*, with *b* the twin axis.

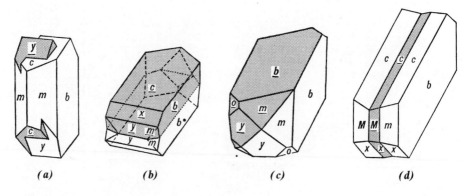

(a) *(b)* *(c)* *(d)*

FIG. 199. Twinning in feldspar. (*a*) Carlsbad twin. (*b*) Manebach twin. (*c*) Baveno twin. (*d*) Albite twins.

Crystal Habit and Crystalline Aggregates

Terms used to express the appearance or habit of individual crystals, or of aggregates of crystals, are given below.

1. Minerals in isolated or distinct crystals may be described as:

    a. *Acicular.* Slender needlelike crystals.
    b. *Capillary and filiform.* Hairlike or threadlike crystals.
    c. *Bladed.* Elongated crystals flattened like a knife blade.

2. For groups of distinct crystals the following terms are used:

    a. *Dendritic.* Arborescent, in slender divergent branches, somewhat plantlike.
    b. *Reticulated.* Latticelike groups of slender crystals.
    c. *Divergent* or *radiated.* Radiating crystals groups.
    d. *Drusy.* A surface covered with a layer of small crystals.

3. Parallel or radiating groups of individual crystals, are described as:

    a. *Columnar.* Stout columnlike individuals.
    b. *Bladed.* An aggregate of many flattened blades.
    c. *Fibrous.* Aggregate of slender fibers, parallel or radiating.
    d. *Stellated.* Radiating individuals forming starlike or circular groups.
    e. *Globular.* Radiating individuals forming spherical or hemispherical groups.
    f. *Botryoidal.* Globular forms resembling, as the word derived from the Greek implies, a "bunch of grapes."
    g. *Reniform.* Radiating individuals terminating in rounded kidney-shaped masses (Fig. 200).

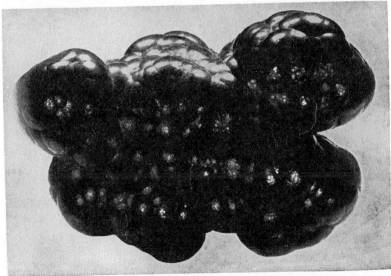

FIG. 200.   Reniform hematite, Cumberland, England.

h. *Mammillary.* Large rounded masses resembling mammae, formed by radiating individuals.

i. *Colloform.* Spherical forms composed of radiating individuals without regard to size; this includes botryoidal, reniform, and mammillary.

4. A mineral aggregate composed of scales or lamellae is described as:

a. *Foliated.* Easily separable into plates or leaves.

b. *Micaceous.* Similar to foliated, but splits into exceedingly thin sheets, as in the micas.

c. *Lamellar* or *tabular.* Flat platelike individuals superimposed upon and adhering to each other.

d. *Plumose.* Fine scales with divergent or featherlike structure.

5. A mineral aggregate composed of grains is *Granular.*

6. Miscellaneous terms:

a. *Stalactitic.* Pendent cylinders or cones. Stalactites form by deposition from mineral-bearing waters dripping from the roofs of caverns.

b. *Concentric.* More or less spherical layers superimposed upon one another about a common center.

c. *Pisolitic.* Rounded masses about the size of peas.

d. *Oölitic.* A mineral aggregate formed of small spheres resembling fish roe.

e. *Banded.* A mineral in narrow bands of different color or texture.

f. *Massive.* Compact material without form or distinguishing features.

g. *Amygdaloidal.* A rock such as basalt containing almond-shaped nodules.

h. *Geode.* A rock cavity lined by mineral matter but not wholly filled. Geodes may be banded as in agate, due to successive depositions of material and the inner surface is frequently covered with projecting crystals.

i. *Concretion.* Masses formed by deposition of material about a nucleus. Some concretions are roughly spherical whereas others assume a great variety of shapes.

X-RAY CRYSTALLOGRAPHY

The application of x-rays to the study of crystals was the greatest single impetus ever given to crystallography. Prior to 1912, crystallographers had correctly deduced from cleavage, optical properties, and the regularity of external form that crystals had an orderly structure; but their thinking concerning the geometry of crystal lattices had only the force of a hypothesis. Since the use of x-rays, it has been possible not only to measure the distance between successive atomic planes but also to tell the positions of the various atoms within the crystal.

X-rays were discovered accidentally by Wilhelm Conrad Röentgen in 1895 while experimenting with the production of cathode rays in sealed discharge tubes wrapped in black paper. The electron stream in the discharge tube, striking the glass of the tube, produced low-intensity x-radiation which caused some fluorescent material nearby to glow in the dark. Röentgen correctly inferred that a new type of radiation was being produced, fittingly called "x-radiation" because of the many mysteries connected with it. Röentgen was unsuccessful in his efforts to measure the wave length of x-rays, and it was this unsolved problem that led to the discovery of the diffraction of x-rays by crystals.

The fact that most substances are more or less transparent to x-rays brought about their almost immediate use in hospitals for medical purposes in the location of fractures, foreign bodies, and diseased tissue in much the same manner in which they are used today.

It was not until 1912, seventeen years after the discovery of x-radiation, that, at the suggestion of Max von Laue, x-rays were used to study crystals. The original experiments were carried out at the University of Munich where von Laue was a lecturer in Professor Sommerfeld's department. Sommerfeld was interested in the nature and excitation of x-rays and Laue in interference phenomena. Also at the University of Munich was Paul Heinrich Groth, a leading crystallographer. With the gathering together of such a group of distinguished scientists having these special interests, the stage was set for the momentous discovery.

In 1912, Paul Ewald was working under Sommerfeld's direction on a doctor's dissertation involving the scattering of light waves on passing through a crystal. In thinking about this problem, von Laue raised the question: what would be the effect if it were possible to use electromagnetic waves having essentially the same wave length as the interatomic distances in crystals? Would a crystal act as a three-dimensional diffraction grating forming spectra that could be recorded? If so, it would be possible to measure precisely the wave length of the x-rays employed, assuming interatomic spacing of the crystal; or assuming a wave length for the x-rays, measure the interplanar spacing of the crystal. Methods for making such an experiment were discussed, and Freidrich and Knipping, two research students, agreed to carry them out. Several experiments in which copper sulfate was used as the object crystal were unsuccessful. Finally they passed a narrow beam of x-rays through a cleavage plate of sphalerite, ZnS, allowing the beam to fall on a photographic plate. The developed plate showed a large number of small spots arranged in geometrical pattern around the large spot produced by the direct x-ray beam. This pattern was shown to be identical with that predictable from the diffraction of x-rays by a regular arrangement of scattering points within the crystal. Thus, a single experiment demonstrated the regular, orderly arrangement of atomic particles within crystals and the agreement as to order of magnitude of the wave length of x-rays with the interplanar spacing of

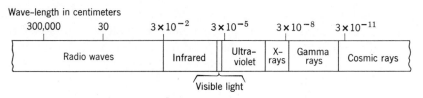

FIG. 201.   The electromagnetic spectrum.

crystals. Although largely replaced by more powerful means of x-ray investigation, this technique, the *Laue method*, is still in use.

Within the next few years, great strides were made as a result of the work of the English physicists, William Henry Bragg and William Lawrence Bragg, father and son. In 1914 the first structure of a compound, halite, NaCl, was worked out by the Braggs; and in the years following many more structures were solved by them. The Braggs also greatly simplified von Laue's mathematical generalizations as to the geometry of x-ray diffraction and popularized the results of their research through their well-written books on the subject.[6]

Electromagnetic waves form a continuous series varying in wave length from long radio waves of wave lengths of the order of thousands of meters to cosmic radiation whose wave lengths are of the order of 10^{-12} meters (a millionth of a millionth of a meter!). All forms of electromagnetic radiation have certain properties in common such as propagation along straight lines at a speed of 300,000 km per second in vacuum, reflection, refraction according to Snell's law, diffraction at edges and by slits or gratings, and a relation between energy and wave length given by Planck's law: $e = hv = hc/\lambda$, where e is energy, v frequency, c velocity of propagation, λ wave length, and h Planck's constant. Thus, the shorter the wave the greater the energy involved in its production and the greater its powers of penetration. X-rays occupy only a small portion of the spectrum, with wave lengths varying between slightly more than 100 Å and 0.02 Å. (See Fig. 201.) X-rays used in investigation of crystals have wave lengths of the order of 1 Å. Visible light has wave lengths between 7200 and 4000 Å, more than 1000 times as great, and hence is less penetrating and energetic than x-radiation.

When electrons moving at a high velocity strike the atoms of any element, x-rays are produced. Orbital electrons of the *K*, *L*, and *M* shells, deep within the extranuclear structure of the element being bombarded, are raised briefly to excited states by the energy contribution of the bombarding electrons. From

[6] W. H. Bragg and W. L. Bragg, *X-rays and Crystal Structures*, G. Bell and Sons, Ltd., London, 1924.

W. H. Bragg and W. L. Bragg, *The Crystalline State*, The Macmillan Company, New York, 1934.

W. H. Bragg, *An Introduction to Crystal Analysis*, G. Bell and Sons, Ltd., London, 1928.

W. L. Bragg, *Atomic Structure of Minerals*, Cornell University Press, Ithaca, New York, 1937.

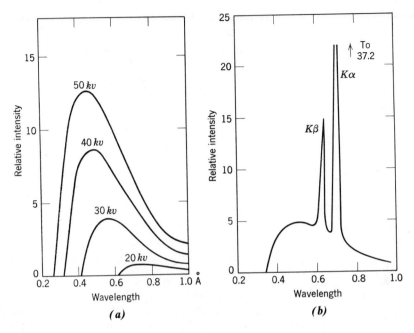

FIG. 202. X-ray spectrum. (a) Distribution of intensity with wavelength in the continuous x-ray spectrum of tungsten at various voltages. (b) Intensity curve showing characteristic wavelength superimposed on the continuous x-ray spectrum of molybdenum. (After Ulrey, *Phys. Rev.*, 11, 401.)

these excited states the orbital electrons fall back to their ground state or to lower energy levels, emitting quanta of energetic radiation in the process. This is x-radiation.

In the modern x-ray tube there is nearly a complete vacuum. The tube is fitted with a tungsten filament as cathode, and an anode composed of one of a number of metals. When the filament is heated by passage of a current, thermal electrons are evolved. Application of a high direct-current voltage accelerates these electrons, and their impact on the target (anode) generates the x-radiation.

The nature of the x-rays depends on the metal of the target and the applied voltage. No x-rays are produced until the voltage reaches a certain minimum value dependent on the target material. At that point a continuous x-ray spectrum is generated. On increasing potential, the intensity of all wave lengths increases, and the value of the minimum wave length becomes progressively less (Fig. 202a). This continuous spectrum containing all wave lengths within a given range is analogous to white light in the visible spectrum and is called *white radiation.*

As the voltage across the tube is increased there becomes superimposed on the white radiation a *line spectrum* or *characteristic radiation* peculiar to the target material. This characteristic radiation, many times more intense than the white radiation, consists of several isolated wave lengths from which a single wave length may be selected by filtering (Fig. 202b). It is thus analogous to mono-chromatic light in the visible spectrum and is called *monochromatic x-radiation*.[7]

THE BRAGG EQUATION

Following the original successful experiment with x-rays at Munich, von Laue worked out three equations (the Laue equations) to explain the observed diffrac-tion phenomena. Three equations were necessary for three-dimensional space. He showed that to produce a spot on a photographic plate three conditions must be simultaneously satisfied. Shortly thereafter, W. L. Bragg, working on x-ray diffraction in England, pointed out that, although x-rays are indeed diffracted by crystals, they act as though they were reflected from planes within the crystal. However, unlike the reflection of light, x-rays are not "reflected" continuously from a given crystal plane. Using a given wave length, λ, Bragg showed that a "reflection" took place from a family of parallel planes only under certain con-ditions. These conditions must satisfy the equation: $n\lambda = 2d \sin \theta$ where n is an integer $(1, 2, 3, \ldots, n)$, λ the wave length, d the distance between successive parallel planes, and θ the angle of incidence and reflection of the x-ray beam from the given atomic plane. This equation, known as the *Bragg law*, expresses in a simpler manner the simultaneous fulfillment of the three Laue equations.

We have seen that the faces most likely to appear on crystals are those parallel to atomic planes having the greatest density of lattice points. Parallel to each face is a family of equispaced identical planes. When an x-ray beam strikes a crystal it penetrates it, and the resulting "reflection" is not from a single plane but from an almost infinite number of parallel planes, each contributing a small bit to the total "reflection." In order that the "reflection" be of sufficient intensity to be recorded, the individual reflections must be in phase with one another. The following conditions necessary for reinforcement were demonstrated by W. L. Bragg.

In Fig. 203 the lines p, p_1, and p_2 represent the trace of a family of atomic planes with spacing d. X-rays striking any of these planes by itself would be reflected at the incident angle θ, whatever the value of θ. However, to reinforce one another in order to give a reflection that can be recorded, these reflected rays

---

[7] The wave lengths of the characteristic x-radiation emitted by the various metals have been accurately determined. The $K\alpha$ wave lengths (weighted averages of $K\alpha_1$ and $K\alpha_2$) of those most commonly used are:

	Å		Å
Molybdenum	0.7107	Cobalt	1.7902
Copper	1.5418	Iron	1.9373
Tungsten	1.4800	Chromium	2.2909

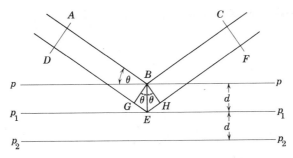

FIG. 203. Geometry of x-ray reflection.

must be in phase. The path of the waves along DEF reflected at E is longer than the path of the wave along ABC reflected at B. If the two sets of waves are to be in phase, the path difference of ABC and DEF must be a whole number of wave lengths ($n\lambda$). In Fig. 203 BG and BH are drawn perpendicular to AB and BC respectively so that $AB = DG$ and $BC = HF$. To satisfy the condition that the two waves be in phase $GE + EH$ must be equal to an integral number of wave lengths. BE is perpendicular to the lines p and p_1 and is equal to the interplanar spacing d. In $\triangle GBE$, $d \sin \theta = GE$; and, in $\triangle HBE$, $d \sin \theta = EH$. Thus for in-phase reflection $GE + EH = 2d \sin \theta = n\lambda$.

This is the Bragg equation. For a given d spacing and given λ, reflections can take place only at those angles of θ which satisfy the equation. Suppose for example, a monochromatic x-ray beam is parallel to a cleavage plate of halite and the plate is supported in such a way that it can be rotated about an axis at right angles to the x-ray beam. As the halite is slowly rotated there is no reflection until the incident beam makes an angle θ which satisfies the Bragg equation, with $n = 1$. On continued rotation there are further reflections only when the equation is satisfied at certain θ angles with $n = 2, 3, 4, 5$, etc. These are known as the first-, second-, third-order, etc., reflections.

We have learned that a crystal is built on a three-dimensional lattice with characteristic periodicities or *identity periods* along the crystallographic axes. We have also seen that this lattice acts as a three-dimensional grating in scattering x-rays. The scattering may be pictured as taking place independently along each of the principal rows of atoms parallel to the crystallographic axes, but, for a diffracted beam to be recorded on the photographic film, the scattering from rows in all three dimensions must be in phase.

Consider a single row of scattering points of periodicity c (Fig. 204). The scattered rays will reinforce each other when they are in phase, that is, when they have a path difference of an integral number of wave lengths. Thus, the in-phase scattering makes definite angles with the row of atoms dependent on the periodicity along the row and the wave length of x-rays.

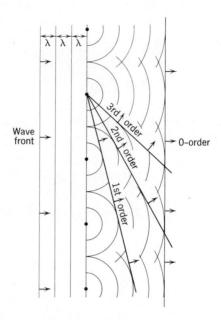

FIG. 204. Scattering of x-rays by a row of atoms.

In Fig. 205 rays 1 and 2 will be in phase only when $n\lambda = c \cos \phi$. For any value of $n\lambda$, ϕ is constant, and the diffracted rays form a cone with the row of points as axis. Since the scattered rays will be in phase for the same angle ϕ on the other side of the incident beam, there will be another similar but inverted

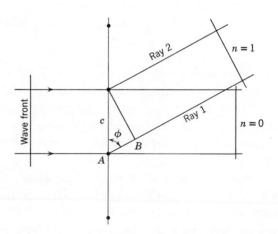

FIG. 205. Conditions for x-ray diffraction from a row of atoms.

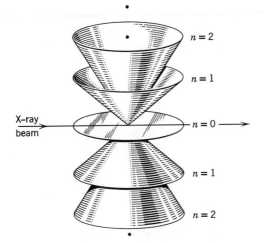

FIG. 206. Diffraction cones from row of atoms.

cone on that side (Fig. 206). When $n = 0$, the cone becomes a plane which includes the incident beam. The greater the value of n the larger the value of ϕ, and hence the narrower the cones. All have the same axis however, and all have their vertices at the same point, the intersection of the incident beam and the row of atoms.

In a three-dimensional lattice there are two other axial directions, each with its characteristic periodicity of scattering points and each capable of generating its own series of nested cones with characteristic apical angles. Diffraction cones from the three rows of scattering atoms intersect each other, but only when all three intersect in a common line is there a diffracted beam (Fig. 207). This line

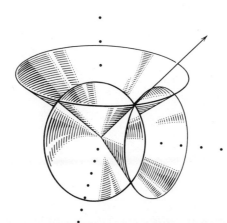

FIG. 207. Diffraction cones from three rows of scattering atoms, intersecting in a common line.

of intersection is the direction of the beam which is recorded on the film. For all other possible directions, destructive interference takes place. When the three cones intersect in a common line, the Bragg law, $n\lambda = 2d \sin \theta$ is also satisfied.

LAUE METHOD

In the Laue method a single crystal which remains stationary is used. A photographic plate or flat film, enclosed in a light-tight envelope is placed a known distance, usually 5 centimeters, from the crystal. A beam of white x-radiation is passed through the crystal at right angles to the photographic plate. The direct beam causes a darkening at the center of the photograph so that a small disc of lead is usually placed in front of the film to intercept and absorb it. The angle of incidence, θ, between the x-ray beam and the various atomic planes with their given d spacings within the crystal is fixed. However, since x-rays of all wave lengths are present, the Bragg law, $n\lambda = 2d \sin \theta$, can be satisfied by each family of atomic planes provided $(2d \sin \theta)/n$ is within the range of wave lengths furnished by the tube. Around the central point of a Laue photograph are arranged diffraction spots, each one resulting from reflection of x-rays from a certain given series of atomic planes (Fig. 208).

The Laue method, although of great historical interest, has largely been replaced by other more powerful methods of x-ray crystal analysis. Its use today is primarily for determining symmetry. If a crystal is so oriented that the x-ray beam is parallel to a symmetry element, the arrangement of spots on the photograph reveals this symmetry. A Laue photograph of a mineral taken with the x-ray beam parallel to the 2-fold axis of a monoclinic crystal will show a 2-fold

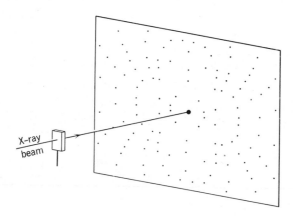

FIG. 208. Laue photograph.

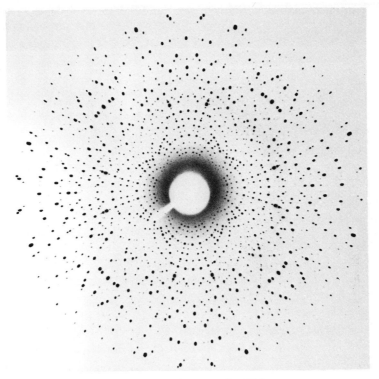

FIG. 209. Laue photograph of idocrase.

arrangement of spots; if the beam is parallel to the symmetry plane, the photograph shows a line of symmetry. A photograph of an orthorhombic crystal with the beam parallel to a crystallographic axis shows a 2-fold arrangement as well as two lines of symmetry. Figure 209 shows a 4-fold arrangement of spots as given by idocrase with the x-ray beam parallel to the 4-fold symmetry axis.

Unfortunately, x-rays do not differentiate between opposite ends of a polar axis and thus diffraction effects always introduce a center of symmetry.

ROTATION, WEISSENBERG, AND PRECESSION METHODS

In the rotation method and all refinements of it, a single crystal is used. The crystal must be oriented so that it can be rotated around one of the principal crystallographic axes. If crystal faces are present orientation is effected most easily by means of an optical goniometer; without crystal faces orientation is possible but laborious. The camera is a cylinder of known diameter, coaxial with the axis of rotation of the crystal, and a photographic film, protected from the light by an envelope, is wrapped around the inside of the cylindrical camera. A beam of monochromatic x-rays enters the camera through a slit and strikes the crystal.

Under these conditions, with a stationary crystal, any reflection that occurs would be fortuitous. However, as the crystal is slowly rotated, various families of atomic planes will be brought into position, so that for them θ has a value that will, with known wave length, λ, satisfy $n\lambda = 2d \sin \theta$. A given family of planes gives rise to separate reflections when $n = 1, 2, 3, 4$, etc.

In taking a rotation photograph, the crystal is rotated about one of the principal lattice rows, usually a crystallographic axis. This lattice row is at right angles to the incident x-ray beam. Consequently, whatever diffracted beams arise must lie along cones having axes in common with the axis of rotation of the crystal. This axis is also the axis of the cylindrical film, and therefore the cones intersect the film in a series of circles (Fig. 210). When the film is developed and laid flat,

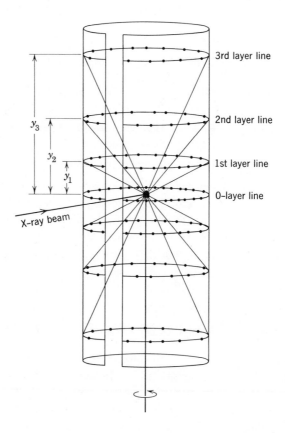

FIG. 210.   Intersection of diffraction cones with cylindrical film.

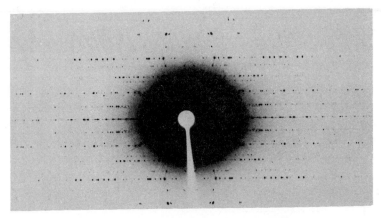

FIG. 211. Rotation photograph of idocrase.

the circles appear as straight parallel lines (Fig. 211). Each of these is a *layer line*, which corresponds to a cone of diffracted rays for which n has some integral value. Thus, the layer line including the incident beam is called the 0-layer, the first layer line is that for which $n = 1$, the next $n = 2$, and so on. The layer lines are not continuous since diffraction spots appear only where all three cones intersect (see Fig. 207).

The separation of the layer lines is determined by the cone angles, which are in turn dependent on the spacing in the lattice row about which the crystal was rotated. Therefore, if we know: (1) the diameter of the cylindrical film ($2r$), (2) the wave length of the x-rays (λ), (3) the distance from the 0-layer to the nth layer on the film (y_n), it is possible to determine the spacing, or *identity period* (I), along the rotation axis of the crystal from the following relations:

$$\frac{y_n}{r} = \tan v \text{ (Fig. 212a)} \qquad I = \frac{n\lambda}{\sin v} \text{ (Fig. 212b)}$$

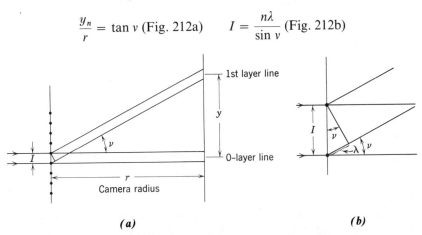

(a)   *(b)*

FIG. 212. Geometry for calculation of identity period, *I*.

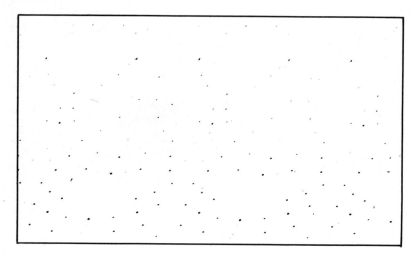

FIG. 213.   Weissenberg photograph of idocrase.

If rotation photographs are taken with the crystal rotating about each of the three crystallographic axes, it is possible to determine the unit-cell dimensions. The identity periods determined by rotating the crystal successively about a, b, and c are the edges of the unit cell. This is true for a crystal of any symmetry. However, in the isometric system one photograph to determine a suffices; in the tetragonal and hexagonal, two photographs are necessary, one about the c axis and one about an a axis. It is frequently desirable to identify specific spots on the film with the atomic planes that gave rise to them. This process of correlation called *indexing* cannot be done with a rotation photograph since the crystal orientation is not completely known. However, several modifications of the rotation method have been devised to permit complete indexing of reflections. Those most commonly used are the *Weissenberg* and the *precession* methods. In the Weissenberg method, the camera is translated to and fro during rotation of the crystal, and the spots of a layer line are spread out in festoons on the film (Fig. 213). Although it is a laborious process, each spot can be indexed. The precession method, devised by M. J. Buerger, yields the same information in a more straightforward way (Fig. 214). In this method a flat film and crystal both move with a complex gyratory motion, mechanically compensating for the distortions produced by the Weissenberg method. Thus a great amount of data is provided in a readily usable form. Further consideration of these methods is beyond the scope of this discussion. However, we have seen that valuable information can be obtained from a rotation photograph without identifying individual reflections.

FIG. 214. Precession photograph of idocrase.

POWDER METHOD

The relative rarity of well-formed crystals and the difficulty of making the precise orientation required by single crystal methods led to the discovery of the *powder method* of x-ray investigation. For the powder method, the specimen is ground as finely as possible and bonded together by an amorphous material, such as flexible collodion, into a needlelike spindle 0.2 to 0.3 millimeter in diameter. This spindle, the *powder mount*, consists ideally of crystalline particles in completely random orientation. To insure randomness of orientation of these tiny particles with respect to the impinging x-ray beam, the mount is generally rotated in the path of the beam during the exposure.

The powder camera is a flat, disc-shaped box with an adjustable pin at the center for attachment of the mount. The cylindrical wall of the camera is pierced diametrically for a demountable slit system and opposing beam catcher. The light-tight lid may be removed to insert and remove the film, which, for the most popular type of camera, is a narrow strip about 14 inches long and 1 inch wide. Two holes are punched in the film and so located that, when the film is fitted

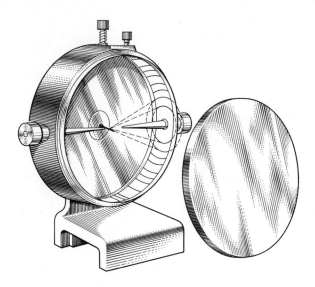

FIG. 215.   Powder camera.

snugly to the inner curve of the camera, the slit tube and beam catcher pass through the holes. This type of mounting is called the Straumanis method (Fig. 215).

A narrow beam of monochromatic x-rays is allowed to pass through the collimating slit and fall on the spindle-shaped mount, which is carefully centered on the short axis of the camera so that the mount remains in the x-ray beam as it rotates during the exposure. The undeviated beam passes through and around the mount and enters the lead-lined beam catcher, through which it leaves the camera.

When the beam of monochromatic x-rays strikes the mount, all possible diffractions take place simultaneously. If the orientation of the crystalline particles in the mount is truly random, for each family of atomic planes with its characteristic d spacing, there are many particles whose orientation is such that they make the proper θ angle with the incident beam to satisfy the Bragg law, $n\lambda = 2d \sin \theta$. The reflections from a given set of planes form cones with the incident beam as axis and the internal angle 4θ. Any set of atomic planes yields a series of nested cones corresponding to reflections of the first, second, third and higher orders ($n = 1, 2, 3 \ldots$). Different families of planes with different d spacings will satisfy the Bragg law at appropriate values of θ for different integral values of n, thus giving rise to separate sets of nested cones of reflected rays.

If the rays forming these cones are permitted to fall on a flat photographic plate at right angles to the incident beam, a series of concentric circles will result (Fig. 216). However, only reflections with small values of the angle 2θ can be recorded in this manner.

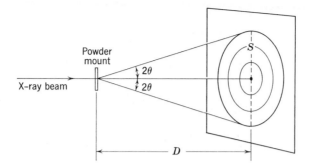

FIG. 216. X-ray diffraction from powder mount recorded on a flat plate.

In order to record reflections of 2θ up to $180°$, the film is fitted snugly into a cylindrical camera at the axis of which the specimen is mounted in the path of the x-ray beam. Under these conditions, the film intercepts the cones of reflected rays along curved lines (Fig. 217). Since the axes of the cones coincide with the x-ray beam, for each cone there will be two curved lines on the film symmetrically disposed on each side of the slit through which the x-rays leave the camera. The angular distance between these two arcs is 4θ.

When the film is developed and laid flat, the arcs are seen to have their centers at the two holes in the film. Reflections of low θ have their center at the exit hole through which the beam catcher passes through the film. Going out from this point, the arcs are of increasing radius until at $2\theta = 90°$ they are straight lines. Lines made by reflections with $2\theta > 90°$ curve in the opposite direction and are concentric about the hole through which the x-rays enter the camera. These are known as *back reflections*.

If a flat film is used and the distance D from the specimen to the film is known, it is possible to calculate θ by measuring S, the diameter of the rings. It may be seen in Fig. 216 that $\tan 2\theta = S/2D$. When a cylindrical camera is used, the distance S is measured with the film laid flat. Under these conditions, $S = R \times 4\theta$ in

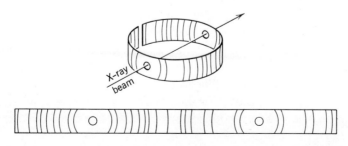

FIG. 217. X-ray diffraction from powder mount recorded on cylindrical film.

radians[8] or $\theta = S/4R$ in radians where R is the radius of the camera and S is measured in the same units as R. Most powder cameras are constructed with a radius such that S, measured in millimeters on the flat film can be converted easily to θ. For example, when the radius of the camera is 57.3 millimeters, the circumference is 360 millimeters. Using such a camera, each millimeter measured along the film is equal to 1°. Hence, a distance S of 60 millimeters measured on the film is equal to $60° = 4\theta$, and θ is accordingly equal to 15°.

It is not possible to measure the symmetrical distance S for values of θ much in excess of 40° ($S = 160$ millimeters) on two-hole films of the Straumanis type. In order to obtain θ for lines of θ higher than about 40°, the center about which the low θ lines are concentric must first be found by measurement of a number of such lines. Then the distance $S/2$ may be measured. It must be borne in mind that, if the camera radius is not 57.3 degrees, a correction factor must be applied. Thus, if a camera having a radius of 28.65 millimeters is used, the correction factor is $57.3/28.65 = 2$, and all S values measured must be divided by 2 to obtain θ in degrees.

The Straumanis method, the type of mounting described with two holes in the film, is now widely used. In older type powder cameras the x-ray beam entered between the ends of the cylindrical film and left through a centered hole. During the process of developing, the film usually shrinks. With the Straumanis camera, the effective diameter of the film can be measured and compared with the true diameter. This provides a shrinkage factor that can be applied to film measurements in making calculations.

When the reflecting angle θ corresponding to a given line on a powder photograph has been determined, it is possible to calculate the interplanar spacing of the family of atomic planes that gave rise to the reflection by use of the Bragg equation $n\lambda = 2d \sin \theta$, or $d = n\lambda/(2 \sin \theta)$. Since it is usually impossible to tell the order of a given reflection, n in the formula above is generally given the value 1, and d is determined in every case as though the line was produced by a first-order reflection. For substances crystallizing in the isometric, and tetragonal systems, it is easy to index the lines of a powder photograph and thus determine cell dimensions. For crystals of the other systems, indexing is more difficult. However, for these crystals, if the unit cell dimensions are known, indexing is performed routinely with high speed computer technique.

The powder method finds its chief use in mineralogy as a determinative tool. One can use it for this purpose without knowing anything of the crystal structure or symmetry. Every crystalline substance produces its own powder pattern, which, since it is dependent on the internal structure, is characteristic for that substance. The powder photograph is often spoken of as the "fingerprint" of a

[8] One radian = 57.3°

5-0490 MINOR CORRECTION

d	3.34	4.26	1.82	4.26	SiO_2	
I/I₁	100	35	17	35	SILICON IV OXIDE	ALPHA QUARTZ

Rad. CuKα₁   λ 1.5405   Filter Ni
Dia.   Cut off   Coll.
I/I₁ G.C. Diffractcmeter   d corr. abs.?
Ref. Swanson and Fuyat, NBS Circular 539, Vol. III (1953)

Sys. Hexagonal   S.G. D_3^4 - P3₂21
a₀ 4.913 b₀   c₀ 5.405 A   C1.10
α   β   γ   Z 3
Ref. Ibid.

εa   n ωβ 1.544ℓ γ 1.553   Sign +
2V   Dx2.647 mp   Color
Ref. Ibid.

Mineral from Lake Toxaway, N.C. Spect. anal.: <0.01% Al; <0.001% Ca,Cu,Fe,Mg.
X-ray pattern at 25°C.

Replaces 1-0649, 2-0458,2-0459, 2-0471, 3-0419, 3-0427, 3-0444

d Å	I/I₁	hkl	d Å	I/I₁	hkl
4.26	35	100	1.228	2	220
3.343	100	101	1.1997	5	213
2.458	12	110	1.1973	2	221
2.282	12	102	1.1838	4	114
2.237	6	111	1.1802	4	310
2.128	9	200	1.1530	2	311
1.980	6	201	1.1408	<1	204
1.817	17	112	1.1144	<1	303
1.801	<1	003	1.0816	4	312
1.672	7	202	1.0636	1	400
1.659	3	103	1.0477	2	105
1.608	<1	210	1.0437	2	401
1.541	15	211	1.0346	2	214
1.453	3	113	1.0149	2	223
1.418	<1	300	0.9896	2	402,115
1.382	7	212	.9872	2	313
1.375	11	203	.9781	<1	304
1.372	9	301	.9762	1	320
1.288	3	104	.9607	2	321
1.256	4	302	.9285	<1	410

FIG. 218. ASTM card for quartz. The *d* spacings, their relative intensities and indices are given. At the top of card are given the three strongest lines and their relative intensities. The fourth *d* is of the greatest spacing.

mineral, since it differs from the powder pattern of every other mineral. Thus, if an unknown mineral is suspected of being the same as a known mineral, powder photographs of both are taken. If the photographs correspond exactly line for line, the two minerals are identical. Many organizations maintain extensive files of standard photographs of known minerals, and by comparison, an unknown mineral may be identified readily if there is some indication as to its probable nature.

However, one is frequently completely at a loss as to the identity of the unknown mineral, and a systematic comparison with the thousands of photographs in the reference file would be too time consuming. When this happens, the investigator turns to the card file of x-ray diffraction data prepared by the American Society for Testing Materials (ASTM) (Fig. 218). On these cards are recorded the *d* spacings for thousands of crystalline substances. In order to use the cards, the investigator must calculate the *d* spacings for the most prominent lines in the powder pattern of his unknown substance and estimate the relative intensity of the lines on a scale where the strongest line is taken as 100. Then he may seek a corresponding series of *d*'s in the ASTM file cards, which are arranged in order of *d* of the most intense line.[9] Since many substances have intense lines corresponding to the same *d* value and since many factors may operate to change the relative intensity of the lines in a powder pattern, all substances are cross-indexed

[9] This compilation of data is available on microfilm as well as on file cards.

for their second and third most intense lines. After the most likely "suspects" have been selected from the file, comparison of the weaker reflections, which are also listed on the ASTM card, will speedily identify the substance in most cases. In this way a completely unknown substance may generally be identified in a short time on a very small volume of sample.

The powder method is of wider usefulness, however, and there are several other applications in which it is of great value. Variations in chemical composition of a known substance involve the substitution of atoms, generally of a somewhat different size, for atoms properly occurring in certain sites in the lattice. As a result of this substitution the cell dimensions and hence the interplanar spacings are slightly changed, and the positions of the lines on the powder photograph corresponding to these interplanar spacings are accordingly shifted. By measuring these small shifts in position of the lines in powder patterns of substances of known structure, changes in chemical composition may often be accurately detected.

Further, the relative proportions of two or more known minerals in a mixture may be often determined conveniently by the comparison of the intensities of the lines in the mixture with the intensities of the same lines in photographs of prepared controls of known composition.

X-RAY DIFFRACTOMETER

In recent years the usefulness of the powder method has been greatly increased and its field of application extended by the introduction of the *x-ray diffractometer*. This powerful research tool uses essentially monochromatic x-radiation and a finely powdered sample, as does the powder-film method, but records the information as to the reflections present as an inked trace on a printed strip chart. The equipment supplied by one manufacturer is shown in Fig. 219.

The sample is prepared for diffractometer analysis by grinding to a fine powder, which powder is then spread uniformly over the surface of a glass slide, using a small amount of adhesive binder. The instrument is so constructed that this slide, when clamped in place, rotates in the path of a collimated x-ray beam while a counting tube, mounted on an arm, rotates about it to pick up the reflected x-ray beams.

When the instrument is set at the zero position, the x-ray beam is parallel to the slide and passes directly into a counting tube. The slide mount and the counter are driven by a motor through separate gear trains so that, while the slide and specimen rotate through the angle θ, the counter rotates through 2θ. The purpose of this arrangement is to maintain such a relation among x-ray source, sample, and counter that no reflections are cut off by the glass slide.

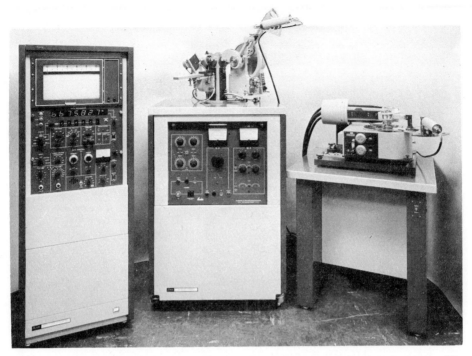

FIG. 219. X-ray diffractometer. (Courtesy of Philips Electronics Co. Inc., Mt. Vernon, N.Y.)

If the specimen has been properly prepared, there will be thousands of tiny crystalline particles on the slide in random orientation. As in powder photography, all possible reflections from atomic planes take place simultaneously. However, instead of recording all of them on a film at one time, the counting tube receives each reflection separately.

In operation, the sample, the counting tube, and the paper drive of the strip chart recorder are all set in motion simultaneously. If an atomic plane has a d spacing such that a reflection occurs at $\theta = 20°$, there is no evidence of this reflection until the counting tube has been rotated through 2θ, or $40°$. At this point the reflected beam enters the tube causing it to conduct. The current pulse thus generated is amplified and causes a deflection on the recording strip chart. Thus, as the counting tube scans, the strip chart records as peaks the reflections from the specimen. The angle 2θ at which the reflection occurred may be read directly from the position of the peak on the strip chart. The heights of the peaks are directly proportional to the intensities of the reflections causing them.

The chart on which the record is drawn is divided in tenths of inches and moves at a constant speed, generally 0.5 inch per minute. At this chart speed and a scanning speed of the counting tube of 1° per minute, 0.5 inch on the chart is equivalent

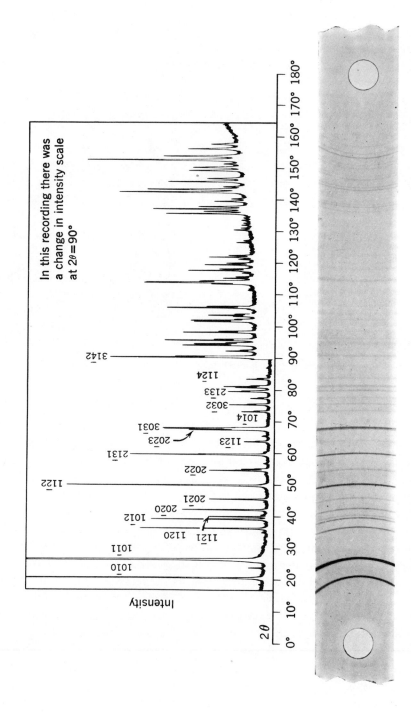

FIG. 220. Comparison of diffractometer record and powder film of quartz. On the diffractometer recording are given the indices of the crystal planes that gave rise to the various reflections. (Courtesy of W. Parrish, Philips Laboratories, Irvington, N.Y.)

to a 2θ of $1°$. The positions of the peaks on the chart can be read easily to a 2θ of $0.05°$, and d spacings of the atomic planes giving rise to them can be determined by use of the equation $n\lambda = 2d \sin \theta$. As in powder photography, the reflections are all considered as first order, unless the record is to be indexed with a view to determining the cell constants.

Although the diffractometer yields data similar to those derived from a powder photograph, it has definite advantages. The powder method requires several hours of exposure plus the time necessary for developing, fixing, washing, and drying the film; a diffractometer run can be made in 1 hour. It is frequently difficult to estimate the intensity of the lines on a powder photograph, whereas the height of peaks on a diffractometer chart can be determined graphically with good precision. The powder photograph must be carefully measured to obtain 2θ values, whereas 2θ can be read directly from the diffractometer strip chart. Figure 220 compares a diffractometer strip chart with a powder photograph of the same mineral and indicates how the data may be used to enter the ASTM card file.

REFERENCES

Geometrical Crystallography

Buerger, M. J., *Elementary Crystallography*, 3rd ed. John Wiley and Sons, New York, 1963.

Phillips, F. C., *An Introduction to Crystallography*, 3rd ed. Longmans, Green and Co., New York, 1962.

Tunnell, G., and J. Murdoch, *Laboratory Manual of Crystallography for Students of Mineralogy and Geology*. Wm. C. Brown Co., Dubuque, 1957.

Wolfe, C. W., *Manual for Geometrical Crystallography*. Edwards Brothers, Ann Arbor, 1953.

X-ray Crystallography

Azaroff, L. V., and M. J. Buerger, *The Powder Method in X-ray Crystallography*. McGraw-Hill Book Co., New York, 1958.

Bragg, W. H., and W. L. Bragg, *X-ray and Crystal Structure*. G. Bell and Sons, London, 1924.

Buerger, M. J., *The Precession Method in X-ray Crystallography*. John Wiley and Sons, New York, 1964.

Buerger, M. J., *Contemporary Crystallography*. McGraw-Hill Book Co., New York, 1970.

Bunn, C. W., *Chemical Crystallography*, 2nd ed. Clarendon Press, Oxford, 1963.

Klug, H. P., and L. E. Alexander, *X-ray Diffraction Procedures for Polycrystalline and Amorphous Material*. John Wiley and Sons, New York, 1954.

Lonsdale, K., *Crystals and X-rays*. G. Bell and Sons, London, 1948.

3

PHYSICAL MINERALOGY

The physical properties are highly important in the rapid determination of minerals, since most of them can be recognized at sight or determined by simple tests.

CLEAVAGE, PARTING, AND FRACTURE

1. *Cleavage.* A mineral has *cleavage* if it breaks along definite plane surfaces. Cleavage may be perfect as in the micas, or more or less obscure as in beryl and apatite. In some minerals, it is completely absent.

Cleavage depends on crystal structure and takes place only parallel to atomic planes. If a family of parallel atomic planes has a weak binding force between them, cleavage is likely to take place along these planes. This weakness may be a result of a weak type of bond, a greater lattice spacing between the planes, or a combination of the two. Graphite has a plate-like cleavage. Within the plates there is a strong bond, but across the plates there is a weak bond giving rise to the cleavage. A weak bond is usually accompanied by a large lattice spacing since the attractive force cannot hold the planes closely together. Diamond has but one bond type, and its excellent cleavage takes place along those lattice planes having the largest spacing. (See Fig. 287.)

Because cleavage is the breaking of a crystal between atomic planes, it is a directional property, and any parallel plane through the crystal is a potential

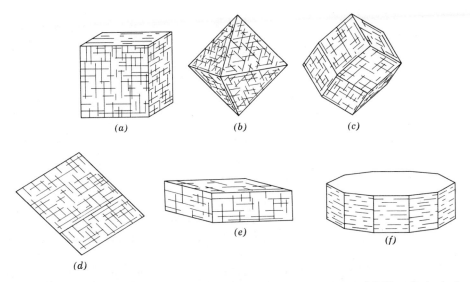

FIG. 221. Cleavage. (*a*) Cubic. (*b*) Octahedral. (*c*) Dodecahedral. (*d*) Rhombohedral. (*e*) Prismatic and pinacoid. (*f*) Pinacoidal.

cleavage plane. Moreover, it is always parallel to crystal faces or possible crystal faces (usually those with simplest indices), since both faces and cleavage reflect the same crystal structure.

In describing a cleavage its quality and crystallographic direction should be given. The quality is expressed as perfect, good, fair, etc. The direction is expressed by the name or indices of the form which the cleavage parallels, as cubic {001}. (Fig. 221), octahedral {111}, rhombohedral {10$\bar{1}$1}, prismatic {110}, pinacoidal {001}, etc. Cleavage is always consistent with the symmetry; thus, if one octahedral cleavage direction is developed, it implies that there must be three other directions similar to it. If one dodecahedral cleavage direction is present, it likewise implies five other, similar directions. Not all minerals show cleavage, and only a comparatively few show it in an eminent degree, but in these it serves as an outstanding diagnostic criterion.

2. *Parting.* When minerals break along planes of structural weakness, they have *parting.* The weakness may result from pressure or twinning; and, since it is parallel to crystallographic directions, it resembles cleavage. However, parting, unlike cleavage, is not shown by all specimens but only those that are twinned or have been subjected to the proper pressure. Even in these specimens there are a limited number of planes in a given direction along which the mineral will break. For example, twinned crystals part along composition planes but between these planes they fracture irregularly. Familiar examples of parting are found in the octahedral parting of magnetite, the basal parting of pyroxene, and the rhombohedral parting of corundum (see Figs. 222 and 223).

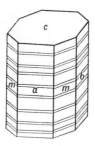

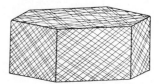

FIG. 222. Basal
parting, pyroxene.

FIG. 223. Rhombohedral
parting, corundum.

3. *Fracture*. The way a mineral breaks when it does not yield along cleavage or parting surfaces is its *fracture*. Different kinds of fracture are designated as follows:

a. Conchoidal. The smooth, curved fracture resembling the interior surface of a shell (see Fig. 224). This is most commonly observed in such substances as glass and quartz. *b. Fibrous or splintery*. *c. Hackly*. Jagged fractures with sharp edges. *d. Uneven or irregular*. Fractures producing rough and irregular surfaces.

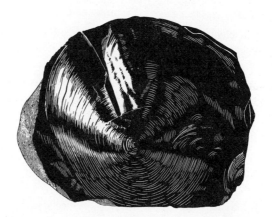

FIG. 224. Conchoidal fracture, obsidian.

HARDNESS

The resistance that a smooth surface of a mineral offers to scratching is its *hardness* (designated by **H**). Like the other physical properties of minerals, hardness is dependent on the crystal structure. The stronger the binding forces between the atoms, the harder the mineral. The degree of hardness is determined by observing the comparative ease or difficulty with which one mineral is scratched by another, or by a file or knife. The hardness of a mineral might then be said to be its "scratchability." A series of ten common minerals has been chosen as a scale, by comparison with which the relative hardness of any mineral can be told. The following minerals arranged in order of increasing hardness comprise what is known as the *Mohs scale of hardness:*

Scale of Hardness

1. Talc	6. Orthoclase
2. Gypsum	7. Quartz
3. Calcite	8. Topaz
4. Fluorite	9. Corundum
5. Apatite	10. Diamond

Talc, number 1 in the scale, has a structure made up of plates so weakly bound to one another that the pressure of the fingers is sufficient to slide one plate over the other. At the other end of the scale is diamond with its constituent carbon atoms so firmly bound to each other that no other mineral can force them apart to cause a scratch.

In order to determine the relative hardness of any mineral in terms of this scale, it is necessary to find which of these minerals it can and which it cannot scratch. In making the determination, the following should be observed: Sometimes when one mineral is softer than another, portions of the first will leave a mark on the second which may be mistaken for a scratch. The mark can be rubbed off, whereas a true scratch will be permanent. Some minerals are frequently altered on the surface to material which is much softer than the original mineral. A fresh surface of the specimen to be tested should therefore be used. The physical nature of a mineral may prevent a correct determination of its hardness. For instance, if a mineral is pulverulent, granular, or splintery, it may be broken down and apparently scratched by a mineral much softer than itself. It is always advisable when making the hardness test to confirm it by reversing the order of procedure; that is, do not only try to scratch mineral A by mineral B, but also try to scratch B by A.

The following materials serve in addition to the above scale: the hardness of the finger nail is a little over 2, a copper coin about 3, the steel of a pocket knife a little over 5, window glass $5\frac{1}{2}$, and the steel of a file $6\frac{1}{2}$. With a little practice, the hardness of minerals under 5 can be quickly estimated by the ease with which they can be scratched with a pocket knife.

Hardness, we have seen (page 10), is a vectorial property. Thus crystals may show varying degrees of hardness depending on the directions in which they are scratched. The directional hardness differences in most common minerals are so slight that, if they can be detected at all, it is only through the use of delicate instruments. Two exceptions are kyanite and calcite. In kyanite, **H** = 5 parallel to the length, but **H** = 7 across the length. The hardness of calcite is 3 on all surfaces except {0001}. On this form, however, it can be scratched by the fingernail and has a hardness of 2.

It can be seen that only within relatively wide limits can one be quantitative in the determination of hardness. Moreover, the interval of hardness between different pairs of minerals in the scale varies. For instance, the hardness difference between corundum and diamond is many times greater than that between topaz and corundum.

TENACITY

The resistance that a mineral offers to breaking, crushing, bending, or tearing—in short, its cohesiveness—is known as tenacity. The following terms are used to describe tenacity in minerals:

1. *Brittle*. A mineral that breaks or powders easily.
2. *Malleable*. A mineral that can be hammered out into thin sheets.
3. *Sectile*. A mineral that can be cut into thin shavings with a knife.
4. *Ductile*. A mineral that can be drawn into wire.
5. *Flexible*. A mineral that bends but does not resume its original shape when the pressure is released.
6. *Elastic*. A mineral that, after being bent, will resume its original position upon the release of the pressure.

SPECIFIC GRAVITY

Specific gravity (**G**) or relative density[1] is a number that expresses the ratio between the weight of a substance and the weight of an equal volume of water at 4°C. Thus a mineral with a specific gravity of 2, weighs twice as much as the same volume of water. The specific gravity of a mineral is frequently an important aid in its identification, particularly in working with fine crystals or gemstones, when other tests would injure the specimens.

The specific gravity of a crystalline substance depends on (1) the kind of atoms of which it is composed, and (2) the manner in which the atoms are packed together. In isostructural compounds (see page 194), in which the packing is constant,

---

[1] *Density* and specific gravity are sometimes used interchangeably. However, *density* requires the citation of units, for example, grams per cubic centimeter or pounds per cubic foot.

those with elements of higher atomic weight will usually have higher specific gravities. This is well shown by the orthorhombic carbonates listed below in which the chief difference is in the cations.

Specific Gravity Change with Change in Cation

Mineral	Composition	Atomic Weight of Cation	Specific Gravity
Aragonite	$CaCO_3$	40.08	2.95
Strontianite	$SrCO_3$	87.62	3.76
Witherite	$BaCO_3$	137.34	4.29
Cerussite	$PbCO_3$	207.19	6.55

In a solid-solution series (see page 198), there is a continuous change in specific gravity with change in chemical composition. For example, the mineral olivine, $(Mg, Fe)_2SiO_4$, is a solid-solution series between forsterite, Mg_2SiO_4 (**G** 3.3), and fayalite, Fe_2SiO_4 (**G** 4.4). Thus from determination of specific gravity one can obtain a close approximation of the chemical composition.

The influence of the packing of atoms on specific gravity is well illustrated in polymorphous compounds (see page 175). In these compounds the composition remains constant, but the packing of the atoms varies. The most dramatic example is given by diamond and graphite, both elemental carbon. Diamond with specific gravity 3.5 has a closely packed structure, giving a high density of atoms per unit volume; whereas in graphite, specific gravity 2.2, the carbon atoms are loosely packed.

Average Specific Gravity. Most people from everyday experience have acquired a sense of relative weight even in regard to minerals. For example, ulexite (**G** 1.96) seems light, whereas barite (**G** 4.5) seems heavy for nonmetallic minerals. This means that one has developed an idea of an average specific gravity or a feeling of what a nonmetallic mineral of a given size should weigh. This average specific gravity can be considered to be between 2.65 and 2.75. The reason for this is that the specific gravities of quartz (**G** 2.65), feldspar (**G** 2.60–2.75), and calcite (**G** 2.72), the most common and abundant nonmetallic minerals, fall mostly within this range. The same sense may be developed in regard to metallic minerals. Graphite (**G** 2.2) seems light, while silver (**G** 10.5) seems heavy. The average specific gravity for metallic minerals is about 5.0, that of pyrite. Thus, with a little practice, one can by merely lifting specimens distinguish minerals that have comparatively small differences in specific gravity.

Determination of Specific Gravity. In order to determine specific gravity accurately, the mineral must be pure, a requirement frequently difficult to fulfill.

It must also be compact with no cracks or cavities within which bubbles or films of air could be imprisoned. For normal mineralogical work, the specimen should have a volume of about one cubic centimeter. If these conditions cannot be met, a specific gravity determination by any rapid and simple method means little.

The necessary steps in making an ordinary specific gravity determination are, briefly, as follows: The mineral is first weighed in air. Let this weight be represented by W_a. It is then immersed in water and weighed again. Under these conditions it weighs less, since in water it is buoyed up by a force equivalent to the weight of the water displaced. Let the weight in water be represented by W_w. Then $W_a - W_w$ equals the apparent loss of weight in water, or the weight of an equal volume of water. The expression $W_a/(W_a - W_w)$ will therefore yield a number which is the specific gravity.

Jolly Balance. Since specific gravity is merely a ratio, it is not necessary to determine the absolute weight of the specimen but merely values proportional to the weights in air and in water. This can be done by means of a *Jolly balance*, (Fig. 225)[2] with which the data for making the calculations are obtained by the stretching of a spiral spring. In using the balance, a fragment is first placed on the upper scale pan and the elongation of the spring noted. This is proportional to the weight in air (W_a). The fragment is then transferred to the lower pan and immersed in water. The elongation of the spring is now proportional to the weight of the fragment in water (W_w).

FIG. 225. Jolly balance.

[2] Manufactured by Eberbach and Son, Ann Arbor, Michigan.

FIG. 226. Berman balance.

A torsion balance was adapted by the late Harry Berman of Harvard University for obtaining specific gravities of small particles weighing less than 25 milligrams (Fig. 226).[3] To the advanced worker interested in accurate determinations this balance is particularly useful, since it is frequently possible to obtain only a tiny mineral fragment free from impurities. In using it, however, one must make correction for temperature and use a liquid with a low surface tension.

Beam Balance. The beam balance is a very convenient and accurate instrument for determining specific gravity. It is easily and inexpensively constructed and can be made in any size. Occasionally it is necessary to obtain the specific gravity of a specimen weighing as much as a kilogram. This can be done easily and quickly using a large beam balance. The balance illustrated in Fig. 227 was devised by S. L. Penfield, whose description of its operation, slightly modified, follows.

The beam made of wood or brass is supported at *b* on a fine wire or needle, which permits it to swing freely. The long arm *bc* is divided into a decimal scale; the short arm carries a

[3] This balance is distributed through Bethlehem Instrument Company, Bethlehem, Pennsylvania.

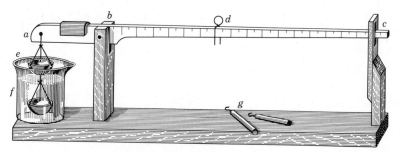

FIG. 227.   Beam balance.

double arrangement of pans so suspended that one of them is in air and the other in water. A piece of lead on the short arm serves to almost counterbalance the long arm. When the pans are empty, the beam is brought to a horizontal position, marked upon the upright near c, by means of a rider d. A number of counterpoises are needed; but, since it is their position on the beam and not their actual weight that is recorded, they need not be of any specific denomination. After the beam is adjusted by means of the rider d, a mineral fragment is placed in the upper pan and a counterpoise is chosen which, when placed near the end of the long arm, will bring it into a horizontal position. A value, W_a, proportional to the weight of the mineral in air, is given by the position of the counterpoise on the scale. The mineral is next transferred to the lower pan, and the same counterpoise is brought nearer the fulcrum b until the beam becomes horizontal again. The position of the counterpoise now gives a value, W_w, proportional to the weight of the mineral in water.

Pycnometer.   When a mineral cannot be obtained in a homogeneous mass large enough to permit use of one of the balance methods, the specific gravity of a powder or an aggregate of mineral fragments can be accurately obtained by means of the *pycnometer*. The pycnometer is a small bottle (Fig. 228) fitted with a ground-glass stopper through which a capillary opening has been drilled.

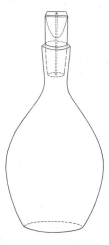

FIG. 228.   Pycnometer.

In making a specific-gravity determination, the dry bottle with stopper is first weighed empty (P). The mineral fragments are then introduced into the bottle and a second weighing (M) is made. The bottle is partially filled with distilled water and boiled for a few minutes to drive off any air bubbles. After cooling, the pycnometer is filled with distilled water and weighed (S), care being taken that the water rises to the top of the capillary opening but that no excess water is present. The last weighing (W) is made after emptying the bottle and refilling with distilled water alone. The specific gravity can thus be determined:

$$G = \frac{M - P}{W + M - P - S}$$

Heavy Liquids. Several liquids with relatively high densities are sometimes used in the determination of the specific gravity of minerals. The two liquids most easily used are bromoform (**G** 2.89) and methylene iodide (**G** 3.33). These liquids are miscible with acetone (**G** 0.79), and thus, by mixing, a solution of any intermediate specific gravity may be obtained. A mineral grain is introduced into the heavy liquid, and the liquid diluted with acetone until the mineral neither rises nor sinks. The specific gravity of the liquid and the mineral are then the same, and that of the liquid may be quickly determined by means of a Westphal balance.

Heavy liquids are frequently used in the separation of grains from mixtures composed of several constituents. For example, a separation of the constituent mineral grains of a sand composed of quartz (**G** 2.65), tourmaline (**G** 3.20), and garnet (**G** 4.25) could be quickly made. In bromoform the quartz would float and the tourmaline and garnet would sink. After removing and washing these "heavy minerals" in acetone, they could be separated from each other in methylene iodide; the tourmaline would float and the garnet would sink.

Calculation of Specific Gravity. If one knows the number of the various kinds of atoms in the unit cell and the volume of the unit cell, the specific gravity can be calculated. The chemical formula of the mineral gives the proportions of the different atoms but not necessarily the exact number, for some minerals have several formula weights per cell. The number, usually small, is indicated by Z. For example, in aragonite, $CaCO_3$, the ratios of the atoms is 1Ca:1C:3O, but there are four formula weights per cell or 4Ca, 4C, 12O. The molecular weight, M, of $CaCO_3$ is 100.09; the molecular weight of the contents of the unit cell ($Z = 4$) is $4 \times 100.09 = 400.36$.

The volume of the unit cell, V, in the orthogonal crystal systems is found by multiplying the cell dimensions, as $a \times b \times c = V$. In the inclined systems the angles between cell edges must also be considered in obtaining the volume. Aragonite is orthorhombic with cell dimensions: $a = 4.95$ Å, $b = 7.96$, $c = 5.73$. Therefore, $V = 225.76$ Å^3.

Converting Å³ to cm³, we divide by $(10^8)^3 = 10^{24}$ or $V = 225.76 \times 10^{-24}$ cm³. Knowing the values M and V, the density, D, can be calculated using the formula:

$$D = \frac{Z \times M}{N \times V}$$

where N is Avogadro's number 6.02338×10^{23}. Substituting values for aragonite,

$$D = \frac{4 \times 100.09}{6.02338 \times 10^{23} \times 225.76 \times 10^{-24}} = 2.945 \text{ gr/cm}^3$$

This value, 2.945, for the calculated density of aragonite is in excellent agreement with the best measured values which are 2.947 ± 0.002.

In the study of new minerals, the numerical value of Z is commonly unknown. Hence it is necessary to make successive trials of the calculations given above, using different values of Z until the best possible agreement with the measured specific gravity is secured. Z is always an integer and generally small.

PROPERTIES DEPENDING UPON LIGHT

PART I: GENERAL

Luster is the general appearance of a mineral surface in reflected light. There are two types of luster, *metallic* and *nonmetallic*, but with no sharp division between them. Minerals with an intermediate luster are said to be submetallic.

A mineral having the brilliant appearance of a metal has a metallic luster. Such minerals are quite opaque to light and, as a result, give a black or very dark streak (see page 138). Galena, pyrite, and chalcopyrite are common minerals with metallic luster.

All minerals without a metallic appearance have a nonmetallic luster. They are, in general, light-colored and transmit light, if not through thick portions, at least through thin edges. The streak of a nonmetallic mineral is either colorless or very light in color. The following terms are used to describe further the luster of nonmetallic minerals:

Vitreous	The luster of glass. Examples—quartz and tourmaline.
Resinous	Having the luster of resin. Examples—sphalerite and sulfur.
Pearly	An iridescent pearl-like luster. This is usually observed on mineral surfaces that are parallel to cleavage planes. Examples—basal plane on apophyllite and cleavage surface of talc.

Greasy	Appears as if covered with a thin layer of oil. This luster results from light scattered by a microscopically rough surface. Examples, nepheline and some specimens of sphalerite and massive quartz.
Silky	Silklike. It is caused by the reflection of light from a fine fibrous parallel aggregate. Examples—fibrous gypsum, malachite, and serpentine (chrysotile).
Adamantine	A hard, brilliant luster like that of a diamond. It is due to the mineral's high index of refraction (see page 142). Examples— the transparent lead minerals, like cerussite and anglesite.

Color

When white light strikes the surface of a mineral, part of it is reflected and part refracted. If the light suffers no absorption, the mineral is colorless, both in reflected and transmitted light. Minerals are colored because certain wave lengths of light are absorbed, and the color results from a combination of those wave lengths that reach the eye. Some minerals show different colors when light is transmitted along different crystallographic directions. This selective absorption known as *pleochroism* is shown by transparent varieties of cordierite and spodumen. If there are only two such directions, as in tourmaline, the property is called dichroism (see page 162).

For some minerals, color is a fundamental property directly related to one of its major constituent elements and is therefore constant and characteristic. For such minerals, called *idiochromatic*, color serves as an important means of identification. Thus malachite is always green, azurite is always blue, and rhodonite and rhodochrosite are always red or pink.

For most metallic minerals, color is constant, as: the brass-yellow of chalcopyrite, the brownish bronze of bornite, and the copper-red of niccolite. Because alteration may produce a colored tarnish on these minerals different from the true color, it is important that a fresh surface be examined. This is particularly true of copper minerals, especially bornite. This mineral is called "peacock copper ore" because on exposure its bronze color is covered by a blue-violet film.

Most minerals are composed of elements that produce no characteristic color and are colorless. Yet color varieties of many of them are common. In some cases the color is attributable to appreciable amounts of an element such as iron that has a strong pigmenting power. In several solid-solution series, color is directly related to the amount of iron present. For example, the amphibole, tremolite, with composition $Ca_2Mg_5(Si_8O_{22})(OH)_2$ is white; but when Fe^2 substitutes for Mg, the mineral is green. With increasing amounts of iron, the color changes from pale to dark green to nearly black. Also in sphalerite, ZnS, the progressive substitution of iron for zinc changes the color from white through yellow and brown to black.

Minerals such as those mentioned that show a variation in color are called *allochromatic*. However, the presence of a major chemical constituent, is only one of the causes of color in minerals. Other factors are small amounts of chemical impurities, defects in crystal structure and finely divided inclusion of other minerals.

The ions of certain elements are strongly light-absorbing and their presence in small, even trace, amounts may cause the mineral to be deeply colored. Chief among these elements known as *chromophores* are: Fe, Mn, Cu, Cr, Co, Ni, and V. Thus the deep green of emerald results from small amounts of chromium in beryl and the purple of amethyst is attributed to trace amounts of iron in quartz. Zoisite, normally white or gray, was found in 1967 in gem quality crystals of a rich sapphire blue; the color results from the presence of vanadium.

Lattice imperfections in minerals having no chromophoric ions may cause white light to be selectively absorbed. The imperfections may be due to structural voids or to the presence of foreign ions. Color in minerals with structural imperfections may be changed or introduced by heat treatment or exposure to high energy radiation. For example, x-radiation turns some colorless quartz smoky, but on heating to 400°C the quartz returns to its colorless state. Not all colorless quartz responds in this way. X-radiation has no effect on some crystals and others are selectively pigmented. It is believed that the coloration is associated with the presence of the foreign ions. Colorless diamonds by exposure to the proper radiation can be colored either green or blue, and the green diamond will turn yellow with heat treatment. The color of many other gem stones can be enhanced by heating.

The color of certain silicate minerals results from the addition of anions such as Cl^{1-}, CO_3^{2-}, SO_4^{3-} to the structure. All these ions may be present in lazurite (blue), Cl^{1-} in sodalite (blue, green), CO_3^{2-} in cancrinite (orange-yellow), SO_4^{3-} in hauynite (blue). Heating causes a change in color in all these minerals.

The mechanical admixture of impurities can give a variety of colors to otherwise colorless minerals. Quartz may be green because of the presence of chlorite and calcite black, colored by manganese oxide or carbon. Hematite, as the most common pigmenting impurity, imparts its red color to many minerals including some feldspar and calcite and the fine-grained variety of quartz, jasper.

Streak. The color of a finely powdered mineral is known as its *streak*. Although the color of a mineral may vary, the streak is usually constant and is thus useful in mineral identification. The streak is determined by rubbing the mineral on a piece of unglazed porcelain, a *streak plate*. The streak plate has a hardness of about 7, and thus it cannot be used with minerals of greater hardness.

Play of Colors.   Interference of light either at the surface or in the interior of a mineral may produce a series of colors as the angle of incident light changes. The striking play of colors seen in precious opal results from the interference of light reflected from submicroscopic layers of differing refractive index. Common opal lacks this layering and the scattered light produces a pearly or milky *opalescence*.

An internal iridescence is caused by light defracted and reflected from closely spaced fractures, cleavage planes, twin lamellae, or minute foreign inclusions in parallel orientation. Some specimens of labradorite show this well with the color changing from blue to green or yellow with changing angle of incident light. The delicate colors of peristerite, an albite feldspar, are of similar origin.

A surface iridescence similar to that produced by soap bubbles or thin films of oil on water is caused by interference of light as it is reflected from thin surface films. It is most commonly seen on metallic minerals, particularly hematite, limonite, and sphalerite.

Chatoyancy and Asterism.   In reflected light some minerals have a silky appearance which results from closely packed parallel fibers or from a parallel arrangement of inclusions or cavities. When a cabochon gem stone is cut from such a mineral, it is crossed by a band of light at right angles to the length of the fibers or direction of the inclusions. This property known as *chatoyancy* is shown particularly well by "satin spar" gypsum, *cat's eye*, a gem variety of chrysoberyl and "tiger eye," fibrous crocidolite replaced by quartz.

In some crystals, particularly those of the hexagonal system, inclusions may be arranged in three crystallographic directions at 120° to each other. A cabochon stone cut from such a crystal shows what might be called a triple chatoyancy, that is, one beam of light at right angles to each direction of inclusions producing a six-pointed star. The phenomenon, seen in star rubies and sapphires, is termed *asterism*. Some phlogopite mica containing rutile needles oriented by the pseudo-hexagonal lattice, show a striking asterism in transmitted light.

Luminescence

Any emission of light by a mineral that is not the direct result of incandescence is *luminescence*. The phenomenon that may be brought about in several ways is usually observed in minerals containing foreign ions called *activators*. Most luminescence is faint and can be seen only in the dark.

Fluorescence and Phosphorescence.   Minerals that luminesce during exposure to ultraviolet light, x-rays, or cathode rays are *fluorescent*. If the luminescence continues after the exciting rays are cut off, the mineral is said to be *phosphorescent*. There is no sharp distinction between fluorescence and phosphorescence, for some minerals that appear only to fluoresce can be shown by refined methods to continue to glow for a small fraction of a second after the removal of the exciting radiation.

Fluorescence is produced when the energy of the short wave radiation is absorbed by the impurity ions and released as longer wave radiation—visible light. Minerals vary in their ability to absorb ultraviolet light of a given wave length. Thus some fluoresce only in short wave u.v., whereas others may fluoresce only in long wave u.v., and still others will fluoresce under either wave length of u.v. The color of the emitted light varies considerably with the wave lengths or source of ultraviolet light.

Fluorescence is an unpredictable property, for some specimens of a mineral show it, whereas other apparently similar specimens even from the same locality, do not. Thus only some fluorite, the mineral from which the property receives its name, will fluoresce. Its usual blue fluorescence may result from the presence of organic material or rare earth ions. Other minerals that frequently, but by no means invariably fluoresce are scheelite, willemite, calcite, eucryptite, scapolite, diamond, hyalite, and autunite. The pale blue fluorescence of most scheelite is ascribed to molybdenum substituting for tungsten. And the brilliant fluorescence of willemite and calcite from Franklin, New Jersey is attributed to the presence of manganese.

With the development of synthetic phosphores, fluorescence has become a commonly observed phenomenon in fluorescent lamps, paints, cloth, and tapes. The fluorescent property of minerals also has practical applications in prospecting and ore dressing. With a portable ultraviolet light one can at night detect scheelite in an outcrop; and underground the miner can quickly estimate the amount of scheelite on a freshly blasted surface. At Franklin, New Jersey ultraviolet light has long been used to determine the amount of willemite that goes into the tailings. Eucryptite is an ore of lithium in the great pegmatite at Bikita, Rhodesia. In white light it is indistinguishable from quartz, but under ultraviolet light it fluoresces a salmon-pink and can be easily separated.

Thermoluminescence is the property possessed by some minerals of emitting visible light when heated to a temperature below that of red heat. It is best shown by anhydrous nonmetallic minerals that contain foreign ions as activators. When a thermoluminescent mineral is heated, the initial visible light, usually faint, is given off at a temperature between 50° and 100°C, and light usually ceases to be emitted at temperatures higher than 475°C. For a long time, fluorite has been known to possess this property; the variety *chlorophane* was named because of the green light emitted. Other minerals which are commonly thermoluminescent are calcite, apatite, scapolite, lepidolite, and feldspar.

Triboluminescence is the property possessed by some minerals of becoming luminous on being crushed, scratched, or rubbed. Most minerals showing this property are nonmetallic, anhydrous, and possess a good cleavage. Fluorite, sphalerite, and lepidolite may be triboluminescent and, less commonly, pectolite, amblygonite, feldspar, and calcite.

PART II: SPECIAL OPTICAL PROPERTIES

Although the optical properties are the least easily determined of the several characteristics of minerals that depend upon light, they are important in mineral identification. Since they are usually observed microscopically, only a very small amount of material is necessary. The following discussion deals with the optical properties that can be determined in transmitted light and thus applies only to nonopaque minerals.

Nature of Light.   To account for all the properties of light it is necessary to resort to two theories: the *wave theory* and the *corpuscular theory*. However, it is the wave theory that we shall consider in explaining the optical behavior of crystals. This theory assumes that visible light, as part of the electromagnetic spectrum, travels in straight lines with a transverse wave motion; that is, it vibrates at right angles to the direction of propagation. The wave motion is similar to that generated by dropping a pebble into still water with waves moving out from the central point. The water merely rises and falls, it is only the wave front that moves forward. The *wave length* (λ) of such a wave motion is the distance between successive crests (or troughs); the *amplitude* is the displacement on either side of the position of equilibrium; the *frequency* is the number of waves per second passing a fixed point; and the velocity is the frequency multiplied by the wave length. Similarly light waves (Fig. 229) have length, amplitude, frequency, and velocity but their transverse vibrations, perpendicular to the direction of propagation, take place in all possible directions.

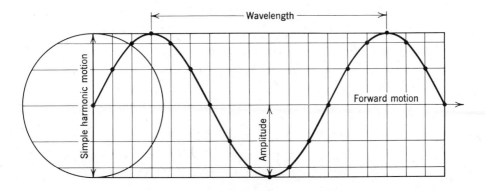

FIG. 229.   Simple sinusoidal wave motion.

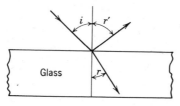

FIG. 230. Reflected and re-fracted light.

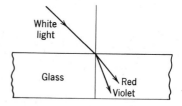

FIG. 231. Different refraction for different wavelengths of light.

Visible light occupies a very small portion of the electromagnetic spectrum (see Fig. 201). The wave length determines the color and varies from slightly more than 7000 Å at the red end to about 4000 Å at the violet end. White light is composed of all wave lengths between these limits, whereas light of a single wave length is called monochromatic.

Reflection and Refraction. When a light ray passes from a rare medium, such as air, into a denser medium, such as glass, part of it is reflected from the surface back into the air and part enters the glass (Fig. 230). The reflected ray obeys the laws of reflection which state: (a) that the angle of incidence (i) equals the angle of reflection (r'), both angles measure from the surface normal, and (b) that the incident and reflected rays lie in the same plane. The light that passes into the glass no longer follows the path of the incident ray but is bent or refracted. The amount of bending depends on the obliquity of the incident ray and the relative densities of the two media: the greater the angle of incidence and the greater the density difference, the greater the refraction.

Index of Refraction. The precise relationship of the angle of incidence (i) to the angle of refraction (r) is given by Snell's law which states that for the same two media the ratio of $\sin i : \sin r$ is a constant. This is usually expressed as $\sin i / \sin r = n$, where the constant, n, is the *index of refraction.*

The refraction of a ray of light on entering a denser medium is accompanied by a decrease in velocity. It can be shown that the index of refraction may also be ex-pressed by the ratio between the velocity of light in air (V) and its velocity in the denser medium (v), that is, $V/v = n$. Since the velocity of light in air may be con-sidered equal to 1, $n = 1/v$, or the index of refraction is equal to the reciprocal of the velocity.[4]

The velocity of light in glass is equal to frequency multiplied by wave length, therefore with fixed frequency, the longer the wave length the greater the velocity. Red light with its longer wave length has a greater velocity than violet light and because of the reciprocal relation between velocity and refractive index, n for red

---

[4] The velocity of light in vacuum is taken as unity, but since the velocity of light in air is 0.9997, it may also be considered as unity.

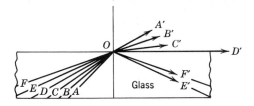

FIG. 232. Critical angle.

light is less than *n* for violet light (Fig. 231). A crystal thus has different refractive indices for different wave lengths of light. This phenomenon is known as *dispersion of the indices*, and because of it, monochromatic light is used for accurate determination of refractive index.

Total Reflection and the Critical Angle

We have seen (Fig. 230) that light is refracted toward the normal when it passes from a lower to a higher refractive index medium. When the conditions are reversed, as in Fig. 232, and light moves from the higher to the lower index medium, it is refracted away from the normal. In Fig. 232, assume that lines *A*, *B*, *C*, etc. represent light rays moving through glass, and into air at point O. The greater the obliquity of the incident ray, the greater the angle of refraction. Finally an angle of incidence is reached, as at ray *D*, for which the angle of refraction is 90°, and the ray then grazes the surface. The angle of incidence at which this takes place is known as the *critical angle*. Rays such as *E* and *F*, striking the interface at a greater angle, are totally reflected back into the higher index medium.

The measurement of the critical angle is a quick and easy method of determining the refractive index of both liquids and solids. The instrument used is a refractometer, of which there are many types. A description of one of these, the Pulfrich refractometer, will suffice to illustrate the underlying principles. This instrument employs a polished hemisphere of high refractive index glass (Fig. 233). A crystal

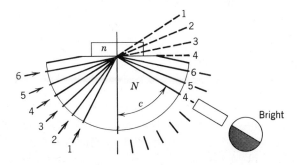

FIG. 233. Pulfrich refractometer and measurement of critical angle *c*.

face or polished surface of the mineral is placed on the equatorial plane of the hemisphere but separated from it by a film of liquid. The liquid is necessary to exclude the air and must have a refractive index higher than that of the mineral but lower than that of the hemisphere. Slightly converging light is directed upward through the hemisphere and, depending on the angle of incidence, is either partly refracted through the unknown or totally reflected back through the hemisphere. If a telescope is placed in a position to receive the reflected rays, we can observe a sharp boundary between the portion of the field intensely illuminated by the totally reflected light and the remainder of the field. When the telescope is moved so that its cross hairs are precisely on the contact, the critical angle c is read on a scale. Knowing this angle and the index of refraction of the hemisphere, N, we can calculate the index of refraction of the mineral: n mineral $=$ sin critical angle $\times$ N hemisphere.

Isotropic and Anisotropic Crystals

For optical considerations all transparent substances can be divided into two groups: isotropic and anisotropic. The isotropic group includes such non-crystalline substances as gases, liquids, and glass, but it also includes crystals that belong to the isometric crystal system. In them light moves in all directions with equal velocity and hence each isotropic substance has a single refractive index. In anisotropic substances, which include all crystals except those of the isometric system, the velocity of light varies with crystallographic direction and thus there is a range of refractive index.

In general, light passing through an anisotropic crystal is broken into two polarized rays vibrating in mutually perpendicular planes. Thus for a given orientation, a crystal has two indices of refraction, one associated with each polarized ray.

Polarized Light

We have seen that light can be considered a wave motion with vibrations taking place in all directions at right angles to the direction of propagation. When the wave motion is confined to vibrations in a single plane, the light is said to be *plane polarized*. The three principal ways of polarizing light are by double refraction, absorption, and reflection.

Polarized Light by Double Refraction.   It has been pointed out that when light passes through an anisotropic crystal it is divided into two polarized rays. The principle on which the first efficient polarizer was based was the elimination of one of these rays. The crystalline material used was the optically clear variety of calcite, Iceland spar, and the polarizer was called the *Nicol prism*, after the inventor William Nicol. Calcite has such a strong double refraction that each ray produces a separate image when an object is viewed through a cleavage fragment (Fig. 234). In making the Nicol prism (Fig. 236), an elongated cleavage rhombohedron of calcite is sawed at a specified angle and the two halves rejoined by

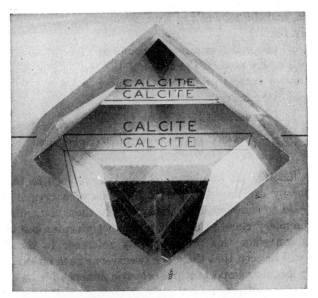

FIG. 234. Calcite, viewed normal to the rhombohedron face showing double refraction. The double repetition of "calcite" at the top of the photograph is seen through a face cut on the specimen parallel to the base.

FIG. 235. Calcite showing no double refraction. Viewed parallel to the c axis.

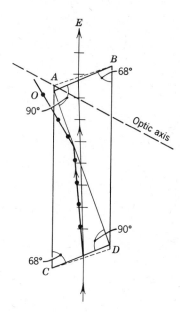

FIG. 236. Nicol prism.

cementing with Canada balsam. Faces are then ground at the ends of the prism which makes angles of 90° with the cemented surface. On entering the prism light is resolved into two rays "*O*" and "*E*." Because of the greater refraction of the *O* ray it is totally reflected at the Canada balsam surface. The *E* ray with refractive index close to that of the balsam proceeds essentially undeviated through the prism and emerges as plane polarized light.

Polarized Light by Absorption.   The polarized rays into which light is divided in anisotropic crystals may be differentially absorbed. If one ray suffers nearly complete absorption and the other very little, the emerging light will be plane polarized. This phenomenon is well illustrated by some tourmaline crystals (Fig. 237). Light passed through the crystal at right angles to [001] emerges essentially plane polarized, with vibrations parallel to the *c* axis. The other ray,

FIG. 237. Polarized light by absorption. (*a*) Tourmaline. (*b*) *Polaroid*.

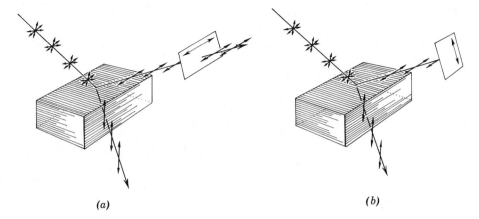

(a)   (b)

FIG. 238.   Polarized light by reflection and refraction.

vibrating perpendicularly to it, is almost completely absorbed. Polarizing sheets, such as *Polaroid*, are made by aligning crystals on an acetate base. These crystals absorb very little light in one vibration direction, but are highly absorptive in the other. The light transmitted by the sheet is thus plane polarized. Because they are thin and can be made in large sheets, manufactured polarizing plates are extensively used in optical equipment, including many polarizing microscopes.

Polarized Light by Reflection.   Light reflected from a smooth nonmetallic surface is partially polarized with the vibration directions parallel to the reflecting surface. The extent of polarization depends on the angle of incidence (Fig. 238) and the index of refraction of the reflecting surface. It is most nearly polarized when the angle between the reflected and refracted ray is 90° (Brewster's law). The fact that reflected light is polarized can be easily demonstrated by viewing it through a polarizing filter. When the vibration direction of the filter is parallel to the reflecting surface the light passes through the filter with only slight reduction in intensity; when the filter is turned 90°, only a small percentage of the light reaches the eye.

The Polarizing Microscope

The polarizing microscope is the most important instrument for determining the optical properties of crystals; with it more information can be obtained easily and quickly than with more specialized devices. Several manufacturers each make a number of models of polarizing microscopes that vary in complexity of design and sophistication and hence in price. A student model made by E. Leitz (SM-Pol) is illustrated in Fig. 239 with the essential parts named.

Although a polarizing microscope differs in detail from an ordinary compound microscope, its primary function is the same: to yield an enlarged image of an

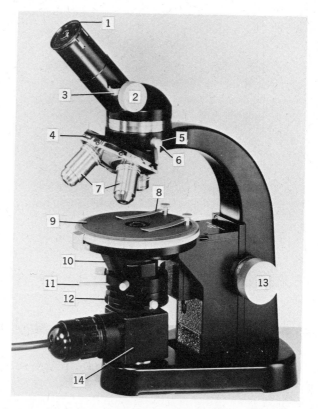

FIG. 239. Polarizing microscope (SM-Pol E. Leitz, Inc.). 1. Ocular. 2. Amici-Bertrand lens. 3. Pinhole stop for auxiliary lens, swings out and in. 4. Revolving nosepiece for oculars. 5. Analyzer. 6. Tube slot for compensators (out of sight). 7. Objectives. 8. Knurled head for swinging out of top part of condenser (out of sight). 9. Rotating stage. 10. Condenser (behind shield). 11. Iris diaphragm. 12. Rotatable polarizer. 13. Focusing controls on both sides. 14. Substage illuminator.

object placed on the stage. The magnification is produced by a combination of two sets of lenses: the objective and the ocular. The function of the objective lens, at the lower end of the microscope tube, is to produce an image that is sharp and clear. The ocular merely enlarges this image including any imperfection resulting from a poor quality objective. For mineralogical work it is desirable to have three objectives: low-, medium-, and high-power. In Fig. 239 these are shown mounted on a revolving nose piece and can be successively rotated into position. The magnification produced by an objective is usually indicated on its housing, such as 2x (low), 10x (medium), and 50x (high). Oculars also have different magnifications such as 5x, 7x, 10x. The total magnification of the image can be determined by multiplying the magnification of the objective by that of the ocular as: 50x · 10x = 500x. Although in routine work the three objectives are frequently interchanged, a single ocular usually suffices. The ocular assembly, which slips into the upper end of the microscope tube, carries cross hairs—one N–S (front–back) the other E–W. These enable us to locate under high power a particular mineral grain that has been brought to the center of the field under low power. They are also essential in aligning cleavage fragments for making angular measurements. A condenser is located below the stage. The upper lens of the condenser, used with high power objectives, makes the light strongly converging and can be rotated easily into or out of the optical system. The iris diaphragm, also located below the stage, can be opened or closed to control the depth of focus and to regulate the intensity of light striking the object.

In addition to the lenses, condenser and diaphragm mentioned above which are common to all compound microscopes, the polarizing microscope has several other features. The *polarizer* below the stage is a polarizing plate, or Nicol prism that transmits plane polarized light vibrating in a N–S (front–back) direction. The *analyzer*, fitted in the tube above the stage, is a similar plate or prism that transmits light vibrating only in an E–W direction. The polarizer and analyzer are collectively called *polars*. When both polars are in position they are said to be crossed and, if no anisotropic crystal is between them, no light reaches the eye. The polarizer remains fixed but the analyzer can be removed from the optical path at will. The Bertrand-Amici lens is an accessory that is used to observe interference figures. In working with crystals it is frequently necessary to change their orientation. This is accomplished by means of a rotating stage, whose axis of rotation is the same as the microscope axis.

Microscopic Examination of Minerals and Rocks

The polarizing microscope is also called the petrographic microscope because it is used in the study of rocks. In examining thin sections of rocks, the textural relationships are brought out and certain optical properties can be determined. It is equally effective in working with powdered mineral fragments. On such loose grains all the optical properties can be determined and in most cases they characterize a mineral sufficiently to permit its identification.

The optimum size of mineral grains for examination with a polarizing microscope is minus 50 mesh − plus 100 mesh, but larger or smaller sizes may be used. To prepare a mount for examination (1) put a few mineral grains on an object glass, a slide 46 mm × 27 mm, (2) immerse the grains in a drop of liquid, and (3) place a cover glass on top of the liquid. Using this type of mount the refractive indices of mineral grains are determined by the *immersion method*.

In using this method there should be available a series of calibrated liquids ranging in refractive index from 1.41 to 1.77, with a difference of 0.01 or less between adjacent liquids. These liquids cover the refractive index range of most of the common minerals. The immersion method is one of trial and error and involves comparing the refractive index of the unknown with that of a known liquid.

Isotropic Crystals and the Becke Line. Since light moves in all directions through an isotropic substance with equal velocities, there is no double refraction and only a single index of refraction. With a polarizing microscope, objects are always viewed in polarized light which conventionally is vibrating N–S. If the object is an isotropic mineral, the light passes through it and continues to vibrate in the same plane. If the analyzer is inserted, darkness results for this polar permits light to pass only if it is vibrating in an E–W direction. Darkness remains as the position of the crystal is changed by rotating the microscope stage. This is a characteristic that distinguishes isotropic from anisotropic crystals.

Let us first consider how the single refractive index of isotropic substances is determined. The first mount may be made using any liquid, but if the mineral is a complete unknown, it is well to select a liquid near the middle of the range. When the grains are brought into sharp focus using a medium power objective and plane polarized light, in all likelihood they will stand in relief; that is, be clearly discernable from the surrounding liquid. This is because the light is refracted as it passes from one medium to another of different refractive index. The farther apart the indices of refraction of mineral and liquid, the greater the relief. But when the indices of the two are the same, there is no refraction at the interface, and the grains are essentially invisible. Relief shows that the index of refraction of the mineral is different from that of the liquid; but is it higher or lower? The answer to this important question can be found by means of the *Becke line* (Fig. 240). If the mineral grain is thrown slightly out of focus by raising the microscope tube (in some microscopes this is accomplished by lowering the

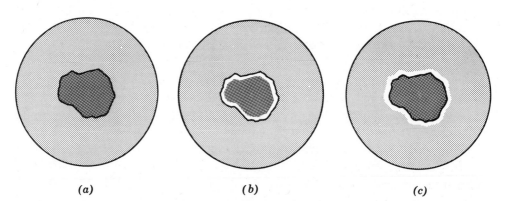

(a)   *(b)*   *(c)*

FIG. 240. The Becke Line. When thrown out of focus by raising microscope tube, white line moves into medium of higher refractive index. (*a*) In focus. (*b*) *n* of grain > *n* of liquid. (*c*) *n* of grain < *n* of liquid.

stage), a narrow line of light will form at its edge and move toward the medium of higher refractive index. Thus if the Becke line moves into the mineral grain, a new mount must be made using a higher refractive index liquid. After several tries it may be found that there is no Becke line and the mineral grains are invisible in a given liquid. The index of refraction of the mineral is then the same as that of the calibrated liquid. More frequently, however, the refractive index of the mineral is found to be greater than that one liquid but less than that of its next higher neighbor. In such cases it is necessary to interpolate. If the Becke line moving into the mineral in the lower index liquid is more intense than the line moving out of the mineral into the next higher liquid, it can be assumed that the refractive index of the mineral is closer to that of the higher liquid. In this way it is usually possible to report a refractive index to ± 0.003.

Frequently the Becke line can be sharpened by restricting the light by means of the substage diaphragm. Most liquids have a greater dispersion than minerals. Thus, if the refractive index of liquid and mineral are matched for a wave length near the center of the spectrum, the mineral has a higher index than the liquid for red light but a lower index for violet light. This is evidenced, when observed in white light, by a reddish line moving into the grain while a bluish line moves out.

Aside from color, the single index of refraction is the only significant optical characteristic of isotropic minerals. It is, therefore, important in mineral identification to consider other properties such as cleavage, fracture, color, hardness, and specific gravity.

Uniaxial Crystals

In hexagonal and tetragonal crystals c is a unique axis and differs in length from the a axes that are at right angles to it. The optical symmetry agrees with the crystal symmetry, for light moves with a different velocity parallel to the c axis than in the plane of the a axes or in any other crystal direction. Therefore, such crystals have two principal indices of refraction, and the difference between them is the *birefringence*.

We have seen that light moves in all directions through an isotropic substance with equal velocity and vibrates in all directions at right angles to the directions of propagation. In hexagonal and tetragonal crystals, there is one and only one direction in which light moves in this way. This is parallel to the c axis, with vibrations in all directions in the basal plane. For this reason the c axis is called the *optic axis*, and hexagonal and tetragonal crystals are called optically *uniaxial*. This distinguishes them from orthorhombic, monoclinic, and triclinic crystals which have two optic axes and are called *biaxial*.

When light moves in uniaxial crystals in any direction other than parallel to the c axis, it is broken into two rays traveling with different velocities. One, the *ordinary ray,* vibrates in the basal plane; the other, the *extraordinary ray,* vibrates at right angles to it and thus in a plane that includes the c axis. Such a plane, of which there is an infinite number, is referred to as the *principal section.* The nature of these two rays can be brought out in the following way. Assume that the direction of the incident beam is varied to make all possible angles with the crystal axes, and that the distance traveled by the resulting rays in any instant can be measured. We would find that:

1. One ray, with waves always vibrating in the basal plane, traveled the same distance in the same time. It's surface can be represented by a sphere, and since it acts much as ordinary light it is the *ordinary ray* (*O* ray).

2. The other ray, with waves vibrating in the plane that includes the c axis, traveled in the same time different distances depending on the orientation of the incident beam. If the varying distances of this, the *extraordinary ray* (*E* ray) were plotted, they would outline an ellipsoid of revolution, with the optic axis the axis of revolution.

Uniaxial crystals are divided into two optical groups: positive and negative. They are *positive* if the *O* ray has the greater velocity, and *negative* if the *E* ray has the greater velocity. Cross sections of the ray velocity surfaces are shown in Fig. 241. Note that in both positive and negative crystals the *O* and *E* rays have the same velocity when traveling along the optic (c) axis. But the difference in their velocities becomes progressively greater as the direction of light propagation moves away from the optic axis, reaching a maximum at 90°.

Since the two rays have different velocities, there are two indices of refraction in uniaxial crystals. Each index is associated with a vibration direction. The index related to vibration along the ordinary ray is designated ω (omega); whereas that associated with the extraordinary ray is ε (epsilon) or ε'. In positive crystals the *O* ray has a greater velocity than the *E* ray, and ω is less than ε. But in negative crystals with the *E* ray having the greater velocity, ω is greater than ε.

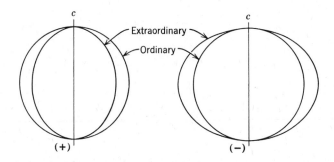

FIG. 241.   Ray velocity surfaces of uniaxial crystals.

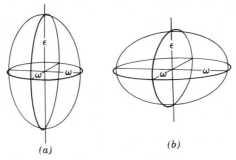

(a)   *(b)*

FIG. 242. Optical indicatrix, uniaxial crystals. (*a*) Positive. (*b*) Negative.

The *uniaxial indicatrix* is a geometrical figure that is helpful in visualizing the relation of the refractive indices and their vibration directions to the direction of propagation of light through a crystal. For positive crystals the indicatrix is a prolate spheroid of revolution; for negative crystals it is an oblate spheroid of revolution (Fig. 242). In their construction the direction of radial lines from the centers represent vibration directions, whereas the length of the lines is proportional to the refractive indices. First consider light moving parallel to the optic axis. It is not doubly refracted but moves through the crystal as the ordinary ray with waves vibrating in all directions in the basal plane. This is why light moving parallel to the *c* axis of calcite (Fig. 235) produces a single image. There is a single refractive index for all these vibrations, proportional to the radius of the equatorial circle of the indicatrix. Now consider light traveling perpendicular to the optic axis. It is doubly refracted. The waves of the ordinary ray vibrate, as always, in the basal plane and the associated refractive index, ω, is again an equatorial radius of the indicatrix. The vibration direction of waves of the extraordinary ray must be at right angles to both the vibration direction of the ordinary waves and the direction of propagation. Thus, in this special case, it is parallel to the optic axis. The axis of revolution of the indicatrix is then proportional to ε, the greatest index in (+) crystals and the least in (−) crystals. It can be seen that light moving through a crystal in a random direction gives rise to two rays: (1) the *O* ray with waves vibrating in the basal section, the associated index, ω, and (2) the *E* ray with waves vibrating in the principal section in a direction at right angles to propagation. The length of the radial line along this vibration direction is an ε', a refractive index lying between ω and ε.

A study of the indicatrix shows that (1) ω, can be determined on *any* crystal grain; and only ω can be measured when light moves parallel to the optic axis, (2) ε can be measured only when light moves normal to the *c* axis, and (3) a randomly oriented grain yields, in addition to ω, an index intermediate to ω and ε, called ε'. The less the angle between the direction of light propagation and the normal to the optic axis, the closer is the value ε' to true ε.

Uniaxial Crystals Between Crossed Polars

Extinction. We have seen that because isotropic crystals remain dark in all positions between crossed polars, they can be distinguished from anisotropic crystals. However, there are special conditions under which uniaxial crystals present a dark field when viewed between crossed polars. One of these conditions is when light moves parallel to the optic axis. Moving in this direction, light from the polarizer passes through the crystal as through an isotropic substance and is completely cut out by the analyzer. The other special condition is when the vibration direction of light from the polarizer coincides exactly with one of the vibration directions of the crystal. In this situation, light passes through the crystal as either the *O* ray or the *E* ray to be completely eliminated by the analyzer, and the crystal is said to be at *extinction*. As the crystal is rotated from this extinction position it becomes progressively lighter, reaching a maximum brightness at 45°. There are four extinction positions in a 360° rotation, one every 90°.

Interference. Let us consider how the crystal effects the behavior of polarized light as it is rotated from one extinction position to another. Figure 243 represents five positions of a tiny quartz crystal elongated on the *c* axis and lying on a prism face. In the diagrams it is assumed that light from the polarizer is moving upward, normal to the page and vibrating in direction *P-P*. The vibration direction of the analyzer is *A-A*. The crystal in (*a*) is at an extinction position and light moves through it as the *E* ray vibrating parallel to the *c* axis. At 90°, position (*b*), the crystal is also at extinction with light moving through it as the *O* ray. When the

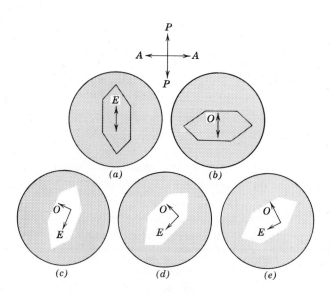

FIG. 243. Quartz crystal between crossed polars.

crystal is turned as in (*c*), (*d*), and (*e*), polarized light entering it is resolved into two components. One moves through it as the *O* ray, vibrating in the basal plane; the other as the *E* ray vibrating in the principal section. In (*c*) most light is transmitted as the *E* ray but in (*e*) it is transmitted mostly as the *O* ray. In (*d*), the 45° position, the amounts of light transmitted by the two rays are equal.

When these rays from the crystal enter the analyzer each is broken up into an *O* and *E* ray conforming in vibration directions to those of the analyzer. Only the components of the rays vibrating in an E–W direction, are permitted to pass. During their passage through the crystal the two rays travel with different velocities and thus on emerging there is a phase difference since one is ahead of the other. The amount it is ahead depends both on the difference in velocities and on the thickness of crystal traversed. Since both rays vibrate in the same plane of the analyzer, they interfere. For monochromatic light, if one ray is an integral number of wave lengths ($n\lambda$) behind the other, the interference results in darkness. On the other hand, if the path difference is $\lambda/2$, $3\lambda/2$ or in general $(2n - 1)\,\lambda/2$, the waves reinforce one another to produce maximum brightness.

Each wave length has its own sets of critical conditions when interference produces darkness. Consequently, when white light is used, "darkness' for one wave length means its elimination from the spectrum and its complimentary color appears. The colors thus produced are called *interference colors*. There are different *orders* of interference depending on whether the color results from a path difference of 1λ, 2λ, 3λ, ..., $n\lambda$. These called 1st-order, 2nd-order, 3rd-order, etc. interference colors are shown in the color plate facing page 156.

Interference colors depend upon three factors: orientation, thickness and birefringence. With a continuous change in direction of light, from parallel to perpendicular to the optic axis, there is a continuous increase in interference colors. For a given orientation, the thicker the crystal and the greater its birefringence the higher the order of interference color. If a crystal plate is of uniform thickness, as a cleavage flake may be or a grain in a rock thin section, it will show a single interference color. As seen in immersion liquids grains commonly vary in thickness and a variation in interference colors reflects this irregularity.

Accessory Plates

The gypsum plate, mica plate, and quartz wedge are accessory plates used with the polarizing microscope; their function is to produce interference of known amounts and thus predetermined colors. The *gypsum plate* is made by cleaving a gypsum crystal to such a thickness that in white light it produces a uniform red interference color: *red of the 1st order*. The *mica plate* is made with a thin mica flake, cleaved to a thickness that for yellow light it yields a path difference of a quarter of a wave length. It is thus also called the *quarter wave plate*. The *quartz wedge* is an elongated wedge-shaped piece of quartz with the vibration direction

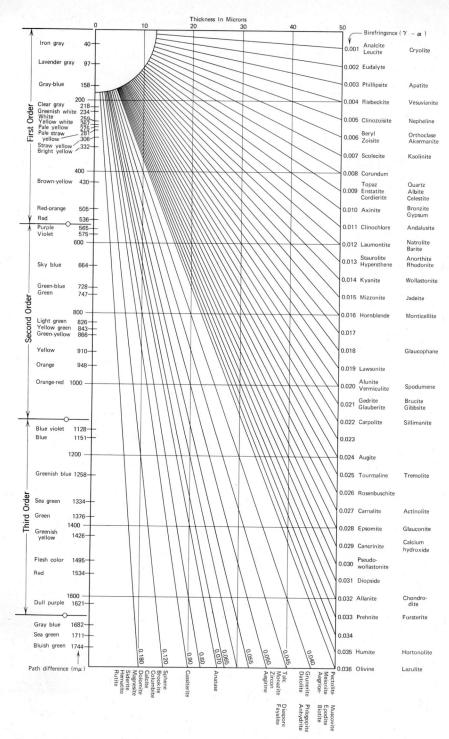

Key to Michel–Lévy Color Chart

Color chart courtesy of Carl Zeiss

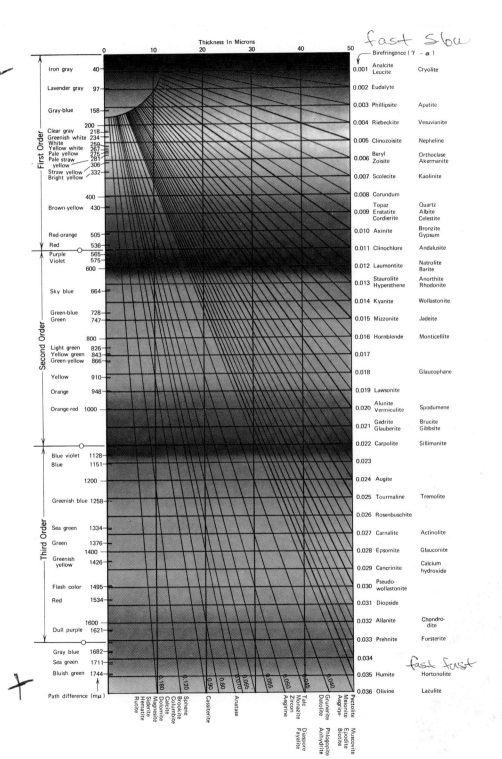

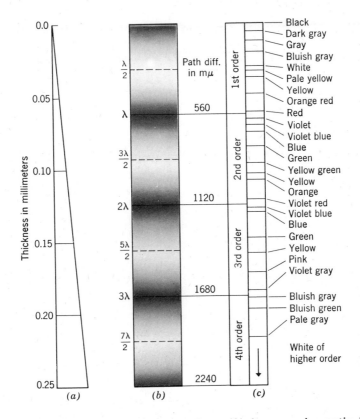

FIG. 244. Quartz wedge between crossed polars: (*b*) In monochromatic light; (*c*) Colors in white light.

of the fast ray parallel to the length and the slow ray across the length. As thicker portions of the wedge are placed in the optical path, the path difference of the rays passing through it also increases producing a succession of interference colors. The number of orders depends on the wedge angle; the greater the angle the more orders per unit of length.

When the quartz wedge is examined between crossed polars in monochromatic light it is crossed by alternating dark and light bands; dark where the path difference is $n\lambda$ and brightest where the path difference is $(2n - 1)\lambda/2$ (Fig. 244). In white light a succession of interference colors is observed. They resemble the colors seen in thin oil films on water, in soap bubbles, and when two slightly curved pieces of glass are pressed together. The colors result from interference phenomena that were described by Sir Isaac Newton, and since then have been called *Newton's Colors*. The succession of colors in the first three orders is given in Plate II.

Uniaxial Crystals in Convergent Polarized Light

What are known as *interference figures* are seen when properly oriented crystal sections are examined in convergent polarized light. To see them, the polarizing microscope (usually used as an orthoscope) is converted to a conoscope by swinging in the upper substage condensing lens, so that the section can be observed in strongly converging light, and using a high power objective. The interference figure then appears as an image just above the upper lens of the objective and can be seen between crossed polars by removing the ocular and looking down the microscope tube. If the Bertrand lens, an accessory lens located above the analyzer, is inserted, an enlarged image of the figure can be seen through the ocular.

The principal interference figure of a uniaxial crystal, the *optic axis figure*, Fig. 245, is seen when one views the crystal parallel to the *c* axis. Only for the central rays from the converging lens is there no double refraction, the others, traversing the crystal in directions not parallel to the *c* axis, are resolved into *O* and *E* rays having increasing path difference as the obliquity to the *c* axis increases. The interference of these rays produces concentric circles of interference colors. The center is black with no interference but moving outward there is a progression from 1st-order to 2nd-order to 3rd-order, etc., interference colors. If the crystal section is of uniform thickness, no change will be noted as it is moved horizontally. If, however, the thickness varies, as in a wedge-shaped fragment, the positions of the colors change with horizontal movement. At the thin edge there may be only gray of the 1st-order, but as the crystal is moved so the light path through it becomes greater, all the 1st-order colors may appear. And with increasing crystal thickness, the path difference of the two rays may be great enough to yield 2nd-, 3rd-, and higher-order interference colors.

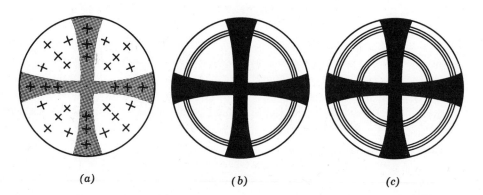

(a)   *(b)*   *(c)*

FIG. 245. Uniaxial optic axis interference figure. (*a*) Radial lines indicate vibration directions of *E* ray; tangential lines indicate vibration directions of *O* ray. (*b*) and (*c*) show isochromatic curves.

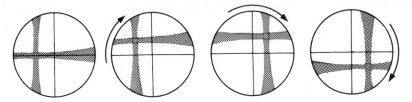

FIG. 246. Positions of off-centered uniaxial optic axis figure on clockwise rotation of microscope stage.

The reason for the black cross superimposed on the rings of interference colors is brought out in Fig. 245a. In this drawing the radial dashes indicate the vibration directions of the E ray and those at right angles the vibration directions of the O ray. It will be seen that where these vibrations directions are parallel or nearly parallel to the vibration directions of the polarizer and analyzer no light passes and thus the formation of the dark cross.

Figure 245 illustrates a centered optic axis figure as obtained on a crystal plate whose c axis coincides with the axis of the microscope; as the stage is rotated, no movement of the figure is seen. If the optic axis of the crystal makes an angle with the axis of the microscope, the black cross is no longer symmetrically located in the field of view (Fig. 246). When the stage is rotated, the center of the cross moves in a circular path, but the bars of the cross remain parallel to the vibration directions of the polarizer and analyzer. Even if the inclination of the optic axis is so great that the center of the cross does not appear, on rotation of the crystal the bars move across the field maintaining their parallelism to the vibration directions of the polars.

The flash figure, is an interference figure produced by a uniaxial crystal when its optic axis is normal to the axis of the microscope. That is, a hexagonal or tetragonal crystal lying on a face in the prism zone. When the crystal is at an extinction position the figure is an ill-defined cross occupying much of the field. On rotation of the stage, the cross breaks into two hyperbolas that rapidly leave the field in those quadrants containing the optic axis. The cross forms because the converging light is broken into O and E rays with vibration directions mostly parallel or nearly parallel to the vibration directions of the polarizer and analyzer. A centered flash figure not only indicates the vibration direction of the E ray, but assures one that in this direction a true value of ε can be obtained in plane polarized light.

Determination of Optic Sign

The mica plate, the gypsum plate, and the quartz wedge may be used with an uniaxial optic axis figure to determine the optic sign, that is, whether the crystal is positive or negative. They are inserted below the analyzer in a slot in the microscope tube so positioned that when the plates are in place, their vibration directions make angles of 45° with the vibration directions of the polars.

From the previous discussion we have learned that in the optic axis interference figure, the E ray vibrates radially and the O ray tangentially. By use of an accessory plate in which vibration directions of the slow and fast rays are known, one can tell whether the E ray of the crystal is slower (positive crystals) or faster (negative crystals) than the O ray and thus determine the optic sign. For most American-made equipment the vibration of the slow ray is at right angles to the length of the plate and is so marked on the metal carrier. However, before using an accessory the vibration directions should be checked. The principle in the use of all the plates is the same: to add to or subtract from the path difference of the O and E rays of the crystal.

If the *mica plate* is superimposed on an uniaxial optic axis figure in which the ordinary ray is slow, (negative crystal) the interference of the plate reinforces the interference colors in the SE and NW quadrants, causing them to shift slightly toward the center. At the same time subtraction causes the colors in the NE and SW quadrants to shift slightly away from the center. The most marked effect produced by the mica plate is the formation of two black spots near the center of the black cross in the quadrants where subtraction occurs (Fig. 247).

The *gypsum plate* is usually used to determine the optic sign when low-order interference colors or no colors at all are seen in the optic axis figure. It has the effect of superimposing red of the 1st-order on the interference figure. If the figure shows several orders of interference colors, one should consider the color effect on the grays of the 1st-order near the center. In the quadrants where there is addition, the red plus the gray gives blue; in the alternate quadrants the red minus the gray gives yellow. The arrangement of colors in positive crystals is: yellow SE–NW; blue NE–SW; and in negative crystals, yellow NE–SW, blue SE–NW. It is suggested that the student insert the colors in Fig. 248.

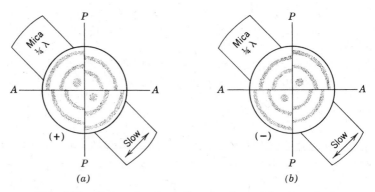

FIG. 247. Determination of optic sign with mica plate.

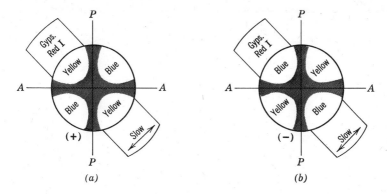

FIG. 248. Determination of optic sign with gypsum plate.

The *quartz wedge* is most effective in determining optic sign when high-order interference colors are present in the optic axis figure. The wedge is usually inserted with the thin edge first. If its retardation is added to that of the crystal, the interference colors in two opposite quadrants will increase progressively as the wedge moves through the microscope tube. If the retardation is subtracted from that produced by the crystal, the order of colors will decrease. Thus as the quartz wedge is slowly inserted over an optic axis figure of a negative crystal, the color bands in the SE–NW quadrants move toward the center and disappear. At the same time in the NE–SW quadrants the colors move outward to the edge of the field. In a positive crystal similar phenomena are observed but the colors move in the opposite directions, that is, away from the center in the SE–NW quadrants and toward the center in the NE–SW quadrants.

Sign of Elongation

Hexagonal and tetragonal crystals are frequently elongated on the *c* axis or have prismatic cleavage that permits them to break into splintery fragments also elongated parallel to *c*. If such an orientation is known, one can determine the optic sign by turning the elongated grain to the 45° position and inserting the gypsum plate. If the interference colors rise, the slow ray of the gypsum has been superimposed on the slow ray of the mineral. If this is also the direction of elongation, it means the *E* ray is slow and the mineral has positive elongation and is optically positive. When the slow ray of the gypsum plate is parallel to the elongation of the mineral grain and the interference colors fall, the mineral has negative elongation and is optically negative (Fig. 249).

Very commonly the interference colors of small grains are grays of the first order. Thus, on superimposing red of the first order, addition gives a blue color and subtraction a yellow color.

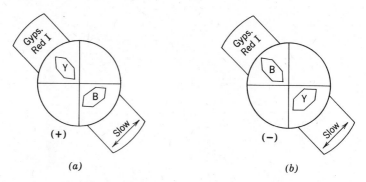

FIG. 249. Determination of sign of elongation with gypsum plate.

Absorption and Dichroism

In the discussion of polarized light (page 146) it was pointed out that in some tourmaline the absorption of one ray is nearly complete but for the other ray it is negligible. Although less striking, many crystals show a similar phenomenon: more light is absorbed in one vibration direction than in the other. In tourmaline where absorption of the O ray is greatest, it is expressed as, absorption: $O > E$ or $\omega > \varepsilon$. In other crystals certain wave lengths may be absorbed in one direction and different wave lengths in the other direction and the complimentary colors are transmitted. Thus the crystal has different colors in different vibration directions and is said to be *dichroic*. Dichroism is expressed by giving the colors, for example, O or ω = yellow, E or ε = pink. Absorption is independent of other properties and is considered, as are refractive indices, a fundamental optical property of crystals.

Biaxial Crystals

Orthorhombic, monoclinic, and triclinic crystals are called optically biaxial for they have two directions, in which light travels with zero birefringence. In uniaxial crystals there is only one such direction.

Light moving through a biaxial crystal, except along an optic axis, travels as two rays with mutually perpendicular vibrations. The velocities of the rays differ from each other and change with changing crystallographic direction. The vibration directions of the fastest ray, X, and the slowest ray, Z, are at right angles to each other. The direction perpendicular to the plane defined by X and Z is designated as Y. For biaxial crystals there are thus three indices of refraction resulting from rays vibrating in each of these principal optical directions. The numerical difference between the greatest and least refractive indices is the *birefringence*. Various letters and symbols have been used to designate the refractive indices, but the most generally accepted are the Greek letters as follows:

Indices of Biaxial Crystals[a]

Index	Direction	Ray Velocity
(alpha) α lowest	X	Highest
(beta) β middle	Y	Intermediate
(gamma) γ highest	Z	Lowest

[a] Other equivalent designations are: $\alpha = nX, n_x, N_x, N_p$; $\beta = nY, n_y, N_y, N_m$; $\gamma = nZ, n_z, N_z, N_g$.

The Biaxial Indicatrix

The biaxial indicatrix is a triaxial ellipsoid with its three axes the mutually perpendicular optical directions X, Y, and Z. The lengths of the semiaxes are proportional to the refractive indices: α along X, β along Y, and γ along Z. Figure 250 shows the three principal sections through the indicatrix; these are the planes XY, YZ, and XZ. They all are ellipses and each has the length of its semimajor and semiminor axes proportional to refractive indices as shown. Of most interest is the XZ section. With its semimajor axis proportional to γ and its semiminor axis proportional to α, there must be points on the ellipse between these extremes where the radius is proportional to the intermediate index, β. In Fig. 250 this radius is marked S. With two exceptions every section passing through the center of a triaxial ellipsoid is an ellipse. The exceptions are *circular sections* of which S is the radius. The two directions normal to these sections are the *optic axes*; and the XZ plane in which they lie is called the *optic plane*. The Y direction perpendicular to this plane is the *optic normal*. Light moving along the optic axes, and vibrating in the circular sections shows no birefringence and gives the constant refractive index, β.

With variation in refractive indices there is a corresponding variation in the axial lengths of the biaxial indicatrix. Some crystals are nearly uniaxial and in these the intermediate index, β, is very close to either α or γ. If β is close to α, the circular sections make only a small angle with the XY plane and the optic axes make the same angle with the Z direction. This is angle V and the angle between the two optic axes, known as the *optic angle*, is $2V$. The optic angle is always acute and since, in this case, it is bisected by Z, Z is called the *acute bisectrix* (Bxa); X is the *obtuse bisectrix* (Bxo) since it bisects the obtuse angle between the optic axes. When Z is the Bxa, the crystal is optically positive.

If β is closer to γ than to α, the acute angle between the optic axes is bisected by X and the obtuse angle bisected by Z. In this case, with X the Bxa, the crystal

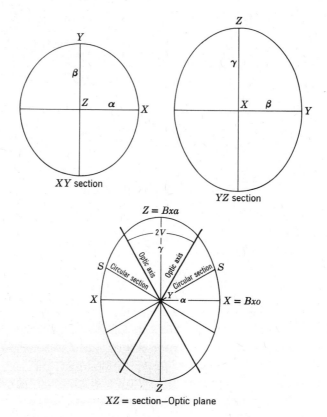

FIG. 250. Principal sections through the biaxial indicatrix of a positive crystal.

is negative. When β lies exactly half way between α and γ, the optic angle is 90°.

The relation between the optic angle and the indices of refraction is expressed by formula (a) below. A close approximation to the optic angle can be made using formula (b).

$$\text{(a) } \text{Cos}^2 \, V_x = \frac{\gamma^2(\beta^2 - \alpha^2)}{\beta^2(\gamma^2 - \alpha^2)} \qquad \text{(b) } \text{Cos}^2 \, V_x = \frac{\beta - \alpha}{\gamma - \alpha}$$

The error using the simplified formula increases with increase of both birefringence and V and always yields values for V' less than true V. It should be noted that in using either formula *half* the optic angle is calculated, and that it is determined with X the bisectrix. Thus, when $V < 45°$ the crystal is negative but when $V > 45°$ the crystal is positive.

Biaxial Crystals in Convergent Polarized Light

Biaxial interference figures are obtained and observed in the same manner as uniaxial figures, that is, with converging light, high power objective, and Bertrand

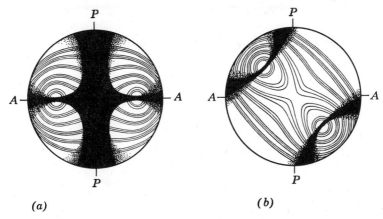

FIG. 251. Acute bisectrix interference figure. (*a*) Parallel position. (*b*) 45° position.

lens. Although interference figures can be observed on random sections of biaxial crystals, the most symmetrical and informative are obtained on sections normal to the optical directions *X*, *Y*, and *Z* and to an optic axis.

The acute bisectrix figure is observed on a crystal plate cut normal to the acute bisectrix. If 2*V* is very small there are four positions during a 360° rotation at which the figure resembles the uniaxial optic axis figure. That is, a black cross is surrounded by circular bands of interference colors. However, as the stage is turned, the black cross breaks into two hyperbolas that have a slight but maximum separation at a 45° rotation; and the color bands, known as *isochromatic curves*, assume an oval shape. The hyperbolas are called isogyres and the dark areas, called melatopes, at their vertices in the 45° position mark the points of emergence of the optic axes. Thus with increasing optic angle the separation of the isogyres increases, and the isochromatic curves are arranged symmetrically about the melatopes as shown in Fig. 251. For most crystals when 2*V* exceed 60° the isogyres leave the field at the 45° position; the larger the optic angle, the faster they leave.

The portion of the interference figure occupied by the isogyres is dark, for here, light as it emerges from the section, has vibration directions parallel to those of the polarizer and analyzer. The dark cross is thus present when the obtuse bisectrix and optic normal coincide with the vibration directions of the polars. The bar of the cross parallel to the optic plane is narrower and better defined than the other bar (Fig. 251). Since light travels along the optic axes with no birefringence, their points of emergence are, of course, dark in all positions of the figure.

The Apparent Optic Angle. The distance between the points of emergence of the optic axes is dependent not only on 2*V* but on *β* as well. The refractive index of the crystal is *β* for light rays moving along the optic axes. These rays are refracted on leaving the crystal, giving an apparent optic angle, 2*E*, which is greater

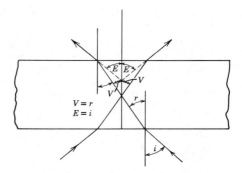

FIG. 252.   Relation of 2V to 2E.

than the real angle, $2V$ (Fig. 252). The higher the refractive index, the greater the refraction. Thus if two crystals have the same $2V$, the one with the higher β index has the larger apparent angle and the farther apart the optic axes emerge in the interference figure.

The *optic axis figure* is observed on mineral grains cut normal to an optic axis. Such grains are easy to select for they remain essentially dark between crossed polars on complete rotation. The figure consists of a single isogyre at the center of which is the emergence of the optic axis. When the optic plane is parallel to the vibration direction of either polar, the isogyre crosses the center of the field as a straight bar. On rotation of the stage it swings across the field forming a hyperbola in the 45° position. In this position the figure can be pictured as half an acute bisectrix figure with the convex side of the isogyre pointing toward the acute bisectrix. As $2V$ increases, the curvature of the isogyre decreases and when $2V = 90°$, the isogyre is straight (Fig. 253).

The *obtuse bisectrix figure* is obtained on a crystal section cut normal to the obtuse bisectrix. When the plane of the optic axes is parallel to the vibration direction of either polar, there is a black cross. On rotation of the stage the cross breaks into two isogyres that move rapidly out of the field in the direction of the acute bisectrix. Although not as informative as an acute bisectrix figure, a centered

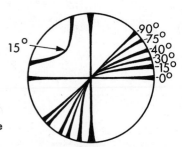

FIG. 253.   Curvature of isogyre in optic axis figure from 0° to 90° 2V.

obtuse bisectrix figure indicates that an accurate determination of β and of either α or γ can be made on the mineral section producing it.

The *optic normal figure* is obtained on sections cut parallel to the plane of the optic axes and resembles the flash figure of a uniaxial crystal. When the X and Z optical directions are parallel to the vibration directions of the polars, the figure is a poorly defined cross. On slight rotation of the stage it splits into hyperbolas that move rapidly out of the field in the quadrants containing the acute bisectrix. An optic normal figure is obtained on sections with maximum birerefringence and indicates that α and γ can be determined on this section.

Determination of Optic Sign of a Biaxial Crystal

The optic sign of biaxial crystals can best be determined on acute bisectrix or optic axis figures with the aid of accessory plates. Let us assume that a in Fig. 254 represents an acute bisectrix figure of a negative crystal in the 45° position. By definition, X is the acute bisectrix and Z the obtuse bisectrix. OP is the trace of the optic plane and Y, the vibration direction of β, at right angles. The velocity is constant for all rays moving along the optic axes, and for them the refractive index of the crystal is β, including those vibrating in the optic plane. Consider the velocities of other rays vibrating in the optic plane. In a negative crystal, those emerging between the isogyres of an acute bisectrix figure have a lesser velocity,

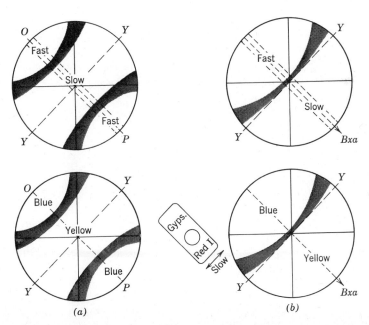

FIG. 254. Optic sign determination of negative crystal with gypsum plate. (*a*) Acute bisectrix figure. (*b*) Optic axis figure.

those emerging outside the isogyres have a greater velocity. If the gypsum plate is superimposed over such a figure, the slow ray of the plate combines with the fast ray of the crystal and subtraction of interference colors produces a yellow color on the convex sides of the isogyres. On the concave sides of the isogyres a blue color is produced by addition, the slow ray of the plate over the slow ray of the crystal. In a positive crystal the reverse color effect is seen, for here Z is the acute bisectrix.

An optic axis figure in the 45° position can be used in a manner similar to the acute bisectrix figure in determining optic sign. Insertion of the gypsum plate yields for ($-$) crystal: convex side yellow, concave side blue (Fig. 254b), for ($+$) crystal: convex side blue, concave side yellow.

The quartz wedge may be used to determine optic sign if several isochromatic bands are present. As the wedge is inserted the colors move out in those portions of the figure where there is subtraction and move in where there is addition. In other words, in the areas that a gypsum plate would render blue, the colors move in; in those that it would render yellow, they move out.

Optical Orientation in Biaxial Crystals

The orientation of the optical indicatrix is one of the fundamental optical properties. It is given by expressing the relationship of the X, Y, and Z optical directions to the crystallographic axes a, b, and c.

In *orthorhombic crystals* each of the crystallographic axes is coincident with one of the principal optical directions. For example, the optical orientation of anhydrite is: $X = c$, $Y = b$, $Z = a$. Usually the optical directions coinciding with only two axes are given for this completely fixes the position of the indicatrix.

It is difficult or impossible to determine the axial directions on microscopic grains of some minerals. To do it one must use a fragment oriented by x-ray study or broken from a faced crystal. However, even in small particles, orientation can be expressed relative to cleavages. Powdered fragments tend to lie on cleavages, which in orthorhombic crystals are commonly pinacoidal or prismatic. For example, barite has {001} and {210} cleavage and most grains lie on faces of these forms. Those lying on {001} will be diamond shaped (Fig. 255a) and have *symmetrical extinction*. That is, the extinction position makes equal angles with the bounding cleavage faces. Fragments lying on {210} will have *parallel extinction* (Fig. 255b). Symmetrical and parallel extinction are characteristic of orthorhombic crystals.

Barite is ($+$), $X = c$, $Y = b$, $Z = a$. Thus, grains lying on {001} yield a centered Bxo figure and one can determine β in the b direction and γ in the a direction. Grains showing parallel extinction do not give a centered interference figure but α can be measured parallel to c.

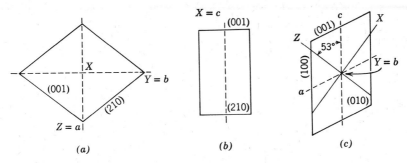

FIG. 255. Optical orientation. (*a*) Barite on {001} showing symmetrical extinction. (*b*) Barite on {210} showing parallel extinction. (*c*) Gypsum on {010} showing extinction angle.

In *monoclinic crystals* one of the principal optical directions (X, Y, or Z) of the indicatrix coincides with the b axis; the other two lie in the a-c plane of the crystal. The orientation is given by stating which optical direction equals b and indicating the *extinction angle*, the angle between the c axis and one of the other optical directions. If the extinction lies between the $+$ ends of the a and c axes, the angle is positive; between $+c$ and $-a$, the angle is negative. In gypsum $Y = b$ and $Z \wedge c = 53°$. Thus a fragment lying on the {010} cleavage would yield an optic normal interference figure and α and γ could be determined; γ at the extinction position $+53°$; α at extinction position $-37°$. A grain lying on the {100} cleavage would show parallel extinction and β could be measured at right angles to the trace of {010}. Crystal fragments lying on any other face in the [001] zone, as on {110} will show an extinction angle but the angle to be recorded, usually that of *maximum extinction* is observed on {010}. Parallel extinction indicates the grain is lying on a face in the [010] zone and the index of the ray vibrating parallel to b can be measured.

In *triclinic crystals* the optical indicatrix can occupy any position relative to the crystallographic axes. Thus a complete optical orientation necessitates giving ϕ and ρ angles of the principal optical directions. But in most cases it suffices to give the extinction angles observed on grains lying on known cleavage faces.

Dispersion

We have seen that the refractive indices of a mineral vary with the wave length of light. This dispersion of the indices means that, with a variation in the color of light, there is a variation in the indicatrix. The resulting change in the positions of the optic axes, and the accompanying change in $2V$, is known as dispersion of the optic axes, but is usually called just *dispersion*. Instead of giving different values for $2V$ for different wave lengths of light, dispersion is usually expressed by stating whether $2V$ is greater or lesser for red light than for violet light.

Using white light, dispersion can be observed in acute bisectrix figures and optic axis figures and is evidenced as a red fringe on one side of an isogyre and a blue fringe on the other side. Let us assume that $2V$ is greater for red than for violet light. Red light moving along the "red" optic axis has zero path difference. Thus where this axis emerges, red has been removed from the white light and a blue color appears. Similarly, violet light has been removed at the point of emergence of the "violet" axis and a red color appears. In this case the red fringe would appear on the convex side of the isogyre, the blue fringe on the concave side and the dispersion is expressed as: $r > v$. If the positions of the color fringes were reversed, the dispersion would be: $r < v$. In most interference figures the color fringes are subtle and the isogyre is essentially black.

The foregoing explanation of dispersion is strictly true only for orthorhombic crystals where the plane of the optic axis is an axial plane of the crystal and the acute bisectrix a crystal axis. In monoclinic crystals there is dispersion of the bisectrices as well as of the optic axes giving rise to three types of dispersion. The dispersion is called: *crossed* when Bxa $= b$; *horizontal* when Bxo $= b$; and *inclined* when the optic normal $= b$. The distribution of the color fringes indicating these special types of dispersion are rarely seen, and monoclinic dispersion is usually expressed as $r >$ or $< v$.

Absorption and Pleochroism. The absorption of light in biaxial crystals may differ in the X, Y, and Z optical directions. If the difference is only in intensity and X has the greatest absorption and Z the least, it is expressed as $X > Y > Z$. If different wave lengths are absorbed in different directions the mineral is said to be pleochroic and the color of the transmitted light is given. For example, in hypersthene the *pleochroism* is: $X =$ brownish red, $Y =$ reddish yellow, $Z =$ green. The term pleochroism is commonly used to denote all differential absorption in both uniaxial and biaxial crystals.

ELECTRICAL AND MAGNETIC PROPERTIES

The conduction of electricity in crystals is related to the type of bonding. Minerals with pure metallic bonding, such as the native metals, are excellent electrical conductors, whereas those in which the bonding is partially metallic, as in some sulfide minerals, are semiconductors. Ionic or covalent bonded minerals are usually nonconductors. For nonisometric minerals, electrical conductivity is a vectorial property varying with crystallographic direction. For example, the hexagonal mineral, graphite, is a far better conductor at right angles to the c axis than parallel to it.

Piezoelectricity. Polar axes are present only in crystals that lack a symmetry center. Of the 32 crystal classes, 21 have no center of symmetry and of these all but one, the gyroidal class, has at least one polar axis with different crystal forms at opposite ends. If pressure is exerted at the ends of a polar axis, a flow of electrons toward one end produces a negative electric charge, while a positive charge is induced at the opposite end. This is *piezoelectricity* and any mineral crystallizing in one of the 20 classes with polar axes should show it. However, in some minerals the charge developed is too weak to be detected.

The property of piezoelectricity was first detected in quartz in 1881 by Pierre and Jacques Curie but it was nearly forty years later before it was used in a practical way. Toward the end of World War I it was found that sound waves produced by a submarine could be detected by the piezoelectric current generated when they impinged on a submerged quartz plate. The device was developed too late to have great value during the war, but it pointed the way to other applications. In 1921 the piezoelectric property of quartz was first used to control radio frequencies, and since then millions of quartz plates have been used for this purpose. When subjected to an alternating current, a properly cut slice of quartz is mechanically deformed and vibrates by being flexed first one way and then the other; the thinner the slice, the greater the frequency of vibration. By placing a quartz plate in the electric field generated by a radio circuit, the frequency of transmission or reception is controlled when the frequency of the quartz coincides with the oscillations of the circuit.

The piezoelectric property of tourmaline has been known almost as long as that of quartz, but compared to quartz tourmaline is a less effective radio oscillator and is extremely rare. Nevertheless, small amounts of it are used today in piezoelectric pressure gauges. Tourmaline is hexagonal with *c* a polar axis. Plates cut normal to this direction will generate an electric current when subjected to a transient pressure. The current generated is proportional to the area of the plate and to the pressure. Tourmaline gauges were developed to record the blast pressure of the first atomic bomb in 1945 and since then have been used by the United States with each atomic explosion. However, lesser pressures also can be recorded by them, such as those generated by firing a rifle or by surf beating on a sea wall.

Pyroelectricity. Temperature changes in a crystal may cause the simultaneous development of positive and negative charges at opposite ends of a polar axis. This property of *pyroelectricity* is observed, as is piezoelectricity, only on crystals with polar axes. Crystals that belong to the ten crystal classes having a unique polar axis are considered to show "true" or *primary* pyroelectricity. For example, tourmaline has a single polar axis, *c*, and falls within this group, whereas quartz with its three polar *a* axes does not. However, a temperature gradient in all other crystals having polar axes such as quartz will produce a pyroelectric effect. In such crystals the polarization is the result of the deformation resulting from unequal thermal expansion that produces piezoelectric effects. If quartz is heated to about 100°C, it will develop on cooling positive charges at three alternate prismatic edges and negative charges at the three remaining edges. These charges have been called *secondary* pyroelectric polarization.

By means of single crystal x-ray photographs some minerals can be assigned to specific point groups, but for others x-ray data are ambiguous. For example, on the basis of x-ray photographs, a mineral might belong to either of the classes $2/m2/m2/m$ or $mm2$. If it could be shown to be piezoelectric or pyroelectric it would definitely belong to $mm2$.

Magnetism.   Magnetite, Fe_3O_4, and pyrrhotite, $Fe_{1-x}S$ are the only common minerals attracted to a small hand magnet. They are *ferromagnetic.* Lodestone, a variety of magnetite, is a natural magnet with the attracting power and polarity of a true magnet. In the field of a powerful electromagnet many other minerals, especially those containing iron, are attracted. These are called *paramagnetic,* whereas those that are repelled are called *diamagnetic.* Because of their different magnetic susceptibilities, minerals can be separated from each other by an electromagnet and magnetic separation of minerals is a standard procedure on both laboratory and commercial scale. Thus the magnetic separator is standard equipment in mineralogical laboratories and is used on a commercial scale to separate ore minerals from gangue.

REFERENCES

Optical Mineralogy

Bloss, F. D., *An Introduction to the Methods of Optical Crystallography.* Holt Rinehart and Winston, New York, 1961.

Hartshorne, N. H., and A. Stuart, *Practical Optical Crystallography.* American Elsevier Pub. Co., New York, 1964.

Kerr, P. F., *Optical Mineralogy*, 3rd ed. McGraw-Hill Book Company, New York, 1958.

Larsen, E. S., and H. Berman, *The Microscopic Determination of the Nonopaque Minerals*, 2nd ed. U.S. Geological Survey Bulletin 848, Government Printing Office, Washington, D.C., 1934.

Wahlstrom, E. E., *Optical Crystallography*, 4th ed. John Wiley and Sons, New York, 1969.

4

CHEMICAL MINERALOGY

INTRODUCTION

The chemical composition of a mineral is of fundamental importance, for all other properties are in great measure dependent upon it. However, these properties depend not only on the chemical composition but also upon the geometry of arrangement of the constituent atoms and the nature of the electrical forces which bind them together.

For more than a century, the classification of minerals has been firmly placed on a chemical basis. Consequently, the final proof of identity of a mineral has been chemical composition. The present-day classification of minerals, however, considers structure as well as gross chemical composition and takes cognizance of the wide latitude in chemical content permitted by the substitution of atoms of one element for those of another in a given structural framework. Considerable clarification of the relationships among minerals results from the introduction of structural concepts into mineral classification, and may have economic significance.

For example, the value of many ore minerals arises from their content of a metal which is a vicarious, rather than an essential, constituent. This is true of thorium in monazite, silver in tetrahedrite, and in general of gallium, germanium, indium, and many other elements. In these cases, a knowledge of the mechanism by which the vicarious constituents come to be present may be of great economic significance.

In this section on chemical mineralogy the general principles relating the chemistry of minerals to their crystallography and physical properties is discussed under the heading of *crystal chemistry*. This is followed by a brief description of the methods of testing for different elements found in minerals. Because of the scope and size of this book, it is necessary to assume that the reader is familiar with at least the essentials of chemical fact and nomenclature.

CRYSTAL CHEMISTRY

That a relation exists between chemical composition and crystal morphology was recognized in the eighteenth century. The ability to determine crystal structure by x-ray diffraction methods added a new dimension to this relationship. As a result of the interest aroused among crystallographers and chemists a new science, *crystal chemistry*, came into being. The goal of this science is the elucidation of the relationships between chemical composition, internal structure, and physical properties in crystalline matter. In mineralogy it serves as a unifying thread upon which the often apparently unrelated facts of descriptive mineralogy may be strung.

Relation of Chemistry to Mineral Classification

Chemical composition is the basis for the modern classification of minerals. According to this scheme, minerals are divided into classes depending on the dominant anion or anionic group. (See page 221.) There are a number of reasons why this criterion is a valid basis for the broad framework of mineral classification. First, minerals having the same anion or anionic group dominant in their composition have unmistakable family resemblances, in general stronger and more clearly marked than those shared by minerals containing the same dominant cation. Thus the carbonates resemble each other more closely than do the minerals of copper. Second, minerals related by dominance of the same anion tend to occur together or in the same or similar geologic environment. Thus the sulfides occur in close mutual association in deposits of vein or replacement type, whereas the silicates make up the great bulk of the rocks of the earth's crust. Third, such a scheme of mineral classification agrees well with the current chemical practice in the naming and classification of inorganic compounds.

It is apparent, however, that so one-sided a view of the nature of minerals leaves many troublesome questions unanswered. Why do minerals deviate so widely from the properties expected on the basis of chemical composition alone? Why do the anionic groups influence the properties of most compounds more than the cations? What uniformity connects those substances of similar crystallography and properties but diverse chemical composition? We shall have to deal with these questions and many more before we can hope to reach an adequate understanding of the nature of mineral substances.

Ordinarily the first task of a student in a course in general chemistry is the discrimination of physical properties and physical change as opposed to chemical properties and chemical change. The most cursory examination reveals that both chemical and physical properties depend to some extent on composition. Lead is heavy, and compounds that contain it generally have notably high specific gravities reflecting the influence of component elements on the physical properties. Likewise, radicals that are groups or combinations of elements, although they cannot exist in the free state like an element, have characteristic properties which they confer on all compounds containing them. Thus the carbonate radical characteristically reacts with acids to yield carbon dioxide, which is liberated from the scene of the reaction as bubbles of gas.

We may therefore specify the chemical properties of the compound lead carbonate, the mineral cerussite, in part by listing those characteristic reactions or tests for the elements contained in it. Such a specification of tests for the elements enables us to place cerussite unambiguously in the scheme of classification of minerals according to chemical composition. We may make the chemical definition of the mineral cerussite more rigorous by specifying not only that it must yield qualitative tests for lead and carbonate but also that *quantitative* chemical analysis must yield proportions by weight of lead and carbonate of 83.5 per cent PbO and 16.5 per cent CO_2.

Polymorphism. From these considerations it seems as though quantitative chemical composition should serve admirably as an exact and rigorous basis of classification. However, if we examine compounds of calcium and carbonate ion, we speedily find that the quantitative chemical classification is not unambiguous. In nature $CaCO_3$ exists in *two* stable compounds, the minerals calcite and aragonite. They are indistinguishable by chemical means, yet differ in almost every other property. Calcite is hexagonal scalenohedral; aragonite is orthorhombic. Calcite has a perfect rhombohedral cleavage; aragonite has prismatic and pinacoidal cleavage. The minerals differ slightly in specific gravity and hardness and, what is more devastating to any notion of their identity, show totally unlike x-ray diffraction patterns.

Further examples of this ambiguity are: pyrite and marcasite, which share the composition FeS_2; graphite and diamond, both elemental carbon; and the silica system, which includes no less than eight physically distinct substances of diverse crystallography all having the composition SiO_2.

Comparison of Dimorphous Minerals

Chemical Substance	Mineral	Crystal System	Hardness	Specific Gravity
C	Diamond	Isometric	10	3.51
	Graphite	Hexagonal	1	2.23
FeS_2	Pyrite	Isometric	6	5.02
	Marcasite	Orthorhombic	6	4.85
$CaCO_3$	Calcite	Rhombohedral	3	2.72
	Aragonite	Orthorhombic	$3\frac{1}{2}$	2.95

This phenomenon, in which the same chemical substance exists in two or more physically distinct forms, is termed *polymorphism* or *allotropy*. Compounds are said to be *dimorphous* if they exist in two modifications, *trimorphous* if they exist in three modifications.

The introduction of crystal structure resolves the problem posed by polymorphous forms. Diamond is harder and denser than graphite simply because its component particles of carbon are more closely packed and more tightly bonded together. Not only does the concept of the importance of structure account for every case of polymorphism, but also it resolves the difficulty presented by the apparently anomalous variation in physical properties of other compounds from those predictable on the basis of chemical composition alone. Thus, on the basis of composition, it is predictable that all compounds of lead will have high densities. This is true, but in fact the compounds of lead do not show a simple proportional relation between density and percentage of lead. In terms of structure, these departures may easily be explained on the basis of differences in packing. A similar consideration accounts for the otherwise baffling fact that corundum, Al_2O_3, entirely made up of light elements, is nearly as dense as chalcopyrite, $CuFeS_2$, a compound largely made up of much heavier elements.

Atoms, Ions, and the Periodic Table

Atoms are the smallest subdivision of matter that retain the characteristics of the elements. They consist of a very small, massive nuclei composed of protons and neutrons surrounded by a much larger region thinly populated by electrons. Although atoms are so small that it is impossible to see them even with the highest magnification of the electron microscope, the sizes have been measured and are generally expressed as atomic radii in angstrom units. For example, the smallest atom, hydrogen, has a radius of only 0.46 Å, whereas the largest, cesium, has a radius of 2.62 Å. (See table, page 177.)

Periodic Table of the Chemical Elements with Atomic and Ionic Radii

Ia	IIa	IIIb	IVb	Vb	VIb	VIIb	VIIIb	VIIIb	VIIIb	Ib	IIb	IIIa	IVa	Va	VIa	VIIa	VIIIa
H 1 / 0.46																	He 2 / 1.78
Li 3 / 1.52 / Li^+ 0.68	Be 4 / 1.12 / Be^{2+} 0.35											B 5 / 0.97 / B^{3+} 0.23	C 6 / 0.77 / C^{4+} 0.16	N 7 / 0.71 / N^{3+} 0.16 / N^{5+} 0.13	O 8 / O^{2-} 1.40 / O 0.60 / O^{6+} 0.10	F 9 / F^- 1.33 / F^{7+} 0.08	Ne 10 / 1.60
Na 11 / 1.86 / Na^+ 0.97	Mg 12 / 1.60 / Mg^{2+} 0.66											Al 13 / 1.43 / Al^{3+} 0.51	Si 14 / Si^{4-} 1.98 / Si 1.17 / Si^{4+} 0.39	P 15 / P^{3+} 0.44 / P^{5+} 0.35	S 16 / S^{2-} 1.74 / S 1.04 / S^{4+} 0.37 / S^{6+} 0.30	Cl 17 / Cl^- 1.81 / Cl 1.07 / Cl^{5+} 0.34 / Cl^{7+} 0.27	Ar 18 / 1.91
K 19 / 2.31 / K^+ 1.33	Ca 20 / 1.96 / Ca^{2+} 0.99	Sc 21 / 1.96 / Sc^{3+} 0.81	Ti 22 / 1.46 / Ti^{3+} 0.76 / Ti^{4+} 0.68	V 23 / V^{2+} 0.88 / V^{3+} 0.74 / V^{4+} 0.63 / V^{5+} 0.59	Cr 24 / 1.25 / Cr^{3+} 0.63 / Cr^{6+} 0.52	Mn 25 / Mn^{2+} 0.80 / Mn^{3+} 0.66 / Mn^{4+} 0.60 / Mn^{7+} 0.46	Fe 26 / 1.24 / Fe^{2+} 0.74 / Fe^{3+} 0.64	Co 27 / 1.25 / Co^{2+} 0.72 / Co^{3+} 0.63	Ni 28 / 1.24 / Ni^{2+} 0.69	Cu 29 / 1.28 / Cu^+ 0.96 / Cu^{2+} 0.72	Zn 30 / 1.33 / Zn^{2+} 0.74	Ga 31 / 1.22 / Ga^{3+} 0.62	Ge 32 / 1.22 / Ge^{2+} 0.73 / Ge^{4+} 0.53	As 33 / 1.25 / As^{3+} 0.58 / As^{5+} 0.46	Se 34 / Se^{2-} 1.91 / Se 1.16 / Se^{4+} 0.50 / Se^{6+} 0.42	Br 35 / Br^- 1.96 / Br 1.19 / Br^{5+} 0.47 / Br^{7+} 0.39	Kr 36 / 2.01
Rb 37 / 2.43 / Rb^+ 1.47	Sr 38 / 2.15 / Sr^{2+} 1.12	Y 39 / 1.81 / Y^{3+} 0.92	Zr 40 / 1.56 / Zr^{4+} 0.79	Nb 41 / 1.43 / Nb^{4+} 0.74 / Nb^{5+} 0.69	Mo 42 / 1.36 / Mo^{4+} 0.70 / Mo^{6+} 0.62	Tc 43 / Tc^{7+} 0.56	Ru 44 / 1.33 / Ru^{4+} 0.67	Rh 45 / 1.34 / Rh^{3+} 0.68	Pd 46 / 1.37 / Pd^{2+} 0.80 / Pd^{4+} 0.65	Ag 47 / 1.44 / Ag^+ 1.26 / Ag^{2+} 0.89	Cd 48 / 1.49 / Cd^{2+} 0.97	In 49 / 1.62 / In^{3+} 0.81	Sn 50 / Sn^{4-} 2.15 / Sn 1.40 / Sn^{2+} 0.93 / Sn^{4+} 0.71	Sb 51 / 1.45 / Sb^{3+} 0.76 / Sb^{5+} 0.62	Te 52 / Te^{2-} 2.11 / Te 1.43 / Te^{4+} 0.70 / Te^{6+} 0.56	I 53 / I^- 2.20 / I 1.36 / I^{5+} 0.62 / I^{7+} 0.50	Xe 54 / 2.20
Cs 55 / 2.62 / Cs^+ 1.67	Ba 56 / 2.17 / Ba^{2+} 1.34	La 57 / 1.86 *) / La^{3+} 1.14	Hf 72 / 1.58 / Hf^{4+} 0.78	Ta 73 / 1.43 / Ta^{5+} 0.68	W 74 / 1.36 / W^{4+} 0.70 / W^{6+} 0.62	Re 75 / 1.36 / Re^{4+} 0.70 / Re^{7+} 0.56	Os 76 / 1.35 / Os^{6+} 0.69	Ir 77 / 1.35 / Ir^{4+} 0.68	Pt 78 / 1.38 / Pt^{2+} 0.80 / Pt^{4+} 0.65	Au 79 / 1.44 / Au^+ 1.37 / Au^{3+} 0.85	Hg 80 / 1.50 / Hg^{2+} 1.10	Tl 81 / 1.70 / Tl^+ 1.47 / Tl^{3+} 0.95	Pb 82 / 1.75 / Pb^{2+} 1.20 / Pb^{4+} 0.84	Bi 83 / 1.55 / Bi^{3+} 0.96 / Bi^{5+} 0.74	Po 84 / Po^{6+} 0.67	At 85 / At^{7+} 0.62	Rn 86
Fr 87 / 1.80 / Fr^+ 1.80	Ra 88 / Ra^{2+} 1.43	Ac 89 / Ac^{3+} 1.18 *)															

Lanthanides *)

58	59	60	61	62	63	64	65	66	67	68	69	70	71
Ce 1.82 / Ce^{3+} 1.07 / Ce^{4+} 0.94	Pr 1.81 / Pr^{3+} 1.06 / Pr^{4+} 0.92	Nd 1.80 / Nd^{3+} 1.04	Pm	Sm / Sm^{3+} 1.00	Eu / Eu^{3+} 0.98	Gd / Gd^{3+} 0.97	Tb / Tb^{3+} 0.93 / Tb^{4+} 0.81	Dy / Dy^{3+} 0.92	Ho / Ho^{3+} 0.91	Er 1.86 / Er^{3+} 0.89	Tm / Tm^{3+} 0.87	Yb / Yb^{3+} 0.86	Lu / Lu^{3+} 0.85

Actinides **)

90	91	92	93	94	95	96	97	98	99	100	101	102	103
Th 1.80 / Th^{4+} 1.02	Pa 1.13 / Pa^{4+} 0.98 / Pa^{5+} 0.89	U 1.38 / U^{4+} 0.97 / U^{6+} 0.80	Np / Np^{3+} 1.10 / Np^{4+} 0.95 / Np^{7+} 0.71	Pu / Pu^{3+} 1.08 / Pu^{4+} 0.93	Am / Am^{3+} 1.07 / Am^{4+} 0.92	Cm	Bk	Cf	Es	Fm	Md	No	Lw

We are indebted to the Danish physicist Niels Bohr for the most widely accepted picture of the atom. In 1912 he developed the concept of the "planetary" atom in which the electrons are visualized as circling the nucleus in "orbits" or energy levels at distances from the nucleus depending on the energies of the electrons. According to this mechanical model, the atom can be considered as a minute solar system. At the center, corresponding to the sun, is the *nucleus* which, except in the hydrogen atom, is made up of *protons* and *neutrons*. The hydrogen *nucleus* is made up of a single proton. Each proton carries a unit charge of positive electricity; the neutron, as the name implies, is electrically neutral. Each electron, which, like a planet of the solar system, moves in an orbit around the nucleus, carries a charge of negative electricity. Since the atom as a whole is electrically neutral, there must be as many electrons as protons. The weight of the atom is concentrated in the nucleus, for the mass of an electron is only $\frac{1}{1850}$ that of the lightest nucleus. Although the electrons and nuclei are both extremely small, the electrons move so rapidly about the nuclei that they give to the atoms relatively large effective diameters—ten to twenty thousand times the diameter of the nucleus.

The simplest atom is that of hydrogen, in which the nucleus has one electron moving around it, as is shown diagrammatically in Fig. 256. Atoms of the other natural elements have from two electrons (helium) to ninety-two electrons (uranium) moving in orbits about their nuclei.

The fundamental difference between atoms of the different elements lies in the electrical charge of the nucleus. This positive charge is the same as the number

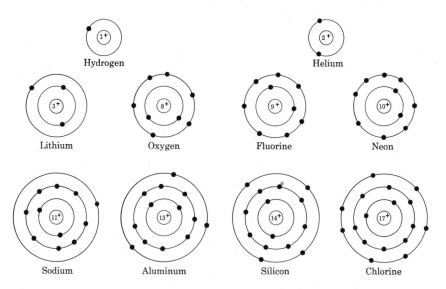

FIG. 256. Schematic diagram of atoms.

Atomic Weights, 1961

	Symbol	Atomic Weight		Symbol	Atomic Weight
Aluminum	Al	26.9815	Neodymium	Nd	144.24
Antimony	Sb	121.75	Neon	Ne	20.183
Argon	A	39.948	Nickel	Ni	58.71
Arsenic	As	74.9216	Niobium	Nb	92.906
Barium	Ba	137.34	Nitrogen	N	14.0067
Beryllium	Be	9.0122	Osmium	Os	190.2
Bismuth	Bi	208.980	Oxygen	O	15.9994
Boron	B	10.811	Palladium	Pd	106.4
Bromine	Br	79.909	Phosphorus	P	30.9738
Cadmium	Cd	112.40	Platinum	Pt	195.09
Calcium	Ca	40.08	Potassium	K	39.102
Carbon	C	12.0111	Praseodymium . . .	Pr	140.907
Cerium	Ce	140.12	Protactinium	Pa	231.
Cesium	Cs	132.905	Radium	Ra	226.
Chlorine	Cl	35.453	Radon	Rn	222.
Chromium	Cr	51.996	Rhenium	Re	186.2
Cobalt	Co	58.9332	Rhodium	Rh	102.905
Copper	Cu	63.54	Rubidium	Rb	85.47
Dysprosium	Dy	162.50	Ruthenium	Ru	101.07
Erbium	Er	167.26	Samarium	Sm	150.35
Europium	Eu	151.96	Scandium	Sc	44.956
Fluorine	F	18.9984	Selenium	Se	78.96
Gadolinium	Gd	157.25	Silicon	Si	28.086
Gallium	Ga	69.72	Silver	Ag	107.870
Germanium	Ge	72.59	Sodium	Na	22.9898
Gold	Au	196.967	Strontium	Sr	87.62
Hafnium	Hf	178.49	Sulfur	S	32.064
Helium	He	4.0026	Tantalum	Ta	180.948
Holmium	Ho	164.930	Tellurium	Te	127.60
Hydrogen	H	1.0079	Terbium	Tb	158.924
Indium	In	114.82	Thallium	Tl	204.37
Iodine	I	126.9044	Thorium	Th	232.038
Iridium	Ir	192.2	Thulium	Tm	168.934
Iron	Fe	55.847	Tin	Sn	118.70
Krypton	Kr	83.80	Titanium	Ti	47.90
Lanthanum	La	138.91	Tungsten	W	183.86
Lead	Pb	207.19	Uranium	U	238.03
Lithium	Li	6.939	Vanadium	V	50.942
Lutetium	Lu	174.97	Xenon	Xe	131.30
Magnesium	Mg	24.312	Ytterbium	Yb	173.04
Manganese	Mn	54.9380	Yttrium	Y	88.905
Mercury	Hg	200.59	Zinc	Zn	65.37
Molybdenum . . .	Mo	95.94	Zirconium	Zr	91.22

of protons, and this number, equal to the number of electrons, is called the *atomic number*. The elements in the Periodic Table, page 177, are arranged according to increasing atomic number. The *atomic weight* of an element is a number expressing its relative weight in terms of the weight of the element oxygen, which is taken as 16. The atomic weights of the elements are listed with the symbols for the elements on page 179.

The chemical attributes of the elements, with which the mineralogist is concerned, depend upon the configuration of the electronic superstructure of their atoms. The electrons are considered to be arranged about the nucleus in energy levels, or shells, commonly alluded to by the numbers 1 to 7.

A close relationship exists between the electronic superstructure of an atom, the chemical properties of the element and the place in the periodic table. In groups Ia, IIa, IIIb, IVb, Vb, VIb, and VIIa the number of electrons in the outermost shell is the same for the atoms of each element in a group and is equal to the group number. The atoms of the alkali-metals, Group Ia, have one electron and atoms of the halogens, Group VIIa, have seven electrons. It is the number of these outer shell or *valance electrons* that largely determines the chemical properties of an element. With the exception of helium, with only one shell occupied by two electrons, the other inert gases of Group VIII have atoms with eight electrons in their outermost shells.

Bonding Forces in Crystals

The forces that bind together the particles of crystalline solids are electrical in nature. Their type and intensity are largely responsible for the physical and chemical properties of minerals. Hardness, cleavage, fusibility, electrical, and thermal conductivity, and the coefficient of thermal expansion are directly related to binding forces. In general, the stronger the bond the harder the crystal, the higher its melting point, and the smaller its coefficient of thermal expansion. The great hardness of diamond is attributed to the very strong electrical forces linking its constituent carbon atoms. The structural patterns of the minerals periclase, MgO, and halite, $NaCl$, are similar, yet periclase melts at $2800°C$, whereas halite melts at $801°C$. The greater amount of heat energy required to separate the atoms in periclase indicates it has a stronger electrical bond than halite.

These electrical forces are *chemical bonds* and are described as belonging to one of four principal *bond types*: ionic, covalent, metallic, and van der Waals'. It should be understood that this classification is one of expediency and that transitions may exist between all types. The electrical interaction of the ions or atoms constituting the structural units determines the properties of the resulting crystal. It is the resemblance in properties among crystals having similar types of electrical interaction that justifies the use of the classification of bonding mechanisms. Thus, the bonding forces linking the atoms of silicon and oxygen in quartz display in almost equal amount the characteristics of the ionic and the covalent bond. In addition, galena, PbS, displays characteristics of the metallic bond, as good electrical conductivity, and some of the ionic bond, as excellent cleavage and brittleness. Furthermore, many crystals, such as mica, contain two or more bond types of different character and strength. Such crystals are called *heterodesmic* in contrast with crystals like diamond, and halite in which all bonds are of the same kind—*homodesmic*.

The Ionic Bond

A comparison of the chemical activity of elements with the configuration of their outer, or valence, electron shells leads to the conclusion that all atoms have a strong tendency to achieve a stable configuration of the outer shell in which all possible electron sites are filled. In the noble gases, helium, neon, argon, krypton, and xenon, which are almost completely inert, this condition is met. We have seen that sodium, for example, has a single valence electron in its outer shell, which it loses readily, leaving the atom with an unbalanced positive charge. Such a charged atom is called an *ion*. Positively charged atoms are *cations*, and an atom bearing a single positive charge, such as sodium, is *monovalent* and is represented by the symbol Na^+. In similar fashion, we will speak of divalent, trivalent, tetravalent, and pentavalent cations (Ca^{2+}, Fe^{3+}, Ti^{4+}, and P^{5+}). On the other hand chlorine and the other elements of the seventh group of the periodic table most easily attain the stable configuration by capturing an electron to fill the single vacancy in their outer valence shells. This produces *monovalent anions* (Cl^-, Br^-, F^-, I^-) with a net unbalanced negative charge. Oxygen, sulfur, and elements of the sixth group form divalent anions ($O^=$, $S^=$, $Se^=$, $Te^=$).

A solution of sodium chloride is thought of as containing free ions of Na^+ and Cl^-. The incessant colliding of molecules of the solvent with the ions keeps them dissociated as long as the free energy of the particles in the solution is kept high. If, however, the temperature is reduced, or the volume of the solution decreased below a certain critical value, the mutual attraction of the opposing electrical charges of sodium and chlorine ions exceeds the disruptive forces of collision, and the ions lock together to form the nucleus of a crystal. When a sufficient number of ions attach themselves to the growing nucleus, it settles out of solution.

A crystal of sodium chloride removed from the solution in which it crystallized has characteristic properties by which it may be recognized. The cubic crystal habit, cleavage, specific gravity, index of refraction, etc., are subject to little variation. These properties in no way resemble those of the shining metal or the greenish, acrid gas which are the elemental constituents of the substance. Touching the crystal to the tongue yields the taste of the solution.

The properties conveyed into the crystal by its constituent particles are the properties of the ions, not of the elements. We may correctly infer that the crystal is made up of ions which require only the presence of a suitable solvent to dissociate into free charged particles. These ions are joined together in the crystal structure by the attraction of their unlike electrostatic charges, and hence such a bonding mechanism is referred to as *ionic* or *electrostatic bond*.

As we have seen in the example of sodium chloride, the ionic bond confers on crystals in which it is dominant the important property of dissolving in polar solvents, such as water, to yield conducting solutions containing free ions.

Physically, ionic-bonded crystals are generally of moderate hardness and specific gravity, have fairly high melting and boiling points, and are very poor conductors of electricity and heat. Because the electrostatic charge constituting the ionic bond is spread over the whole surface of the atom, this bond type is not highly directional and the symmetry of the resultant crystal is generally high.

Ionic Radius. Atoms and ions do not possess definite surfaces and must be thought of as tiny, very dense, and highly charged nuclei surrounded by space which is sparsely inhabited by clouds of electrons whose density varies with distance from the nucleus, falling finally to zero. Hence, the radius of an ion cannot be defined precisely except in terms of its interaction with other ions.

Between any pair of oppositely charged ions there exists an attractive electrostatic force directly proportional to the product of their charges and inversely proportional to the square of the distance between their centers (Coulomb's law). When ions approach each other under the influence of these forces, repulsive forces are set up. These repulsive forces arise from the interaction of the negatively charged electron clouds and from the opposition of the positively charged nuclei and increase rapidly with diminishing internuclear distance. The distance at which these repulsive forces balance the attractive forces is the characteristic interionic spacing for this pair of ions. In the simplest case, when both cations and anions are fairly large and feebly charged and both have numerous symmetrically disposed neighbors of opposite sign, ions may be regarded as spheres in contact. Sodium chloride, in which both cation and anion are monovalent, fairly large, and surrounded by six neighbors of opposite polarity, is a good example. In such crystals, the interionic distance may be regarded as the sum of the radii of the two ions in contact.

The interionic distance may be measured as an interplanar spacing from x-ray diffraction data. Thus, if we assume the radius of one of the ions, the radius of the other may be found. It was in this way that the ionic radii given on page 177 were obtained. The values shown are based in part on a radius of 1.40 Å for the oxygen ion but contain numerous correction factors. These ionic radii apply only to ionic bonded crystals and strictly only to those in which both cations and anions have six closest neighbors. Atoms and ions are not rigid bodies but respond to external electrical forces by dilatation and deformation. A larger number of neighboring ions tends to distend the central ion; a smaller number allows it to collapse a little. Some distortion of shape may accompany the distension of ions. These effects are collectively called *polarization* and are of great importance in crystal structures.

If the bond is not predominantly ionic, the interionic spacing will not agree with the sum of the radii as taken from the table, even after applying correction for polarization. Thus, the length of the silicon-oxygen bond in silicates is characteristically about 1.6 Å, much less than the sum of the radii taken from the table (1.79 Å). The large discrepancy in the case of the silicon-oxygen bond, which is regarded as about 50 per cent ionic and 50 per cent covalent, may be contrasted with the good agreement of the sum of the tabulated radii with the measured interionic spacing in halite. (Sum of the radii from table = 2.78 Å; measured interionic spacing from x-ray data is 2.81 Å.)

In summation, the size of ions in the same family in the periodic table increases with increasing atomic number up to element 57, lanthanum. Above 57 the size of ions of given charge diminishes in the rare-earth sequence, giving rise to the *lanthanide contraction*, after which the size again increases. For any atomic number, the size of the ion depends on the state of ionization. Cations are in general smaller, more rigid, and less extensible than anions, and for a given element the positive ionization states are always markedly smaller than the negative. Thus, sulfur forms both divalent negative ions and tetra- and hexavalent positive ions. The contrast between the ionic radii characterizing the different valence states is typical

$S^=$	S^0	S^{4+}	S^{6+}
1.74	1.04	0.37	0.30

Anions are generally larger, more easily polarized, and less sensitive to variation in size with change in valence than cations.

Bond Strength and Physical Properties. The strength of the ionic bond, that is, the amount of energy required to break it, depends on two factors: the center-to-center spacing between the two ions and their total charge. In a crystal with a "pure" ionic-bonding mechanism, such as sodium chloride, this law holds rigorously.

The effect of increased interionic distance on the strength of the ionic bond is readily seen in the halides of sodium. The table, page 184, shows the melting points and interionic distances for these compounds. It is apparent that the strength of the bond, as measured by the melting temperature, is inversely proportional to the length of the bond. The melting temperatures of the fluorides of the alkali metals illustrate that it does not matter whether it is the size of the anion or cation that is varied; bond strength varies in inverse proportion to bond length. LiF offers an intriguing exception to this generalization, explained by anion-anion repulsion in a structure having a very small cation.

Melting Point vs. Interionic Distance in Ionic-Bonded Compounds[a]

Compound	Interionic Distance (Å)	M.P. (°C)	Compound	Interionic Distance (Å)	M.P. (°C)
NaF	2.31	980–997	SrO	2.55	2430
NaCl	2.81	801	BaO	2.75	1923
NaBr	2.97	755			
NaI	3.23	651	LiF	2.01	870
			NaF	2.31	980–997
MgO	2.10	2800	KF	2.67	880
CaO	2.40	2580	RbF	2.82	760

[a] Data from *Handbook of Chem. and Phys.*, 37th ed., Chem. Rubber Publishing Co.

Hardness vs. Interionic Distance in Ionic-Bonded Compounds[a]

Compound	Interionic Distance (Å)	Hardness (Mohs)
BeO	1.65	9.0
MgO	2.10	6.5
CaO	2.40	4.5
SrO	2.55	3.5
BaO	2.75	3.3

Hardness vs. Charge on Coordinated Ions

Compound	Interionic Distance (Å)	Hardness (Mohs)
Na^+F^-	2.31	3.2
$Mg^{2+}O^{-2}$	2.10	6.5
$Sc^{3+}N^{-3}$	2.23	7–8
$Ti^{4+}C^{-4}$	2.23	8–9

[a] Data from *Crystal Chemistry*, R. C. Evans, Cambridge University Press, 1952.

The charge on the coordinated ions has an even more powerful effect on the strength of the bond. Comparison of the absolute values of melting temperature for the alkali-earth oxides, which are divalent compounds, with the absolute values for the monovalent alkali fluorides, in which the interionic spacings are closely comparable, reveals the magnitude of this effect. Although the interionic distance, or bond length, is almost the same for corresponding oxides and fluorides, the bonds uniting the more highly charged ions are obviously much stronger. The table shows the effect of interionic spacing and charge on the hardness for a number of substances.

Covalent Bond

We have seen that ions of chlorine may enter into ionic-bonded crystals as stable units of structure because, by taking a free electron from the environment, they achieve a filled outer valence shell. A single atom of chlorine, with a void in its valence shell is in a very highly reactive condition. It seizes upon and combines with almost anything in its neighborhood. Generally, the nearest neighbor of such a single chlorine atom is another chlorine atom, and the two promptly unite in such a way that one electron does double duty in the outer valence shells of both atoms, and both thus achieve the stable inert-gas configuration. As a result of this sharing of an electron the two atoms of chlorine are bound together in an extremely tight, intimate relation.

This electron-sharing or *covalent* bond is the strongest of the chemical bonds. Minerals so bonded are characterized by general insolubility, great stability, and very high melting and boiling points. They do not yield ions to such solutions as they form and hence are nonconductors of electricity both in the solid state and in solution. Because the electrical forces constituting the bond are sharply localized in the vicinity of the shared electron, the bond is highly directional and the symmetry of the resulting crystals is likely to be lower than where ionic bonding occurs. In chlorine, the bonding energy of the atom is entirely consumed in linking to one neighbor, and stable Cl_2 molecules result, which show little tendency to link together. Certain other elements—in general, those near the middle of the periodic table, such as carbon, silicon, aluminum, and sulfur—have two, three, and four vacant spaces in the outer electron shell, and hence not all the bonding energy is consumed in linking to one neighboring atom. Such elements tend, therefore, to unite in covalent bonding with a number of adjacent atoms, forming very stable groups of atoms of rigidly fixed shape and dimensions, which may then link together to form aggregates or groups.

Carbon is an outstanding example of such an atom. Carbon atoms have four vacancies in their valence shells which they may fill by electron sharing with four other carbon atoms, forming a very stable, firmly bonded configuration having the shape of a tetrahedron with carbon atoms at the four apices. Every carbon atom is linked in this way to four others to form a continuous network. The energy of the bond is strongly localized in the vicinity of the shared electrons, producing a very rigid and highly polarized structure—that of diamond, the hardest natural substance.

Covalent Atomic Radii. In covalent bonded structures, the interatomic distance is generally equal to the arithmetic mean of the interatomic distances in crystals of the elemental substances. Thus, in diamond, the C–C spacing is 1.54 Å; in metallic silicon the Si–Si distance is 2.34 Å. We may therefore suppose that if these atoms unite to form a compound, SiC, the silicon-carbon distance will be near 1.94 Å, the arithmetic mean of the elemental spacings. X-ray measurement determines this spacing in the familiar synthetic abrasive, silicon carbide, as 1.93 Å.

Estimation of the Character of the Bonding Mechanism. It is now generally recognized that there is some electron sharing in most ionic-bonded crystals, whereas atoms in covalent-bonded substances often display some electrostatic charges. Assessment of the relative proportions of ionic to covalent character is based in part on the polarizing power and polarizability of the ions involved.

Electronegativity of Elements

Li 1.0	Be 1.5	B 2.0	C 2.5		N 3.0	O 3.5	F 4.0
Na 0.9	Mg 1.2	Al 1.5	Si 1.8		P 2.1	S 2.5	Cl 3.0
K 0.8	Ca 1.0	Sc 1.3	Ti 1.6	Ge 1.8	As 2.0	Se 2.4	Br 2.8
Rb 0.8	Sr 1.0	Y 1.3	Zr 1.6	Sn 1.7	Sb 1.8	Te 2.1	I 2.5
Cs 0.7	Ba 0.9						

Compounds of a highly polarizing ion with an easily polarized ion, such as AgI, may show strongly covalent character. In contrast, AgF, because of the lower polarizability of the smaller fluorine ion, is a dominantly ionic-bonded compound. In general, the presence of ions of small size and high charge tends to favor electron sharing as a bonding mechanism.

Bonds between elements of the first and seventh groups of the periodic table and between the second and the sixth groups are dominantly ionic. Examples are the alkali halides and the alkali-earth oxides. Bonds between like atoms or atoms close together in the periodic table will be covalent. The silicon-oxygen bond is 50 per cent ionic, the aluminum-oxygen bond 63 per cent, but the boron-oxygen bond only 44 per cent ionic. Pauling has generalized this concept and rendered it quantitative by assigning to each element a numerical value of *electronegativity*. The greater the difference in electronegativity between any two elements, the more ionic the bond between them. (See table above and Fig. 257.)

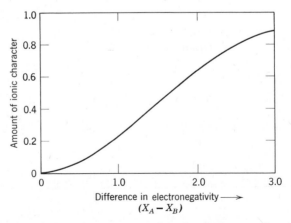

FIG. 257. Curve relating the amount of ionic character of a bond *A-B* to the difference in electronegativity $X_A - X_B$ of the atoms. (After Linus Pauling, *The Nature of the Chemical Bond*, Cornell University Press, Ithaca, 1948.)

Van der Waals' Bond

Returning to the chlorine molecules, if we take energy from these Cl_2 molecules by cooling the gas, the molecules will ultimately collapse into the close-packed, chaotic liquid state. If still more heat energy is taken away, the amplitude of vibration of the Cl_2 molecules is further reduced, and ultimately the minute, stray electrical fields existing about the essentially electrically neutral atoms will serve to lock the sluggishly moving molecules into the orderly structure of the solid state. This phenomenon of the solidification of chlorine takes place at very low temperatures, and warming above $-102°C$ will permit the molecules to break the very weak bonds and return to the disordered state of a liquid. This weak bond which ties neutral molecules and essentially uncharged structure units into a lattice by virtue of the small residual charges on their surfaces is called the *van der Waals'* or stray-field bond and is the weakest of the chemical bonds. Common only in organic compounds and solidified gases, it is not often encountered in minerals, but, when it is, it generally defines a zone of ready cleavage and low hardness. An example is the mineral graphite, which consists of covalent bonded sheets of carbon atoms linked only by van der Waals' bonds.

Metallic Bond

Metallic sodium is soft, lustrous, opaque, and sectile and conducts heat and electric current well. X-ray diffraction analysis reveals that it has the regular repetitive pattern of a true crystalline solid. The properties of the metal differ so from those of its salts that we are led to suspect a different mechanism of bonding. Sodium, like all true metals, conducts electricity, which means that electrons are free to move readily through the structure. So prodigal with their electrons are sodium and its close relatives, cesium, rubidium, and potassium, that the impact of the radiant energy of light knocks a considerable number entirely free of the structure. This photoelectric effect, on which such instruments as exposure meters depend, shows that the electrons are very weakly tied into the metal structure. We may thus postulate that the structural units of true metals are really the atomic nuclei bound together by the aggregate electric charge of a cloud of electrons that surrounds the nuclei. An electron owes no allegiance to any particular nucleus and is free to drift through the structure or even out of it entirely without disrupting the bonding mechanism. This type of bond is fittingly called the *metallic bond*. To it metals owe their high plasticity, tenacity, ductility, and conductivity, as well as their generally low hardness, melting point, and boiling point. Among minerals, only the native metals display pure metallic bonding.

Crystals with More than One Bond Type

Among naturally occurring substances, with their tremendous diversity and complexity, the presence of only one type of bonding is rare, and two or more bond types coexist in most minerals. Where this is so, the crystal shares in the properties of the different bond types represented, and often strongly directional properties result. Thus, in the mineral graphite, the cohesion of the thin sheets in which the mineral generally occurs is the result of the strong covalent bonding in the plane of the sheets, whereas the excellent cleavage reflects the weak van der Waals' bond joining the sheets together. The micas, which consist of sheets of strongly bonded silica tetrahedra (see page 430), with the relatively weak ionic bond joining the sheets together through the cations, similarly reflect in their remarkable cleavage the difference in strength of the two bond types. The prismatic habit and cleavage of the pyroxenes and amphiboles, and the chunky, blocky habit and cleavage of the feldspars, we shall see, may likewise be traced to the influence of relatively weak bonds joining together more strongly bonded structure units having a chain, band, or block shape.

The Coordination Principle

When oppositely charged ions unite to form a crystal structure in which the binding forces are dominantly electrostatic, each ion tends to gather to itself, or to *coordinate*, as many ions of opposite sign as size permits. When the atoms are linked by simple electrostatic bonds, they may be regarded as spheres in contact, and the geometry is simple. The coordinated ions always cluster about the central coordinating ion in such a way that their centers lie at the apices of a polyhedron. Thus, in a stable crystal structure, each cation lies at the center of a *coordination polyhedron* of anions. The number of anions in the polyhedron is the coordination number (**C.N.**) of the cation with respect to the given anion, and is determined by their relative sizes. Thus, in NaCl each Na has six closest Cl neighbors and is said to be in 6 coordination with Cl (**C.N.** 6). In fluorite, CaF_2, each calcium ion is at the center of a coordination polyhedron consisting of eight fluorine ions and hence is in 8 coordination with respect to fluorine (**C.N.** 8).

Anions may also be regarded as occupying the centers of coordination polyhedra formed of cations. In NaCl each chloride ion has six sodium neighbors and hence is in 6 coordination with respect to sodium. Since both sodium and chlorine are in 6 coordination, there must be equal numbers of both, in agreement with the formula, NaCl. On the other hand, examination of the fluorite structure (see Fig. 331) reveals that each fluorine ion has four closest calcium neighbors and hence is in 4 coordination with respect to calcium (**C.N.** 4). Although these four calcium ions do not touch each other, they form a definite coordination polyhedron about the central fluorine ion in such a way that the calcium ions lie at the apices of a regular tetrahedron. Since each calcium ion has eight fluorine neighbors, while each fluorine ion has only four calcium neighbors, it is obvious that there are twice as many fluorine as calcium ions in the structure. This accords with the formula CaF_2, and with the valences for calcium and fluorine.

It is easily seen that the relative sizes of the calcium and fluorine ions would permit a structure containing equal numbers of each with both ions in 8 coordination. The fact that in fluorite only half the possible calcium sites are filled calls attention to an important restriction on crystal structures; viz., *the total numbers of ions of all kinds in any stable crystal structure must be such that the crystal as a whole is electrically neutral.* That is, the total number of positive charges must equal the total number of negative charges; hence, in fluorite there may be only half as many divalent positive calcium ions as there are monovalent negative fluorine ions.

Radius Ratio

Although each ion in a crystal affects every other ion to some extent, the strongest forces exist between ions which are nearest neighbors. These are said to constitute the *first coordination shell*. The geometry of arrangement of this shell and hence the coordination number are dependent on the relative sizes of the coordinated ions. The relative size of ions is generally expressed as a *radius ratio*, $R_A : R_X$, where R_A is the radius of the cation and R_X the radius of the anion in angstrom units. The radius ratio of sodium and chlorine in halite, $NaCl$, is therefore

$$R_{Na} = 0.97 \text{ Å} \qquad R_{Cl} = 1.81 \text{ Å}$$
$$R_{Na} : R_{Cl} = 0.97/1.81 = 0.54$$

The radius ratio of calcium and fluorine in fluorite, CaF_2, is

$$R_{Ca} = 0.99 \text{ Å} \qquad R_F = 1.33 \text{ Å}$$
$$R_{Ca} : R_F = 0.99/1.33 = 0.74$$

When two or more cations are present in a structure, coordinated with the same anion, separate radius ratios may be computed for each. Thus, in spinel, $MgAl_2O_4$, both magnesium and aluminum coordinate oxygen anions. Hence,

$$R_{Mg} = 0.66 \text{ Å} \qquad R_{Al} = 0.51 \text{ Å} \qquad R_O = 1.40 \text{ Å}$$
$$R_{Mg} : R_O = 0.66/1.40 = 0.47 \qquad R_{Al} : R_O = 0.51/1.40 = 0.36$$

When coordinating and coordinated ions are the same size, the radius ratio is 1. Trial with a tray of identical spheres, such as pingpong balls, reveals that identical spherical units may be packed stably together in either of two equally economical ways, called cubic closest packing (Fig. 258) and hexagonal closest packing (Fig. 259). Either way each sphere has its twelve closest neighbors all mutually in contact (**C.N.** 12). Twelve coordination is rare in minerals, occurring in the native metals of the gold group, which are not, however, ionic bonded.

When the coordinating cation is slightly smaller than the anions, 8 coordination

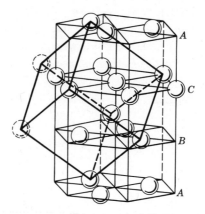

FIG. 258. Cubic closest packing. Atoms are in 12 coordination with a sequence of layers *ABC, ABC* giving rise to a face-centered cubic arrangement.

is stable. This is also called *cubic coordination* for the centers of the anions lie at the eight corners of a cube (Fig. 260). If we consider a cubic coordination polyhedron in which the anions touch each other as well as the central cation, we may compute the limiting value of radius ratio for **C.N.** = 8. Allowing the radius of the anion to equal unity, the radius of the cation for this limiting condition must be 0.732 (Fig. 261). Hence, cubic coordination has maximum stability for radius ratios between 0.732 and 1.000.

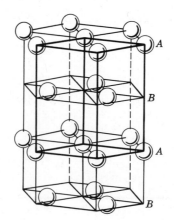

FIG. 259. Hexagonal closed packing. Atoms are in 12 coordination with a sequence of layers *AB, AB*, etc.

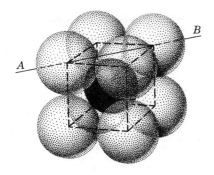

FIG. 260. Cubic or 8 coordination of X ions about, an A ion. $R_A:R_X > 0.732$.

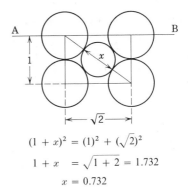

$$(1 + x)^2 = (1)^2 + (\sqrt{2})^2$$
$$1 + x = \sqrt{1 + 2} = 1.732$$
$$x = 0.732$$

FIG. 261. Limiting condition for cubic coordination.

For values of radius ratio less than 0.732, 8 coordination is not as stable as 6, in which the centers of the coordinated ions lie at the apices of a regular octahedron. Six coordination is accordingly called *octahedral coordination* (Fig. 262). We may, as before, calculate the limiting value of radius ratio for the condition in which the six coordinated anions touch each other and the central cation. The lower limit of radius ratio for stable 6 coordination is found to be 0.414 (Fig. 263). Hence, we may expect six to be the common coordination number when the radius ratio lies between 0.732 and 0.414. Na and Cl in halite, Ca and CO_3 in calcite, the *B*-type cations in spinel and the *Y*-type ions in silicates are examples of 6 coordination.

Figure 264 illustrates the change from 6 to 8 coordination in the alkali chlorides with increasing ionic radius of the cation. It is interesting to note that rubidium may be in both 6 and 8 coordination and thus rubidium chloride is polymorphous.

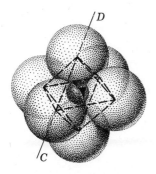

FIG. 262. Octahedral or 6 coordination of X ions about an A ion. $R_A:R_X = 0.732 - 0.414$.

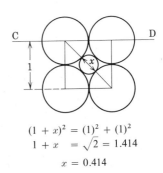

$$(1 + x)^2 = (1)^2 + (1)^2$$
$$1 + x = \sqrt{2} = 1.414$$
$$x = 0.414$$

FIG. 263. Limiting condition for octahedral coordination.

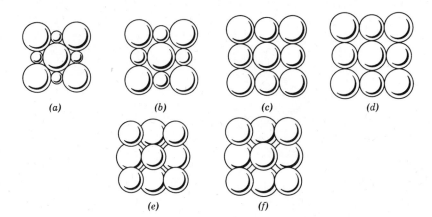

FIG. 264. Packing of ions. (*a*) Lithium chloride. (*b*) Sodium chloride. (*c*) Potassium chloride. (*d*) Rubidium chloride. (*e*) Rubidium chloride. (*f*) Cesium chloride. (*a*), (*b*), (*c*), and (*d*) have sodium structure with 6 coordination; (*e*) and (*f*) have cesium chloride structure with 8 coordination. In each figure the larger circles represent anions, the smaller circles cations.

It may be proved in similar fashion that 4 or *tetrahedral* coordination, in which the centers of the coordinated ions lie at the apices of a regular tetrahedron, has maximum stability between radius ratio limits of 0.414 and 0.225. Tetrahedral coordination is typified by the SiO_4 group in silicates, by the *A*-type ion in spinel, and by the ZnS structure (Fig. 265).

Triangular or 3 coordination is stable between limits of 0.225 and 0.155 and is common in nature in the CO_3, NO_3, and BO_3 ionic groups (Fig. 266).

Two coordination is very rare in ionic-bonded crystals. Examples are the uranyl group $(UO_2)^{2+}$, the nitrite group $(NO_2)^=$, and copper with respect to oxygen in cuprite, Cu_2O (Fig. 267).

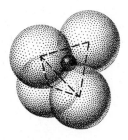

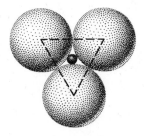

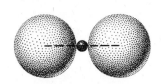

FIG. 265. Tetrahedral or 4 coordination of *X* ions about an *A* ion. $R_A:R_X = 0.414 - 0.225$.

FIG. 266. Triangular coordination of *X* ions about an *A* ion. $R_A:R_X = 0.255 - 0.155$.

FIG. 267. Linear or 2 coordination of *X* ions about an *A* ion. $R_A:R_X < 0.155$.

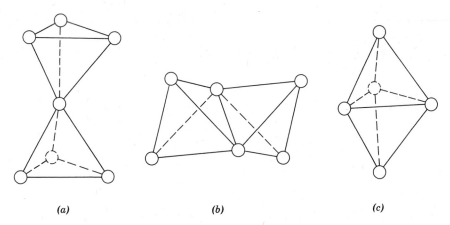

(a)   (b)   (c)

FIG. 268.   Tetrahedrons sharing corners, edges, and faces.

Although rare, examples of 5-, 7-, 9-, and 10-fold coordination are known. Such coordination numbers are possible only in complex structures and result from the filling of interstices between other coordination polyhedra.

The coordination polyhedra are frequently distorted. The smaller and more strongly polarizing the coordinating cation, or the larger and more polarizable the anion, the greater the distortion and the wider the departure from the theoretical radius-ratio limits. Also, if the bonding mechanism is not purely ionic, radius-ratio considerations may not be safely used to determine the coordination number.

Obviously, every ion in a crystal structure has some effect on every other ion—attractive if the charges are opposite, repulsive if the same. Hence, ions tend to group themselves in space to form crystal lattices in such a way that cations are as far apart as possible yet consistent with the coordination of the anions that will result in electrical neutrality. Thus, when cations share anions, they do so in such a way as to place themselves as far apart as possible. Hence, the coordination polyhedra formed around each are linked more commonly through corners than through edges or faces (Fig. 268). Cations tend to share as small a number of anions as possible, and sharing of as many as three or four anions is rare.

Pauling's Rules.   Every stable crystal, in its ordered internal architecture, bears witness to the operation of certain universal principles which determine the structure of solid matter. These principles were enunciated in 1929 by Pauling in the form of the following five rules:

Rule 1.   A coordinated polyhedron of anions is formed about each cation, the cation-anion distance being determined by the radius sum and the coordination number of the cation by the radius ratio.

Rule 2. *The electrostatic valency principle.* In a stable coordination structure the total strength of the valency bonds which reach an anion from all the neighboring cations is equal to the charge of the anion.

Rule 3. The existence of edges, and particularly of faces, common to two anion polyhedra in a coordinated structure decreases its stability; this effect is large for cations with high valency and small coordination number and is especially large when the radius ratio approaches the lower limit of stability of the polyhedron.

Rule 4. In a crystal containing different cations, those of high valency and small coordination number tend not to share polyhedron elements with each other.

Rule 5. *The principle of parsimony.* The number of essentially different kinds of constituents in a crystal tends to be small.

These rules constitute a formal statement of principles that have been discussed elsewhere. Rule 5, the principle of parsimony, refers not to chemically different kinds of constituents but to structurally distinct types of atomic sites. Thus, in crystals of very complex composition, a number of different ions may occupy the same structural position. These ions must be considered as a single "constituent" in the sense of rule 5.

Structure Type

Although there seems to be little in common between uraninite, UO_2, and fluorite, CaF_2, their x-ray powder patterns show analogous lines, although different in spacing and intensity. Structure analysis reveals that the uranium atoms in uraninite are in 4 coordination with respect to oxygen, whereas eight oxygens are grouped about each uranium. In fluorite, four calcium ions are grouped about each fluorine, and eight fluorines are packed about each calcium. Uraninite and fluorite have structures that are analogous in every respect, although the cell dimensions are different and the properties are, of course, utterly unlike. These two substances are said to be *isostructural*, or *isotypous*, and belong to the same *structure type*. All crystals in which the centers of the constituent atoms occupy geometrically similar positions, regardless of the size of the atoms or the absolute dimensions of the structure, are said to belong to the same structure type. For example, all crystals in which there are equal numbers of cations and anions in 6 coordination belong to the NaCl structure type. A large number of minerals of diverse composition belong to this structure type, including KCl, sylvite; MgO, periclase; NiO, bunsenite; PbS, galena; MnS, alabandite; AgCl, cerargyrite; TiN, osbornite; and many others.

Of great importance in mineralogy is the concept of the *isostructural group*: a group of minerals related to each other by analogous structures, generally having a common anion, and frequently displaying extensive ionic substitution. Many groups of minerals are isostructural, of which the barite group of sulfates, the calcite group of carbonates, and the aragonite group of carbonates are perhaps the best examples. The extremely close relationship that exists among the members of many groups is illustrated by the aragonite group listed herewith.

Aragonite Group

Mineral	Chemical Composition	Axial Ratios $a:b:c$	$110 \wedge 1\bar{1}0$	Cleavage	
Aragonite	$CaCO_3$	0.622:1:0.720	63° 48′	{110}	{010}
Witherite	$BaCO_3$	0.594:1:0.740	62° 12′	{110}	{010}
Strontianite	$SrCO_3$	0.609:1:0.723	62° 41′	{110}	
Cerussite	$PbCO_3$	0.608:1:0.721	62° 46′	{110}	{021}

Electrostatic Valency

A fundamental principle of organization of ionic-bonded crystals is that the sum of the strength of all bonds reaching an ion must equal the valence of that ion. Therefore, we may compute the relative strength of any bond in a crystal structure by dividing the total charge on the coordinating ion by the number of nearest neighbors to which it is linked. The resulting number, called the *electrostatic valency* (e.v.) is a measure of the strength of any of the bonds reaching the coordinating ion from its nearest neighbors. For instance, in sodium chloride, each sodium ion has a single positive charge and has six nearest neighbors; hence the electrostatic valency equals 1/6. This number is a measure of the strength of the bond reaching any sodium ion from any neighboring chlorine ion. The chlorine ions have also a single charge and are in 6 coordination with respect to sodium, hence the e.v. for chlorine is also 1/6. Crystals in which all bonds are of equal strength are called *isodesmic*. This generalization is so simple that it seems trivial, but in some cases unexpected results emerge from calculation of electrostatic valencies. For example, minerals of the spinel group have formulas of the type AB_2O_4, where A is a divalent cation such as magnesium or ferrous iron and B is a trivalent cation such as aluminum or ferric iron. Such compounds have often been called aluminates and ferrates, by analogy with such compounds as borates and oxalates. This nomenclature suggests that ionic clusters or radicals are present in the structure. X-ray data reveal that the A ions are in 4 coordination whereas the B ions are in 6 coordination. Hence, for the A ions, e.v. = 2/4 = 1/2; and, for B ions, e.v. = 3/6 = 1/2. In spite of the appearance of the formula, all bonds are the same strength. Such crystals are isodesmic and are multiple oxides, not oxysalts.

When small, highly charged cations coordinate larger and less strongly charged anions, compact, firmly bonded groups result, as the carbonates and nitrates. If the strength of the bonds within such groupings is computed the numerical value of the e.v. is always greater than one-half the total charge on the anion. This means that in such groups, the anions are more strongly bonded to the central coordinating cation than they can possibly be bonded to any other ion. For example, in the carbonate group, tetravalent carbon is in 3 coordination with divalent oxygen and hence we may compute the e.v. $= 4/3 = 1\frac{1}{3}$. This is greater than one-half the charge on the oxygen ion, and hence a functional group, or radical, exists. This is the carbonate triangle, the basic structural unit of carbonate minerals. Another example is the sulfate group. Oxygen is in 4 coordination with hexavalent positive sulfur; hence the e.v. $= 6/4 = 1\frac{1}{2}$. Since this is greater than one-half the charge on the oxygen ion, the sulfate radical forms a tightly knit group, and oxygen is bonded to sulfur more strongly than it can be bound to any other ion in the structure. This is the tetrahedral unit that is the fundamental basis of the structure of all sulfates. Such compounds as sulfates and carbonates are said to be *anisodesmic*.

Of course, it must be understood that, if the radicals are regarded as single structural units, then in a compound such as $CaCO_3$, calcite, all calcium-carbonate bonds are of equal strength and resemble those of an isodesmic crystal. Likewise, in simple sulfates like barite, all barium-sulfate bonds may be equal in strength. The crystals are, however, called anisodesmic because the presence of the strong carbon-oxygen and sulfur-oxygen bonds in addition to the weaker calcium-carbonate and barium-sulfate bonds.

Logic requires a third case: that in which the strength of bonds joining the central coordinating cation to its coordinated anions equals exactly half the bonding energy of the anion. In this case each anion may be bonded to some other unit in the structure just as strongly as it is to the coordinating cation. The other unit may be an identical cation, and the anion shared between two cations may enter into the coordination polyhedra of both. Let us consider the case of boron-oxygen groupings. Boron is trivalent with ionic radius 0.23. Hence the radius ratio with oxygen is 0.164, indicating triangular coordination, and the e.v. will hence be $3/3 = 1$. This is half the bonding strength of the oxygen ion. Consequently, a BO_3 triangle may link to some other ion just as strongly as to the central B ion. If this ion is another B^{3+}, two triangles may result, linked through a common oxygen to form a single $B_2O_5^{-4}$ group. In similar fashion, BO_3 triangles may join, or *polymerize*, to form chains, sheets, or boxworks by sharing oxygen ions. Such crystals are called *mesodesmic*, and the most important example is the silicates.

All ionic crystals may now be classified on the basis of the relative strength of their bonds into isodesmic, mesodesmic, and anisodesmic crystals.

Compositional Variation in Minerals

Minerals crystallize from solutions of complex composition, and hence ample opportunity is afforded for the substitution of one ion for another. As a result, practically all minerals display variation in chemical composition between localities and even between one specimen and another from the same locality.

Compositional variation takes place by substitution, in a given structure, of an ion or ionic group for another ion or ionic group. We may call this relationship *ionic substitution*. There are several factors that determine the extent to which it may take place. The most important is the size of the ion. Ions of two elements can readily substitute for each other only if their ionic radii are similar or differ by less than 15 per cent. If the radii of two ions differ by 15 to 30 per cent, substitution is limited and rare; if they differ by more than 30 per cent, little substitution is likely. These limits are only approximate but are based on empirical data derived from study of a number of mineral systems displaying ionic substitution. Another factor of great importance is the temperature at which the crystal is grown. The higher the temperature, the greater the amount of thermal disorder, and the less stringent the space requirements of the lattice. Hence, crystals grown at high temperatures may display extensive ionic substitution that would not have taken place at lower temperatures. An example is the high-temperature potassium feldspar sanidine, which contains much larger amounts of sodium than could be accommodated stably in crystals grown at lower temperatures.

In some cases the extent of substitution is limited. Thus Fe^2 may enter the lattice of sphalerite, ZnS, but iron cannot completely replace zinc, for FeS, troilite, is hexagonal, whereas sphalerite is isometric. Experimental work has shown that temperature of crystallization and environment are important factors in determining the extent to which iron can substitute for zinc. These investigations indicate that sphalerite may contain up to 50 mole per cent FeS in environments relatively rich in iron as indicated by the presence of pyrrhotite. The equilibrium

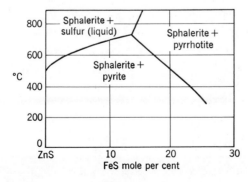

FIG. 269. Substitution of Fe^2 for Zn in sphalerite.

curve, Fig. 269, shows the FeS content of sphalerite possible at various temperatures in the presence of sulfur, pyrite, or pyrrhotite.

Coupled Ionic Substitution

It may seem that, for one ion to substitute for another, the two must have the same charge. This is not necessarily so, for frequently an ion may substitute for another of different charge. Electrical neutrality of the crystal is maintained by the simultaneous substitution elsewhere in the lattice of another ion whose charge balances out the deficit or excess caused by the first substitution. Thus, Ca^2 may readily substitute for Na^+ because the ions are almost identical in size. The deficit of one electron caused by the substitution of Ca for Na may be balanced by the simultaneous substitution, in another type of ionic site of Al^{3+} for Si^{4+}. Thus, $Ca^{2+} + Al^{3+} = Na^+ + Si^{4+}$, and the growing crystal retains its electrical neutrality. This is the mechanism of compositional variation in the soda-lime feldspars and is called *coupled substitution*. Coupling may involve two cations as in the example cited; a cation and an anion, two anions, or even the substitution of a neutral atom or a lattice vacancy for a cation or an anion in order to achieve electrical neutrality. Neutral atoms in lattice vacancies introduced in this way may have important effects on the electrical and optical properties of the crystal.

Solid Solution or Isomorphism

Numerous examples of complete ionic substitution are offered by isostructural mineral groups. In the calcite group ferrous iron may enter the lattice of magnesite, $MgCO_3$, in any proportion, and magnesium is likewise capable of entering the lattice of siderite, $FeCO_3$, in all proportions. It is apparent that the magnesite and siderite lattices are sufficiently close in their space requirements so that either ion may fill those requirements with equal ease. Since the charge of both ions is the same, the substitution is simple. In such a case, the relation between siderite and magnesite may be pictured as being one of *solid solution*, and the minerals of intermediate composition regarded as though they were homogeneous solutions of $MgCO_3$ in $FeCO_3$ or vice versa. Of course, it is clearly understood that there are no actual molecules of either of these pure compounds and that ionic substitution is completely random within the framework of the lattice (see Fig. 270). However, the idea of solid solution is useful and may be used to express the relationships in any mineral series in which the compositions lie between two pure compounds, called *end members*, as limits. Examples of such solid-solution series are the plagioclase feldspars, in which the end members are albite, $Na(AlSi_3O_8)$ and anorthite, $Ca(Al_2Si_2O_8)$, and the garnets in which complete solid-solution relations exist between most of the named varieties.

FIG. 270. Random substitution of anions in sodium chloride structure. Cations—small dark; anions—large white and black.

The term *isomorphism*, originally proposed in 1819 to describe crystals which had the same outward form, is used today by many mineralogists synonymously with *solid solution*. It is felt wiser to restore the term isomorphism to its original meaning and to use solid solution for those cases of complete ionic substitution within the framework of an isostructural group.

Exsolution

Another aspect of crystal growth to which radius ratio and geometry of packing afford a clear approach is that of disordered, nonequilibrium crystals. What happens to a rapidly grown alkali halide crystal growing from a solution containing all the alkali metals and all the halogens? If crystallization is at a sufficiently high temperature and growth is sufficiently rapid, there is a good deal of "play" in the structure, organization is loose, and ions whose sizes do not meet the equilibrium requirements for radius ratio may be incorporated in the structure, as, for example, Rb and Cs in the NaCl structure. When such a crystal is cooled, large stresses which tend to force them from the lattice exist as a result of the poor fit of the outsize ions. If sufficient time is allowed, these misfit ions migrate through the lattice and tend to group locally. Such concentrations of alien ions then arrange themselves into coordinations better suited to their size and form nuclei of a different structure which grow in the solid body of the host crystal.

This process of segregation and growth of rejected ions into crystal domains in the solid state from a disordered crystal is called *exsolution*. It is important in the interpretation of many natural mineral intergrowths. Thus, thin lamellae of sodium feldspar in potassium feldspar are interpreted as the result of exsolution of excess sodium from the potassium feldspar after crystallization. These intergrowths, called *perthite*, are very common in rocks, and this crystal chemical explanation of their formation aids materially in interpreting the history of the rock. Similarly, lenticular bodies of ilmenite in magnetite are interpreted as exsolved crystals.

Homeomorphism

Although similarities in crystals of different minerals usually result from chemically identical or similar ions in the structure, some minerals closely resemble others in crystal habit although they are quite unlike chemically. If the geometry of the arrangement of dissimilar ions is the same, similar-appearing crystals may result, as, for example, rutile, TiO_2, and zircon, $ZrSiO_4$. Both these minerals are tetragonal with similar crystal forms and axial ratios in spite of their unlike chemical compositions. Such minerals are called *homeomorphs*.

Pseudomorphs

If a crystal of a mineral is altered so that the internal structure is changed but the external form is preserved, it is called a *pseudomorph*, or *false form*. The chemical composition and structure of a pseudomorph belong to one mineral species, whereas the crystal forms correspond to another. For example, pyrite may change to limonite but preserve all the external features of the pyrite. Such a crystal is described as a pseudomorph of *limonite after pyrite*. Pseudomorphs are usually further defined according to the manner in which they were formed, as by:

1. *Substitution.* In this type of pseudomorph there is a gradual removal of the original material and a corresponding and simultaneous replacement of it by another with no chemical reaction between the two. A common example of this is the substitution of silica for the wood fiber to form petrified wood. Another example is quartz, SiO_2, after fluorite, CaF_2.

2. *Incrustation.* In the formation of this type of pseudomorph a crust of one mineral is deposited over crystals of another. A common example is quartz incrusting cubes of fluorite. The fluorite may later be entirely carried away by solution, but its former presence is indicated by the casts left in the quartz.

3. *Alteration.* In this type of pseudomorph there has been only a partial addition of new material, or a partial removal of the original material. The change of anhydrite, $CaSO_4$, to gypsum, $CaSO_4 \cdot 2H_2O$, and the change of galena, PbS, to anglesite, $PbSO_4$, are examples of alteration pseudomorphs. A core of the unaltered mineral may be found in such pseudomorphs.

4. *Paramorphism.* The name paramorph is given to a crystal whose internal structure has changed to that of a polymorphous form without producing any change in external form. Thus aragonite frequently goes over to calcite, and rutile changes to brookite.

Mineraloids

There are a number of amorphous natural solids which show no sign of crystallinity. They have formed by the solidification of a colloidal solution and are called *gel minerals* or *mineraloids*. Mineraloids are usually formed under conditions of low pressure and low temperature and commonly originate during the process of weathering. They characteristically occur in mammillary, botryoidal, and stalactitic masses. The power of mineraloids to absorb other substances accounts for their often wide variations in chemical composition. Limonite and opal are familiar examples of gel minerals. Lechatelierite, natural silica glass, is a mineraloid.

There are other natural noncrystalline solids known as *metamict minerals*. These originally formed as crystalline solids but the crystal structure has been destroyed by radiation from radioactive elements. As one expects of amorphous substance, metamict minerals do not diffract x-rays and are optically isotropic. In addition they have no cleavage but break with a concoidal fracture. Many metamict minerals are bounded by crystal faces and are thus amorphous pseudomorphs after an earlier crystalline phase. When a metamict mineral is heated its crystal structure is reconstituted and its density is increased. The recrystallization is frequently accompanied by the internal generation of so much heat that the mineral becomes incandescent.

All metamict minerals are radioactive and it is believed that the structural breakdown results from the bombardment of alpha particles from contained uranium or thorium. However, the process is not well understood for some minerals containing less than 0.5 per cent of these elements may be metamict, whereas others containing high percentages of them may be crystalline.

CHEMICAL ANALYSES AND FORMULAS

Some elements as gold, arsenic, and sulfur occur in the native state and their formulas are the chemical symbols of the elements. But most minerals are compounds composed of two or more elements and their formulas indicate the proportions of the elements present. Thus in PbS there is one atom of sulfur for each atom of lead; and in $CuFeS_2$, there are two atoms of sulfur for each atom of copper and iron. The chemical formulas of minerals have been derived from quantitative chemical analyses. These analyses give the weight percentages of the elements as reported by the chemist which usually total slightly more or less than 100. Consider the following analysis of chalcopyrite.

	1 Weight Per Cent	2 Weight Per Cent Recalculated	3 Atomic Weights	4 Atomic Proportions	5 Atomic Ratios
Cu	34.30	34.40	63.54	0.54139	1
Fe	30.59	30.68	55.85	0.54932	1 } approx.
S	34.82	34.92	32.07	1.08886	2
Total	99.71	100.00			

The percentages in column 1 reported by the chemist, are recalculated to 100, column 2, by dividing them by the total 99.71, and multiplying by 100. Since the elements have different atomic weights, the percentages do not represent the ratios of the different atoms. To arrive at their relative proportions, the percentage in each case is divided by the atomic weight of the element. This gives a series of numbers, (column 4) the atomic proportions, from which the atomic ratios can be quickly derived. In the analysis of chalcopyrite, these ratios are $S:Cu:Fe = 2:1:1$; thus $CuFeS_2$ is the chemical formula.

Most minerals are oxidized compounds, and by convention the analyses are reported as percentages of oxides. By calculations similar to those outlined above, the ratios of the oxides are determined. But in this case the percentages are divided by the molecular weights (the sum of the atomic weights of the elements in the oxides). As an example consider the following analysis of gypsum.

	1 Weight Per Cent	2 Weight Per Cent Recalculated	3 Molecular Weights	4 Molecular Proportions	5 Molecular Ratios
CaO	32.44	32.51	56.08	0.57970	1
SO_3	46.61	46.71	80.07	0.58336	1 } approx.
H_2O	20.74	20.78	18.0	1.15444	2
Total	99.79	100.00			

From the molecular ratios in column 5 we see $CaO:SO_3:H_2O = 1:1:2$ and the composition can be written as $CaO \cdot SO_3 \cdot 2H_2O$ or $CaSO_4 \cdot 2H_2O$.

The solutions and melts from which minerals crystallize contain many elements providing ample opportunity for atomic substitution. Thus in nature the pure compound is rare and several different ions may occupy a given site in the crystal structure giving rise in some cases to complicated formulas.

For example, in the mineral olivine $(Mg, Fe)_2SiO_4$, Mg and Fe can substitute for each other in all proportions and thus a complete solid solution series exists between forsterite, Mg_2SiO_4 and fayalite, Fe_2SiO_4. In addition, small amounts of Mn and Ca are commonly reported in analyses. Consider the following analysis of olivine from New Hampshire.

	1	2	3	4	5	6
	Weight Per Cent	Weight Per Cent Recalculated	Molecular Weights	Molecular Proportions	Atomic Proportions	with O = 4
SiO_2	34.96	35.21	60.09	0.58595	Si 0.5860	0.99
FeO	36.77	37.03	71.85	0.51537	Fe 0.5134	0.87 ⎫
MnO	0.52	0.53	70.94	0.00747	Mn 0.0075	0.01 ⎬ 2.02
MgO	27.04	27.23	40.31	0.67551	Mg 0.6755	1.14 ⎭
					O 2.3684	4.0
Total	99.29	100.00				

To arrive at an empirical formula from an analysis it is frequently desirable to determine the atomic proportions (columns 5 and 6). In this example of olivine, the proportions of the metal atoms are the same as the corresponding molecular proportions. For each atom of the divalent metals there is one oxygen atom, but for each atom of silicon there are two oxygen atoms. Therefore, the atomic proportion of oxygen equals the atomic proportions of Fe + Mn + Mg + 2Si. Assuming 4 oxygens, the atomic proportions are given in column 6, and the formula can be written as: $Mg_{1.14}Fe_{0.87}Mn_{0.01}Si_{0.99}O_4$; or neglecting the minor amount of Mn, this is very close to $(Mg_{0.57}Fe_{0.43})_2SiO_4$. In the olivine solid solution series, the ratio of (Mg + Fe):Si is always 2:1. Thus a general formula can be written as: $(Mg_{1-x}Fe_x)_2SiO_4$, where x can range from 0 to 1.

In solid solution series there is usually a continuous change in many of the properties with change in chemical composition. In the olivine series the specific gravity, refractive index, and unit cell dimensions all increase progressively as Fe^2 substitutes for Mg; and by their determination one can obtain a close approximation of the chemical composition.

It is frequently desirable to calculate the theoretical composition of a mineral from its chemical formula. This requires the determination of the proportions by weight of the different elements in 100 parts of the mineral. Take for example, chalcopyrite, $CuFeS_2$. The atomic weights of the elements divided by their sum, the molecular weight, yields the weight percentages as follows:

	Atomic Weight		Weight Per Cent
Cu	63.54×1	63.54	34.62
Fe	55.85×1	55.85	30.43
S	32.07×2	64.14	34.95
	Molecular weight =	183.53	100.00

For the example of olivine $(Mg_{0.57}Fe_{0.43})_2SiO_4$ given above the theoretical composition is obtained as follows:

	Molecular Weights		Weight Per Cent
MgO	$40.32 \times 0.57 \times 2 =$	45.96	27.39
FeO	$71.85 \times 0.43 \times 2 =$	61.79	36.82
SiO_2	$60.06 \times 1 \quad\quad =$	60.06	35.79
	Molecular weight =	167.81	100.00

BLOWPIPE AND CHEMICAL TESTS

During the nineteenth century a number of quick and easy tests, known as the *blowpipe tests* were developed for making qualitative determinations of most of the common elements found in minerals. In spite of the sophisticated apparatus available for modern mineralogical investigation, the blowpipe remains a serviceable tool in determinative mineralogy.

The ordinary blowpipe consists essentially of a tapering tube ending in a small opening through which air can be forced in a thin stream. When this current of air is directed into a luminous flame, combustion takes place more rapidly and completely, producing a very hot flame.

Figure 271 represents a common type of blowpipe. The mouthpiece, either c or c', is fitted into the upper end of the tube, and air from the lungs forced into it issues from the small opening at the other end. The tip of the blowpipe, b, is placed just within a flame rich in carbon such as is obtained from a candle or ordinary household gas. When gas is available, the flame may be produced using a Bunsen burner in which an inner tube has been inserted to cut off the supply of air at the base and thus yield a luminous flame. The upper end of the tube is flattened and cut at an angle (Fig. 272). The gas flame should be adjusted to about one inch in height. The resulting blowpipe flame should be nonluminous, narrow, sharp-pointed, and clean-cut (Fig. 272). If illuminating gas in not available, a candle can be used.

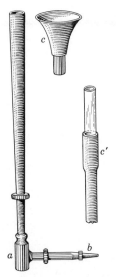

FIG. 271. Common blowpipe.

Figure 273 illustrates a more elaborate blowpipe, but is constructed with a gas fitting, *c*, so that air forced into the tube through the mouthpiece *a*, and gas admitted through tube *d*, issue together from the opening *e*.

It usually requires some practice before one can produce a steady and continuous blowpipe flame. Some blowpipe tests can be completed before it is necessary to replenish the supply of air in the lungs. Frequently, however, an operation takes a longer time than this permits, and an interruption in order to fill the lungs afresh materially interferes with the success of the experiment. Consequently, it is important to be able to maintain a steady stream of air from the blowpipe for a considerable time. This is accomplished by distending the cheeks so as to form a reservoir of air in the mouth. When the supply of air in the lungs must be replenished, the passage from the mouth into the throat is closed by lifting the root of the tongue; thus, while breathing in through the nose, a steady stream of air is also being forced out of the reservoir in the mouth. In this way a constant flame may be obtained.

Fusion by Means of the Blowpipe Flame. The hottest part of a blowpipe flame is just beyond the visible portion, at *a* Fig. 272, and may reach a temperature as

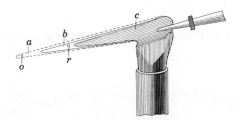

FIG. 272. Blowpipe flame. (*a*) Oxidizing flame. (*b*) Reducing flame. (*c*) Unburned gas. *o* and *r* indicate positions of assay in oxidizing and reducing flames respectively.

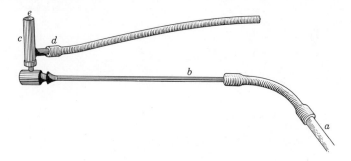

FIG. 273. Improved blowpipe.

high as 1500°C, but the temperature varies depending on the type of gas and the mixture of gas and air. The ease with which a mineral fuses is an important aid to its identification. To make this test the *assay*, a small fragment of the mineral, is held in forceps and placed in the hottest portion of the flame. It may melt completely or only the sharp edges may round over. In either case it is said to be fusible.

Minerals can therefore be divided into two classes: fusible and infusible. The fusible minerals can be further classified according to the ease with which they fuse. To assist in this classification, six minerals with different degrees of fusibility have been chosen as a scale of reference. For instance, a mineral has a fusibility of 3, if it fuses as easily as the mineral listed as 3 in the scale. In making such comparative tests, it is necessary to use fragments of uniform size, preferably a splinter, not larger than 1 mm in cross section. The scale of fusibility is given below.

Scale of Fusibility

Number	Mineral	Approximate Fusing Point	Remarks
1	Stibnite	525°C	Very easily fusible in the candle flame
2	Chalcopyrite	800°	A small fragment will fuse easily in the Bunsen burner flame
3	Garnet (Almandite)	1050°	Infusible in the Bunsen burner flame but fuses easily in the blowpipe flame
4	Actinolite	1200°	A sharp-pointed splinter fuses with little difficulty in the blowpipe flame
5	Orthoclase	1300°	The edges of fragments are rounded with difficulty in the blowpipe flame
6	Bronzite	1400°	Practically infusible in the blowpipe flame. Only the fine ends of splinters are rounded
7	Quartz	1710°	Infusible in the blowpipe flame

Reducing and Oxidizing Flames. Reduction consists essentially in taking oxygen away from a chemical compound, and oxidation consists in adding oxygen to it. These two opposite chemical reactions can be accomplished by the flame of a Bunsen burner or of a blowpipe. Cone b, Fig. 272, contains CO, carbon monoxide. This is a reducing agent, since, because of its strong tendency to take up oxygen to become CO_2, it will, if possible, take oxygen away from another substance in contact with it. For instance, if a small fragment of ferric oxide, Fe_2O_3, is held in this part of the blowpipe flame, it will be reduced by the removal of oxygen to ferrous oxide, FeO. In Fig. 272 cone b is the reducing part of the blowpipe flame, and if a reduction test is to be performed the mineral fragment is placed at r.

If oxidation is to be accomplished, the mineral must be placed entirely outside of the flame, at o in Fig. 272, where the oxygen of the air can have free access to it, but where it is heated by the flame. Under these conditions, if the reaction is possible, oxygen will be added to the mineral and the substance will be oxidized. Pyrite, FeS_2, for instance, if placed in the oxidizing flame, would be converted into ferric oxide, Fe_2O_3, and sulfur dioxide, SO_2.

Metallic Globules Reduced on Charcoal

Element	Color and Character of Globule	Remarks
Gold Au	Yellow, soft, no coating, remains bright	Metallic gold can be easily reduced from the gold tellurides without a flux
Silver Ag	White, soft, no coating, remains bright	Usually necessary to use reducing mixture. To distinguish from other globules, dissolve in nitric acid, add hydrochloric acid to obtain white silver chloride precipitate
Tin Sn	White, soft, becomes dull on cooling. White coating of oxide film	Globules form with difficulty even with reducing mixture. The metallic globule is oxidized in nitric acid to a white hydroxide
Copper Cu	Red, soft, surface black when cold, difficultly fusible	Copper minerals should be roasted to drive off sulfur, arsenic, and antimony before mixing with reducing mixture
Lead Pb	Gray, soft, fusible. Bright in reducing flame, iridescent in oxidizing flame	The incandescent charcoal will reduce the lead. To distinguish from other globules dissolve in nitric acid, and from clear solution precipitate white lead sulfate by addition of sulfuric acid

Note. Easily fusible metallic beads are often obtained on heating metallic compounds containing sulfur, antimony, or arsenic. Such beads are always very brittle and often magnetic. Magnetic masses or globules are obtained when compounds of iron, nickel, and cobalt are heated on charcoal.

Sublimates on Charcoal

Element	Composition of Coating	Color and Character of Coating on Charcoal	Remarks
As	Arsenious oxide As_2O_3	White and volatile, depositing at some distance from the assay	Usually accompanied by garlic odor
Sb	Antimony oxides Sb_2O_3, Sb_2O_4	White and volatile, depositing close to the assay	Less volatile than arsenic oxide
Se	Selenium oxide SeO_2	Volatile white, tinged with red on outside; to gray near assay	Accompanied by a peculiar odor. Coating touched with reducing flame gives blue flame
Te	Tellurium oxide TeO_2	Dense white; volatile. On outside gray to brownish	In reducing flame coating gives bluish green flame color
Zn	Zinc oxide ZnO	If mixed with sodium carbonate on charcoal gives nonvolatile sublimate, yellow when hot, white when cold near assay	Coating moistened with cobalt nitrate and ignited turns green
Sn	Tin oxide SnO_2	Faint yellow when hot, white when cold. Nonvolatile in the oxidizing flame	Coating moistened with cobalt nitrate and ignited turns bluish green
Mo	Molybdenum oxide MoO_3	Pale yellow when hot, white when cold. May be crystalline. Volatile in the oxidizing flame. Red MoO_2 under assay	Coating touched for a moment by a reducing flame becomes dark blue
Pb	Lead oxide PbO	Yellow near the mineral and white farther away. Volatile	Coating may be composed of white sulfite and sulfate of lead in addition to the oxide
Pb	Lead iodide PbI_2	Chrome yellow. Volatile	This reaction when lead minerals heated with iodide flux
Bi	Bismuth oxide Bi_2O_3	Yellow near the mineral and white farther away. Volatile	To be told from the lead oxide coating by iodide test
Bi	Bismuth iodide BiI_3	Bright red with yellow ring near assay	This reaction when bismuth minerals heated with iodide flux

FIG. 274. Charcoal block with antimony oxide coating.

Charcoal in Blowpiping. Charcoal blocks measuring about $4'' \times 1'' \times \frac{1}{2}''$ are used to reduce metals from their minerals and to obtain characteristic oxide coatings on the surface (Fig. 274). Although it is impossible to extract the metal from some minerals by blowpipe methods, a few can be reduced merely by heating on charcoal and others can be reduced by means of a flux. A mixture of sodium carbonate and charcoal in equal proportions, called the *reducing mixture*, serves as a good flux for most reductions. In all reduction experiments the mineral should be powdered and placed in a small depression in the charcoal. When the flux is used, it should be mixed with the mineral in the proportion, 2 flux to 1 mineral.

Sublimates on Plaster

Element	Composition of Coating	Color and Character of Coating	Remarks
Se	SeO_2	Red to crimson. Volatile	Volatilizes giving reddish fumes and characteristic odor
Te	TeO_2	Dark brown. Volatile	In reducing flame coating gives bluish green flame color
Cd	CdO	Greenish yellow with brown both near assay and at a distance	Nonvolatile
Pb	PbI_2	Chrome yellow with iodide flux	
Bi	BiI_3	Chocolate-brown with underlying red with iodide flux	Subjected to ammonia fumes coating becomes first orange-yellow then red
Mo	MoO_3	White in oxidizing flame. Red MoO_2 under assay	Coating turns deep ultramarine blue if touched with reducing flame
Mo	MoI_4	Ultramarine blue with iodide flux	
Sb	SbI_3	Orange to red with iodide flux	Disappears when subjected to ammonia fumes

Sublimates in the Open Tube

Element	Products of Oxidation Composition	Color and Character	Remarks
S	Sulfur dioxide SO_2	SO_2, a colorless gas, issues from the upper end of the tube	The gas has a pungent and irritating odor. Moistened blue litmus paper placed at the upper end of the tube becomes red, owing to the acid reaction of sulfurous acid
As	Arsenious oxide As_2O_3	White, highly volatile, and crystalline	The sublimate condenses at a considerable distance above the heated portion in small octahedral crystals
Sb	Antimonious oxide Sb_2O_3	White, volatile, and crystalline	The sublimate forms a white ring closer to the heated portion of the tube than the arsenious oxide. Is obtained from antimony compounds which do not contain sulfur
Sb	Antimony tetraoxide Sb_2O_4	Pale yellow when hot, white when cold. Dense, nonvolatile, amorphous	Obtained from antimony sulfide and sulfantimonites. Settles mostly on the bottom of the tube, usually is accompanied by Sb_2O_3
Mo	Molybdenum trioxide MoO_3	Pale yellow to white crystals form a network near heated portion	If crystals are touched with reducing flame they turn blue

The table on page 208 gives a list of the elements which yield sublimates when their minerals are heated in the oxidizing flame on charcoal. In some cases more characteristic coatings are obtained when the assay has had a chemical reagent added to it. The most important reagent is the so-called *iodide-flux* or *bismuth-flux*, a mixture of potassium iodide and sulfur. When this reagent is used colored iodide coatings may result.

Sublimates on Plaster. In some cases it is preferable to collect sublimates on the surface of a plaster of Paris tablet rather than on charcoal. The material to be tested is placed in a small depression made near one end of the tablet and then heated before the blowpipe exactly as on charcoal. The plaster tablet is used to bring out the color of sublimates poorly displayed on the black background of charcoal. The iodide coatings are especially marked on the plaster tablet.

Open-Tube Test. Tubing of hard glass is used in making what are known as open-tube tests. The tubing, with an internal diameter of $\frac{1}{4}$ inch, should be cut

Closed-Tube Tests

Element	Substance	Color and Character	Remarks
	H_2O	Colorless liquid, easily volatile	Minerals containing water of crystallization or hydroxyl give on moderate heating a deposit of drops of water on the cold upper walls of the tube. The water may be acid from hydrochloric, hydrofluoric, sulfuric, or other volatile acid
S	S	Red when hot, yellow when cold. Volatile	Given only by native sulfur and those sulfides which contain a high percentage of sulfur
As	As	Two rings around tube: one black amorphous material; the other near the bottom a silver gray crystalline material, the "arsenic mirror"	Given by native arsenic and some arsenides
As	AsS As_2S_3	Deep red liquid when hot, reddish yellow solid when cold	Given by realgar, AsS, and orpiment, As_2S_3, and some arsenic sulfides
Sb	Sb_2S_2O	Slight reddish brown coating near the bottom of the tube	Given by antimony sulfide and some antimony sulfides
Hg	HgS	Black amorphous sublimate	This test given when cinnabar, HgS, is heated alone
Hg	Hg	Gray, metallic globules	Metallic mercury is obtained when cinnabar is mixed with sodium carbonate and heated

into approximately 5-inch lengths. A small amount of the mineral to be tested is powdered and placed in the tube at a point about one-third of its length from one end. The tube is then inclined at as sharp an angle as possible, with the mineral lying nearer the lower end. The tube is held over a Bunsen burner flame in such a way that at first the flame plays on the upper part of the tube. This serves to convert the inclined tube into a chimney, up which a current of air flows. After a moment the tube is shifted so that the flame heats it at a point just above the mineral, or in some cases the flame may be directly beneath the mineral. The mineral is being heated under these conditions in a steady current of air, and it will be oxidized if such a reaction is possible. Various oxides may come off as gases and either escape at the end of the tube or be condensed as sublimates upon its walls. The table on page 210 lists those elements which yield most characteristic reactions in open tubes.

Closed-Tube Test. The closed tube made of soft glass should have a length of about $3\frac{1}{2}$ inches and an internal diameter from $\frac{1}{8}$ to $\frac{3}{16}$ inch. Two closed tubes can be made by fusing the center of a piece of tubing 7 inches in length and pulling it apart. The closed-tube test is used to determine what takes place when a mineral is heated in the absence of oxygen. In performing the test the mineral is broken into small fragments or powdered, placed in the closed end of the tube, and heated in the Bunsen burner flame.

Flame Test. Some elements may volatilize and impart characteristic colors to the flame, when minerals containing them are heated. A flame test is made by heating a small fragment of the mineral held in the forceps, but a more decisive test is usually obtained when the fine powder of the mineral is introduced into the Bunsen burner flame on a piece of platinum wire. Some minerals contain elements which normally color the flame but because the compound is nonvolatile they fail to give the characteristic flame until broken down by an acid or a flux.

Frequently a flame color is masked by the presence of a sodium flame. Although a mineral may contain no sodium, it may be coated with dust of a sodium compound ever present in the laboratory, and a yellow flame may result. A filter of blue glass held in front of the flame will completely absorb the yellow sodium flame and allow the characteristic flame colors of other elements to be observed.

Bead Tests. Some elements, when dissolved in fluxes, give a characteristic color to the fused mass. The fluxes that are most commonly used are borax, $Na_2B_4O_7 \cdot 10H_2O$ and salt of phosphorus, $HNaNH_4PO_4 \cdot 4H_2O$. Sodium carbonate, Na_2CO_3 is used in the test for manganese. The operation is most satisfactorily performed by first fusing the flux on a small loop of platinum wire into the form of a lens-shaped bead. For best results the loop on the wire should have the shape and size shown in Fig. 275. After the flux has been fused into a bead on the wire, but before it cools, a small amount of the powdered mineral is introduced into it

FIG. 275. Loop in platinum wire.

Flame Colorations

Element	Color of Flame	Remarks
Strontium Sr	Crimson	Strontium minerals which give the flame color also give alkaline residues after being heated
Lithium Li	Crimson	Lithium minerals which give the flame color do not give alkaline residues after being heated. (Difference from strontium)
Calcium Ca	Orange	In the majority of cases a distinct calcium flame will be obtained only after the assay has been moistened with HCl
Sodium Na	Intense yellow	A very delicate reaction. The flame should be very strong and persistent to indicate the presence of sodium in the mineral as an essential constituent
Barium Ba	Yellow-green	Minerals which give the barium flame also give alkaline residues after ignition
Molybdenum Mo	Yellow-green	Obtained from the oxide or sulfide of molybdenum
Boron B	Yellow-green	Minerals giving a boron flame rarely give alkaline residues after ignition. Many boron minerals will give a green flame only after they have been broken down by sulfuric acid or the "boron flux"
Copper Cu	Emerald-green	Obtained from copper oxide
Copper Cu	Azure-blue	Obtained from copper chloride. Any copper mineral will give the copper chloride flame after being moistened with hydrochloric acid
Chlorine Cl	Azure-blue (copper chloride flame)	If a mineral containing chlorine is mixed with copper oxide and introduced into the flame, the copper chloride flame results
Phosphorus P	Pale bluish green	A phosphorus mineral may not give the flame color until moistened with sulfuric acid. Not a decisive test
Zinc Zn	Bluish green	Appears usually as bright streaks in the flame
Antimony Sb	Pale green	Flame best observed when mineral is fused on charcoal. Color plays about assay
Lead Pb	Pale azure-blue	Flame may be observed to play about assay when a lead mineral is fused on charcoal
Potassium K	Violet	It may be necessary to decompose mineral with a flux of gypsum, $CaSO_4 \cdot 2H_2O$, to obtain flame color

Bead Tests[a]

Oxides of	Borax Bead		Phosphorus Salt Bead	
	Oxidizing Flame	Reducing Flame	Oxidizing Flame	Reducing Flame
Titanium	Colorless to white	Brownish violet	Colorless	Violet
Tungsten	Colorless or white	Yellow to yellowish brown	Colorless	Fine blue
Molybdenum	Colorless or white	Brown	Colorless	Fine green
Chromium	Yellowish green	Fine green	Fine green	Fine green
Vanadium	Yellowish green almost colorless	Fine green	Yellow	Fine green
Uranium	Yellow	Pale green to nearly colorless	Pale greenish yellow	Fine green
Iron	Yellow	Pale bottle-green	Yellow to almost colorless	Almost colorless
Copper	Blue-green	Opaque red with much oxide	Blue	Opaque red
Cobalt	Blue	Blue	Blue	Blue
Nickel	Reddish brown	Opaque gray	Yellow to reddish yellow	Yellow to reddish yellow
Manganese	Reddish violet	Colorless	Violet	Colorless

[a] The colors given are those obtained after the bead is cold.

and is dissolved by further heating.[1] The color of the resulting bead may depend upon whether it was heated in the oxidizing or reducing flame.

Etch Tests. A series of *etch tests* is used for the determination of fine-grained mixtures of metallic minerals. The specimen must first be polished and set in an appropriate mount so that the polished surface can be examined microscopically. When thus examined many minerals have a characteristic appearance and may be recognized by inspection. However, minerals which resemble one another on the polished surface can be differentiated by etching their surfaces with appropriate reagents. This method is used particularly by the economic geologist in the study of ore minerals. The technique has become so highly developed and specialized that one must refer to specific books on the subject for an adequate treatment.

---

[1] All metallic minerals should be roasted before being introduced into the bead, for it is the oxide of the metal that is desired. It is especially desirable to rid the mineral of arsenic, for even small amounts of this element make the platinum brittle and cause the end of the wire to break off.

Microchemical Reactions. Another method for the determination of the elements in minerals uses the polarizing microscope to observe chemical reactions and the products of such reactions. This microchemical method is particularly applicable to small quantities, and thus has a definite advantage over other methods if a limited amount of material is available. A discussion of this method is beyond the scope of the present book, and the reader is referred to specialized books on the subject.

Spectroscopic Analysis. Optical spectrographic analysis has for years been standard procedure in mineralogical research, but the necessary equipment is costly and can be operated only by the skilled technician. Recently several small, relatively low-priced spectroscopes and spectrographs have appeared on the market. Because of their simple operation, they can be used as a determinative tool in the elementary mineralogy laboratory. One such instrument, the Vreeland Spectroscope, is illustrated in Fig. 276.

We have seen that when certain elements are volatilized they impart characteristic colors to the flame. The spectroscope analyzes these colors by identifying the wave lengths producing them. In most cases the color is found to result from a combination of several wave lengths.

When atoms are heated their electrons are excited and rise to higher energy levels; the greater the temperature, the higher the state of excitation. The fall of the electrons to lower energy levels is accompanied by the emission of radiant energy. The energy difference between the two levels determines the wave length of the emitted radiation. Since the electronic configuration of the atom is different for each element, every element has a characteristic spectrum.

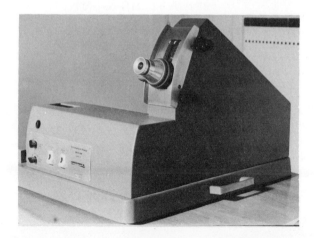

FIG. 276. Vreeland spectroscope. (Courtesy of Spectrex Co., Redwood City, California.)

In the high temperature of a carbon arc as used in the spectroscope and spectrograph, minerals are vaporized and the spectrum emitted by the incandescent vapor can be analyzed. The spectrum is made up of lines which correspond to single wave lengths or range in wave lengths. In a *spectrograph* the lines are recorded photographically, but in the spectroscope they are observed visually. In the Vreeland instrument (Fig. 276), while the spectrum from the vaporizing mineral is being observed, standard spectra of many different elements can be brought in the field of view for comparison, and the elements composing the mineral can be identified.

1. DRY REAGENTS

Sodium carbonate (anhydrous) Na_2CO_3. Used as a flux to decompose minerals by fusion on charcoal, and in some bead tests.

Borax, $Na_2B_4O_7 \cdot 10H_2O$. Used for bead tests.

Salt of phosphorus, $HNaNH_4PO_4 \cdot 4H_2O$. Used for bead tests.

Cupric oxide, CuO. Used for making a flame test for chlorine.

Potassium iodide and sulfur in equal parts (bismuth flux) used in a test for bismuth, lead, molybdenum, etc.

Metallic zinc or tin to make reduction tests in HCl solutions.

2. WET REAGENTS

Hydrochloric acid, HCl. Used for solution of minerals. The concentrated acid is diluted with three parts of water for ordinary use.

Nitric acid, HNO_3. Commonly used in concentrated form.

Sulfuric acid, H_2SO_4. For normal use the concentrated acid is diluted with four parts of water. *Note.* Dilute by pouring acid into the water, not vice versa.

Ammonium hydroxide, NH_4OH.

Ammonium oxalate $(NH_4)_2C_2O_4$.

Sodium phosphate, $Na_2HPO_4 \cdot 12H_2O$. Used as aqueous solution.

Ammonium molybdate, $(NH_4)_2MoO_4$, for phosphate test.

Silver nitrate, $AgNO_3$, for test for chlorine.

Cobalt nitrate, $Co(NO_3)_2$, in dilute solution.

Dimethylglyoxine, used in test for nickel.

Hydrogen peroxide, H_2O_2, used in titanium test.

TESTS FOR THE ELEMENTS

A few of the important blowpipe and chemical tests for the determination of specific elements in minerals are given on the following pages. For a more complete treatment the student is referred to *The Manual of Determinative Mineralogy*.[2]

Aluminum. Light-colored infusible aluminum minerals turn blue when moistened with $Co(NO_3)_2$ and ignited. Zinc silicates also turn blue under similar conditions. A flocculent white precipitate of $Al(OH)_3$ forms when an excess of NH_4OH is added to an acid solution of an aluminum-bearing mineral.

Antimony. When an antimony mineral is heated on charcoal, a white sublimate forms near the assay, page 208. Sublimates are obtained in the open tube, page 210.

Arsenic. Unoxidized arsenic minerals give a white volatile coating on charcoal at some distance from the assay, page 208. Arsenic also gives characteristic open- and closed-tube tests, pages 210 and 211.

Barium. Yellowish-green flame is characteristic of barium, page 213. White $BaSO_4$ is precipitated from an acid solution by the addition of dilute H_2SO_4.

Beryllium. There is no simple test for beryllium.

Bismuth. When heated with reducing mixture on charcoal, a bismuth mineral yields a metallic globule and an oxide coating. The metal is easily fusible but only imperfectly malleable and when hammered breaks into grains. When the assay is heated with bismuth flux, the sublimate is yellow near the mineral and red farther away (pages 208, 209).

Boron. A yellow green flame is characteristic of boron, page 213.

Calcium. Calcium is precipitated from alkaline solutions as white calcium oxalate by the addition of ammonium oxalate. Calcium imparts an orange color to the flame, page 213.

Carbon. All carbonate minerals dissolve with effervescence in HCl. Some dissolve only in hot acid; and others (for example, cerussite, $PbCO_3$) form an insoluble chloride which quickly brings the reaction to an end.

Chlorine. A copper chloride flame test is obtained from a mixture of a chloride with CuO, page 213. When $AgNO_3$ is added to a dilute HNO_3 solution of a chloride, a white precipitate of AgCl forms. The precipitate is soluble in NH_4OH.

Chromium. The green bead tests for chromium are characteristic, page 214.

Cobalt. Cobalt minerals yield a distinctive dark blue color to the borax bead, page 214.

Copper. When a copper mineral is moistened with HCl and heated, it yields an intense blue flame tinged with green. Copper sulfides should first be roasted before moistening with acid, page 213. Copper can be reduced from some mineral on charcoal, page 207.

---

[2] George J. Brush and Samuel L. Penfield, *The Manual of Determinative Mineralogy with an Introduction on Blowpipe Analysis*, 16th ed., John Wiley and Sons, New York, 1926.

Fluorine.   Fluorine can be detected by converting it to HF and observing the etching effects on glass as follows: Cover a watch glass with paraffin and then remove the coating in places. Place the powdered mineral in the watch glass with a few drops of concentrated H_2SO_4. If the mineral is a soluble fluoride, HF will be liberated and etch the glass where it is exposed.

Gold.   Gold can be reduced from a telluride mineral on charcoal, page 207.

Iron.   Minerals containing an appreciable amount of iron become magnetic when heated in the reducing flame. Cobalt and nickel minerals react similarly. A reddish-brown flocculent precipitate of $Fe(OH)_3$ forms when NH_4OH is added to an acid solution of an iron mineral.

Lead.   Lead minerals when powdered, mixed with reducing mixture and heated on charcoal, yield a metallic globule, page 207. A coating of PbO forms on the charcoal, yellow near the assay and white farther away, page 208.

Lithium.   Lithium imparts a strong crimson color to the flame, page 213.

Magnesium.   The only satisfactory test for magnesium is to precipitate it as NH_4MgPO_4. This is done by adding $Na_2HPO_4 \cdot 12H_2O$ to a strongly ammoniacal solution of the mineral.

Manganese.   Manganese gives a bluish-green color to a Na_2CO_3 bead. This is a more sensitive test than the borax bead test, page 214.

Mercury.   A mercury mineral mixed with anhydrous Na_2CO_3 and heated in the closed tube yields a deposit of metallic mercury on the upper part of the tube, page 211.

Molybdenum.   A white sublimate of MoO_3 forms when an unoxidized molybdenum mineral is heated by the oxidizing flame on plaster. Touched with the reducing flame, the sublimate turns a deep ultramarine blue, page 209. In the open tube a network of MoO_3 crystals form near the assay, page 210.

Nickel.   When an alcohol solution of dimethylglyoxine is added to an ammoniacal solution containing nickel, a scarlet precipitate forms. This is a very sensitive test. Nickel gives a brownish color to the borax bead in the oxidizing flame and an opaque gray in the reducing flame, page 214.

Niobium.   To test for niobium: Fuse the powdered mineral on charcoal with anhydrous Na_2CO_3, dissolve the resulting mass in HCl, add a few grains of metallic tin or zinc and boil. If niobium is present, the solution turns dark blue then brown but becomes colorless on dilution with water.

Phosphorus.   To test for phosphorus: Dissolve the mineral in nitric acid and add a few drops of the solution to an excess of ammonium molybdate solution. A yellow precipitate of ammonium phosphomolybdate indicates phosphorus. Phosphates give a bluish-green flame color, page 213.

Silicon.   When a powdered silicate mineral is heated in a salt of phosphorus bead, the bases are dissolved, leaving the silica as a translucent skeleton.

Silver.   When HCl is added to a nitric acid solution of a silver mineral, white curdy AgCl precipitates. This is a very sensitive test. Many silver minerals can be reduced to silver by heating them with reducing mixture on charcoal. The resulting globules are white and malleable, page 207.

Sodium. Sodium minerals give a strong persistent yellow flame, page 213.

Strontium. Strontium minerals give a strong persistent crimson flame, page 213.

Sulfur. Sulfides heated in the open tube give off SO_2 that may be detected by its pungent odor. Moistened blue litmus paper inserted into the upper end of the tube turns red, page 210. Sulfur in sulfates can be detected as follows: Fuse the mineral with reducing mixture on charcoal; place the fused mass on a clean silver surface with a drop of water. A dark brown stain of silver sulfide results.

Tellurium. A tellurium mineral heated on charcoal gives a white volatile sublimate, page 208.

Tin. Fragments of tin minerals placed in dilute HCl with metallic zinc and warmed become coated with metallic tin. Tin can be reduced from its minerals by heating with reducing mixture on charcoal, page 207.

Titanium. To test for titanium: Fuse the mineral with anhydrous sodium carbonate, and dissolve the fused mass in equal amounts of concentrated H_2SO_4 and water. The cold solution turns yellow on addition of hydrogen peroxide.

Tungsten. To test for tungsten, powder the mineral and fuse with anhydrous sodium carbonate. When the fused mass is dissolved in HCl, a yellow precipitate, WO_3, forms. The solution turns blue when boiled with grains of metallic tin or zinc. If the solution is diluted, it remains blue, whereas in the similar test for niobium dilution renders the solution colorless.

Uranium. Uranium colors the salt of phosphorus bead yellow-green in the oxidizing flame, page 214. This bead will fluoresce yellow in ultraviolet light.

Vanadium. The amber color of the salt of phosphorus bead in the oxidizing flame is characteristic, page 214.

Zinc. When fused with reducing mixture on charcoal, zinc minerals yield a nonvolatile coating, yellow when hot, white when cold. The coating, moistened with cobalt nitrate and heated, turns green.

Zirconium. There is no simple test for zirconium.

REFERENCES: Crystal Chemistry

Bragg, W. L., and G. F. Claringbull, *Crystal Structures of Minerals.* Cornell University Press, New York, 1965.

Bunn, C. W., *Chemical Crystallography*, 2nd ed. Clarendon Press, Oxford, 1963.

Evans, R. C., *An Introduction to Crystal Chemistry*, 2nd ed. The University Press, Oxford, 1966.

Fyfe, W. S., *Geochemistry of Solids.* McGraw-Hill, New York, 1963.

Mason, B., *Principles of Geochemistry*, 3rd ed. John Wiley and Sons, New York, 1966.

Pauling, L., *The Nature of the Chemical Bond.* Cornell University Press, Ithaca, 1948.

Strunz, H., *Mineralogische Tabellen*, 4th ed. Akademische Verlagsgesellschaft, Leipzig, 1966.

5

DESCRIPTIVE
MINERALOGY

INTRODUCTION

This section includes descriptions of all the common minerals and those rarer ones of most economic importance. The list, numbering about 200, is relatively small, since nearly 2000 minerals are recognized as valid species. The names and chemical compositions of some other minerals are given.

The descriptions of all minerals follow a common scheme of presentation. The headings used and the data given under each are as follows:

Crystallography. Under this heading are given the following crystallographic information:

The crystal system and symbol of the crystal class.

Crystallographic features usually observed by inspection, such as habit, twinning and crystal forms.

"*Angles*," interfacial angles for those minerals frequently found in well developed crystals.

Other pertinent data are listed without subheadings.

As an example consider the following data for cerussite:

Pmcn (space group), $a = 5.15$, $b = 8.47$, $c = 6.11$ Å (unit cell dimensions),

$a:b.c = 0.608:1:0.721$ (axial ratios),[1] $Z = 4$ (formula weights in the unit cell). *d's:* 3.59(10), 3.50(4), 3.07(2), 2.49(3), 2.08(3) (strongest x-ray lines in angstrom units with relative intensities in parentheses).

Physical Properties. The physical properties: *Cleavage*, **H** (hardness), **G** (specific gravity), *Luster, Color*, and *Optics:* (brief summary of optical data) are listed.

Composition. Chemical composition; frequently with the percentage of elements or oxides and the elements that may substitute for those given in the chemical formula.

Diagnostic Features. The outstanding properties and tests that aid one in recognizing the mineral and distinguishing it from others.

Occurrence. A brief statement of the mode of occurrence and characteristic mineral associations are given. The localities where a mineral is or has been found is notable amount or quality are mentioned. For minerals found abundantly the world over, emphasis is on American localities.

Use. For a mineral of economic value, there is a brief statement of its uses.

Similar Species. The similarity of the species listed to the mineral whose description precedes may be either on the basis of chemical composition or crystal structure. Consequently, the names of these minerals may not appear in the proper places in the scheme of classification followed in this book.

Mineral Classification

Minerals have been classified in several ways but classification on the basis of chemical composition has nearly universal acceptance and is used in this book.[2] The broadest divisions of the classification are:

1. Native elements
2. Sulfides
3. Sulfosalts
4. Oxides
   a. Simple and multiple
   b. Hydroxides
5. Halides
6. Carbonates
7. Nitrates
8. Borates
9. Phosphates
10. Sulfates
11. Tungstates
12. Silicates

[1] The axial ratios are derived from the cell dimensions and thus may be slightly inconsistent with the interfacial angles obtained from morphological measurements (see p. 116).

[2] The classification in this book is based upon that used in the following works:

a. C. Palache, H. Berman, C. Frondel, *Dana's System of Mineralogy*, 7th ed., Vol. I (1944), Vol. II (1952). John Wiley and Sons, New York. A critical discussion of all the minerals with the exception of the silicates.

b. H. Strunz, *Mineralogische Tabellen*, 4th ed., Akademische Verlags., Leipzig, 1966. A listing and classification on crystal-chemical grounds of all the minerals and a brief summary of properties.

The above classes are subdivided into *families* on the basis of chemical types, and the family in turn may be divided into *groups* which show a close crystallographic and structural similarity. A group is made up of *species*, which may form *series* with each other, and finally a species may have several *varieties*. In each of the classes, the mineral with the highest ratio of metal to nonmetal is given first, followed by those containing progressively less metal. Such a relatively small number of minerals are described in this book that often only one member of a group or family is represented, and thus a rigorous adherence to division and subdivision is impractical.

As an introduction to each of the chemical classes a few prefatory remarks are made regarding the crystal chemistry of the class. These discussions are in no way complete but are given as a basis for understanding the reasons for the similarities and differences of members of the class.

NATIVE ELEMENTS

With the exception of the free gases of the atmosphere, only about twenty elements are found in the native state. These elements can be divided into (1) metals; (2) semimetals; (3) nonmetals. The more common native metals constitute three isostructural groups: the *gold group*, gold, silver, copper, and lead; the *platinum group*, platinum, palladium, iridium, and osmium; and the *iron group*, iron and nickel-iron. In addition, mercury, tantalum, tin, and zinc have been found. The native semimetals form two isostructural groups: arsenic, antimony, and bismuth crystallize in the hexagonal-scalenohedral class, and the less common selenium and tellurium crystallize in the trigonal-trapezohedral class. The important nonmetals are carbon, in the form of diamond and graphite, and sulfur.

Native Elements

Metals		Semimetals	
Gold Group		Arsenic Group	
Gold	Au	Arsenic	As
Silver	Ag	Bismuth	Bi
Copper	Cu	Nonmetals	
Platinum Group		Sulfur	S
Platinum	Pt	Diamond	C
Iron Group		Graphite	C
Iron	Fe		

Native Metals

It is fitting that *descriptive mineralogy* begin with a discussion of the gold group, for man's knowledge of the properties and usefulness of metals arose

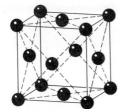

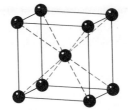

FIG. 277. Face-centered cubic lattice.

FIG. 278. Body-centered cubic lattice.

from the chance discovery of nuggets and masses of these minerals. Many relatively advanced early cultures, such as the Amerindian, were restricted in their use of metal to that found in the native state.

The elements of the gold group belong to the same family in the periodic classification of the elements and, hence, their atoms have somewhat similar chemical properties; all are sufficiently inert to occur free in nature. When uncombined with other elements, the atoms of these metals are united into crystal structures by the rather weak metallic bond. The minerals are isostructural and are built on the face-centered cubic lattice with identical atoms in 12 coordination. (See Figs. 277 and 279.)

FIG. 279. Copper, Cu, packing model. The photograph shows copper atoms in 12 coordination (cubic closest packing) arranged in a face-centered cubic lattice. This structure is shared by silver, gold, and many other metals.

The similar properties of the members of this group arise from the common structure. All are rather soft, malleable, ductile, and sectile. All are excellent conductors of heat and electricity, display metallic luster and hackly fracture, and have rather low melting points. These are properties conferred by the metallic bond. All are isometric hexoctahedral and have unusually high densities. These properties arise from the very economical packing, called *cubic closest packing*.

Those properties with respect to which the minerals of this group differ arise from the properties of the atoms of the individual elements. Thus, the yellow of gold, the red of copper, and the white of silver are atomic properties. The specific gravities likewise depend on atomic properties and show a rough proportionality to the atomic weights.

The members of this group are isostructural, and the dominant factor controlling solid solution between them is atomic radius. Thus, gold and silver, which have the same atomic radii (1.44 Å), display complete mutual solid solubility but usually contain only small amounts of copper (atomic radius 1.28 Å). Conversely, native copper carries only traces of gold and silver in solid solution.

Although only platinum is discussed here, the Platinum Group also includes the rarer minerals palladium, platiniridium, and iridosmine. The last two are respectively alloys of iridium and platinum and iridium and osmium. Their structure is hexagonal close-packed; whereas the structure of platinum and paladium is cubic close-packed, similar to metals of the Gold Group. The platinum metals are harder and have higher melting points than metals of the gold group.

Crystals of the iron group metals, although isometric, have a body-centered cubic lattice in which each atom touches eight others. Their structure is thus similar to that of cesium chloride (Fig. 278). Iron and nickel both have atomic radii of 1.24 Å, and thus nickel can and usually does substitute for some of the iron. The natural solid solution, nickel-iron, is particularly characteristic of iron meteorites and is believed to constitute a large part of the earth's core.

GOLD—Au

Crystallography. Isometric; $4/m\bar{3}2/m$. Crystals are commonly octahedral (Fig. 280), rarely showing the faces of the dodecahedron, cube, and trapezohedron $\{113\}$. Often in arborescent crystal groups with crystals elongated in the direction of a 3-fold symmetry axis, or flattened parallel to an octahedron face. Crystals are irregularly formed; passing into filiform, reticulated, and dendritic shapes (Fig. 281). Seldom shows crystal forms; usually in irregular plates, scales, or masses.

$Fm3m;\ a = 4.079\ \text{Å};\ Z = 4.\ d's:\ 2.36(10),\ 2.04(7),\ 1.443(6),\ 1.229(8),\ 0.785(5).$

FIG. 280.   Malformed gold octahedron.

Physical Properties.   H $2\frac{1}{2}$–3. **G** 19.3 when pure. The presence of other metals decreases the specific gravity, which may be as low as 15. *Fracture* hackley. Very malleable and ductile. Opaque. *Color* various shades of yellow depending on the purity, becoming paler with increase of silver.

Composition.   A complete solid-solution series exists between gold and silver, and most gold contains some silver. California gold carries 10 to 15 per cent silver. When silver is present in amounts greater than 20 per cent, the alloy is known as *electrum*. Small amounts of copper and iron may be present as well as traces of bismuth, lead, tin, zinc, and the platinum metals. The purity or *fineness* of gold is expressed in parts per 1000. Most gold contains about 10 per cent of other metals and thus has a fineness of 900.

Diagnostic Features.   Gold is distinguished from the yellow sulfides pyrite and chalcopyrite and from yellow flakes of altered micas by its sectility and its high specific gravity. It fuses at 3 (1063°C) and is soluble only in aqua regia.

Occurrence.   Although gold is a rare element, it occurs in nature widely distributed in small amounts. It is found most commonly in veins that bear a genetic relation to silicic types of igneous rocks. In places it has been found intimately associated with igneous rocks as in the Bushveld igneous complex in the Transvaal, South Africa. Most gold occurs as the native metal; tellurium and possibly selenium are the only elements that are combined with it in nature.

The chief source of gold are hydrothermal gold-quartz veins where, together with pyrite and other sulfides, gold was deposited from ascending mineral-bearing solutions. The gold is merely mechanically mixed with the sulfides and

FIG. 281.   Dendritic gold.

is not in chemical combination. At and near the surface of the earth oxidation of the gold-bearing sulfides sets the gold free making its extraction easy. Such ores are called "free-milling" because their gold content can be recovered by amalgamation. Finely crushed ore is washed over copper plates coated with mercury. When sulfides are present in any quantity, not all the gold can be recovered by amalgamation, and either the cyanide or chlorination process must be used. In the cyanide process finely crushed ore is treated with a solution of potassium or sodium cyanide, forming a soluble cyanide. The gold is then recovered by precipitation with zinc or by electrolysis. The chlorination process renders the gold in a soluble form by treating the crushed and roasted ore with chlorine. In the majority of veins the gold is so finely divided and uniformly distributed that its presence in the ore cannot be detected with the eye. With the chemical processes of gold extraction, ores carrying values as low as $1.00 per ton can be worked at a profit. It is interesting to note that, with the value of gold at $35 per troy ounce, such ore would contain about 0.0001 per cent gold by weight.

When gold-bearing veins are weathered, the gold liberated either remains in the soil mantle as an eluvial deposit, or is washed into the neighboring streams to form a placer deposit. Because of its high specific gravity, gold works its way through the lighter sands and gravels to lodge behind irregularities or to be caught in crevices in the bedrock. The rounded or flattened nuggets of placer gold can be recovered by panning, a process of washing away all but the heavy concentrate from which the gold can be easily separated. On a larger scale, gold-bearing sand is washed through sluices where the gold collects behind cross-bars or riffles and amalgamates with mercury placed behind the riffles. Most placer mining is today carried on with dredges, some of which are gigantic and can extract the gold from thousands of cubic yards of gravel a day. Some dredges can operate at a profit handling gravels which average no more than 10 cents in gold per cubic yard.

Of the estimated world production of gold in 1967 of 45,600,000 ounces, approximately two-thirds come from the Republic of South Africa. The principal sources of South African gold are the Precambrian Witwatersrand conglomerate "the Rand," in the Transvaal and similar conglomerates in the Orange Free State. The reefs in these conglomerates in which the gold is concentrated are thought to be fossil placers. Although no accurate figures are available, the U.S.S.R. is believed to be the second-ranking country in gold production. Gold is mined in the Ural Mountains but it is estimated that two-thirds of the production is from placers, mostly in Siberia. Canada ranks third in gold production followed by the United States, Ghana, Australia, and the Philippines.

The most important gold-producing states of the United States, in order of the amounts of gold produced at present are: South Dakota, Nevada, Utah, Arizona, and California. The Homestake mine, South Dakota, and the Carlin mine,

Nevada account for over 60 per cent of the country's production. The gold from Arizona and Utah is essentially a biproduct of copper mining and that from California is largely from placer operations. For 100 years following the discovery of gold in the streams of California in 1848, that state was the leading producer. The most important gold-producing districts were those of the Mother Lode, gold-quartz veins along the western slope of the Sierra Nevada.

Use. The principal use of gold is as a monetary standard. Other uses are in jewelry, scientific instruments, electroplating, gold leaf, and dental appliances.

SILVER—Ag

Crystallography. Isometric; $4/m\bar{3}2/m$. Crystals are commonly malformed and in branching, arborescent, or reticulated groups. Found usually in irregular masses, plates, and scales; in places as coarse or fine wire. (Fig. 282).

$Fm3m$; $a = 4.09$ Å; $Z = 4$. d's: 2.34(10), 1.439(6), 1.228(8), 0.936(7), 0.934(8).

Physical Properties. H $2\frac{1}{2}$–3. G 10.5 when pure, 10–12 when impure. *Fracture* hackly. Malleable and ductile. *Luster* metallic. *Color* and *streak* silver-white, often tarnished to brown or gray-black.

Composition. Native silver frequently contains alloyed gold, mercury, and copper, more rarely traces of platinum, antimony, and bismuth. *Amalgam* is a solid solution of silver and mercury.

FIG. 282. Silver, Chañarcillo, Chile.

Diagnostic Features. Silver can be distinguished by its malleability, color on a fresh surface, and high specific gravity. Fusible at 2 (960°C) to bright globule which is soluble in nitric acid and gives on addition of HCl a curdy white precipitate of silver chloride.

Occurrence. Native silver is widely distributed in small amounts in the oxidized zone of ore deposits. However, native silver in larger deposits is the result of deposition from primary hydrothermal solutions. There are three types of primary deposits: (1) Associated with sulfides, zeolites calcite, barite, fluorite, and quartz as typified by the occurrence in Kongsberg, Norway. Mines at this locality were worked for several hundred years and produced magnificent specimens of crystallized wire silver. (2) With arsenides and sulfides of cobalt, nickel and silver and with native bismuth. Such was the type at the old silver mines at Freiberg and Schneeberg in Saxony and at Cobalt, Ontario. (3) With uraninite and cobalt-nickel minerals. Deposits at the old but still active mines of Joachimsthal, Bohemia and at Great Bear Lake, Northwest Territories are of this type.

In the United States small amounts of primary native silver are associated with native copper on the Keweenaw Peninsula, Michigan. Elsewhere in the United States native silver is largely secondary.

Use. An ore of silver, although most of the world's supply comes from other minerals. Silver has many uses chief of which are in photographic film emulsions, plating, brazing alloys, tableware, and electronic equipment. Because of declining production and increased price, silver in coinage is being replaced by other metals.

COPPER—Cu

Crystallography. Isometric; $4/m\bar{3}2/m$. Tetrahexahedron faces are common, as well as the cube, dodecahedron, and octahedron. Crystals are usually malformed and in branching and arborescent groups (Figs. 283 and 284). Usually occurs in irregular masses, plates, and scales and in twisted and wirelike forms.

$Fm3m$; $a = 3.615$ Å, $Z = 4$, $d's$: 2.09(10), 1.81(10), 1.28(10), 1.09(10), 1.05(8).

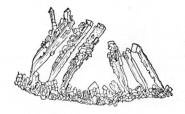

FIG. 283.   Dendritic copper.

FIG. 284. Native copper, Keweenaw Peninsula, Michigan.

Physical Properties. H $2\frac{1}{2}$–3. G 8.9. Highly ductile and malleable. *Fracture* hackly. *Luster* metallic. *Color* copper-red on fresh surface, usually dark with dull luster because of tarnish.

Composition. Native copper often contains small amounts of silver, bismuth, mercury, arsenic, antimony.

Diagnostic Features. Native copper can be recognized by its red color on fresh surfaces, hackly fracture, high specific gravity, and malleability. Fuses at 3 (1083°C) to a globule. It dissolves readily in nitric acid, and the solution is colored a deep blue on addition of an excess of ammonium hydroxide.

Occurrence. Small amounts of native copper have been found at many localities in the oxidized zones of copper deposits associated with cuprite, malachite, and azurite.

Most primary deposits of native copper are associated with basaltic lavas, where deposition of copper resulted from the reaction of hydrothermal solutions with iron minerals. The only major deposit of this type is on the Keweenaw Peninsula, Michigan, on the southern shore of Lake Superior. The rocks of the area are composed of Precambrian basic lava flows interbedded with conglomerates. The series, exposed for 100 miles along the length of the peninsula, dips to the northwest beneath Lake Superior to emerge on Isle Royale 50 miles away. Some copper is found in veins cutting these rocks, but it occurs principally either in the lava flows filling cavities or in the conglomerates filling interstices or replacing pebbles. Associated minerals are prehnite, datolite, epidote, calcite, and zeolites. Small amounts of native silver are present. Michigan copper had long been used by the American Indians but it was not until 1840 that its source was located. Exploitation began shortly thereafter and during the next 75 years there was active mining throughout the length of the Keweenaw peninsula. Most of the copper in the district was in small irregular grains, but notable large masses were found; one found in 1857 weighed 420 tons.

Sporadic occurrences of copper similar to the Lake Superior district have been found in the sandstone areas of the eastern United States, notably in New Jersey, and in the glacial drift overlying a similar area in Connecticut. In Bolivia at Corocoro, southwest of La Paz, there is a noted occurrence in sandstone. Native copper occurs in small amounts associated with the oxidized copper ores of Arizona, New Mexico, and northern Mexico.

Use. A minor ore of copper; copper sulfides are today the principal ores of the metal. The greatest use of copper is for electrical purposes, mostly as wire. It is also extensively used in alloys, such as brass (copper and zinc), bronze (copper and tin with some zinc), and German silver (copper, zinc, and nickel). These and many other minor uses make copper second only to iron as a metal essential to modern civilization.

PLATINUM—Pt

Crystallography. Isometric; $4/m\overline{3}2/m$. Cubic crystals are rare and commonly malformed. Usually found in small grains and scales. In some places it occurs in irregular masses and nuggets of larger size.

$Fm3m$; $a = 3.923$ Å, $Z = 4$. $d's$: 2.27(9), 1.180(10), 1.956(8), 1.384(8).

Physical Properties. H $4-4\frac{1}{2}$. (Unusually high for a metal.) G 21.45 when pure; 14–19 when native. Malleable and ductile. *Color* steel-gray, with bright luster. Magnetic when rich in iron.

Composition. Platinum is usually alloyed with several per cent iron and with smaller amounts of iridium, osmium, rhodium, palladium; also copper, gold, nickel.

Diagnostic Features. Determined by its high specific gravity, malleability, infusibility in the blowpipe flame, and insolubility except in aqua regia.

Occurrence. Most platinum occurs as the native metal in ultra-basic rocks, especially dunites, associated with olivine, chromite, pyroxene, and magnetite. It has been mined extensively in placers which are usually close to the platinum-bearing igneous rock.

Platinum was first discovered in the United States of Colombia, South America. It was taken to Europe in 1735 where it received the name *platina* from the word *plata* (Spanish for silver) because of its resemblance to silver. A small amount of platinum is still produced in Colombia from placers in two districts near the Pacific Coast. In 1822 platinum was discovered in placers on the Upper Tura River on the eastern slope of the Ural Mountains, U.S.S.R. From that time until 1934 most of the world's supply of platinum came from placers of that district, which centered around the town of Nizhne Tagil. In 1934, because of the large amount of platinum recovered from the copper-nickel ore of Sudbury, Ontario, Canada became the leading producer. At Sudbury platinum occurs in sperrylite, $PtAs_2$. In 1954 the Republic of South Africa moved into first place. Part of the South African production is as a by-product from gold mining on the Rand, but the chief source is the Merensky Reef in the ultrabasic rocks of the Bushveld igneous complex. The Merensky Reef is a horizon of this layered intrusive about 12 inches thick extending for many miles with a uniform platinum content of about one-half ounce per ton of ore. The chief mining operations are near Rustenburg in the Transvaal. Although Canadian and South African production have steadily increased, Russia once again took the lead in 1963. It is now estimated that over 50 per cent of the world's platinum metals comes from this country. In smaller amounts platinum has come from Borneo, New South Wales, New Zealand, Brazil, Peru, and Madagascar.

In the United States small amounts of platinum have been recovered from the gold-bearing sands of North Carolina and from black-sand placers in California, Oregon, and Alaska.

Use. The uses of platinum depend chiefly upon its high melting point ($1755°C$), resistance to chemical attack, and superior hardness. It is used as a catalyst in the chemical and petroleum industry, chemical apparatus, electrical equipment, and jewelry. It is also used in dentistry, surgical instruments, pyrometry, and photography.

Similar species. *Iridium; iridosmine*, an alloy of iridium and osmium; and *palladium* are rare minerals of the platinum group associated with platinum.

Iron—Fe

Crystallography. Isometric; $4/m\bar{3}2/m$. Crystals are rare. Terrestrial: in blebs and large masses; meteoric: in plates and lamellar masses, and in regular intergrowth with nickel-iron.

Ordinary iron (α-Fe) body-centered cubic, *Im3m*, $a = 2.86$ Å, $Z = 2$. *d's:* 2.02(9), 1.168(10), 1.430(7), 1.012(7). Nickel-iron, face-centered cubic, *Fm3m*, $a = 3.56$ Å, $Z = 4$. *d's:* 2.06(10), 1.073(4), 1.78(3), 1.27(2).

Physical Properties. (α-Fe). *Cleavage* {010} poor. **H** $4\frac{1}{2}$. **G** 7.3–7.9. *Fracture* hackly. Malleable. Opaque. *Luster* metallic. *Color* steel-gray to black. Strongly magnetic.

Composition. (α-Fe) always with some nickel and frequently small amounts of cobalt, copper, manganese, sulfur, carbon. *Nickel-iron*, (Ni,Fe), contains up to 76 per cent nickel.

Diagnostic Features. Iron can be recognized by its strong magnetism, its malleability, and the oxide coating usually on its surface. It is infusible but soluble in HCl.

Occurrence. Occurs sparingly as terrestrial iron and in meteorites. Terrestrial iron is regarded as a primary magmatic constituent or a secondary product formed by the reduction of iron compounds by assimilated carbonaceous material. The most important locality is Disko Island, Greenland where fragments ranging from small grains to masses of many tons are included in basalts. Masses of nickel-iron have been found in Josephine and Jackson counties, Oregon.

Iron meteorites are composed largely of α-Fe, *kamacite*, containing about 7 per cent nickel and nickel-iron, *taenite*, with 30 to 75 per cent nickel. When polished sections of iron meteorites are etched, the regular intergrowth of these two phases is shown by the Widmanstatten pattern.

Native Semimetals

The native semimetals form an isostructural group with similar properties. They belong to the hexagonal-scalenohedral class and have good basal cleavage. They are rather brittle and much poorer conductors of heat and electricity than the native metals. These properties reflect a bond type that is intermediate between true metallic and covalent; hence it is stronger and more directive in its properties, leading to lower symmetry. The structure may be pictured as a modified type of 6 coordination in which each atom is more strongly bonded to three of its neighbors than to the other three (Fig. 285). The structure is thus made up of sheets of atoms

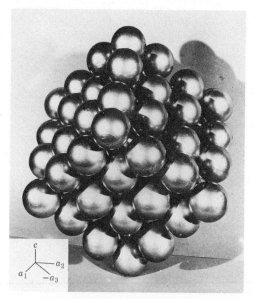

FIG. 285. Bismuth, Bi, packing model. The model shows the layered nature of the structure resulting from the fact that each atom is somewhat closer to three of its neighbors than to the other three.

strongly bonded to one adjacent sheet and weakly bonded to the other. The sheets are parallel to the base, and the weak bonding between sets of double sheets gives rise to the cleavage.

Antimony belongs in the Arsenic Group. It also is found in the native state but is less common than arsenic and bismuth.

Arsenic—As

Crystallography. Hexagonal-R; $\bar{3}2/m$. Pseudocubic crystals rare. Usually granular massive, reniform, and stalactitic.

Physical Properties. *Cleavage* $\{0001\}$ perfect. **H** $3\frac{1}{2}$. **G** 5.7. *Luster* nearly metallic on fresh surface. *Color* tin-white on fresh fracture, tarnishes to dark gray on exposure. Streak gray. Brittle.

$R\bar{3}m$; $a = 3.76$, $c = 10.55$ Å; $a:c = 1:2.805$, $Z = 6$. *d's:* 3.52(4), 2.77(10), 2.05(6), 1.88(7), 1.768(4).

Composition. Native arsenic often contains antimony and traces of iron, silver, gold, and bismuth.

Diagnostic Features. Before the blowpipe on charcoal it volatilizes without fusing and gives a white volatile coating of arsenious oxide and the odor of garlic.

Occurrence. Arsenic is a comparatively rare mineral found in veins in crystalline rocks associated with silver, cobalt, or nickel ores. Found in the silver mines of Freiberg, in Saxony; at Andreasberg in the Harz Mountains; in Bohemia; Rumania; and Alsace. Found sparingly in the United States.

Use. Very minor ore of arsenic; most arsenic of commerce is obtained from other arsenic-bearing minerals. (See arsenopyrite, page 266).

Name. The name arsenic is derived from a Greek word meaning *masculine* from the belief that metals were of different sexes. The term was first applied to the sulfide of arsenic because of its potent properties.

Similar Species. Allemontite, AsSb, an intermediate compound of arsenic and antimony.

Bismuth—Bi

Crystallography. Hexagonal-R; $\bar{3}2/m$. Distinct crystals are rare. Usually laminated and granular; may be reticulated or arborescent. Artificial crystals pseudocubic $\{01\bar{1}2\}$.

$R\bar{3}m$; $a = 4.54$, $c = 11.85$ Å; $a:c = 1:2.609$, $Z = 6$. *d's:* 3.28(10), 2.35(5), 2.27(5), 1.86(3), 1.44(3).

Physical Properties. *Cleavage* $\{0001\}$ perfect. **H** $2–2\frac{1}{2}$. **G** 9.8. Sectile. Brittle. *Luster* metallic. *Color* silver-white with a decided reddish tone. *Streak* silver-white, shining.

Composition. Small amounts of arsenic, sulfur, tellurium, antimony may be present.

Diagnostic Features. Bismuth is recognized chiefly by its laminated nature, its reddish-silver color, its perfect cleavage, and its sectility. Fusible at 1 (271°C). Before the blowpipe on charcoal it gives a brittle metallic globule and yellow to white coating of bismuth oxide. When mixed with potassium iodide and sulfur and heated on charcoal it gives a brilliant yellow to red coating.

Occurrence. Bismuth is a comparatively rare mineral, occurring with ores of silver, cobalt, nickel, lead, and tin. Found in the silver veins of Saxony; in Norway and Sweden; and at Cornwall, England. Important deposits are found in Australia, but the most productive deposits are in Bolivia. It is found in small veins associated with silver and cobalt minerals at Cobalt, Ontario, Canada.

Use. The chief ore of bismuth. Bismuth forms low-melting alloys with lead, tin, and cadmium, which are used for electric fuses and safety plugs in water sprinkling systems. About 75 per cent of the bismuth produced is for medicine and cosmetics. Bismuth nitrate is relatively opaque to x-rays and is taken internally when the digestive organs are to be photographed.

Name. Etymology in dispute; possibly from the Greek meaning *lead white*.

SULFUR—S

Crystallography. Orthorhombic; $2/m2/m2/m$. Pyramidal habit common (Fig. 286a), often with two dipyramids, first-order prism and base in combination (Fig. 286b). Commonly found in irregular masses imperfectly crystallized. Also massive, reniform, stalactitic, as incrustations, earthy.

Angles: $c(001) \wedge n(011) = 62° 18'$, $c(001) \wedge p(111) = 70° 40'$, $c(001) \wedge s(113) = 45° 10'$, $p(111) \wedge p(1\bar{1}1) = 73° 34'$.

Fddd; $a = 10.45$, $b = 12.84$, $c = 24.37$ Å; $a:b:c = 0.814:1:1.898$; $Z = 128$. *d's:* 7.76(4), 5.75(5), 3.90(10), 3.48(4), 3.24(6).

There are three polymorphic forms of sulfur. The ordinary natural sulfur is orthorhombic; the other two are monoclinic and very rare as minerals.

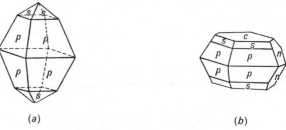

(a)   (b)

FIG. 286.   Sulfur crystals.

Physical Properties. *Fracture* conchoidal to uneven. Brittle. **H** $1\frac{1}{2}$–$2\frac{1}{2}$. **G** 2.05–2.09. *Luster* resinous. *Color* sulfur-yellow, varying with impurities to yellow shades of green, gray, and red. Transparent to translucent. *Optics:* (+); $\alpha = 1.957$, $\beta = 2.037$, $\gamma = 2.245$, 2V = 69°.

Sulfur is a poor conductor of heat. When a crystal is held in the hand close to the ear it will be heard to crack. This is due to the expansion of the surface layers because of the heat from the hand, while the interior, on account of the slow heat conductivity, is unaffected. Crystals of sulfur should, therefore, be handled with care.

Composition. Native sulfur may contain small amounts of selenium.

Diagnostic Features. Sulfur can be told by its yellow color and the ease with which it burns. The absence of a good cleavage distinguishes it from orpiment. Fusible at 1 (112.8°C) and burns with a blue flame to sulfur dioxide. Sublimates in closed tube, giving a red to dark yellow liquid when hot, and a yellow solid when cold.

Occurrence. Sulfur often occurs at or near the crater rims of active or extinct volcanoes where it has been derived from the gases given off in fumaroles. These may furnish sulfur as a direct sublimation product or by the incomplete oxidation of hydrogen sulfide gas. It is also formed from sulfates, by the action of sulfur-forming bacteria. Sulfur may be found in veins associated with metallic sulfides and formed by the oxidation of the sulfides. It is most commonly found in the Tertiary sedimentary rocks and most frequently associated with gypsum and limestone; often in clay rocks; frequently with bituminous deposits. The large deposits near Girgenti, Sicily, are noteworthy for the fine crystals associated with celestite, gypsum, calcite, aragonite. Sulfur is also found associated with the volcanoes of Mexico, Hawaii, Japan, Argentina, and at Ollague, Chile, where it is mined at an elevation of 19,000 feet.

In the United States the most productive deposits are in Texas and Louisiana where sulfur is associated with anhydrite, gypsum, and calcite in the cap rock of salt domes. There are ten or twelve producing localities, but the largest are Boling Dome in Texas and Grand Ecaille, Plaquemines Parish, Louisiana. Sulfur is also recovered from salt domes in Mexico and offshore in the shallow waters of the Gulf of Mexico. Sulfur is obtained from these deposits by the Frasch method. Superheated water is pumped down to the sulfur horizon, where it melts the sulfur; compressed air then forces the sulfur to the surface.

About half of the world's sulfur is produced as the native element; the remainder is recovered as a by-product in smelting sulfide ores, from sour natural gas and from pyrite.

Use. Sulfur is used in the chemical industry chiefly in the manufacture of sulfuric acid. It is also used in fertilizers, insecticides, explosives, coal-tar products, rubber, and in the preparation of wood pulp for paper manufacture.

Diamond and Graphite

Diamond and graphite, both elemental carbon, afford the most spectacular example of polymorphism. Both may be wholly burned to carbon dioxide at sufficiently high temperatures. Despite this chemical identity, diamond and graphite differ in almost every other regard, and the differences are directly traceable to the structures. Diamond has an exceptionally close-knit and strongly bonded structure in which each carbon atom is bound by powerful and highly directive covalent bonds to four carbon neighbors at the apices of a regular tetrahedron. The resulting structure, illustrated in Fig. 287, although strongly bonded, is not close packed, and only 34 per cent of the available space is filled. The presence in the structure of rather widely spaced sheets of carbon atoms parallel to the (111) planes may be observed in Fig. 287. These sheets are the planes of maximum atomic population and account for the prominent {111} cleavage of diamond.

The structure of graphite, illustrated in Fig. 288, consists of sheets of six-membered rings in which each carbon atom has three near neighbors arranged at the apices of an equilateral triangle. Three of the four valence electrons in each carbon atom may be considered to be locked up in tight covalent bonds with its three close neighbors in the plane of the sheet. The fourth is free to wander over the surface of the sheet, creating a dispersed electrical charge which bestows on graphite its relatively high electrical conductivity. In contrast, diamond, in which all four valence electrons are locked up in covalent bonds, is among the best of electrical insulators.

The sheets composing the graphite crystal are stacked in such a way that alternate sheets are in identical position, with the intervening sheet translated a distance

FIG. 287. Diamond packing model.

FIG. 288. Graphite packing model.

of one-half the identity period in the plane of the sheets. The distance between sheets is much greater than one atomic diameter, and van der Waals' binding forces perpendicular to the sheets are very weak. This wide separation and weak binding give rise to the perfect basal cleavage and easy gliding parallel to the sheets. Because of this open structure only about 21 per cent of the available space in graphite is filled, and the specific gravity is proportionately less than that of diamond. The relation

$$\frac{\% \text{ filling of space in graphite}}{\% \text{ filling of space in diamond}} = \frac{\text{specific gravity of graphite}}{\text{specific gravity of diamond}}$$

is very nearly numerically exact, indicating that the carbon-carbon bonds in the two structures are closely similar.

The synthesis of diamond in 1955 was the realization of an age-old dream and established firmly the stability relations between the polymorphs of carbon over a wide range of pressure and temperature (Fig. 289). Diamond, as is to be expected from its high specific gravity and fairly close packing, is the high-pressure polymorph. At low pressures or temperatures it is unstable with respect to graphite and may be converted to graphite at moderate temperatures in sealed containers. The reason that diamond and graphite can coexist at room temperatures and pressures is because the reaction is very sluggish. In order to permit inversion of graphite to diamond, extremely high temperatures are needed to cause the carbon atoms in the graphite lattice to break loose by thermal agitation and to make them available for building the diamond lattice. Such temperatures also increase the pressure required to bring about the inversion. Therefore, diamonds were not

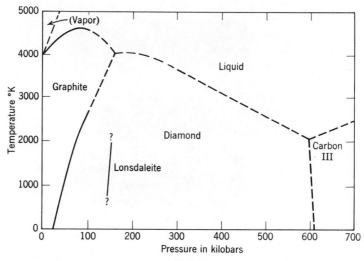

FIG. 289. Carbon phase diagram. Composite diagram based on experimental data of several workers.

synthesized until an apparatus could be built that would simultaneously exert a very high pressure and withstand the high temperature. Such a press was built by General Electric engineers, and the first diamonds were synthesized in 1955 using pressures of 600,000 to 1,500,000 pounds per square inch and temperatures of 750 to 2750°C!

The success of the diamond syntheses encouraged experimentation with boron nitride, BN, whose structure is similar to graphite. In 1956 a high-pressure, high-temperature dimorph of boron nitride having the structure and hardness of diamond was prepared. This compound, known by the trade name of *Borazon* is expected to be of great industrial use. Borazon is not a mineral, but its synthesis illustrates the application to industrial problems of crystal chemical concepts first evolved and tested in minerals.

After the initial synthesis of diamond, scientists at the General Electric Company continued experimentation. In 1967 they produced a hexagonal transparent polymorph of carbon with specific gravity and refractive indices close to those of diamond. Almost simultaneously the hexagonal diamond was found in the Canyon Diablo, Arizona iron meteorite. The name *lonsdaleite* is given to this naturally occurring mineral.

DIAMOND—C

Crystallography. Isometric; $4/m\bar{3}2/m$. Crystals usually octahedral but may be cubic or dodecahedral. Curved faces, especially of the hexoctahedron are frequently observed. Crystals may be flattened on {111}. Twins on {111} (spinel

(a)

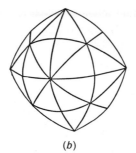
(b)

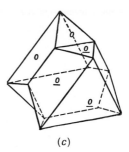
(c)

FIG. 290.   Diamond crystals.

law) are common, usually flattened parallel to the twin plane (Fig. 290c). *Bort*, a variety of diamond, has rounded forms and rough exterior resulting from a radial or cryptocrystalline aggregate. The term is also applied to badly colored or flawed diamonds without gem value.

Fd3m; a = 3.567 Å, *Z* = 8. *d's:* 2.06(10), 1.26(8), 1.072(7), 0.813(6), 0.721(9).

Physical Properties.   *Cleavage* {111} perfect. **H** 10 (hardest known mineral). **G** 3.51. *Luster* adamantine; uncut crystals have a characteristic greasy appearance. The very high refractive index, 2.42, and the strong dispersion of light account for the brilliancy and "fire" of the cut diamond. *Color* usually pale yellow or colorless; also pale shades of red, orange, green, blue, and brown. Deeper shades are rare. *Carbonado* or *carbon* is black or grayish black bort. It is noncleavable, opaque, and less brittle than crystals.

Composition.   Pure carbon.

Diagnostic Features.   Diamond is distinguished from similar-appearing minerals by its great hardness, adamantine luster, and cleavage. Insoluble in acids and alkalis. At a high temperature in oxygen will burn to CO_2 gas, leaving no ash.

Occurrence.   Diamonds have been discovered in many different localities but in only a few in notable amount. Most commonly the diamond is found in alluvial deposits, where it accumulates because of its inert chemical nature, its great hardness, and its fairly high specific gravity. In several countries in Africa and more recently in Siberia diamonds have been found *in situ*. The rock in which they occur is an altered peridotite called *kimberlite*. Although these intrusive bodies vary in size and shape, many are roughly circular with a pipe-like shape and are referred to as "diamond pipes." In the deepest mine, the Kimberley, South Africa, the diameter of the pipe decreased with depth and mining stopped at a depth of 3500 feet although the pipe continued. At the surface the kimberlite is weathered to a soft yellow rock, "yellow ground," that in depth gives way to a harder "blue ground." The ratio of diamonds to barren rock varies from one pipe to another. In the Kimberley mine it was 1:8,000,000 but in some it may be as high as 1:30,000,000.

Diamonds were first found in India which remained virtually their only source until they were discovered in Brazil in 1725. The early diamonds came from stream gravels in southern and central India, and it is estimated that twelve million carats were produced from this area. Today the Indian productions is only a few hundred carats a year.

Following tl.eir discovery in stream gravels in the state of Minas Gerais, Brazil, diamonds were also found in the states of Bahia, Goyaz, and Mato Grasso. Brazil today produces about 160,000 carats annually chiefly from the stream gravels near the city of Diamantina, Minas Gerais. Extensive upland deposits of diamond-bearing gravels and clays are also worked. The black carbonado comes only from Bahia.

In 1866 diamonds were discovered in the gravels of the Vaal River, South Africa, and in 1871 in the yellow ground of several pipes located near the present city of Kimberley. Although some diamonds are still recovered from gravels, the principal South African production is from kimberlite pipes. The deposits were originally worked as open pits, but as they deepened, underground methods were adopted. The world's largest and most productive diamond mine is the Premier, 24 miles east of Pretoria, South Africa. Since mining began there in 1903, nearly 30 million carats or six tons of diamonds have been produced. It was at the Premier mine in 1905 that the world's largest diamond, the Cullinan, weighing 3024 carats, was found. Prospecting for diamonds has located in South Africa over 700 kimberlite pipes, dikes, and sills most of which are barren. As recently as 1966 the Finsch Mine, eighty miles west of Kimberley came into production. The most notable pipe outside of South Africa is the Williamson Diamond Mine in Tanzania, discovered in 1940; and because of it, this country is a major diamond producer.

The early method of treatment was to crush the blue ground into coarse fragments and spread it out on platforms to disintegrate gradually under atmospheric influences. The present method is to crush the rock fine enough to permit immediate concentration. The diamonds are finally separated on tables coated with grease, to which the diamonds adhere, whereas the rest of the material is washed away.

In Cape Province, South Africa, on the desert coast just south of the mouth of the Orange River, terrace deposits containing high-quality stones were discovered in 1927. Later similar deposits were found along the coast north of the Orange River in South-West Africa extending 50 to 60 miles up the coast. Elsewhere in Africa alluvial diamonds have been found in the Congo, Angola, Ghana, and Sierra Leone. About 95 per cent of the world's output of diamonds comes at present from the African continent. The Congo is by far the largest producer and furnishes from placer deposits over 50 per cent of the world's supply. These Congo diamonds are mostly of industrial grade and represent only about 13 per cent of the total value of diamonds produced. There is a considerable but unknown production of diamonds in the Soviet Union from both pipes and placer deposits. British Guiana, Venezuela, and Australia are small producers.

Diamonds have been found sparingly in various parts of the United States. Small stones have occasionally been discovered in the stream sands along the eastern slope of the Appalachian Mountains from Virginia south to Georgia. Diamonds have also been reported from the gold sands of northern California and southern Oregon and sporadic occurrences have been noted in the glacial drift in Wisconsin, Michigan, and Ohio. In 1906 diamonds were found in kimberlite in a pipe near Murfreesboro, Pike County, Arkansas. This locality, resembling the diamond pipes of South Africa, has yielded about 40,000 stones but is at present unproductive. In 1951 the old mine workings were opened to tourists who, for a fee, were permitted to look for diamonds.

Use. *In Industry.* Fragments of diamond crystals are used to cut glass. The fine powder is employed in grinding and polishing diamonds and other gem stones. Wheels are impregnated with diamond powder for cutting rocks and other hard materials. Steel bits are set with diamonds, especially the cryptocrystalline variety, carbonado, to make diamond drills used in exploratory mining work. The diamond is also used in wire drawing and in tools for the truing of grinding wheels.

In Gems. The diamond is the most important of the gem stones. Its value depends upon its hardness, its brilliancy, which results from its high index of refraction, and its "fire," resulting from its strong dispersion. In general, the most valuable are those flawless stones which are colorless or possess a "blue-white" color. A faint straw-yellow color, which diamond often shows, detracts from its value. Diamonds colored deep shades of yellow, red, green, or blue, known as fancy stones, are greatly prized, and bring very high prices. Diamonds can be colored deep shades of green by irradiation with high-energy nuclear particles, neutrons, deuterons, and alpha particles, and blue by exposing them to fast-moving electrons. A stone colored green by irradiation can be made a deep yellow by proper heat treatment. These artificially colored stones are difficult to distinguish from those of natural color.

The value of a cut diamond depends upon its color and purity, upon the skill with which it has been cut, and upon its size. A stone weighing one carat (0.2 g) cut in the form of a brilliant would be 6.25 millimeters in diameter and 4 millimeters in depth. A 2-carat stone of the same quality would have a value three or four times as great.

Artificial. For many years experiments have been carried out with the hope of synthesizing diamond and many false claims made of its synthesis. However, it was not until 1955 that the General Electric Company made authenticated diamonds. So far the synthetic diamonds are small and not suitable for cutting into gems, but several million carats are produced each year for industrial purposes.

Name. The name diamond is a corruption of the Greek word *adamas*, meaning *invincible*.

GRAPHITE—C

Crystallography. Hexagonal; $6/m2/m2/m$. In tabular crystals of hexagonal outline with prominent basal plane. Distinct faces of other forms very rare. Triangular markings on the base are the result of gliding along an undetermined second-order pyramid. Usually in foliated or scaly masses, but may be radiated or granular.

$C6_3/mmc$. $a = 2.46$, $c = 6.74$ Å, $Z = 4$. $a:c = 1:2.740$. *d's:* 3.36(10), 2.03(5), 1.675(8), 1.232(3), 1.158(5).

Physical Properties. *Cleavage.* $\{0001\}$ perfect. **H** 1–2 (readily marks paper and soils the fingers). **G** 2.23. *Luster* metallic, sometimes dull earthy. *Color* and *streak*, black. Greasy feel. Folia flexible but not elastic.

Composition. Carbon. Some graphite impure with iron oxide, clay, or other minerals.

Diagnostic Features. Graphite is recognized by its color, foliated nature, and greasy feel. Distinguished from molybdenite by its black color (molybdenite has a blue tone), and black streak on glazed porcelain. Infusible, but may burn to CO_2 at a high temperature. Unattacked by acids.

Occurrence. Graphite most commonly occurs in metamorphic rocks such as crystalline limestones, schists, and gneisses. It may be found as large crystalline plates enclosed or disseminated in small flakes in sufficient amount to form a considerable proportion of the rock. In these cases, it has probably been derived from carbonaceous material of organic origin which has been converted into graphite during metamorphism. Metamorphosed coal beds may be partially converted into graphite, as the graphite coals of Rhode Island and in the coal fields of Sonora, Mexico. Graphite also occurs in hydrothermal veins associated with quartz, biotite, orthoclase, tourmaline, apatite, pyrite, and sphene, as in the deposits at Ticonderoga, New York. The graphite in these veins may have been formed from hydrocarbons introduced into them during the metamorphism of the region and derived from the surrounding carbon-bearing rocks. Graphite occurs occasionally as an original constituent of igneous rocks as in the basalts of Ovifak, Greenland, in a nepheline syenite in India, in a granite pegmatite in Maine. It is also found in some meteorites.

The principal countries producing natural graphite are: North and South Korea, U.S.S.R., Mexico, Austria, China, Malagasy Republic, and Ceylon. The chief deposits in the United States are in the Adirondack region of New York, in Essex, Warren, and Washington counties, particularly at Ticonderoga.

Artificial. Graphite is manufactured on a large scale in electrical furnaces using anthracite coal or petroleum coke as the raw materials. The use of artificial graphite in the United States is considerably in excess of that of the natural mineral.

Use. Used in the manufacture of refractory crucibles for the steel, brass, and bronze industries. Flake graphite for crucibles comes mostly from Ceylon and Malagasy Republic. Mixed with oil, graphite is used as a lubricant, and mixed with fine clay, it forms the "lead" of pencils. It is employed in the manufacture of protective paint for structural steel and is used in foundry facings, batteries, electrodes, generator brushes, and in electrotyping.

Name. Derived from the Greek word meaning *to write*, in allusion to its use in pencils.

SULFIDES

The sulfides form an important class of minerals which includes the majority of the ore minerals. With them are classed the similar but rarer selenides, tellurides, arsenides, and antimonides.

Most of the sulfide minerals are opaque with distinctive colors and characteristically colored streaks. Those that are nonopaque, as cinnabar, realgar, and orpiment, have high refractive indices and transmit light only on thin edges.

The sulfides can be divided into small structural groups, but no broad generalizations can be made regarding their structure. Many of the sulfides have ionic bonding, whereas others, displaying most of the properties of metals, have metallic bonding at least in part. Sphalerite has a structure similar to diamond and like diamond has a covalent bond.

The general formula for the sulfides is given as $A_m X_n$ in which A represents the metallic elements and X the nonmetallic element. The order of listing of the various minerals is in a decreasing ratio of $A:X$.

Sulfides

Argentite	Ag_2S	Cinnabar	HgS
Chalcocite	Cu_2S	Realgar	AsS
Bornite	Cu_5FeS_4	Orpiment	As_2S_3
Galena	PbS	Stibnite	Sb_2S_3
Sphalerite	ZnS	Bismuthinite	Bi_2S_3
Chalcopyrite	$CuFeS_2$	Pyrite	FeS_2
Stannite	Cu_2FeSnS_4	Cobaltite	$(Co,Fe)AsS$
Greenockite	CdS	Marcasite	FeS_2
Pyrrhotite	$Fe_{1-x}S$	Arsenopyrite	$FeAsS$
Niccolite	$NiAs$	Molybdenite	MoS_2
Millerite	NiS	Calaverite	$AuTe_2$
Pentlandite	$(Fe,Ni)S$	Sylvanite	$(Au,Ag)Te_2$
Covellite	CuS	Skutterudite	$(Co,Ni)As_3$

Argentite—Ag$_2$S

Crystallography. Isometric, $4/m\bar{3}2/m$ (above 179°C); monoclinic, $2/m$ (acanthite) at ordinary temperatures. Crystals, paramorphs of the high-temperature form, commonly show the cube, octahedron, and dodecahedron but frequently are arranged in branching or reticulated groups. Most commonly massive or as a coating.

Argentite: *Im3m;* a = 4.89 Å, Z = 2.

Acanthite: *P2$_1$/n;* a = 4.23, b = 6.93, c = 7.86 Å, β = 99° 35′, $a:b:c$ = 0.610:1:1.134. Z = 4. *d's:* 2.60(10), 2.45(8), 2.38(5), 2.22(3), 2.09(4).

Physical Properties. **H** 2–2$\frac{1}{2}$. **G** 7.3. Very sectile; can be cut with a knife like lead. *Luster* metallic. *Color* black. *Streak* black, shining. Opaque. Bright on fresh surface but on exposure becomes dull black, owing to the formation of an earthy sulfide.

Composition. Ag 87.1, S 12.9 per cent.

Diagnostic Features. Argentite can be distinguished by its color, sectility, and high specific gravity. Fusible at 1$\frac{1}{2}$ with intumescence. When fused on charcoal in the oxidizing flame it gives off the odor of sulfur dioxide and yields a globule of silver.

Occurrence. Argentite is an important primary silver mineral found in veins associated with native silver, the ruby silvers, polybasite, stephanite, galena, and sphalerite. It may also be of secondary origin. It is found in microscopic inclusions in argentiferous galena. Argentite is an important ore in the silver mines of Guanajuato and elsewhere in Mexico; in Peru, Chile, and Bolivia. Important European localities are Freiberg, Saxony; Joachimsthal, Bohemia; Schemnitz and Kremnitz, Czechoslovakia, and Kongsberg, Norway. In the United States it has been an important ore mineral in Nevada, notably at the Comstock Lode and at Tonopah. It is also found in the silver districts of Colorado, and in Montana at Butte associated with copper ores.

Use. An important ore of silver.

Name. The name argentite comes from the Latin *argentum*, meaning silver.

CHALCOCITE—Cu$_2$S

Crystallography. Orthorhombic; $2/m2/m2/m$ (below 105°C); above 105°C, hexagonal. Crystals are very rare, usually small and tabular with hexagonal outline; striated parallel to the a axis (Fig. 291). Commonly fine grained and massive.

Abm2; a = 11.92, b = 27.33, c = 13.44 Å. $a:b:c$ = 0.436:1:0.492, Z = 96. *d's:* 3.39(3), 2.40(7), 1.969(8), 1.870(10), 1.695(4).

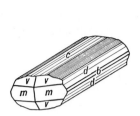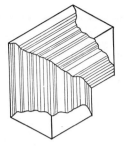

FIG. 291. Chalcocite crystals.

Physical Properties. *Cleavage* {110} poor. *Fracture* conchoidal. **H** $2\frac{1}{2}$–3. **G** 5.5–5.8. *Luster* metallic. Imperfectly sectile. *Color* shining lead-gray, tarnishing to dull black on exposure. *Streak* grayish black. Some chalcocite is soft and sooty.

Composition. Cu 79.8, S 20.2 per cent. May contain small amounts of Ag and Fe.

Diagnostic Features. Chalcocite is distinguished by its lead-gray color and sectility. Fusible at 2–$2\frac{1}{2}$. When heated on charcoal it gives odor of sulfur dioxide and is easily reduced to metallic copper. Roasted mineral, moistened with HCl, gives azure-blue flame.

Occurrence. Chalcocite is one of the most important copper-ore minerals. Fine crystals are rare but have been found at Cornwall, England and Bristol, Connecticut. Chalcocite may occur as a primary mineral in veins with bornite, chalcopyrite, enargite and pyrite. But its principal occurrence is as a supergene mineral in enriched zones of sulfide deposits. Under surface conditions the primary copper sulfides are oxidized; the soluble sulfates formed move downward reacting with the primary minerals to form chalcocite and thus enriching the ore in copper. The water table is the lower limit of the zone of oxidation and here a "chalcocite blanket" may form. Many famous copper mines owe their greatness to this process of secondary enrichment as: Rio Tinto, Spain; Ely, Nevada; Morenci, Miami, Clifton, Arizona; and Butte, Montana.

Much of the world's copper is today produced from what is called "porphyry copper" ore. In these deposits primary copper minerals disseminated through the rock, usually a porphyry, have been altered, at least in part, to chalcocite and thus enriched to form a workable ore body. The amount of copper in such deposits is small, rarely greater than 1 or 2 per cent and may be as low as 0.50 per cent. The largest copper producer in the United States, is a porphyry copper deposit at Bingham, Utah.

Use. An important copper ore.

Similar Species. *Digenite*, Cu_9S_5, is blue to black, associated with chalcocite. *Stromeyerite*, $(Ag,Cu)_2S$, is a steel-gray mineral found in copper-silver veins.

BORNITE—Cu_5FeS_4

Crystallography. Isometric; $4/m\overline{3}2/m$. Rarely in rough cubic and less commonly in dodecahedral and octahedral crystals. Usually massive.

$Fd3m$. $a = 10.93$ Å, $Z = 8$. $d's$: 3.30(8), 3.17(8), 2.74(8), 1.93(10), 1.37(8).

Physical Properties. H 3. G 5.06–5.08. *Luster* metallic. *Color* brownish-bronze on fresh fracture but quickly tarnishing to variegated purple and blue (hence called *peacock ore*) and finally to almost black on exposure. *Streak* grayish-black.

Composition. Cu 63.3, Fe 11.2, S 25.5 per cent. Microscopic admixed blebs of other minerals cause the composition of what appears to be bornite to vary considerably, but analyses of pure material agree with the above formula.

Diagnostic Features. Bornite is distinguished by its characteristic bronze color on the fresh fracture and by the purple tarnish. Fusible at $2\frac{1}{2}$. When heated on charcoal it gives off the odor of sulfur dioxide and becomes magnetic. If, after roasting, it is moistened with hydrochloric acid and heated, it gives an azure-blue flame.

Alteration. Bornite alters readily to chalcocite and covellite.

Occurrence. Bornite is a widely occurring copper ore usually found associated with other copper minerals in hypogene deposits. It is much less frequently found as a supergene mineral, in the upper, enriched zone of copper veins. It is found disseminated in basic rocks, in contact metamorphic deposits, in replacement deposits, and in pegmatites. Bornite frequently occurs in intimate mixtures with chalcopyrite and chalcocite. It is not as important an ore of copper as chalcocite and chalcopyrite.

Good crystals of bornite have been found associated with crystals of chalcocite at Bristol, Connecticut, and at Cornwall, England. Found in large masses in Chile, Peru, Bolivia, and Mexico. In the United States it is found at Magma mine, Pioneer, Arizona; Butte, Montana; Engels mine, Plumas County, California; Halifax County, Virginia; and Superior, Arizona.

Use. An ore of copper.

Name. Bornite was named after the German mineralogist von Born (1742–1791).

GALENA—PbS

Crystallography. Isometric; $4/m\overline{3}2/m$. The most common form is the cube, sometimes truncated by the octahedron (Figs. 292 and 293). Dodecahedron and trisoctahedron rare. Galena has a NaCl type of structure with Pb in place of Na and S in place of Cl.

FIG. 292. Galena crystals.

Fm3m; a = 5.936 Å, *Z* = 4. *d's:* 3.44(9), 2.97(10), 2.10(10), 1.780(9), 1.324(10).

Physical Properties. *Cleavage* perfect {001}. **H** 2½. **G** 7.4–7.6. *Luster* bright metallic. *Color* and *streak* lead-gray.

Composition. Pb 86.6, S 13.4 per cent. Silver is usually present, probably as admixtures of silver minerals such as argentite or tetrahedrite. Inclusions probably also account for the small amounts of zinc, cadmium, antimony, arsenic, and

FIG. 293. (*a*) Galena crystals, Joplin, Missouri. (*b*) Galena on sphalerite and dolomite crystals.

bismuth that may be present. Selenium may substitute for sulfur and a complete series from PbS–PbSe has been reported.

Diagnostic Features. Galena can be easily recognized by its good cleavage, high specific gravity, softness, and lead-gray streak. Fusible at 2. Reduced on charcoal to lead globule with formation of yellow to white coating of lead oxide and odor of sulfur dioxide.

Alteration. By oxidation galena is converted into the sulfate, anglesite, and the carbonate, cerussite.

Occurrence. Galena is a very common metallic sulfide, found in veins associated with sphalerite, pyrite, marcasite, chalcopyrite, cerussite, anglesite, dolomite, calcite, quartz, barite, and fluorite. When found in hydrothermal veins galena is frequently associated with silver minerals; it often contains silver itself and so becomes an important silver ore. A large part of the supply of lead comes as a secondary product from ores mined chiefly for their silver. In a second type of deposit typified by the lead-zinc ores of the Mississippi Valley, galena, associated with sphalerite, is found in veins, open space filling, or replacement bodies in limestones. These are low temperature deposits, located at shallow depths and usually contain little silver. Galena is also found in contact metamorphic deposits, in pegmatites and as disseminations in sedimentary rocks.

Famous world localities are Freiberg, Saxony; the Harz Mountains; Westphalia and Nassau; Přibram, Bohemia; Cornwall, Derbyshire, and Cumberland, England; Sullivan mine, British Columbia; and Broken Hill, Australia.

In the United States there are many lead-producing districts; only the most important are mentioned here. In the Tri-State district of Missouri, Kansas, and Oklahoma centering around Joplin, Missouri, galena is associated with zinc ores and is found in irregular veins and pockets in limestone and chert. It is found in a similar manner but in smaller amount in Illinois, Iowa, and Wisconsin. The deposits of southeast Missouri where galena is disseminated through limestone are particularly productive. In the Coeur d'Alene district, Idaho, galena is the chief ore mineral in the lead-silver vein deposits. Lead is produced in Utah at Brigham and from the silver deposits of the Tintic and Park City districts; and in Colorado, chiefly from the lead-silver ores of the Leadville district.

Use. Practically the only source of lead and an important ore of silver. Metallic lead is used chiefly as follows: for conversion into white lead (a basic lead carbonate), which is the principal ingredient of many white paints, or into the oxides (*litharge*, PbO, and *minium*, Pb_3O_4) used in making glass and in giving a glaze to earthenware; as pipe and sheets; and for shot. It is a principal ingredient of several alloys as solder (lead and tin), type metal (lead and antimony), and low-fusion alloys (lead, bismuth, and tin). Large amounts of metallic lead are used in storage batteries and as shielding in working with uranium and other radioactive substances.

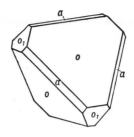

FIG. 294. Sphalerite crystals.

Name. The name galena is derived from the Latin *galena*, a name originally given to lead ore.

Similar Species. *Altaite*, PbTe, and *alabandite*, MnS, like galena, have a NaCl type of structure.

SPHALERITE—ZnS

Zinc Blende

Crystallography. Isometric; $\bar{4}3m$. Tetrahedron, dodecahedron, and cube common forms (Fig. 294), but the crystals, frequently highly complex and usually malformed or in rounded aggregates, often show polysynthetic twinning on $\{111\}$. Usually found in cleavable masses, coarse to fine granular. Compact, botryoidal, cryptocrystalline.

The sphalerite structure is similar to that of diamond with one-half of the carbon atoms of diamond replaced by zinc and the other half by sulfur. Each zinc atom is surrounded by and bonded to four sulfur atoms and in turn each sulfur atom is bonded to four zinc atoms (Fig. 295a).

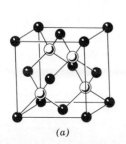

(a)

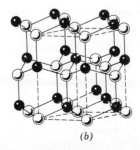

(b)

FIG. 295. Structure of ZnS dimorphs. (*a*) Sphalerite. (*b*) Wurtzite.

FIG. 296.   Sphalerite, ZnS, packing model. Zn (black) and S (white) are both in 4 coordination. Note sheets of atoms parallel to (111). If they are rotated 180° alternately, the wurtzite structure results.

$F\bar{4}3m$; $a = 5.41$ Å; $Z = 4$. $d's$: 3.12(10), 1.910(8), 1.631(7), 1.240(4), 1.106(5). The dimorphic form of ZnS, *wurtzite*, is hexagonal, *6mm* (Fig. 295b).

Physical Properties.   *Cleavage*, {011} perfect but some sphalerite is too fine grained to show cleavage. **H** $3\frac{1}{2}$–4. **G** 3.9–4.1. *Luster* nonmetallic and resinous to submetallic; also adamantine. *Color* white when pure, and green when nearly so. Commonly yellow, brown to black, darkening with increase in iron. Also red (ruby zinc). Transparent to translucent. *Streak* white to yellow and brown.

Composition.   Zn 67, S 33 per cent when pure. Nearly always contains iron with the amount of iron dependent on the temperature and environment. If iron is in excess as indicated by the presence of pyrrhotite, FeS can reach nearly 50 mole per cent (see Fig. 269). If sphalerite and an iron sulfide crystallize together, the amount of iron is an indication of the temperature of formation, and sphalerite becomes a geologic thermometer. Manganese and cadmium are usually present in small amounts in solid solution.

Diagnostic Features.   Sphalerite can be recognized by its striking resinous luster and perfect cleavage. The dark varieties (black jack) can be told by the reddish-brown streak, always lighter than the massive mineral. Pure zinc sulfide is infusible; it becomes fusible, with difficulty, with an increase in the amount of iron. When heated on charcoal it gives the odor of sulfur dioxide and with the reducing mixture, gives a coating of zinc oxide (yellow when hot, white when cold).

Occurrence. Sphalerite, the most important ore mineral of zinc, is extremely common. Its occurrence and mode of origin are similar to those of galena, with which it is commonly found. In the shallow seated lead-zinc deposits of the Tri-State district of Missouri, Kansas, and Oklahoma these minerals are associated with marcasite, chalcopyrite, calcite, and dolomite. Sphalerite with only minor galena occurs in hydrothermal veins and replacement deposits associated with pyrrhotite, pyrite, and magnetite. Sphalerite is also found in veins in igneous rocks and in contact metamorphic deposits.

Zinc is mined in significant amounts in over 40 countries. Although in a few places the ore minerals are hemimorphite and smithsonite and at Franklin, New Jersey willemite, zincite, and franklinite, most of the world's zinc comes from sphalerite. The principal producing countries are: Canada, U.S.S.R., United States, Australia, Peru, Mexico, and Japan. In the United States nearly 60 per cent of the zinc is produced east of the Mississippi River with Tennessee, New York, Pennsylvania, and New Jersey the principal producing states. In western United States, Idaho, Colorado, and Utah are the chief producers. The mines in the Tri-State district, formerly major zinc producers, are now largely exhausted.

Use. The most important ore of zinc. The chief uses for metallic zinc, or *spelter*, are in galvanizing iron; making brass, an alloy of copper and zinc; in electric batteries; and as sheet zinc. Zinc oxide, or zinc white, is used extensively for making paint. Zinc chloride is used as a preservative for wood. Zinc sulfate is used in dyeing and in medicine. Sphalerite also serves as the most important source of cadmium, indium, gallium, and germanium.

Name. Sphalerite comes from the Greek meaning *treacherous*. Blende because, although often resembling galena, it yielded no lead; from the German word meaning *blind* or *deceiving*.

CHALCOPYRITE—$CuFeS_2$

Crystallography. Tetragonal; $\bar{4}2m$. Commonly tetrahedral in aspect with the disphenoidal $p\{112\}$ dominant. Other forms shown in Fig. 297 are rare. Usually massive.

Angles: $p(112) \wedge p'(1\bar{1}2) = 108° 40'$, $c(001) \wedge z(011) = 63° 06'$, $c(001) \wedge p(112) = 54° 20'$, $c(001) \wedge e(102) = 44° 34'$.

$I\bar{4}2d$; $a = 5.25$, $c = 10.32$ Å, $a:c = 1:1.966$; $Z = 4$. d's: 3.03(10), 1.855(10), 1.586(10), 1.205(8), 1.074(8).

Physical Properties. H $3\frac{1}{2}$–4. G 4.1–4.3. *Luster* metallic. *Color* brass-yellow; often tarnished to bronze or iridescent. *Streak* greenish black. Brittle.

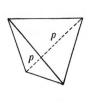
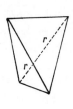
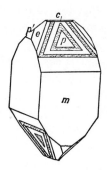

FIG. 297. Chalcopyrite crystals.

Composition. Cu 34.6, Fe 30.4, S 35.0 per cent. Analyses often show variations from the percentages given because of mechanical admixtures of other sulfides, chiefly pyrite.

Diagnostic Features. Recognized by its brass-yellow color and greenish-black streak. Distinguished from pyrite by being softer than steel and from gold by being brittle. Known as "fool's gold," a term also applied to pyrite. On charcoal it fuses at 2 to a magnetic globule, and gives off the odor of sulfur dioxide. Decrepitates and gives sulfur in the closed tube. After roasting, and moistening with HCl, gives the azure-blue copper chloride flame.

Occurrence. Chalcopyrite is the most widely occurring copper mineral and one of the most important sources of that metal. Most sulfide ores contain some chalcopyrite but the most important economically are the hydrothermal vein and replacement deposits. In the low-temperature deposits as in the Tri-State district, it occurs as small crystals associated with galena, sphalerite, and dolomite. Associated with pyrrhotite and pentlandite, it is the chief copper mineral in the ores of Sudbury, Ontario and similar high-temperature deposits. Chalcopyrite is the principal primary copper mineral in the "porphyry-copper" deposits. Also occurs as an original constituent of igneous rocks; in pegmatite dikes; in contact metamorphic deposits; and disseminated in schistose rocks. It may carry gold or silver and become an ore of those metals. Often in subordinate amount with large bodies of pyrite, making them serve as low-grade copper ores.

A few of the localities at which chalcopyrite is the chief ore of copper are: Cornwall, England; Falun, Sweden; Schemnitz, Czechoslovakia; Schlaggenwald, Bohemia; Freiberg, Saxony; Rio Tinto, Spain; South Africa, Zambia and Chile. Found widely in the United States but usually with other copper minerals in equal or greater amount; found at Butte, Montana; Bingham, Utah; Jerome, Arizona; Ducktown, Tennessee; and various districts in California, Colorado, and New

Mexico. In Canada the most important occurrences of chalcopyrite are at Sudbury, Ontario and at Rouyn district, Quebec.

Alteration. Chalcopyrite is the principal source of copper for the secondary minerals malachite, azurite, covellite, chalcocite, and cuprite. Concentrations of copper in the zone of supergene enrichment are often the result of such alteration and removal of copper in solution with its subsequent deposition.

Use. Important ore of copper.

Name. Derived from Greek word meaning *brass* and from *pyrites*.

Greenockite—CdS

Crystallography. Hexagonal; *6mm*. Crystals rare and small, showing prism faces, and terminated usually below with pedion and above with pyramids. Usually pulverulent, and as powdery incrustations.

$P6_3mc$; $a = 4.15$, $c = 6.73$ Å, $a:c = 1:1.622$. $Z = 2$. $d's$: 3.59(8), 3.167(10), 2.071(8), 1.900(8), 1.764(8).

Physical Properties. *Cleavage* $\{11\bar{2}2\}$, $\{0001\}$ poor. **H** $3-3\frac{1}{2}$. **G** 4.9. *Luster* adamantine to resinous, earthy. *Color* various shades of yellow and orange. *Streak* between orange-yellow and brick-red. Greenockite containing zinc shows a strong orange-yellow fluorescence.

Composition. Cd 77.8, S 22.2 per cent. Wurtzite, ZnS, and greenockite are isostructural, and a complete solid-solution series exists between the two minerals (see **Fig. 295b**).

Diagnostic Features. Characterized by its yellow color and pulverulent form and association with zinc ores. Infusible. Yields odor of sulfur dioxide when heated in the open tube, and a reddish-brown coating of cadmium oxide when heated with sodium carbonate on charcoal.

Occurrence. Greenockite is the most common mineral containing cadmium but it is found only in a few localities and in small amounts, usually as an earthy coating on zinc ores, especially sphalerite.

Found in crystals at Bishopton, Renfrew, Scotland; Tsumeb, South-West Africa; and also in Bohemia and Carinthia. In the United States it is found with the zinc ores of the Tri-State district, in Arkansas, and in small amounts at Franklin, New Jersey.

Use. A source of cadmium. Cadmium is used in alloys for antifriction bearings and low-melting alloys. Small amounts, less than 1.5 per cent, will harden copper and silver. The largest use is in electroplating other metals to form a coating resistant to chemical attack. Cadmium is also used in many pigments and chemicals.

Name. Named after Lord Greenock (later Earl Cathcart). The first crystal was found about 1810. It was over one-half inch across and was mistaken for sphalerite.

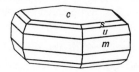

FIG. 298. Pyrrhotite.

PYRRHOTITE—Fe$_{1-x}$S

Magnetic Pyrites

Crystallography. Hexagonal; $6/m2/m2/m$. Crystals usually tabular, in some cases pyramidal (Fig. 298). Usually massive with granular or lamellar habit.

Angles: $c\{0001\} \wedge s\{10\bar{1}2\} = 45°\ 8'$, $c\{0001\} \wedge u\{20\bar{2}1\} = 76°\ 0'$.

$C6/mmc$; $a = 3.44$, $c = 5.73$ Å, $a:c = 1:1.666$; $Z = 2$. $d's$: 2.97(6), 2.63(8), 2.06(10), 1.718(6), 1.045(8).

Physical Properties. H 4. G 4.58–4.65. *Luster* metallic. *Color* brownish bronze. *Streak* black. Magnetic, but varying in intensity; the greater the amount of iron, the lesser the magnetism. Opaque.

Composition. Most pyrrhotite has a deficiency of iron with respect to sulfur, as indicated by the formula Fe$_{1-x}$S, with x between 0 and 0.2. The mineral *troilite* is close to FeS.

Diagnostic Features. Recognized usually by its massive nature, bronze color, and magnetism. Fusible at 3. When heated on charcoal, it gives off the odor of sulfur dioxide and becomes strongly magnetic.

Occurrence. Pyrrhotite is commonly associated with basic igneous rocks, particularly norites. It occurs in them as disseminated grains or, as at Sudbury, Ontario, as large masses associated with pentlandite, chalcopyrite, or other sulfides. At Sudbury vast tonnages of pyrrhotite are mined principally for the copper, nickel, and platinum that are extracted from associated minerals. Pyrrhotite is also found in contact metamorphic deposits, in vein deposits, and in pegmatites.

Large quantities are known in Finland, Norway, and Sweden; in Germany from Andreasberg in the Harz Mountains; at Schneeberg, Saxony; and Bodenmais, Bavaria. In the United States in crystals from Standish, Maine; at the Gap mine, Lancaster County, Pennsylvania; and in considerable amount at Ducktown, Tennessee.

Use. It is mined for its associated nickel, copper, and platinum. At Sudbury, Ontario it is also a source of sulfur and an ore of iron.

Name. The name pyrrhotite comes from the Greek meaning *reddish*.

Niccolite—NiAs

Crystallography. Hexagonal; $6/m2/m2/m$. Rarely in tabular crystals. Usually massive, reniform with columnar structure.

$C6/mmc$; $a = 3.61$, $c = 5.02$ Å; $a:c = 1:1.391$; $Z = 2$. $d's$: 2.66(10), 1.961(9), 1.811(8), 1.328(3), 1.071(4).

Physical Properties. H $5-5\frac{1}{2}$. **G** 7.78. *Luster* metallic. *Color* pale copper-red (hence called *copper nickel*), with gray to blackish tarnish. *Streak* brownish-black. Opaque.

Composition. Ni 43.9, As 56.1 per cent. Usually with a little iron, cobalt, and sulfur. Arsenic frequently replaced in part by antimony.

Diagnostic Features. Characterized by its copper-red color. Fusible at 2. When heated on charcoal it yields a white sublimate and a garliclike odor is given off. Gives nickel test with dimethylglyoxime.

Alteration. Quickly alters to annabergite (green nickel bloom) in moist atmosphere.

Occurrence. Niccolite, with other nickel arsenides and sulfides, pyrrhotite, and chalcopyrite, frequently occurs in, or is associated with, norites. Also found in vein deposits with cobalt and silver minerals.

Found in Germany in the silver mines of Saxony, the Harz Mountains, in Hessen-Nassau; and at Cobalt, Ontario.

Use. A minor ore of nickel.

Name. The first name of this mineral, *kupfernickel*, gave the name *nickel* to the metal. Niccolite is from the Latin for nickel.

Similar Species. Breithauptite, NiSb, is isostructural with niccolite with similar occurrence and association.

Millerite—NiS

Crystallography. Hexagonal—R; $\bar{3}2/m$. Usually in hair-like tufts and radiating groups of slender to capillary crystals (capillary pyrites). In velvety incrustations. Rarely in coarse cleavable masses.

$R3m$; $a = 9.62$, $c = 3.16$ Å; $a:c = 1:0.328$; $Z = 9$. $d's$: 4.77(8), 2.75(10), 2.50(6), 2.22(6), 1.859(10).

Physical Properties. *Cleavage* $\{10\bar{1}1\}$, $\{01\bar{1}2\}$ good. H $3-3\frac{1}{2}$. **G** 5.5 ± 0.2. *Luster* metallic. *Color* pale brass-yellow; with a greenish tinge when in fine hairlike masses. *Streak* black, somewhat greenish.

Composition. Ni 64.7, S 35.3 per cent.

Diagnostic Features. Characterized by its capillary crystals and distinguished from minerals of similar color by nickel tests. Fusible at $1\frac{1}{2}-2$ to a magnetic globule. Gives odor of sulfur dioxide when heated on charcoal.

Occurrence. Millerite forms as a low-temperatu mineral often in cavities and as an alteration of other nickel minerals, or as crystal inclusions in other minerals.

Occurs in various localities in Saxony, Westphalia, and Hessen-Nassau and in Bohemia. In the United States, it is found with hematite and ankerite at Antwerp,

New York; with pyrrhotite at the Gap mine, Lancaster County, Pennsylvania; in geodes in limestone at St. Louis, Missouri; Keokuk, Iowa; and Milwaukee, Wisconsin. In coarse cleavable masses it is a major ore mineral at the Marbridge Mine, Lamotte Township, Quebec.

Use. A subordinate ore of nickel.

Name. In honor of the mineralogist, W. H. Miller (1801–1880), who first studied the crystals.

Pentlandite—$(Fe,Ni)_9S_8$

Crystallography. Isometric; $4/m\bar{3}2/m$. Massive, usually in granular aggregates with octahedral parting.

$Fm3m$; $a = 10.07$ Å; $Z = 4$. d's: 5.84(2), 3.04(6), 2.92(2), 2.31(3), 1.781(10).

Physical Properties. Parting on $\{111\}$. **H** $3\frac{1}{2}$–4. **G** 4.6–5.0. Brittle. *Luster* metallic. *Color* yellowish-bronze. *Streak* light bronze-brown. Opaque. Nonmagnetic.

Composition. $(Fe,Ni)_9S_8$. Usually the ratio of Fe:Ni is close to 1:1. Commonly contains small amounts of cobalt.

Diagnostic Features. Pentlandite closely resembles pyrrhotite in appearance but can be distinguished from it by the octahedral parting and lack of magnetism. Fusible at $1\frac{1}{2}$–2. On heating gives odor of sulfur dioxide and becomes magnetic. Gives nickel test with dimethylglyoxime.

Occurrence. Pentlandite usually occurs in basic igneous rocks where it is commonly associated with other nickel minerals, pyrrhotite, and chalcopyrite, and has probably accumulated by magmatic segregation.

Found at widely separated localities in small amounts but its chief occurrences are in Canada where, associated with pyrrhotite, it is the principal source of nickel at Sudbury, Ontario, and the Lynn Lake area, Manitoba. It is also an important ore mineral in similar deposits in the Petsamo district of U.S.S.R.

Use. The principal ore of nickel. The chief use of nickel is in steel. Nickel steel contains $2\frac{1}{2}$–$3\frac{1}{2}$ per cent nickel, which greatly increases the strength and toughness of the alloy, so that lighter machines can be made without loss of strength. Nickel is also an essential constituent of stainless steel. The manufacture of Monel metal (68 per cent nickel, 32 per cent copper) and Nichrome (38–85 per cent nickel) consumes a large amount of the nickel produced. Other alloys are German silver (nickel, zinc, and copper); metal for coinage—the 5-cent coin of the United States is 25 per cent nickel and 75 per cent copper, low-expansion metals for watch springs and other instruments. Nickel is used in plating; although chromium now largely replaces it for the surface layer, nickel is used for a thicker under-layer.

Name. After J. B. Pentland, who first noted the mineral.

Covellite—CuS

Crystallography. Hexagonal; $6/m2/m2/m$. Rarely in tabular hexagonal crystals. Usually massive as coatings or disseminations through other copper minerals.

$P6_3/mmc$; $a = 3.80$, $c = 16.36$ Å; $a:c = 1:4305$; $Z = 6$. $d's$: 3.06(4), 2.83(6), 2.73(10), 1.899(8), 1.740(5).

Physical Properties. *Cleavage* {0001} perfect giving flexible plates. **H** $1\frac{1}{2}$–2. **G** 4.6–4.76. *Luster* metallic. *Color* indigo-blue or darker. *Streak* lead-gray to black. Often iridescent. Opaque.

Composition. Cu 66.4, S 33.6 per cent. A small amount of iron may be present.

Diagnostic Features. Characterized by the indigo-blue color, micaceous cleavage yielding flexible plates, and association with other copper sulfides. Fusible at $2\frac{1}{2}$. Gives off the odor of sulfur dioxide in the open tube, and much sulfur in the closed tube. The roasted mineral, moistened with HCl and ignited, gives the blue copper chloride flame.

Occurrence. Covellite is not an abundant mineral but is found in most copper deposits as a supergene mineral, usually as a coating, in the zone of sulfide enrichment. It is associated with other copper minerals, principally chalcocite, chalcopyrite, bornite, and enargite, and is derived from them by alteration. Primary covellite is known but uncommon.

Found at Bor, Serbia, Yugoslavia; and Leogang, Austria. In large iridescent crystals from the Calabona mine, Alghero, Sardinia. In the United States covellite is found in appreciable amounts at Butte, Montana; Summitville, Colorado; and La Sal district, Utah. Formerly found at Kennecott, Alaska.

Use. A minor ore of copper.

Name. In honor of N. Covelli (1790–1829), the discoverer of the Vesuvian covellite.

CINNABAR—HgS

Crystallography. Hexagonal—R; 32. Crystals usually rhombohedral, often in penetration twins. Trapezohedral faces rare. Usually fine granular massive; also earthy, as incrustations and disseminations through the rock.

$P3_121$; $a = 4.146$, $c = 9.497$ Å; $a:c = 1:2.291$; $Z = 3$. $d's$: 3.37(10), 3.16(8), 2.87(10), 2.07(8), 1.980(8).

Physical Properties. *Cleavage* {$10\bar{1}0$} perfect. **H** $2\frac{1}{2}$. **G** 8.10. *Luster* adamantine when pure to dull earthy when impure. *Color* vermilion-red when pure to brownish-red when impure. *Streak* scarlet. Transparent to translucent. *Hepatic cinnabar* is an inflammable variety with liver-brown color and in some cases a brownish streak, usually granular or compact.

Composition. Hg 86.2, S 13.8 per cent. Frequently impure from admixture of clay, iron oxide, bitumen.

Diagnostic Features. Recognized by its red color and scarlet streak, high specific gravity, and cleavage. Wholly volatile before the blowpipe. In the closed tube it yields (1) black sublimate of HgS when heated alone, and (2) globules of metallic mercury when heated with sodium carbonate.

Occurrence. Cinnabar is the most important ore of mercury but is found in quantity at comparatively few localities. Occurs as impregnations and as vein fillings near recent volcanic rocks and hot springs and evidently deposited near the surface from solutions which were probably alkaline. Associated with pyrite, marcasite, stibnite, and sulfides of copper in a gangue of opal, chalcedony, quartz, barite, calcite, and fluorite.

The important localities for the occurrence of cinnabar are at Almaden, Spain; Idria, Yugoslavia; Huancavelica in southern Peru; and the provinces of Kweichow and Hunan, China. In the United States the important deposits are in California at New Idria in San Benito County, in Napa County, and at New Almaden in Santa Clara County. It also occurs in Nevada, Utah, Oregon, Arkansas, Idaho, and Texas.

Use. The only important source of mercury. The most important use of mercury has been in the amalgamation process for recovering gold and silver from their ores, but other methods of extraction have lessened its demand for this purpose. It is used in thermometers, barometers, and various scientific and electrical equipment, including the mercury cell, in drugs, and in the form of an amalgam with silver for dental work and with tin in "silvering" mirrors. Several plants in the United States utilize mercury vapor instead of steam for the generation of power. This is a great potential use for mercury. Important military applications include the manufacture of fulminate of mercury for detonating high explosives and of paint for ship bottoms.

Name. The name cinnabar is supposed to have come from India, where it is applied to a red resin.

Similar Species. Metacinnabar is a black isometric polymorph of HgS.

REALGAR—AsS

Crystallography. Monoclinic; $2/m$. Found in short, vertically striated, prismatic crystals (Fig. 299). Frequently coarse to fine granular and often earthy and as an incrustation.

Angles: $b(010) \wedge m(110) = 56° 36'$, $b(010) \wedge l(120) = 37° 10'$, $c(001) \wedge n(011) = 24° 56'$, $c(001) \wedge z(101) = 29° 25'$.

$P2_1/n$; $a = 9.29$, $b = 13.53$, $c = 6.57$ Å; $a:b:c = 0.687:1:0.486$, $\beta = 106° 33'$; $Z = 16$. $d's$: 5.40(10), 3.19(9), 2.94(8), 2.73(8), 2.49(5).

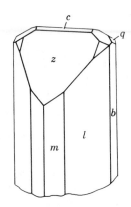

FIG. 299.   Realgar.

Physical Properties.   *Cleavage* {010} good. **H** $1\frac{1}{2}$–2. **G** 3.48. Sectile. *Luster* resinous. *Color* and *streak* red to orange. Translucent to transparent.

Composition.   As 70.1; S 29.9 per cent.

Diagnostic Features.   Realgar is distinguished by its red color, resinous luster, orange-red streak and almost invariable association with orpiment. Fusible at 1. When heated on charcoal it yields a volatile white sublimate with a garlic odor. In the open tube it gives a volatile crystalline sublimate of arsenious oxide and the odor of sulfur dioxide.

Alteration.   On long exposure to light disintegrates to a reddish-yellow powder.

Occurrence.   Realgar is found in veins of lead, silver, and gold ores associated with orpiment, other arsenic minerals, and stibnite. It also occurs as a volcanic sublimation product and as a deposit from hot springs.

Realgar is found associated with silver and lead ores in Hungary, Bohemia, and Saxony. Found in good crystals at Nagyág, Transylvania; Binnenthal, Switzerland; and Allchar, Macedonia. In the United States realgar is found at Mercur, Utah; at Manhattan, Nevada; and deposited from the geyser waters in the Norris Geyser Basin, Yellowstone National Park.

Use.   Realgar was used in fireworks to give a brilliant white light when mixed with saltpeter and ignited. Artificial arsenic sulfide is used for this purpose at present. It was formerly used as a pigment.

Name.   The name is derived from the Arabic, Rahj al ghar, *powder of the mine*.

ORPIMENT—As_2S_3

Crystallography.   Monoclinic; $2/m$. Crystals small, tabular or short prismatic, and rarely distinct; many pseudo-orthorhombic. Usually in foliated or columnar masses (Fig. 300).

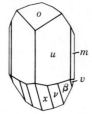

Allchar, Macedonia   **FIG. 300.   Orpiment.**

Angles: $b(010) \wedge m(110) = 39° 59'$, $b(010) \wedge u(210) = 59° 12'$, $b(010) \wedge x(\bar{3}11) = 73° 27'$.

$P2_1/n$; $a = 11.49$, $b = 9.59$, $c = 4.25$ Å; $a:b:c = 1.198:1:0.443$, $\beta = 90° 27'$; $Z = 4$. *d's:* 4.78(10), 2.785(4), 2.707(6), 2.446(6), 2.085(4).

Physical Properties. *Cleavage* {010} perfect; cleavage laminae flexible but not elastic. Sectile. **H** $1\frac{1}{2}$–2. **G** 3.49. *Luster* resinous, pearly on cleavage face. *Color* lemon-yellow. *Streak* pale yellow. Translucent.

Composition. As 61, S 39 per cent.

Diagnostic Features. Characterized by its yellow color and foliated structure. Distinguished from sulfur by its perfect cleavage. Gives the same tests as realgar.

Occurrence. Orpiment is a rare mineral, associated usually with realgar and formed under similar conditions. Found in various places in Rumania, Kurdistan, Peru, Japan, etc. In the United States it occurs at Mercur, Utah, and at Manhattan, Nevada. Deposited with realgar from geyser waters in the Norris Geyser Basin, Yellowstone National Park.

Use. Used in dyeing and in a preparation for the removal of hair from skins. Artificial arsenic sulfide is largely used in place of the mineral. Both realgar and orpiment were formerly used as pigments but this use has been discontinued because of their poisonous nature.

Name. Derived from the Latin, *auripigmentum*, "golden paint," in allusion to its color and because the substance was supposed to contain gold.

STIBNITE—Sb$_2$S$_3$

Crystallography. Orthorhombic; $2/m2/m2/m$. Slender prismatic habit, prism zone vertically striated. Crystals often steeply terminated (Fig. 301). Crystals sometimes curved or bent (Fig. 302). Often in radiating crystal groups or in bladed forms with prominent cleavage. Massive, coarse to fine granular.

Angles: $b(010) \wedge m(110) = 45° 12'$; $b(010) \wedge n(210) = 63° 36'$; $s(111) \wedge m(110) = 64° 17'$; $p(331) \wedge m(110) = 34° 42'$.

Pbnm; $a = 11.22$, $b = 11.30$, $c = 3.84$ Å; $a:b:c = 0.993:1:0.340$; $Z = 4$. *d's:* 5.07(4), 3.58(10), 2.76(3), 2.52(4), 1.933(5).

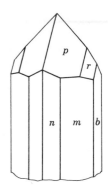

FIG. 301. Stibnite.

Physical Properties. *Cleavage* {010} perfect, showing striations parallel to [100]. **H** 2.0. **G** 4.52–4.62. *Luster* metallic, splendent on cleavage surfaces. *Color* and *streak* lead-gray to black. Opaque.

Composition. Sb 71.4, S 28.6 per cent. May carry small amounts of gold, silver, iron, lead, and copper.

Diagnostic Features. Characterized by its easy fusibility, bladed habit, perfect cleavage in one direction, lead-gray color, and soft black streak. Fusible at 1. When heated on charcoal it gives a dense white coating and the odor of sulfur dioxide. When roasted in the open tube it gives a nonvolatile white sublimate near the bottom of tube and a white volatile sublimate as a ring around tube.

Occurrence. Stibnite is found in low-temperatures hydrothermal veins or replacement deposits and in hot spring deposits. It is associated with other antimony minerals which have formed as the product of its decomposition, and with galena, cinnabar, sphalerite, barite, realgar, orpiment, and gold.

The finest crystals of stibnite have come from the province of Iyo, Island of Shikoku, Japan. The world's most important producing district is in the province of Hunan, China. It occurs also in Algeria, Borneo, Bolivia, Peru, and Mexico.

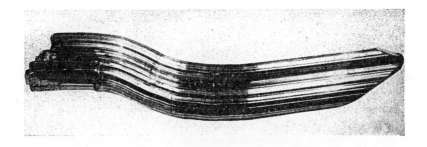

FIG. 302. Curved stibnite crystal, Ischinokowa, Japan.

Found in quantity at only a few localities in the United States, the chief deposits being in California, Nevada, and Idaho.

Use. The chief ore of antimony. The metal is used in various alloys, as antimonial lead for storage batteries, type metal, pewter, babbitt, britannia metals, and antifriction metal. The sulfide is employed in fireworks, matches, percussion caps, vulcanizing rubber and in medicines. Antimony trioxide is used as a pigment and for making glass.

Name. The name stibnite comes from an old Greek word that was applied to the mineral.

Similar Species. Bismuthinite, Bi_2S_3 is a rare mineral isostructural with stibnite with similar crystallographic and physical properties.

PYRITE—FeS_2

Crystallography. Isometric; $2/m\bar{3}$. Frequently in crystals (Fig. 303). The most common forms are the cube, the faces of which are usually striated; the pyritohedron and the octahedron. Figure 303f shows a penetration twin, known as the *iron cross* with [011] the twin axis. Also massive, granular, reniform, globular, and stalactitic.

$Pa3$; $a = 5.42$ Å, $Z = 4$. d's: 2.70(7), 2.42(6), 2.21(5), 1.917(4), 1.632(10).

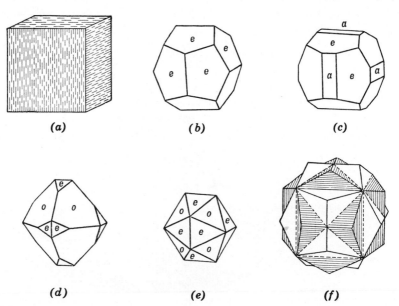

FIG. 303.   Pyrite crystals. (*a*) Striated cube. (*b*) Pyritohedron {210}. (*c*) Cube and pyritohedron. (*d*) and (*e*) Octahedron and pyritohedron. (*f*) Twinned pyritohedrons, iron cross.

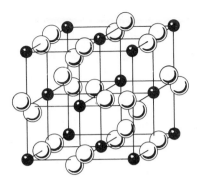

FIG. 304.   Pyrite structure.

Pyrite has a modified type of NaCl structure (Fig. 304) with Fe occupying the position of Na and with S_2 groups occupying the position of Cl. The sulfur pairs are joined along the 3-fold axes, and each sulfur of a pair touches three iron atoms. Each iron atom is surrounded by six sulfur atoms. It will be noted that only one sulfur pair in four lies along a given 3-fold axis (Fig. 305).

Physical Properties.   *Fracture* conchoidal. Brittle. **H** $6-6\frac{1}{2}$ (unusually hard for a sulfide). **G** 5.02. *Luster* metallic, splendent. *Color* pale brass-yellow; may be darker because of tarnish. *Streak* greenish or brownish black. Opaque. Paramagnetic.

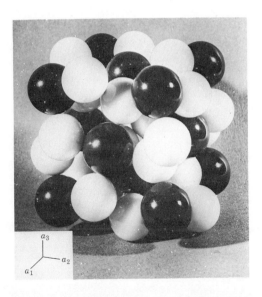

FIG. 305.   Pyrite, FeS_2, packing model. Fe black; S white. Note that the sulfur pairs are aligned along the 3-fold symmetry axes.

Composition. Fe 46.6, S 53.4 per cent. May contain small amounts of nickel and cobalt. Some analyses show considerable nickel, and a complete solid-solution series exists between pyrite and *bravoite*, (Ni, Fe)S$_2$. Frequently carries minute quantities of gold and copper but presumably as microscopic impurities.

Diagnostic Features. Distinguished from chalcopyrite by its paler color and greater hardness, from gold by its brittleness and hardness, and from marcasite by its deeper color and crystal form. Fusible at $2\frac{1}{2}$–3 to a magnetic globule. Yields much sulfur in the closed tube, and sulfur dioxide when heated in the open tube.

Alteration. Pyrite is easily altered to oxides of iron, usually limonite. In general, however, it is much more stable than marcasite. Pseudomorphic crystals of limonite after pyrite are common. Pyrite veins are usually capped by a cellular deposit of limonite, termed *gossan*. Rocks that contain pyrite are unsuitable for structural purposes because the ready oxidation of pyrite would serve both to disintegrate the rock and to stain it with iron oxide.

Occurrence. Pyrite is the most common and widespread of the sulfide minerals. It has formed at both high and low temperatures, but the largest masses probably at high temperature. It occurs as a magmatic segregation, as an accessory mineral in igneous rock, and in contact metamorphic deposits and hydrothermal veins. Pyrite is a common mineral in sedimentary rocks, being both primary and secondary. It is associated with many minerals but found most frequently with chalcopyrite, sphalerite, galena.

Large and extensively developed deposits occur at Rio Tinto and elsewhere in Spain and also in Portugal. Important deposits of pyrite in the United States are in Prince William, Louisa, and Pulaski counties, Virginia, where it occurs in large lenticular masses which conform in position to the foliation of the inclosing schists; in St. Lawrence County, New York; at the Davis Mine, near Charlemont, Massachusetts; and in various places in California, Colorado, and Arizona.

Use. Pyrite is often mined for the gold or copper associated with it. Because of the large amount of sulfur present in the mineral it is used as an iron ore only in those countries where oxide ores are not available. It's chief use is a source of sulfur for sulfuric acid and *copperas* (ferrous sulfate). Copperas is used in dyeing, in the manufacture of inks, as a preservative of wood, and as a disinfectant.

Name. The name *pyrite* is from a Greek word meaning *fire*, in allusion to the brilliant sparks emitted when struck by steel.

COBALTITE—(Co,Fe)AsS

Crystallography. Isometric; $2/m\bar{3}$. Commonly in cubes or pyritohedrons with the faces striated as in pyrite. Also granular.

$P2_13$; $a = 5.57$ Å; $Z = 4$. d's = 2.78(5), 2.48(10), 2.27(7), 1.676(9), 1.488(5).

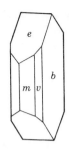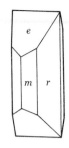

FIG. 306. Marcasite crystals.

Physical Properties. *Cleavage* {001} perfect. Brittle. **H** $5\frac{1}{2}$. **G** 6.33. *Luster* metallic. *Color* silver-white, inclined to red. *Streak* grayish black.

Composition. Usually contains considerable iron (maximum about 10 per cent) and lesser amounts of nickel. *Gersdorffite*, NiAsS, and cobaltite form a complete solid-solution series, but intermediate members are rare.

Diagnostic Features. Although in crystal form cobaltite resembles pyrite, it can be distinguished by its silver color and cleavage. Fusible at 2–3. On charcoal it gives a volatile white sublimate and characteristic garlic odor. In the oxidizing flame in borax bead it gives deep blue color (cobalt).

Occurrence. Cobaltite is usually found in high-temperature deposits, as disseminations in metamorphosed rocks, or in vein deposits with other cobalt and nickel minerals. Notable occurrences of cobaltite are at Tunaberg, Sweden, and Cobalt, Ontario. The largest producer of cobalt today is the Congo, where oxidized cobalt and copper ores are associated.

Use. An ore of cobalt.

MARCASITE—FeS$_2$

Crystallography. Orthorhombic; $2/m2/m2/m$. Crystals commonly tabular {010}; less commonly prismatic [001] (Fig. 306). Often twinned, giving cockscomb and spear-shaped groups (Fig. 307). Usually in radiating forms. Often stalactitic,

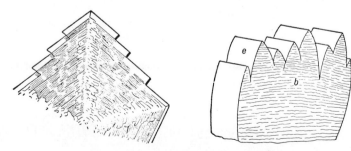

FIG. 307. "Cockscomb" marcasite.

having an inner core with radiating structure and covered with irregular crystal groups. Also globular and reniform.

Angles: $b(010) \wedge m(110) = 50° 40'$, $b(010) \wedge r(140) = 16° 15'$, $e(101) \wedge e'(\bar{1}01) = 105° 21'$.

Pmnn; $a = 4.45$, $b = 5.42$, $c = 3.39$ Å; $a:b:c = 0.821:1:0.625$. $Z = 2$. *d's:* 2.70(10), 2.41(6), 2.32(6), 1.911(5), 1.755(9).

Physical Properties. **H** $6–6\frac{1}{2}$. **G** 4.89. *Luster* metallic. *Color* pale bronze-yellow to almost white on fresh fracture, hence called *white iron pyrites.* Yellow to brown tarnish. *Streak* grayish-black. Opaque.

Composition. Dimorphous with pyrite but of more constant composition.

Diagnostic Features. Usually recognized and distinguished from pyrite by its pale yellow color, its crystals or its fibrous habit, and the following chemical test. When fine powder is treated with cold nitric acid, and the solution is allowed to stand until vigorous action ceases and is then boiled, the mineral is decomposed with separation of sulfur. Pyrite treated in the same manner would have been completely dissolved. Blowpipe tests same as for pyrite.

Alteration. Marcasite usually disintegrates more easily than pyrite with the formation of ferrous sulfate and sulfuric acid. The white powder that forms on marcasite is *melanterite*, $FeSO_4 \cdot 7H_2O$.

Occurrence. Marcasite is found in metalliferous veins, frequently with lead and zinc ores. It is less stable than pyrite, being easily decomposed, and is much less common. It is deposited at low temperatures from acid solutions and commonly formed under surface conditions as a supergene mineral. Marcasite most frequently occurs as replacement deposits in limestone, and often in concretions imbedded in clays, marls, and shales.

Found abundantly in clay near Carlsbad and elsewhere in Bohemia; in various places in Saxony; and in the chalk marl of Folkestone and Dover, England. In the United States marcasite is found with zinc and lead deposits of the Joplin, Missouri district; at Mineral Point, Wisconsin; and at Galena, Illinois.

Use. Marcasite is used to a slight extent as a source of sulfur.

Name. Derived from an Arabic word, at one time applied generally to pyrite.

ARSENOPYRITE—FeAsS

Crystallography. Monoclinic; $2/m$, pseudo-orthorhombic. Crystals are commonly prismatic elongated on c and less commonly on b (Fig. 308). Twinning (1) on $\{100\}$ and $\{001\}$ produces pseudo-orthorhombic crystals; (2) on $\{110\}$ as contact or penetration twins; may be repeated, as in marcasite.

Angles: $q(210) \wedge q'(2\bar{1}0) = 80° 10'$, $u(120) \wedge u'(1\bar{2}0) = 146° 55'$, $n(101) \wedge n'(\bar{1}01) = 68° 13'$.

$P2_1c;\ a = 9.53,\ b = 5.66,\ c = 6.43\ \text{Å},\ \beta = 90°;\ a{:}b{:}c = 1.684{:}1{:}1.136;$
$Z = 8.\ d\text{'s} = 2.66(9),\ 2.43(10),\ 1.820(8),\ 1.634(4),\ 1.348(4).$

Physical Properties. *Cleavage* {101} poor. **H** $5\frac{1}{2}$–6. **G** 6.07. *Luster* metallic. *Color* silver-white. *Streak* black. Opaque.

Composition. Essentially iron arsenide sulfide. Fe 34.3, As 46, S 19.7 per cent. Cobalt may replace part of the iron and a series extends to *glaucodot*, (Co, Fe)AsS.

Diagnostic Features. Distinguished from marcasite by its silver-white color. Its crystal form and lack of cobalt test distinguish it from skutterudite. Fusible at 2 to magnetic globule. When heated on charcoal it gives a volatile coating and garlic odor.

Occurrence. Arsenopyrite is the most common mineral containing arsenic. It occurs with tin and tungsten ores in high-temperature hydrothermal deposits, associated with silver and copper ores, galena, sphalerite, pyrite, chalcopyrite. Frequently associated with gold. Often found sparingly in pegmatites, in contact metamorphic deposits, disseminated in crystalline limestones.

Arsenopyrite is a widespread mineral and is found in considerable abundance in many localities, as at Freiberg and Munzig, Saxony; with tin ores in Cornwall, England; from Tavistock, Devonshire; in various places in Bolivia. In the United States it is found in fine crystals at Franconia, New Hampshire; Roxbury, Connecticut; and Franklin, New Jersey. It is associated with gold at Lead, South Dakota. Large quantities occur at Deloro, Ontario.

Use. The principal source of arsenic. Most of the arsenic produced is recovered in the form of the oxide, as a by-product in the smelting of arsenical ores for copper, gold, lead, and silver. Metallic arsenic is used in some alloys, particularly with lead in shot metal. Arsenic is used chiefly, however, in the form of white arsenic or arsenious oxide in medicine, insecticides, preservatives, pigments, and glass. Arsenic sulfides are used in paints and fireworks.

Name. Arsenopyrite is a contraction of the older term *arsenical pyrites.*

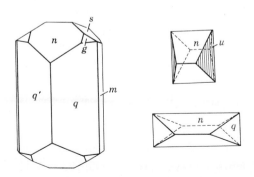

FIG. 308. Arsenopyrite crystals.

MOLYBDENITE—MoS$_2$

Crystallography. Hexagonal; $6/m2/m2/m$. Crystals in hexagonal-shaped plates or short, slightly tapering prisms. Commonly foliated, massive, or in scales.

$P6_3/mmc$; $a = 3.16$, $c = 12.32$ Å; $a:c = 1:3.889$; $Z = 2$. d's: 6.28(10), 2.28(9), 1.824(6), 1.578(4), 1.530(4).

Physical Properties. *Cleavage* {0001} perfect, laminae flexible but not elastic. Sectile. **H** 1–1$\frac{1}{2}$. **G** 4.62–4.73. Greasy feel. *Luster* metallic. *Color* lead-gray. *Streak* grayish-black. Opaque.

Composition. Mo 59.9, S 40.1 per cent.

Diagnostic Features. Resembles graphite but is distinguished from it by higher specific gravity; by a blue tone to its color, whereas graphite has a brown tinge. On glazed porcelain, it gives a greenish streak, graphite a black streak. Infusible. Roasted in the open tube gives odor of sulfur dioxide and deposit of thin plates of molybdic oxide, crossing the tube. When heated with potassium iodide and sulfur on plaster gives a deep blue sublimate.

Alteration. Molybdenite alters to yellow ferrimolybdite, $Fe_2(MoO_4)_3 \cdot 8H_2O$.

Occurrence. Molybdenite forms as an accessory mineral in certain granites; in pegmatites and aplites. Commonly in high-temperature vein deposits associated with cassiterite, scheelite, wolframite, and fluorite. Also in contact metamorphic deposits with lime silicates, scheelite, and chalcopyrite.

Occurs with the tin ores of Bohemia, from various places in Norway and Sweden, from England, China, Mexico, and New South Wales, Australia. In the United States molybdenite is found in many localities: at Blue Hill, Maine; Westmoreland, New Hampshire; in Okanogan County, Washington; Questa, New Mexico. From various pegmatites in Ontario, Canada. The bulk of the world's supply comes from Climax, Colorado, where molybdenite occurs in quartz veinlets in silicified granite with fluorite and topaz. Much molybdenum is produced at Bingham Canyon, Utah as a by-product of the copper mining.

Use. The principal ore of molybdenum.

Name. The name molybdenite comes from the Greek word meaning *lead*.

Calaverite—AuTe$_2$

Crystallography. Monoclinic; $2/m$. Rarely in distinct crystals which are elongated parallel to b; faces of this zone are deeply striated. Terminated at the ends of the b-axis with a large number of faces. Twinning on {101}, {310}, {111} frequent. Usually granular.

$C2/m$; $a = 7.19$, $b = 4.41$, $c = 5.08$ Å; $\beta = 90° 8'$; $a:b:c = 1.630:1:1.152$; $Z = 2$. d's: 3.02(10), 2.93(3), 2.20(4), 2.09(8), 1.758(3).

Physical Properties. **H** 2$\frac{1}{2}$. **G** 9.35. *Luster* metallic. *Color* brass-yellow to silver-white. *Streak* yellowish to greenish-gray. Opaque. Very brittle.

Composition. Au 44.03, Te 55.97 per cent. Silver usually replaces gold to a small extent.

Diagnostic Features. Distinguished from sylvanite by the presence of only a small amount of silver and by the lack of cleavage. Fusible at 1. On charcoal it fuses with a bluish-green flame, yielding globules of metallic gold. When decomposed in hot concentrated sulfuric acid the solution assumes a deep red color (tellurium), and a spongy mass of gold separates.

Occurrence. Calaverite is usually found in low-temperature hydrothermal veins, but also in some higher-temperature deposits. It is associated with sylvanite and other tellurides at Cripple Creek, Colorado, at Kalgoorlie, West Australia and Nagyág, Transylvania.

Use. An ore of gold.

Name. Named from Calaveras County, California, where it was originally found in the Stanislaus mine.

Similar Species. Other rare tellurides are *krennerite*, $AuTe_2$; *altaite*, $PbTe$; *hessite*, Ag_2Te; *petzite*, $(Ag, Au)_2Te$; and *nagyagite*, a sulfotelluride of lead and gold.

Sylvanite—$(Au,Ag)Te_2$

Crystallography. Monoclinic; $2/m$. Distinct crystals rare. Usually bladed or granular. Often in skeleton forms on rock surfaces resembling writing in appearance.

$P2/a$; $a = 14.62$, $b = 4.49$, $c = 8.96$ Å, $\beta = 145° 26'$; $a:b:c = 3.256:1:1.995$; $Z = 2$. d's: 3.05(10), 2.25(3), 2.15(5), 1.989(3), 1.797(2).

Physical Properties. *Cleavage* $\{010\}$ perfect. **H** $1\frac{1}{2}$–2. **G** 8–8.2. *Luster* brilliant metallic. *Color* silver-white. *Streak* gray. Opaque.

Composition. The ratio of gold to silver varies somewhat; when Au:Ag = 1:1, Te 62.1, Au 24.5, Ag 13.4 per cent.

Diagnostic Features. Fusible at 1. When heated in concentrated sulfuric acid, the solution assumes a deep red color (tellurium). When heated on charcoal, it gives a gold-silver globule and blue-green flame of tellurium. Distinguished from calaverite by its good cleavage.

Occurrence. Sylvanite is a rare mineral associated with calaverite and other tellurides, pyrite and other sulfides in small amounts, gold, quartz, chalcedony, fluorite, and carbonates. Usually in veins formed at low temperatures but may be in higher-temperature veins.

It is found at Offenbanya and Nagyág in Transylvania; at Kalgoorlie and Mulgabbie, West Australia. In the United States it is found sparingly at several localities in California and Colorado, but the most notable occurrence is at Cripple Creek, Colorado.

Use. An ore of gold and silver.

Name. Derived from Transylvania, where it was first found, and in allusion to *sylvanium*, one of the names first proposed for the element tellurium.

SKUTTERUDITE—$(Co,Ni)As_3$

Crystallography. Isometric; $2/m\overline{3}$. Common crystal forms are cube and octahedron, more rarely dodecahedron and pyritohedron. Usually massive, dense to granular.

$Im3$; $a = 8.21–8.29$ Å, $Z = 8$; d's: 2.61(10), 2.20(8), 1.841(9), 1.681(7), 1.616(9).

Physical Properties. H $5\frac{1}{2}$–6. G 6.5 ± 0.4. Brittle. *Luster* metallic. *Color* tin-white to silver-gray. *Streak* black. Opaque.

Composition. Essentially cobalt and nickel arsenide but iron usually substitutes for some nickel or cobalt. The high nickel varieties are called *nickel-skutterudite*. *Smaltite*, $(Co,Ni)As_{3-x}$ and *cloanthite*, $(Ni,Co)As_{3-x}$ (with $x = 0.5–1.0$) have skutterudite structure but a deficiency of As.

Diagnostic Features. Fusible at $2–2\frac{1}{2}$. Before a blowpipe on charcoal it gives a volatile coating of arsenious oxide with garlic odor. In borax bead gives blue color (cobalt). It is necessary to make the test for cobalt to distinguish skutterudite from massive arsenopyrite.

Occurrence. Skutterudite is usually found with cobaltite and niccolite in veins formed at moderate temperature. Native silver, bismuth, arsenopyrite, and calcite are also commonly associated with it.

Notable localities are Skutterude, Norway; Annaberg, Schneeberg, and Freiberg, in Saxony; and Cobalt, Ontario, where skutterudite is associated with silver ores.

Use. An ore of cobalt and nickel. Cobalt is chiefly used in alloys for making permanent magnets and high-speed tool steel. Cobalt oxide is used as blue pigment in pottery and glassware.

Name. Skutterudite from the locality, Skutterude, Norway.

Similar Species. *Linnaeite*, Co_3S_4, associated with cobalt and nickel minerals.

SULFOSALTS

The term sulfosalt was originally proposed to indicate that a compound was a salt of one of a series of acids in which sulfur had replaced the oxygen of an ordinary acid. Since such acids may be purely hypothetical, it is perhaps misleading to endeavor to thus explain this class of minerals. Nevertheless, the term sulfosalt is serviceable and is retained for indicating a certain type of unoxidized sulfur mineral distinct from a sulfide.

The sulfides are those minerals in which a metal or a semimetal is combined with sulfur. If both a semimetal and a metal are present, the semimetal takes the place of sulfur in the structure, as in arsenopyrite, FeAsS, and thus acts as an electronegative element. In the sulfosalts, the semimetals play a role more or less like that of metals in the structure, and thus, in a sense, the sulfosalts may be considered double sulfides. Enargite Cu_3AsS_4, may be considered as $3Cu_2S \cdot As_2S_5$. There are nearly 100 sulfosalts, but only a few are important enough to warrant description here.

Sulfosalts

Pyrargyrite	Ag_3SbS_3	Enargite	Cu_3AsS_4
Proustite	Ag_3AsS_3	Bournonite	$PbCuSbS_3$
Tetrahedrite	$Cu_{12}Sb_4S_{13}$	Jamesonite	$Pb_4FeSb_6S_{14}$
Tennantite	$Cu_{12}As_4S_{13}$		

Pyrargyrite—Ag_3SbS_2; Proustite—Ag_3AsS_2

Dark Ruby Silver *Light Ruby Silver*

These minerals, the *ruby silvers*, are isostructural with similar crystal forms, physical properties, and occurrence; thus the general description that follows applies to both.

Crystallography. Hexagonal—R; $3m$. Crystals commonly prismatic with hemimorphic development with ditrigonal pyramid $\{21\overline{3}1\}$ and trigonal pyramid $\{10\overline{1}1\}$. Usually distorted with complex development. Commonly massive, compact, in disseminated grains.

$R3c$. Pyrargyrite: $a = 11.06$, $c = 8.74$ Å; $a:c = 1:0.790$; $Z = 6$. $d's$: 3.35(5), 3.21(8), 2.79(10), 2.58(5), 2.54(5). Proustite: $a = 10.79$, $c = 8.69$ Å; $a:c = 1:0.805$; $Z = 6$. $d's$: 3.28(8), 3.18(8), 2.76(10), 2.56(8), 2.48(8).

Physical Properties. *Cleavage* $\{10\overline{1}1\}$ distinct. *Luster* adamantine. Translucent. **H** $2-2\frac{1}{2}$. **G** 5.85 (pyrargyrite), 5.57 (proustite). *Color* and *streak* red (pyrargyrite), scarlet-vermilion (proustite).

Composition. Pyrargyrite: Ag 59.7, Sb 22.5, S 17.8 per cent. Proustite: Ag 65.4, As 15.2, S 19.4 per cent. There is very little solid solution between the two minerals.

Diagnostic Features. Fusible at 1. Characterized by brilliant luster and red color. Pyrargyrite is distinguished from proustite by darker red color and tests for antimony on charcoal (dense white coating) and in open tube. Proustite when heated on charcoal gives a volatile sublimate of arsenious oxide with characteristic garlic odor.

Occurrence. Pyrargyrite is the more common of these two minerals which are found in low-temperature silver veins as late minerals to crystallize in the sequence of primary deposition. They are associated with other silver sulfosalts, argentite, tetrahedrite, and native silver.

Notable localities are: Andreasberg, Harz Mountains; Freiberg, Saxony; Přibram, Bohemia; Guanajuato, Mexico; Chañarcillo, Chile; and in Bolivia. In the United States they are found in various silver veins in Colorado, Nevada, New Mexico, and Idaho. In Canada they occur in the silver veins at Cobalt, Ontario.

Use. Ores of silver.

Name. Pyrargyrite derived from two Greek words meaning *fire* and *silver*, in allusion to its color and composition. Proustite named in honor of the French chemist, J. L. Proust (1755–1826).

Similar Species. *Polybasite*, $Ag_{16}Sb_2S_{11}$, and *stephanite*, Ag_5SbS_4, are rare minerals of primary origin deposited at low to moderate temperatures. They are associated with other silver sulfosalts, argentite, tetrahedrite and the commoner sulfides.

Tetrahedrite—$Cu_{12}Sb_4S_{13}$;   Tennantite—$Cu_{12}As_4S_{13}$

These two isostructural minerals form a complete solid solution series. They are so similar in crystallographic and physical properties that it is impossible to distinguish them by inspection.

Crystallography. Isometric; $\bar{4}3m$. Habit tetrahedral (Fig. 309), may occur in groups of parallel crystals. Tetrahedron, tristetrahedron, dodecahedron, and cube are the common forms. Frequently in crystals. Contact and penetration twins on [111]. Also massive, coarse, or fine granular.

$\bar{I}43m$. Tetrahedrite: $a = 10.37$ Å; $Z = 2$; d's: 3.00(10), 2.60(6), 2.45(4), 1.839(10), 1.568(8). With increase in silver a increases. Tennantite: $a = 10.21$ Å; $Z = 2$; d's: 2.94(10), 2.55(6), 2.40(4), 1.803(10), 1.537(8).

Physical Properties. H $3-4\frac{1}{2}$. G 4.6–5.1 (tennantite harder but of lower specific gravity than tetrahedrite). *Luster* metallic to submetallic. *Color* grayish-black to black. *Streak* black to brown. Opaque.

Composition. Although Cu is the predominant metal, Fe is always present (1 to 13 per cent) and Zn usually (0 to 8 per cent). Less commonly Ag, Pb, and Hg substitute for Cu. The argentiferous variety, *freibergite*, may contain as much as 18 per cent Ag.

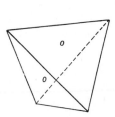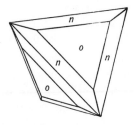

FIG. 309.   Tetrahedrite crystals.

Diagnostic Features. Recognized by their tetrahedral crystals and when massive by their brittleness, luster, and gray color. Fusible at $1\frac{1}{2}$. On charcoal or in open tube they usually give tests for both arsenic and antimony. Therefore, a quantitative analysis for these elements may be necessary to determine at which end of the series a specimen belongs. After roasting and moistening with hydrochloric acid, they give azure-blue copper chloride flame.

Occurrence. Tetrahedrite, the most common member of the sulfosalt group, is widespread in occurrence and varied in association. Tennantite is less widely distributed. Commonly found in hydrothermal veins of copper, silver, lead, and zinc minerals formed at low to moderate temperatures. Rarely in higher-temperature veins or in contact metamorphic deposits. Usually associated with chalcopyrite, pyrite, sphalerite, galena, and various other silver, lead, and copper minerals. May carry sufficient silver to become an important ore of that metal.

Notable localities: Cornwall, England; the Harz Mountains, Germany; Freiberg, Saxony; Příbram, Bohemia; various places in Rumania; and the silver mines of Mexico, Peru, and Bolivia. Found in the United States in various silver and copper mines in Colorado, Montana, Nevada, Arizona, and Utah.

Use. An ore of silver and copper.

Name. Tetrahedrite in allusion to the tetrahedral form of the crystals. Tennantite after the English chemist, Smithson Tennant (1761–1815).

ENARGITE—Cu_3AsS_4

Crystallography. Orthorhombic; $mm2$. Crystals elongated parallel to c and vertically striated, also tabular parallel to $\{001\}$. Columnar, bladed, massive.

Pnm; $a = 6.41$, $b = 7.42$, $c = 6.15$ Å; $a:b:c = 0.864:1:0.829$; $Z = 2$. d's: 3.22(10), 2.87(8), 1.859(9), 1.731(6), 1.590(5).

Physical Properties. *Cleavage* $\{110\}$ perfect, $\{100\}$ and $\{010\}$ distinct. H 3. G 4.45. *Luster* metallic. *Color* and *streak* grayish black to iron-black. Opaque.

Composition. In Cu_3AsS_4: Cu 48.3, As 19.1, S 32.6 per cent. Sb substitutes for As up to 6 per cent by weight, and some Fe and Zn are usually present.

Diagnostic Features. Characterized by its color and cleavage. Distinguished from stibnite by a test for copper. Fusible at 1. On charcoal gives volatile white sublimate of arsenious oxide and characteristic garlic odor. Roasted on charcoal, then moistened with hydrochloric acid and again ignited, gives azure-blue copper chloride flame.

Occurrence. Enargite is a comparatively rare mineral, found in vein and replacement deposits formed at moderate temperatures associated with pyrite, sphalerite, bornite, galena, tetrahedrite, covellite, chalcocite.

Notable localities: Bor, near Zaječar, Yugoslavia. Found abundantly at Morococha and Cerro de Pasco, Peru; also from Chile and Argentina; Island of Luzon, Philippines. In the United States is an important ore mineral at Butte, Montana, and to a lesser extent at Bingham Canyon, Utah. Occurs in the silver mines of the San Juan Mountains, Colorado.

Use.   An ore of copper. Arsenic oxide also obtained from it at Butte, Montana.

Name.   From the Greek meaning *distinct*, in allusion to the cleavage.

Similar Species.   *Famatinite*, Cu_3SbS_4, is the antimony analogue of enargite, but the two minerals are not isostructural.

Bournonite—PbCuSbS₃

Crystallography.   Orthorhombic; $2/m2/m2/m$. Crystals usually short prismatic to tabular. May be complex with many vertical prism and pyramid faces. Frequently twinned on {110}, giving tabular crystals with recurring reentrant angles in the [001] zone (Fig. 310), whence the common name of *cogwheel ore*. Also massive; granular to compact.

Angles: $b(010) \wedge m(110) = 46° 50'$, $(001) \wedge (101) = 43° 45'$.

Pnmm; $a = 8.16$, $b = 8.71$, $c = 7.81$ Å; $a:b:c = 0.937:1:0.897$; $Z = 2$.

d's: 4.37(4), 3.90(8), 2.99(4), 2.74(10), 2.59(5), 1.765(6).

Physical Properties.   *Cleavage* {010} imperfect. **H** $2\frac{1}{2}$–3. **G** 5.8–5.9. *Luster* metallic. *Color* and *streak* steel-gray to black. Opaque.

Composition.   The percentages of the elements in $PbCuSbS_3$: Pb 42.4, Cu 13.0, Sb 24.9, S 19.7. Arsenic may substitute for antimony to about Sb: As = 4:1.

Diagnostic Features.   Recognized by its characteristic crystals, high specific gravity, and the following tests. Fusible at 1. When roasted in the open tube it gives sublimates of antimony oxides. When heated on charcoal with a mixture of potassium iodide and sulfur it gives a chrome-yellow coating of lead iodide.

Occurrence.   Bournonite, one of the commonest of the sulfosalts, occurs typically in hydrothermal veins formed at moderate temperature. It is associated with galena, tetrahedrite, chalcopyrite, sphalerite, pyrite. Frequently noted as microscopic inclusions in galena.

Notable localities are: the Harz Mountains; Kapnik and elsewhere in Rumania; Liskeard, Cornwall; also found in Australia, Mexico, and Bolivia. In the United

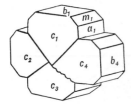

FIG. 310.   Bournonite.

States, if has been found in various places in Arizona, Utah, Nevada, Colorado, and California, but not in notable amount or quality.

Use. An ore of copper, lead, and antimony.

Name. After Count J. L. de Bournon (1751–1825), French crystallographer and mineralogist.

Jamesonite—$Pb_4FeSb_6S_{14}$

Crystallography. Monoclinic; $2/m$. Usually in acicular crystals or in capillary forms with a featherlike appearance; hence called *feather ore*. Also fibrous to compact massive. Several different compounds are often designated "feather ore."

$P2_1/a$; $a = 15.71$, $b = 19.05$, $c = 4.04$ Å; $\beta = 91° 48'$; $a:b:c = 0.825:1:0.212$; $Z = 2$. d's: 3.44(10), 3.18(5), 3.09(5), 2.84(9), 2.75(8).

Physical Properties. *Cleavage* {001} good. Brittle. **H** 2–3. **G** 5.63. *Luster* metallic. *Color* and streak steel-gray to grayish-black. Opaque.

Composition. In $Pb_4FeSb_6S_{14}$: Pb 40.16, Fe 2.71, Sb 35.39, S 21.74 per cent. Copper and zinc may be present in small amounts.

Diagnostic Features. Recognized by its characteristic fibrous (feathery) appearance, and distinguished from stibnite by lack of good lengthwise cleavage. Difficult to distinguish from similar species (see below).

Occurrence. Jamesonite is found in ore veins formed at low to moderate temperatures, associated with other lead sulfosalts, with galena, stibnite, tetrahedrite, sphalerite.

Found in Cornwall, England, and from various localities in Czechoslovakia, Rumania, and Saxony; from Tasmania and Bolivia. In the United States from Sevier County, Arkansas, and at Silver City, South Dakota.

Use. A minor ore of lead.

Name. After the mineralogist Robert Jameson (1774–1854), of Edinburgh.

Similar Species. A number of minerals similar to jamesonite in composition and general physical characteristics are included under the term *feather ore*. These include such minerals as *zinkenite*, $Pb_6Sb_{14}S_{27}$; *boulangerite*, $Pb_5Sb_4S_{11}$; and *meneghinite*, $Pb_{13}Sb_7S_{23}$. Other minerals of similar composition but different habit are *plagionite*, $Pb_5Sb_8S_{17}$; *semseyite*, $Pb_9Sb_8S_4$; and *geocronite*, $Pb_5(Sb,As)_2S_8$.

OXIDES

The oxide minerals include those natural compounds in which oxygen is combined with one or more metals. They are here grouped as: simple oxides, multiple oxides, and hydroxides. The simple oxides, compounds of one metal and oxygen, are of several types with different $A:O$ ratios (the ratio of metal to oxygen) as A_2O, AO, A_2O_3. Although not described on the following pages, ice H_2O, is a simple oxide of the A_2O type in which hydrogen is the cation. The multiple oxides have two nonequivalent metal atoms (A and B). When in some of these compounds hydrogen takes the place of a metal as in goethite, $HFeO_2$, they are grouped with the hydroxides. Within the oxide class are several minerals of great economic importance. These include the chief ores of iron (hematite and magnetite), chromium (chromite), manganese (pyrolusite, manganite, psilomelane), tin (cassiterite), and aluminum (bauxite).

Oxides

A_2O Type

| Cuprite | Cu_2O |
| Zincite | ZnO |

A_2O_3 Type

HEMATITE GROUP

Corundum	Al_2O_3
Hematite	Fe_2O_3
Ilmenite	$FeTiO_3$

AO_2 Type

RUTILE GROUP

Rutile	TiO_2
Pyrolusite	MnO_2
Cassiterite	SnO_2
Uraninite	UO_2

AB_2O_4 Type

SPINEL GROUP

Spinel	$MgAl_2O_4$
Gahnite	$ZnAl_2O_4$
Magnetite	Fe_3O_4
Franklinite	$(Zn,Fe,Mn)(Fe,Mn)_2O_4$
Chromite	$FeCr_2O_4$
Chrysoberyl	$BeAl_2O_4$
Columbite	$(Fe,Mn)(Nb,Ta)_2O_6$

Hydroxides

Brucite	$Mg(OH)_2$
Manganite	$MnO(OH)$
Psilomelane	$(Ba,H_2O)Mn_5O_{10}$

GOETHITE GROUP

Diaspore	$HAlO_2$
Goethite	$HFeO_2$
Bauxite	Al hydrates

CUPRITE—Cu_2O

Crystallography. Isometric; $4/m\bar{3}2/m$. Commonly occurs in crystals showing the cube, octahedron, and dodecahedron, frequently in combination (Fig. 311). Sometimes in elongated capillary crystals, known as "plush copper" or *chalcotrichite*.

$Pn3m$; $a = 4.27$ Å; $Z = 2$. d's: 2.46(10), 2.13(6), 1.506(5), 1.284(4), 0.978(3).

Physical Properties. H $3\frac{1}{2}$–4. G 6.1. *Luster* metallic-adamantine in clear crystallized varieties. *Color* red of various shades; ruby-red in transparent crystals, called "ruby copper." *Streak* brownish-red.

Composition. Cu 88.8, O 11.2 per cent. Usually pure, but iron oxide may be present as an impurity.

Diagnostic Features. Usually distinguished from other red minerals by its crystal form, high luster, streak, and association with limonite. Fusible at 3.

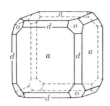 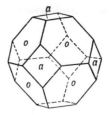

FIG. 311. Cuprite crystals.

Gives azure-blue copper chloride flame if moistened with hydrochloric acid and then heated. Gives globule of copper on charcoal in the reducing flame. When dissolved in concentrated hydrochloric acid and solution diluted with cold water, it gives a white precipitate of cuprous chloride (tests for cuprous copper).

Occurrence. Cuprite is a supergene copper mineral and in places an important copper ore. It is found in the upper oxidized portions of copper veins, associated with limonite and the other secondary copper minerals as native copper, malachite, azurite, and chrysocolla.

Noteworthy foreign countries where cuprite is an ore are Chile, Bolivia, Australia, and the Congo. Fine crystals have been found at Cornwall, England; Chessy, France; and the Ural Mountains. Found in the United States in excellent crystals in the copper deposits at Bisbee, Arizona. Also found at Clifton and Morenci, Arizona.

Use. A minor ore of copper.

Name. Derived from the Latin *cuprum*, copper.

Similar Species. *Tenorite* or *melaconite*, the cupric oxide, CuO, is a black supergene mineral.

ZINCITE—ZnO

Crystallography. Hexagonal; 6*mm*. Crystals are rare, terminated at one end by faces of a steep pyramid and the other by a pedion (see Fig. 112). Usually massive with platy or granular appearance.

$P6_3mc$; $a = 3.25$, $c = 5.19$ Å; $a:c = 1:1.625$; $Z = 2$. *d's:* 2.83(7), 2.62(5), 2.49(10), 1.634(6), 1.486(6).

Physical Properties. *Cleavage* $\{10\overline{1}0\}$ perfect; $\{0001\}$ parting. **H** 4. **G** 5.68. *Luster* subadamantine. *Color* deep red to orange-yellow. *Streak* orange-yellow. Translucent. *Optics:* (+); $\omega = 2.013$, $\varepsilon = 2.029$.

Composition. Zn 80.3, O 19.7 per cent. Divalent manganese often present and probably colors the mineral; pure ZnO is white.

Diagnostic Features. Distinguished chiefly by its red color, orange-yellow streak, and the association with franklinite and willemite. Infusible. Soluble in hydrochloric acid. When mixed with reducing mixture and intensely heated on charcoal it gives a nonvolatile coating of zinc oxide, yellow when hot, white when cold. Usually gives a blue color to the sodium carbonate beads (manganese).

Occurrence. Zincite is confined almost exclusively to the zinc deposits at Franklin, New Jersey, where it is associated with franklinite and willemite in calcite, often in an intimate mixture. Reported only in small amounts from other localities.

Use. An ore of zinc, particularly used for the production of zinc white (zinc oxide).

Oxides of the A_2O_3 Type—Hematite Group

In the hematite group are the isostructural minerals corundum and hematite with symmetry $\overline{3}2/m$, and ilmenite with symmetry $\overline{3}$. The oxygen ions are essentially in hexagonal closest packing with layers parallel to $\{0001\}$. The cations lie between the oxygen layers in octahedral coordination with oxygen but with only two-thirds of the possible sites occupied. One can thus think of the structure as layers of oxygen ions alternating with layers of cations. In ilmenite the symmetry planes and 2-fold axes are not present since the cation layers contain a mixture of iron and titanium ions.

The long-established morphological unit with c one-half the length of c of the structural unit is retained here and the indices of forms of these minerals are expressed according to it.

CORUNDUM—Al₂O₃

Crystallography. Hexagonal—R; $\overline{3}2/m$. Crystals commonly tabular on $\{0001\}$ or prismatic $\{11\overline{2}0\}$. Often tapering hexagonal pyramids (Fig. 312b), rounded into barrel shapes with deep horizontal striations. May show rhombohedral faces. Usually rudely crystallized or massive; coarse or fine granular. Polysynthetic twinning is common on $\{10\overline{1}1\}$ and $\{0001\}$.

Angles: $c(0001) \wedge r(10\overline{1}1) = 57° 35'$, $c(0001) \wedge n(22\overline{4}3) = 61° 12'$, $r(10\overline{1}1) \wedge r'(\overline{1}101) = 93° 56'$, $n(22\overline{4}3) \wedge n'(\overline{2}4\overline{2}3) = 51° 58'$.

$R\overline{3}c$; $a = 4.76$, $c = 12.98$ Å; $a:c/2 = 1:1.363$, $Z = 6$. *d's:* 2.54(6), 2.08(9), 1.738(5), 1.599(10), 1.374(7).

Physical Properties. Parting $\{0001\}$ and $\{10\overline{1}1\}$, the latter giving nearly cubic angles; more rarely prismatic parting. **H** 9. Corundum may alter to mica, and care should be exercised in obtaining a fresh surface for hardness test. **G** 4.02.

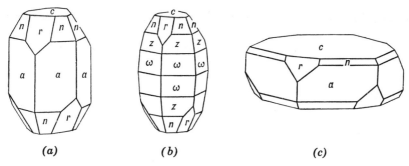

FIG. 312. Corundum crystals.

Luster adamantine to vitreous. Transparent to translucent. *Color* various; usually some shade of brown, pink, or blue. May be white, gray, green, red (ruby), or blue (sapphire). Rubies and sapphires having a stellate opalescence when viewed in the direction of the *c* crystal axis are termed *star ruby* or *star sapphire. Optics:* $(-)$, $\omega = 1.769$, $\varepsilon = 1.760$.

Emery is a black granular corundum intimately mixed with magnetite, hematite, or hercynite.

Composition. Al 52.9, O 47.1 per cent.

Diagnostic Features. Characterized chiefly by its great hardness, high luster, specific gravity, and parting. Infusible. Insoluble. Finely pulverized material moistened with cobalt nitrate and intensely ignited becomes blue (aluminum).

Occurrence. Corundum is common as an accessory mineral in the metamorphic rocks, such as crystalline limestone, mica-schist, gneiss. Found also as an original constituent of certain igneous rocks; usually those deficient in silica, as syenites and nepheline syenites. May be found in large masses in the zone separating peridotites from adjacent country rocks. It is disseminated in small crystals through certain lamprophyric dikes and is found in large crystals in pegmatites. Found frequently in crystals and rolled pebbles in detrital soil and stream sands, where it has been preserved through its hardness and chemical inertness. Associated minerals are commonly chlorite, micas, olivine, serpentine, magnetite, spinel, kyanite, and diaspore.

The finest rubies have come from Burma; the most important locality is near Mogok, 90 miles north of Mandalay. Some stones are found here in the metamorphosed limestone that underlies the area, but most of them have been recovered from the overlying soil and associated stream gravels. Darker, poorer quality rubies have been found in alluvial deposits near Bangkok, Thailand and at Battambang, Cambodia. In Ceylon rubies of relatively inferior grade, associated with more abundant sapphires and other gem stones are recovered from stream gravels. In the United States a few rubies have been found associated with the large corundum deposits of North Carolina.

Sapphires are found in the alluvial deposits of Thailand, Ceylon, and Cambodia associated with rubies. The stones from Cambodia of a cornflower-blue color are most highly prized. Sapphires occur in Kashmir, India, and are found over an extensive area in central Queensland, Australia. In the United States small sapphires of fine color are found in various localities in Montana. They were first discovered in the river sands east of Helena during placer operations for gold, and later found imbedded in the rock of a lamprophyre dike at Yogo Gulch. The rock was formerly mined for sapphires but mining operations ceased in 1929.

Common abrasive corundum is found in large crystals in the Transvaal, Republic of South Africa. These crystals found loose in the soil or in pegmatites constitute a major source of abrasive corundum. Common corundum is found in the United States in various localities along the eastern edge of the Appalachian Mountains in North Carolina and Georgia. At one time it was extensively mined in south-western North Carolina. It occurs here in large masses lying at the edges of intruded masses of an olivine rock (dunite) and is thought to have been a separation from the original magma. Found as an original constituent of a nepheline syenite in Ontario, Canada, where in places it forms more than 10 per cent of the rock mass.

Emery is found in large quantities on Cape Emeri on the island of Naxos and in various localities in Asia Minor, where it has been extensively mined for centuries. In the United States emery has been mined at Chester, Massachusetts, and Peekskill, New York.

Artificial. Artificial corundum is manufactured from bauxite on a large scale. This synthetic material, together with other manufactured abrasives, notably silicon carbide, has largely taken the place of natural corundum as an abrasive.

Synthetic rubies and sapphires, colored with small amounts of chromium and titanium are synthesized by fusing alumina powder in an oxy-hydrogen flame which on cooling forms single crystal "boules." This, the Verneuil process, has been in use since 1902. In 1947 the Linde Air Products of the United States succeeded in synthesizing star rubies and star sapphires. This was accomplished by introducing titanium which, during proper heat treatment, exsolved as oriented rutile needles to produce the star. The artificial rival the natural stones in beauty, and it is difficult for the untrained person to distinguish them.

Use.   As a gem stone and abrasive. The deep red ruby is one of the most valuable of gems, second only to emerald. The blue sapphire is also valuable, and stones of other colors may command good prices. Stones of gem quality are used as watch jewels and as bearings in scientific instruments. Corundum is used as an abrasive, either ground from the pure massive material or in its impure form as emery.

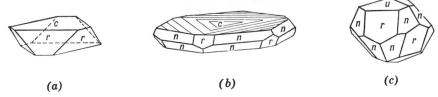

FIG. 313. Hematite crystals.

HEMATITE—Fe₂O₃

Crystallography. Hexagonal—R; $\bar{3}2/m$. Crystals are usually thick to thin tabular on {0001}, basal planes often show triangular markings and the edges of the plates may be beveled with rhombohedral forms (Fig. 313b). Thin plates may be grouped in rosettes (iron roses) (Fig. 314). More rarely crystals are distinctly rhombohedral, often with nearly cubic angles and may be polysynthetically twinned on {0001} and {10$\bar{1}$1}. Also in botryoidal to reniform shapes with radiating structure, *kidney ore* (see Fig. 200, page 103). May also be micaceous and foliated, *specular*. Usually earthy; called *martite* when in octahedral pseudomorphs after magnetite.

Angles: $c(0001) \wedge r(10\bar{1}1) = 57° 37'$, $c(0001) \wedge n(22\bar{4}3) = 61° 13'$, $r(10\bar{1}1) \wedge r'(\bar{1}101) = 94°$, $n(22\bar{4}3) \wedge n'(\bar{2}4\bar{2}3) = 51° 59'$.

$R\bar{3}c$; $a = 5.04$, $c = 13.76$ Å; $a:c/2 = 1:1.365$; $Z = 6$. *d's:* 2.69(10), 2.52(8), 2.21(4), 1.843(6), 1.697(7).

FIG. 314. Hematite, iron rose.

Physical Properties. *Parting* on $\{10\bar{1}1\}$ with nearly cubic angles, also on $\{0001\}$. **H** $5\frac{1}{2}$–$6\frac{1}{2}$, **G** 5.26 for crystals. *Luster* metallic in crystals and dull in earthy varieties. *Color* reddish-brown to black. Red earthy variety is known as *red ocher*. *Streak* light to dark red which becomes black on heating. Translucent.

Composition. Fe 70, O 30 per cent. May contain titanium.

Diagnostic Features. Distinguished chiefly by its characteristic red streak. Infusible. Becomes strongly magnetic on heating in the reducing flame. Slowly soluble in hydrochloric acid; solution with potassium ferrocyanide gives dark blue precipitate (test for ferric iron).

Occurrence. Hematite is widely distributed in rocks of all ages and forms the most abundant and important ore of iron. It may occur as a sublimation product in connection with volcanic activities. Occurs in contact metamorphic deposits and as an accessory mineral in feldspathic igneous rocks such as granite. Found from microscopic scales to enormous masses in regionally metamorphosed rocks where it may have originated by the alteration of limonite, siderite, or magnetite. It is found in red sandstones as the cementing material that binds the quartz grains together. The hematite in a deposit that can be mined economically must be measured in tens of millions of tons. Such major accumulations are largely sedimentary and by leaching of associated silica by meteoric waters many of them have been enriched to high-grade ores (over 50 per cent Fe). Like limonite, it may be formed in irregular masses and beds as the result of the weathering of iron-bearing rocks. The oölitic ores are of sedimentary origin and may occur in beds of considerable size.

Noteworthy localities for hematite crystals are the island of Elba; St. Gothard, Switzerland, in "iron roses"; in the lavas of Vesuvius; at Cleator Moor, Cumberland, England; Minas Gerais, Brazil.

In the United States the columnar and earthy varieties are found in enormous bedded deposits that have furnished a large proportion of the iron ore of the world. The chief iron-ore districts of the United States are grouped around the southern and northwestern shores of Lake Superior in Michigan, Wisconsin, and Minnesota. The chief districts, which are spoken of as iron ranges, are, from east to west, the Marquette in northern Michigan; the Menominee in Michigan to the southwest of the Marquette; the Penokee-Gogebic in northern Wisconsin. In Minnesota the Mesabi, northwest of Duluth; the Vermilion, near the Canadian boundary; and the Cuyuna, southwest of the Mesabi. The iron ore of these different ranges varies from the hard specular variety to the soft red earthy type. Of the several ranges the Mesabi is the largest, and since 1892 has yielded over 2.5 billion tons of high-grade ore, over twice the total production of all the other ranges.

Oölitic hematite is found in the United States in the rocks of the Clinton formation, which extends from central New York south along the line of the

Appalachian Mountains to central Alabama. The most important deposits of the series lie in eastern Tennessee and northern Alabama, near Birmingham. Hematite has been found at Iron Mountain and Pilot Knob in southeastern Missouri. Deposits of considerable importance are located in Wyoming in Laramie and Carbon counties.

Although the production of iron ore within the United States remains large, the rich deposits are being rapidly worked out. In the future much of the iron must come from low-grade deposits or must be imported. The low-grade, silica-rich iron formation from which the high-grade deposits have been derived is known as *taconite* which contains about 25 per cent iron. The iron reserves in taconite are far greater than were the original reserves of high-grade ore, and at present many millions of tons of iron are being concentrated from it each year.

Exploration outside the United States has in recent years been successful in locating several ore bodies with many hundreds of millions of tons of high-grade ore. These are notably in Venezuela, Brazil, and Canada. Brazil's iron mountain, Itabira, is estimated to have 15 billion tons of very pure hematite. Poor transportation over the 325 miles between the deposit and the sea has thus far prevented its major exploitation. In 1947 Cerro Bolivar in Venezuela was discovered as an extremely rich deposit of hematite and by 1954 ore was being shipped from there to the United States. In Canada several new iron ore deposits have been located, but the major ones lie along the boundary between Quebec and Labrador. A 350-mile railroad built into this previously inaccessible area began in 1954 to deliver iron ore to a St. Lawrence River port.

Use. Most important ore of iron. Also used in pigments, *red ocher*, and as polishing powder.

Name. Derived from a Greek word meaning *blood*, in allusion to the color of the powdered mineral.

ILMENITE—FeTiO₃

Crystallography. Hexagonal—R; $\bar{3}$. Crystals are usually thick tabular with prominent basal planes and small rhombohedral truncations. Crystal constants close to those for hematite. Often in thin plates. Usually massive, compact; also in grains or as sand.

Angles: $(0001) \wedge (10\bar{1}1) = 57° 59'$, $(0001) \wedge (22\bar{4}3) = 61° 33'$, $(10\bar{1}1) \wedge (\bar{1}101) = 94° 29'$.

$R\bar{3}$; $a = 5.09$, $c = 14.06$ Å; $a:c/2 = 1:1.381$. $Z = 6$. d's: 2.75(10), 2.54(7), 1.867(5), 1.726(8), 1.507(4).

Physical Properties. H $5\frac{1}{2}$–6. G 4.7. *Luster* metallic to submetallic. *Color* iron-black. *Streak* black to brownish-red. May be magnetic without heating. Opaque.

Composition. Fe 36.8, Ti 31.6, O 31.6 per cent. Some ferric oxide may be present, but if there is over 6 per cent it is probably a result of minute inclusions of hematite. Mg may substitute for Fe grading into *geikielite*, $(Mg, Fe)TiO_3$. Mn may also substitute for Fe.

Diagnostic Features. Ilmenite can be distinguished from hematite by its streak and from magnetite by its lack of strong magnetism. Infusible. Magnetic after heating. After fusion with sodium carbonate it can be dissolved in sulfuric acid and on the addition of hydrogen peroxide the solution turns yellow.

Occurrence. Ilmenite is a common accessory mineral in igneous rocks. It may be present in large masses in gabbros, diorites, and anorthosites as a product of magmatic segregation intimately associated with magnetite. It is also found in some pegmatites and vein deposits. As a constituent of black sands it is associated with magnetite, rutile, zircon, and monazite.

Found in large quantities at Krägerö and other localities in Norway; in Finland; and in crystals at Miask in the Ilmen Mountains, U.S.S.R. It is mined in considerable quantities from beach sands, notably in India and Brazil. In the United States found at Washington, Connecticut; in Orange County, New York; and with many of the magnetite deposits of the Adirondack region, notably at Tahawus, Essex County, where it is actively mined. A large ilmenite-hematite deposit is mined at Allard Lake, Quebec.

Use. The major source of titanium. World production of ilmenite is approximately 3 million tons. It is used principally in the manufacture of titanium dioxide which is being used in increasingly large amounts as a paint pigment replacing older pigments, notably lead compounds. Also the use of metallic titanium as a structural material is increasing. Because of its high strength-to-weight ratio, titanium is proving to be a desirable metal for aircraft construction in both frames and engines. Ilmenite cannot be used as an iron ore because of difficulties in smelting it, but mixtures of ilmenite-magnetite and ilmenite-hematite are separated so that both titanium and iron can be recovered.

Name. From the Ilmen Mountains, U.S.S.R.

Oxides of the AO_2 Type

In general, dioxides fall into two structure types. One has the fluorite structure (see Fig. 330) in which each oxygen has four cation neighbors arranged about it at the apices of a more or less regular tetrahedron, whereas each cation has eight oxygens surrounding it at the corners of a cube. The oxides of 4-valent uranium, thorium, and cerium, now of considerable interest because of their connection with nuclear chemistry, have this structure. In general, all dioxides in which the radius ratio of cation to oxygen $(R_A : R_O)$ lies within or close to the limits for 8

Radius Ratios in the AO_2 Type Oxides (R_O = 1.4 Å)

R_A	R_A/R_O	Element	Mineral	Structure Type
0.60	0.43	Mn	Pyrolusite	Rutile
0.68	0.49	Ti	Rutile	Rutile
0.71	0.51	Sn	Cassiterite	Rutile
0.94	0.67	Ce	Cerianite	Fluorite
0.97	0.69	U	Uraninite	Fluorite
1.02	0.73	Th	Thorianite	Fluorite

coordination (0.732–1) may be expected to have this structure and be isometric hexoctahedral.

The second common AO_2 structure type is that typified by rutile. In this structure, the cation is smaller, with radius ratios of $R_A:R_O$ lying between the limits 0.732 to 0.414 and hence having 6 coordination. Accordingly, there are six oxygens grouped about each cation. Since the requirements for electrical neutrality dictate that there must be half as many cations as oxygens, only half the possible A sites are filled, and there are only three cations grouped about each oxygen (Fig. 315). The effect of this reduction in number of cations is to warp the usual octahedral arrangement characteristic of 6 coordination into a configuration of lower symmetry. Hence the minerals of the rutile group are tetragonal with the prismatic habit reflecting the chainlike structure.

RUTILE—TiO_2

Crystallography. Tetragonal; $4/m2/m2/m$. Usually in prismatic crystals with dipyramid terminations common and vertically striated prism faces. Frequently in elbow twins, often repeated (Fig. 316) with twin plane $\{011\}$. Crystals frequently slender acicular. Also compact massive.

Angles: $e(101) \wedge e'(011) = 45° 2'$, $e(101) \wedge e''(\bar{1}01) = 65° 35'$, $s(111) \wedge s'(\bar{1}11) = 56° 26'$, $s(111) \wedge s''(\bar{1}\bar{1}1) = 84° 40'$.

$P4_2/mnm$; $a = 4.59$, $c = 2.96$ Å; $a:c = 1:0.645$. $Z = 2$. $d's:$ 3.24(10), 2.49(5), 2.18(3), 1.687(7), 1.354(4).

Physical Properties. *Cleavage* $\{110\}$ distinct. **H** $6-6\frac{1}{2}$. **G** 4.18–4.25. *Luster* adamantine to submetallic. *Color* red, reddish-brown to black. *Streak* pale brown. Usually subtranslucent, may be transparent. *Optics:* (+); $\omega = 2.612$, $\varepsilon = 2.899$.

Composition. Ti 60, O 40 per cent. Usually contains Fe^2 which may amount to 10 per cent; also Fe^3, Nb, and Ta may be present. When Fe^2 substitutes for Ti^4, electrical neutrality is maintained by twice as much Nb^5 and Ta^5 entering the compound. The formula can be written: $Fe_x(Nb,Ta)_{2x}Ti_{1-3x}O_2$.

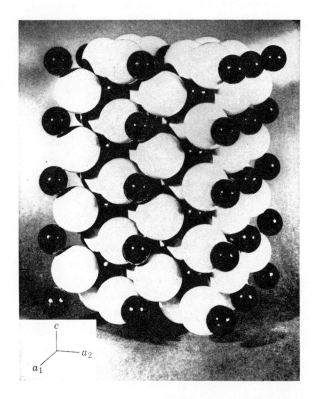

FIG. 315.   Rutile, TiO_2, packing model. O (white) is in 6 coordination about Ti (black) and Ti is in 3 coordination about O. This structure is stable for AX_2 compounds in which the radius is $0.414 - 0.732$. Cassiterite SnO_2 and pyrolusite MnO_2 also have this structure. Compare Fig. 330 (fluorite).

FIG. 316.   Rutile crystals.

Diagnostic Features. Characterized by its peculiar adamantine luster and red color. Lower specific gravity distinguishes it from cassiterite. Infusible. Insoluble. After fusion with sodium carbonate it can be dissolved in sulfuric acid; the solution turns yellow on addition of hydrogen peroxide (titanium).

Occurrence. Rutile is found in granite, granite pegmatites, gneiss, mica schist, metamorphic limestone, and dolomite. It may be present as an accessory mineral in the rock, or in quartz veins traversing it. Often occurs as slender crystals in quartz and micas. Is found in considerable quantities in black sands associated with magnetite, zircon, and monazite.

Notable European localities are: Krägerö, Norway; Yrieix, near Limoges, France; in Switzerland; and the Tyrol. Rutile from beach sands of northern New South Wales and southern Queensland makes Australia the largest producer of rutile. In the United States remarkable crystals have come from Graves Mountain, Lincoln County, Georgia. Rutile is found in Alexander County, North Carolina, and at Magnet Cove, Arkansas. It has been mined in Amherst and Nelson counties, Virginia, and derived in commercial quantities from the black sands of north-eastern Florida.

Artificial. Single crystals of rutile have been manufactured by the Verneuil process. With the proper heat treatment they can be made transparent and nearly colorless, quite different from the natural mineral. Because of its high refractive index and dispersion, this synthetic material makes a beautiful cut gem stone with only a slight yellow tinge. It is sold under a variety of names, some of the better known are *titania*, *kenya gem*, and *miridis*.

Use. Most of the rutile produced is used as a coating of welding rods. Some titanium derived from rutile is used in alloys; for electrodes in arc lights; to give a yellow color to porcelain and false teeth. Manufactured oxide is used as a paint pigment. (See ilmenite.)

Name. From the Latin *rutilus*, red, in allusion to the color.

Similar Species. *Anatase* (tetragonal) and *brookite* (orthorhombic) are rare polymorphous forms of TiO_2.

Perovskite, $CaTiO_3$, is an isometric titanium mineral found usually in metamorphic rocks.

PYROLUSITE—MnO_2

Crystallography. Tetragonal; $4/m2/m2/m$. Rarely in well-developed crystals, *polianite*. Usually in radiating fibers or columns. Also granular massive; often in reniform coats and dendritic shapes (Fig. 317). Frequently as pseudomorphs after manganite.

$P4_2/mnm$; $a = 4.39$, $c = 2.86$ Å; $a:c = 1:0.651$; $Z = 2$. $d's$: 3.11(10), 2.40(5), 2.11(4), 1.623(7), 1.303(3).

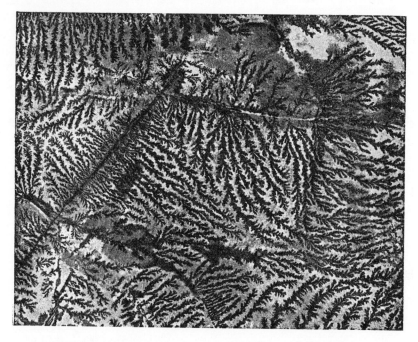

FIG. 317. Dendrites of pyrolusite on limestone, Sardinia.

Physical Properties. *Cleavage* {110} perfect. **H** 1–2 (often soiling the fingers). For coarsely crystalline polianite the hardness is 6–6½. **G** 4.75. *Luster* metallic. *Color* and streak iron-black. Fracture splintery. Opaque.

Composition. Mn 63.2, O 36.8 per cent. Commonly contains a little water.

Diagnostic Features. Characterized by and distinguished from other manganese minerals by its black streak, low hardness, and small amount of water. Infusible. Gives a bluish-green opaque bead with sodium carbonate. When heated in the closed tube it gives oxygen which causes a splinter of charcoal to ignite when placed in tube above the mineral.

Occurrence. Manganese is present in small amounts in most crystalline rocks. When dissolved from these rocks, it may be redeposited as various minerals, but chiefly as pyrolusite. Nodular deposits of pyrolusite are found in bogs, on lake bottoms, and on the floors of shallow seas. Nests and beds of manganese ores are found inclosed in residual clays, derived from the decay of manganiferous limestones. Also found in veins with quartz and various metallic minerals.

Pyrolusite is the most common manganese ore and is widespread in its occurrence. The chief manganese-producing countries are the U.S.S.R., Ghana, India, Republic of South Africa, French Morocco, Brazil, and Cuba. In the United States, manganese ores in small amount are found in Virginia, West Virginia,

Georgia, Arkansas, Tennessee, with the hematite ores of the Lake Superior districts, and in California.

Use. Most important manganese ore. Manganese is used with iron in the manufacture of *spiegeleisen* and *ferromanganese*, employed in making steel. This is its principal use for about 13.2 pounds of manganese are consumed in the production of one ton of steel. Also used in various alloys with copper, zinc, aluminum, tin, and lead. Pyrolusite is used as an oxidizer in the manufacture of chlorine, bromine, and oxygen; as a disinfectant in potassium permanganate; as a drier in paints; as a decolorizer of glass; and in electric dry-cells and batteries. Manganese is also used as a coloring material in bricks, pottery, and glass.

Name. *Pyrolusite* is derived from two Greek words meaning *fire* and *to wash*, because it is used to free glass through its oxidizing effect of the colors due to iron.

Similar Species. *Alabandite*, MnS, is comparatively rare, associated with other sulfides in veins.

Wad is the name given to manganese ore composed of an impure mixture of hydrous manganese oxides.

CASSITERITE—SnO_2

Crystallography. Tetragonal; $4/m2/m2/m$. The common forms are the prisms $\{110\}$ and $\{010\}$ and the dipyramids $\{111\}$ and $\{011\}$ (Fig. 318a). Frequently in elbow-shaped twins with a characteristic notch, giving rise to the miner's term *visor tin* (Fig. 318b); the twin plane is $\{011\}$. Usually massive granular; often in reniform shapes with radiating fibrous appearance, *wood tin*.

Angles: $m(110) \wedge s(111) = 46° 27'$, $a(100) \wedge e(101) = 56° 05'$, $s(111) \wedge s'(\bar{1}\bar{1}1) = 87° 7'$, $e(101) \wedge e'(\bar{1}01) = 67° 50'$.

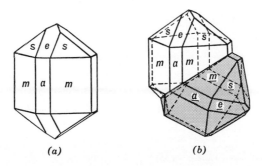

(a)   (b)

FIG. 318.   Cassiterite crystals.

FIG. 319. Cassiterite.

$P4_2/mnm$; $a = 4.73$, $c = 3.18$ Å; $a:c = 1:0.672$, $Z = 2$. $d's$: 2.36(8), 2.64(7), 1.762(10), 1.672(4), 1.212(5).

Physical Properties. Cleavage {010} imperfect. **H** 6–7. **G** 6.8–7.1 (unusually high for a nonmetallic mineral.) *Luster* adamantine to submetallic and dull. *Color* usually brown or black; rarely yellow or white. *Streak* white. Translucent, rarely transparent. *Optics:* (+); $\omega = 1.997$, $\varepsilon = 2.093$.

Composition. Close to SnO_2 with Sn 78.6, O 21.4 per cent. Small amounts of Fe^3 may be present and lesser amounts of Nb and Ta substituting for Sn.

Diagnostic Features. Recognized by its high specific gravity, adamantine luster, and light streak. Infusible. Insoluble. Gives globule of tin with coating of white tin oxide when finely powdered mineral is fused on charcoal with reducing mixture. When cassiterite is placed in dilute hydrochloric acid with metallic zinc the surface of the cassiterite is reduced and the specimen becomes coated with a dull gray deposit of metallic tin that becomes bright on rubbing.

Occurrence. Cassiterite is widely distributed in small amounts but is produced on a commercial scale in only a few localities. It has been noted as an original constituent of igneous rocks and pegmatites, but it is more commonly found in high-temperature hydrothermal veins in or near granitic rocks. Tin veins usually have minerals that contain fluorine or boron, such as tourmaline, topaz, fluorite, and apatite, and the minerals of the wall rocks are commonly much altered. Frequently associated with wolframite, molybdenite and arsenopyrite. Cassiterite is commonly found in the form of rolled pebbles in placer deposits, *stream tin*.

Most of the world's supply of tin comes from alluvial deposits chiefly in Malaysia, Bolivia, U.S.S.R., Thailand, and China. Today Bolivia is the only country where a significant production comes from vein deposits; but in the past the mines of Cornwall, England were major producers. In the United States cassiterite is not found in sufficient quantities to warrant mining but is present in small amounts in numerous pegmatites.

Use. Principal ore of tin. The chief use of tin is in the manufacture of *tin plate* and *tern plate* for food containers. Tern plate is made by applying a coating of tin and lead instead of pure tin. Tin is also used with lead in solders, in babbitt metal with antimony and copper, and in bronze and bell-metal with copper. "Phosphor bronze" contains 89 per cent copper, 10 per cent tin, and 1 per cent phosphorus. Artificial tin oxide is a polishing powder.

Name. From the Greek word meaning *tin*.

Uraninite—UO_2

Crystallography. Isometric; $4/m\overline{3}2/m$. The rare crystals are usually octahedral with subordinate cube and dodecahedron faces. More commonly as the variety *pitchblende*: massive or botryoidal with a banded structure.

$Fm3m$; $a = 5.46$ Å; $Z = 4$. $d's$: 3.15(7), 1.926(6), 1.647(10), 1.255(5), 1.114(5).

Physical Properties. $H\ 5\frac{1}{2}$. G 7.5–9.7 for crystals; 6.5–9 for pitchblende. The specific gravity decreases with oxidation of U^4 to U^6. *Luster* submetallic to pitchlike, dull. *Color* black. *Streak* brownish-black.

Composition. Uraninite is always partially oxidized, and thus the actual composition lies between UO_2 and U_3O_8. Thorium can substitute for uranium, and a complete series between uraninite and thorianite, ThO_2, has been observed in artificial preparations. In addition to thorium, analyses usually show the presence of small amounts of lead, radium, cerium, yttrium, nitrogen, helium, and argon. The lead is present as the stable end product of the radioactive disintegration of both uranium and thorium. Two isotopes of lead are found: Pb^{206} results from the radioactive disintegration of the isotope of uranium U^{238}; and Pb^{207} results from the breakdown of U^{235}. In the disintegration ionized helium atoms (α particles) and electrons (β particles) are emitted. Helium is always found in uraninite. Since radioactive disintegration proceeds at a uniform rate, the accumulation of both helium and lead can be used as a measure of the time elapsed since the mineral crystallized. Both the lead-uranium ratio and the helium-uranium ratio have been used by geologists to determine the age of rocks.

It was in uraninite that helium was first discovered on earth, having been previously noted in the sun's spectrum. Also radium was discovered in uraninite.

Diagnostic Features. Characterized chiefly by its pitchy luster, high specific gravity, color, and streak. Infusible. Gives a yellow-green color to the salt of phosphorus bead. The bead fluoresces under ultraviolet light; a sensitive test for uranium. Because of its radioactivity, uraninite, as well as other uranium compounds, can be detected in small amounts by Geiger-Müller counters, ionization chambers, and similar instruments.

Occurrence. Urananite occurs as a primary constituent of granitic rocks and pegmatites. It also is found in high-temperature hydrothermal veins associated with cassiterite, chalcopyrite, pyrite, and arsenopyrite as at Cornwall, England; or with medium temperature veins with native silver and Co-Ni-As minerals as at Joachimsthal, Czechoslovakia, and Great Bear Lake, Canada. The world's most productive uranium mine is the Shinkolobwe Mine in the Congo. Here uraninite and secondary uranium minerals are in vein deposits associated with cobalt and copper minerals. The ores also carry significant amounts of Mo, W, Au, and Pt.

Wilberforce, Ontario has been a famous Canadian locality, but the present major Canadian sources are the mines at Great Bear Lake, the Beaverlodge region, Saskatchewan; and the Blind River area, Ontario. At Blind River uraninite occurs as detrital grains in a Precambrian quartz conglomerate. In the same type of deposit it is found in the gold-bearing Witwatersrand conglomerates, Republic of South Africa.

In the United States uraninite was found long ago in isolated crystals in pegmatites at Middletown, Glastonbury, and Branchville, Connecticut, and the mica mines of Mitchell County, North Carolina. A narrow vein of it was mined near Central City, Gilpin County, Colorado. Recent exploration has located many workable deposits of uraninite and associated uranium minerals on the Colorado Plateau in Arizona, Colorado, New Mexico, and Utah.

Use. Uraninite is the chief ore of uranium although other minerals are important sources of the element such as carnotite (page 367), tyuyamunite (page 368), torbernite (page 367), and autunite (page 366).

Uranium has assumed an important place among the elements because of its susceptibility to nuclear fission, a process by which the nuclei of uranium atoms are split apart with the generation of tremendous amounts of energy. This energy, first demonstrated in the atomic bomb, is now produced by nuclear-power reactors for generating electricity.

Uraninite is also the source of radium but contains it in very small amounts. Roughly, 750 tons of ore must be mined in order to furnish 12 tons of concentrates; chemical treatment of these concentrates yields about 1 gram of a radium salt. In the form of various compounds uranium has a limited use in coloring glass and porcelain, in photography, and as a chemical reagent.

Name. Uraninite in allusion to the composition.

Similar Species. *Thorianite*, ThO_2, is dark gray to black with submetallic luster. Found chiefly in pegmatites and as water-worn crystals in stream gravels.

Spinel Group, AB_2O_4

The spinel minerals form a closely related isostructural group with the general formula AB_2O_4. In the A position there is substantially complete solid solution

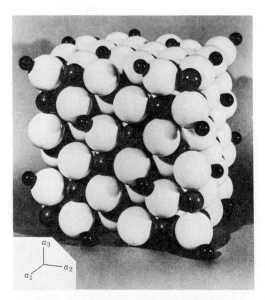

FIG. 320. Spinel packing model. Members of the spinel group have this structure in which Mg, Fe^2 or Zn (large dark) are in 6 coordination and Al, Fe^3 or Cr (small dark) are in 4 coordination with O (white).

of the divalent cations, Mg, Fe, Zn, and Mn. However, since pyroxying of the trivalent B ions, Al, Fe, Cr, for each other is more sensitive to ionic size and polarizing power, there is only incomplete solid solution. Because of the extensive solid solution the minerals of the spinel group show a wide range in color and specific gravity which depend on chemical composition. However, the crystal habit that depends on the geometry of the structure and nature of the bonding is remarkably constant throughout (Fig. 320). The following table shows the relationship of the principal members of the group.

A		$B = Al$		$B = Fe$		$B = Cr$
Mg	Spinel	$MgAl_2O_4$	Magnesio-ferrite	$MgFe_2O_4$	Magnesio-chromite	$MgCr_2O_4$
Fe	Hercynite	$FeAl_2O_4$	Magnetite	$FeFe_2O_4$	Chromite	$FeCr_2O_4$
Zn	Gahnite	$ZnAl_2O_4$	Franklinite	$ZnFe_2O_4$		
Mn	Galaxite	$MnAl_2O_4$	Jacobsite	$MnFe_2O_4$		

SPINEL—$MgAl_2O_4$

Crystallography. Isometric; $4/m\bar{3}2/m$. Usually in octahedral crystals or in twinned octahedrons (spinel twins) (Fig. 321a, b). Dodecahedron may be present

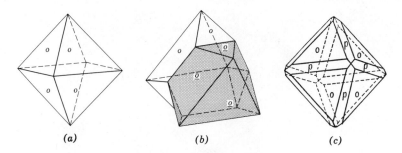

(a) (b) (c)

FIG. 321. Spinel crystals.

as small truncations (Fig. 321c), but other forms rare. Also massive and as irregular grains.

$Fd3m; a = 8.10$ Å; $Z = 8; d's:$ 2.44(9), 2.02(9), 1.552(9), 1.427(10), 1.053(10).

Physical Properties. **H** 8. **G** 3.5–4.1. **G** 3.55 for the composition as given. Nonmetallic. *Luster* vitreous. *Color* various: white, red, lavender, blue, green, brown, black. *Streak* white. Usually translucent, may be clear and transparent. *Optics:* $n = 1.718$.

Composition. MgO 28.2, Al_2O_3 71.8 per cent. Fe^2, Zn, and less commonly Mn^2 substitute for Mg in all proportions. Fe^3 and Cr may substitute in part for Al. The clear red, nearly pure magnesium spinel is known as *ruby spinel*. *Pleonaste* is the iron spinel, dark green to black, and *picotite* is the chrome spinel, yellowish to greenish-brown.

Diagnostic Features. Recognized by its hardness (8), its octahedral crystals, and its vitreous luster. The iron spinel can be distinguished from magnetite by its nonmagnetic character and white streak. Infusible. The finely powdered mineral dissolves completely in the salt of phosphorus bead (proving the absence of silica).

Occurrence. Spinel is a common metamorphic mineral occurring in crystalline limestone, gneisses, and serpentine. Occurs also as an accessory mineral in many dark igneous rocks. Spinel is frequently formed as a contact metamorphic mineral associated with phlogopite, pyrrhotite, chondrodite, and graphite. Found frequently as rolled pebbles in stream sands, where it has been preserved because of its resistant physical and chemical properties. The ruby spinels are found in this way, often associated with the gem corundum, in the sands of Ceylon, Thailand, Upper Burma, and Madagascar. Ordinary spinel is found in various localities in New York and New Jersey.

Use. When transparent and finely colored, spinel is used as a gem. Usually red and known as *ruby spinel* or *balas ruby*. Some stones are blue. The largest cut stone known weighs about 80 carats. The stones are usually comparatively inexpensive.

Artificial. Synthetic spinel has been made by the Verneuil process (see corundum) in various colors rivaling the natural stones in beauty. Synthetic spinel is also used as a refractory.

Similar Species. *Hercynite*, $FeAl_2O_4$, an iron spinel, is associated with corundum in some emery; also found with andalusite, sillimanite, and garnet. *Galaxite*, $MnAl_2O_4$, has not been found pure in nature; but *ferroan galaxite*, $(Mn, Fe)Al_2O_4$ has been reported.

Gahnite—$ZnAl_2O_4$

Crystallography. Isometric; $4/m\bar{3}2/m$. Commonly octahedral with faces striated parallel to the edge between the dodecahedron and octahedron. Less frequently showing well-developed dodecahedrons and cubes.

$Fd3m$; $a = 8.12$ Å; $Z = 8$. *d's:* 2.85(7), 2.44(10), 1.65(4), 1.48(6), 1.232(8).

Physical Properties. H $7\frac{1}{2}$–8. G 4.55. *Luster* vitreous. *Color* dark green. *Streak* grayish. Translucent. *Optics:* $n = 1.80$.

Composition. Fe^2 and Mn^2 may substitute for Zn; and Fe^3 for Al.

Diagnostic Features. Characterized by crystal form (striated octahedrons) and hardness. Infusible. The fine powder fused with sodium carbonate on charcoal gives a white nonvolatile coating of zinc oxide.

Occurrence. Gahnite is a rare mineral. It occurs in zinc deposits and also as a contact mineral in crystalline limestones. Found in large crystals at Bodenmais, Bavaria; in a talcose schist near Falun, Sweden. In the United States found at Charlemont, Massachusetts, and Franklin, New Jersey.

Name. After the Swedish chemist J. G. Gahn, the discoverer of manganese.

MAGNETITE—Fe_3O_4

Crystallography. Isometric; $4/m\bar{3}2/m$. Magnetite is frequently in octahedral crystals (Fig. 322a), occasionally twinned on {111}. More rarely in dodecahedrons (Fig. 322b). Dodecahedrons may be striated parallel to the intersection with the octahedron (Fig. 322c). Other forms rare. Usually granular massive, coarse, or fine grained.

$Fd3m$; $a = 8.40$ Å; $Z = 8$. *d's:* 2.96(6), 2.53(10), 1.611(8), 1.481(9), 1.094(8).

Physical Properties. Octahedral parting on some specimens. H 6. G 5.18. *Luster* metallic. *Color* iron-black. *Streak* black. Strongly magnetic; may act as a natural magnet, known as *lodestone*. Opaque.

Composition. Fe 72.4, O 27.6 per cent. The composition of magnetic usually corresponds closely to Fe_3O_4. However, some analyses show a few per cent of Mg and Mn^2 substituting for Fe^2 and Al, Cr, Mn^3 substituting for Fe^3.

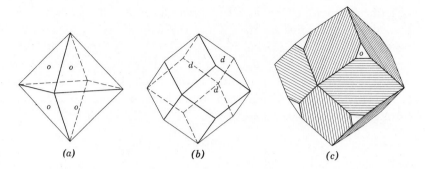

FIG. 322. Magnetite crystals.

Diagnostic Features. Characterized chiefly by its strong magnetism, its black color, and its hardness (6). Can be distinguished from magnetic franklinite by streak. Infusible. Slowly soluble in HCl, and solution reacts for both ferrous and ferric iron.

Occurrence. Magnetite is a common mineral found disseminated as an accessory through most igneous rocks. In certain types through magmatic segregation it becomes one of the chief constituents, and may thus form large ore bodies. Such bodies are often highly titaniferous. Most commonly associated with crystalline metamorphic rocks, also frequently in rocks that are rich in ferromagnesian minerals, such as diorite, gabbro, and peridotite. Occurs also in immense beds and lenses, inclosed in old metamorphic rocks. Found in the black sands of the seashore. Occurs as thin plates and dendritic growths between plates of mica. Often intimately associated with corundum, forming *emery*.

The largest magnetite deposits in the world are in northern Sweden at Kiruna and Gellivare associated with apatite, and are believed to have formed by magmatic segregation. Other important deposits are in Norway, Rumania, and U.S.S.R. The most powerful natural magnets are found in Siberia, in the Harz Mountains, on the Island of Elba, and in the Bushveld igneous complex, Republic of South Africa.

In the United States magnetite has been found in commercial quantities in several localities in the Adirondack region in New York, Utah, California, New Jersey, and Pennsylvania. Found as lodestone and in crystals at Magnet Cove, Arkansas.

Use. An important iron ore.

Name. Probably derived from the locality Magnesia, bordering on Macedonia. A fable, told by Pliny, ascribes its name to a shepherd named Magnes, who first discovered the mineral on Mount Ida by noting that the nails of his shoes and the iron ferrule of his staff adhered to the ground.

Similar Species. *Magnesioferrite*, $MgFe_2O_4$, is a rare mineral found chiefly in fumaroles. *Jacobsite*, $MnFe_2O_4$, is a rare mineral found at Långban, Sweden.

FRANKLINITE—$(Zn,Fe,Mn)(Fe,Mn)_2O_4$

Crystallography. Isometric; $4/m\overline{3}2/m$. Crystals octahedral with dodecahedral truncations; often rounded. Also massive, coarse, or fine granular, in rounded grains.

$Fd3m$; $a = 8.42$ Å; $Z = 8$. *d's:* 2.51(10), 1.610(10), 1.480(9), 1.278(6), 1.091(6).

Physical Properties. **H** 6. **G** 5.15. *Luster* metallic. *Color* iron-black. *Streak* reddish-brown to dark brown. Slightly magnetic.

Composition. Dominately $ZnFe_2O_4$ but always with substitution of Fe^2 and Mn^2 in the *A* position and Mn^3 in the *B* position. Analyses show a wide variation in the proportions of the different elements.

Diagnostic Features. Resembles magnetite but is only slightly magnetic and has a dark brown streak. Usually identified by its characteristic association with willemite and zincite. Infusible. Becomes strongly magnetic on heating in the reducing flame. Gives a bluish green color to sodium carbonate bead in the oxidizing flame (manganese).

Occurrence. Franklinite, with only minor exceptions, is confined to the zinc deposits at Franklin, New Jersey where, inclosed in a granular limestone, it is associated with zincite and willemite.

Use. As an ore of zinc and manganese. The zinc is converted into zinc white, ZnO, and the residue is smelted to form an alloy of iron and manganese, *spiegeleisen*, used in the manufacture of steel.

Name. From Franklin, New Jersey.

CHROMITE—$FeCr_2O_4$

Crystallography. Isometric; $4/m\overline{3}2/m$. Habit octahedral but crystals small and rare. Commonly massive, granular to compact.

$Fd3m$; $a = 8.36$ Å; $Z = 8$. *d's:* 4.83(4), 2.51(10), 2.08(5), 1.602(6), 1.473(8).

Physical Properties. **H** $5\frac{1}{2}$. **G** 4.6. *Luster* metallic to submetallic; frequently pitchy. *Color* iron-black to brownish black. *Streak* dark brown. Subtranslucent. *Optics: n* = 2.16.

Composition. For $FeCr_2O_4$, FeO 32.0, Cr_2O_3 68.0 per cent. Some Mg is always present substituting for Fe^2 and some Al and Fe^3 may substitute for chromium. A complete series probably extends to $MgCr_2O_4$.

Diagnostic Features. The submetallic luster usually distinguishes chromite; the green borax bead is diagnostic. Infusible. When fused on charcoal with sodium carbonate it gives a magnetic residue.

Occurrence. Chromite is a common constituent of peridotites and of serpentines derived from them. One of the first minerals to separate from a cooling magma; large chromite ore deposits are thought to have been derived by such magmatic differentiation. Associated with olivine, serpentine, and corundum.

The important countries for its production are U.S.S.R., Republic of South Africa, Turkey, the Philippines, and Rhodesia. Found only sparingly in the United States. Pennsylvania, Maryland, North Carolina, and Wyoming have produced it in the past. California, Alaska, and Oregon are small producers. During World War II, bands of chromite were mined in the Stillwater igneous complex, Montana.

Uses. The only ore of chromium. Chromite ores are grouped into three categories—metallurgical, refractory, and chemical—on the basis of the chrome content and the chrome-iron ratio. As a metal, chromium is used as a ferroalloy to give steel the combined properties of high hardness, great toughness, and resistance to chemical attack. Chromium is a major constituent in stainless steel. *Nichrome*, an alloy of nickel and chromium, is used for resistance in electrical heating equipment. Chromium is widely used in plating plumbing fixtures, automobile accessories, etc.

Because of its refractory character, chromite is made into bricks for the linings of metallurgical furnaces. The bricks are usually made of crude chromite and coal tar but sometimes of chromite with kaolin, bauxite, or other materials. Chromium is a constituent of certain green, yellow, orange, and red pigments and in K_2Cr_2O and $Na_2Cr_2O_7$, which are used as mordants to fix dyes.

Similar Species. *Magnesiochromite*, $MgCr_2O_4$, is in both occurrence and appearance similar to chromite.

CHRYSOBERYL—$BeAl_2O_4$

Crystallography. Orthorhombic; $2/m2/m2/m$. Usually in crystals tabular on {001}, the faces of which are striated parallel to [100]. Commonly twinned on {130} giving pseudohexagonal appearance (Fig. 323).

From its formula, $BeAl_2O_4$, it would appear that chrysoberyl is a member of the spinel group. However, because of the small size of the beryllium ion, chrysoberyl has a structure of lower symmetry. It consists of oxygen atoms in hexagonal closest packing with Be in 4 coordination with O and Al in 6 coordination with O. The pseudohexagonal lattice is reflected in the pseudohexagonal angles and twinning.

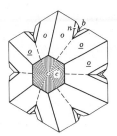

FIG. 323. Chrysoberyl.

Angles: $c(001) \wedge o(111) = 43° 5'$, $b(010) \wedge m(110) = 59° 48'$, $c(001) \wedge n(121) = 51° 8'$.

Pmnb; $a = 5.47$, $b = 9.39$, $c = 4.42$ Å; $a:b:c = 0.583:1:0.471$; $Z = 4$. *d's:* 3.24(8), 2.33(8), 2.57(8), 2.08(10), 1.61(10).

Physical Properties. *Cleavage* {110}. **H** $8\frac{1}{2}$. **G** 3.65–3.8. *Luster* vitreous. *Color* various shades of green, brown, yellow; may be red by transmitted light. *Optics:* (+); $\alpha = 1.746$, $\beta = 1.748$, $\gamma = 1.756$; 2V = 45°; X = c, Y = b.

Alexandrite is a gem variety, emerald-green in daylight, but red by transmitted light and usually red by artificial light. *Cat's eye*, or *cymophane*, is a chatoyant variety which, when cut as an oval or round *cabochon* gem, shows a narrow band of light on its surface. The effect results from minute tubelike cavities or needle-like inclusions with parallel orientation.

Composition. BeO 19.8, Al_2O_3 80.2 per cent. Be 7.1 per cent.

Diagnostic Features. Characterized by its high hardness, its yellowish to emerald-green color, and its twin crystals. Infusible. Insoluble. Wholly soluble in the salt of phosphorus bead (absence of silica).

Occurrence. Chrysoberyl is a rare mineral. It occurs in granitic rocks and pegmatites and in mica schists. Frequently in river sands and gravels. The outstanding alluvial gem deposits are found in Brazil and Ceylon; the *alexandrite* variety comes from the Ural Mountains. In the United States chrysoberyl of gem quality is rarely found. It has been found in Oxford County and elsewhere in Maine; Haddam, Connecticut; and Greenfield, New York. Recently found in Colorado.

Use. As a gem stone. The ordinary yellowish-green stones are inexpensive; the varieties alexandrite and cat's eye are highly prized as gems.

Name. *Chrysoberyl* means *golden beryl. Cymophane* is derived from two Greek words meaning *wave* and *to appear*, in allusion to the chatoyant effect. *Alexandrite* was named in honor of Alexander II of Russia.

COLUMBITE–TANTALITE——$(Fe,Mn)Nb_2O_6$–$(Fe,Mn)Ta_2O_6$

Crystallography. Orthorhombic; $2/m2/m2/m$. Commonly in crystals. The habit is short prismatic or thin tubular on {010}; often in square prisms because

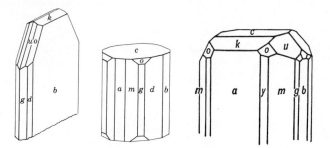

FIG. 324. Columbite-tantalite crystals.

of prominent development of {100} and {010} (Fig. 324). Also in heart-shaped twins, twinned on {201}.

Angles: $b(010) \wedge m(110) = 68° 05'$, $b(010) \wedge g(130) = 39° 39'$, $c(001) \wedge u(111) = 43° 49'$.

Pcan; $a = 5.74$, $b = 14.27$, $c = 5.09$ Å; $a:b:c = 0.402:1:0.357$ (for 17 per cent Ta_2O_5); $Z = 4$. *d's* for columbite: 3.66(7), 2.97(10), 1.767(6), 1.735(7), 1.712(8).

Physical Properties. *Cleavage* {010} good. **H** 6. **G** 5.2–7.9, varying with the composition, increasing with rise in percentage of Ta_2O_5 (Fig. 325). *Luster* submetallic. *Color* iron-black, frequently iridescent. *Streak* dark red to black. Subtranslucent.

Composition. $(Fe,Mn)(Nb,Ta)_2O_6$, which varies in composition from *columbite* $(Fe,Mn)Nb_2O_6$, to *tantalite* $(Fe,Mn)Ta_2O_6$. Often contains small amounts of tin and tungsten. A variety known as *manganotantalite* is essentially tantalite with divalent manganese substituting for most of the ferrous iron.

Diagnostic Features. Recognized usually by its black color with lighter-colored streak, and high specific gravity. Distinguished from wolframite by lower

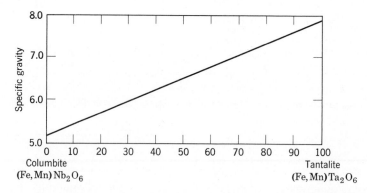

FIG. 325. Columbite-tantalite. Variation of specific gravity with composition.

specific gravity and less distinct cleavage. Difficulty fusible $(5-5\frac{1}{2})$. Fused with borax; the bead dissolved in hydrochloric acid; the solution boiled with tin gives a blue color (niobium). Generally gives a bead test for manganese.

Occurrence. The minerals occur in granitic rocks and pegmatites, associated with quartz, feldspar, mica, tourmaline, beryl, spodumene, cassiterite, samarskite, wolframite, microlite, and monazite.

Notable localities for their occurrence are the Congo; Nigeria; Brazil; near Moss, Norway; Bodenmais, Bavaria; Ilmen Mountains, U.S.S.R., western Australia (*manganotantalite*); and Madagascar. In the United States it is found at Standish, Maine; Haddam, Middletown, and Branchville, Connecticut; in Amelia County, Virginia; Mitchell County, North Carolina; Black Hills, South Dakota; and near Canon City, Colorado.

Use. Source of tantalum and niobium. Because of its resistance to acid corrosion, tantalum is employed in chemical equipment, in surgery for skull plates and sutures, also in some tool steels and in electronic tubes. Niobium has its chief use in alloys in weldable high-speed steels, stainless steels, and alloys resistant to high temperatures, such as used in the gas turbine of the aircraft industry.

Name. *Columbite* from Columbia, a name for America, whence the original specimen was obtained. *Tantalite* from the mythical Tantalus in allusion to the difficulty in dissolving in acid.

Similar Species. *Microlite*, $CaTa_2O_6$, is found in pegmatites; *pyrochlore* and *fergusonite* are oxides of niobium, tantalum, and rare earths found associated with alkalic rocks.

Brucite—$Mg(OH)_2$

Crystallography. Hexagonal—R; $\bar{3}2/m$. Crystals usually tabular on $\{001\}$ and may show small rhombohedral truncations. Commonly foliated, massive.

$C\bar{3}m$; $a = 3.13$, $c = 4.74$ Å; $a:c = 1:1.514$; $Z = 1$. $d's$: 4.74(8), 2.37(10), 1.793(10), 1.372(7), 1.189(9).

Physical Properties. *Cleavage* $\{0001\}$ perfect. Folia flexible but not elastic. Sectile. **H** $2\frac{1}{2}$. **G** 2.39. *Luster* on base pearly, elsewhere vitreous to waxy. *Color* white, gray, light green. Transparent to translucent.

Optics. $(+)$, $\omega = 1.566$, $\varepsilon = 1.581$.

Composition. For $Mg(OH)_2$: MgO 69.0, H_2O 31.0 per cent. Iron and manganese may substitute for magnesium.

Diagnostic Features. Recognized by its foliated nature, light color, and pearly luster on cleavage face. Distinguished from talc by its greater hardness and lack of greasy feel, and from mica by being inelastic. Infusible. Glows before the blowpipe. Gives water in the closed tube. Easily soluble in hydrochloric acid and gives test for magnesium.

Occurrence. Brucite is found associated with serpentine, dolomite, magnesite, and chromite; as an alteration product of periclase and magnesium silicates, especially serpentine. It is also found in crystalline limestone.

Notable foreign localities for its occurrence are at Unst, one of the Shetland Islands, and Aosta, Italy. In the United States found at Tilly Foster Iron Mine, Brewster, New York; at Wood's Mine, Texas, Pennsylvania; and Gabbs, Nevada.

Use. Brucite is used as a raw material for magnesia refractories and is a minor source of metallic magnesium.

Name. In honor of the early American mineralogist Archibald Bruce.

MANGANITE—MnO(OH)

Crystallography. Monoclinic; $2/m$ (pseudo-orthorhombic). Crystals usually prismatic parallel to c and vertically striated (Fig. 326). Often columnar to coarse fibrous. Twinned on {011} as both contact and penetration twins.

Angles: $m(210) \wedge m(2\bar{1}0) = 99° 40'$, $u(101) \wedge u'(\bar{1}01) = 65° 41'$.

$B2_1/d$; $a = 8.84$, $b = 5.23$, $c = 5.74$ Å, $\beta = 90°$, $a:b:c = 1.690:1:1.098$; $Z = 8$. *d's:* 3.38(10), 2.62(9), 2.41(6), 2.26(7), 1.661(9).

Physical Properties. *Cleavage* {010} perfect. {110} and {001} good. **H** 4. **G** 4.3. *Luster* metallic. *Color* steel-gray to iron-black. *Streak* dark brown. Opaque.

Composition. Mn 62.4, O 27.3, H_2O 10.3 per cent.

Diagnostic Features. Recognized chiefly by its black color and prismatic crystals. Hardness (4), and brown streak distinguish it from pyrolusite. Infusible. Powdered mineral gives a bluish-green bead with sodium carbonate. Much water when heated in the closed tube.

Occurrence. Manganite is found associated with other manganese oxides in deposits formed by meteoric waters. Found often in low-temperature hydrothermal veins associated with barite, siderite, and calcite. It frequently alters to pyrolusite.

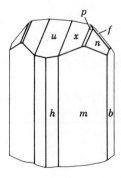

FIG. 326. Manganite.

Occurs at Ilfeld, Harz Mountains, in fine crystals; also at Ilmenau, Thuringia, and Cornwall, England. In the United States at Negaunee, Michigan. In Nova Scotia.

Use.   A minor ore of manganese.

PSILOMELANE—$(Ba,H_2O)_2Mn_5O_{10}$

Crystallography.   Monoclinic, $2/m$. Massive, botryoidal, stalactitic. Appears amorphous.

$A2/m$; $a = 9.56$, $b = 2.88$, $c = 13.85$ Å, $\beta = 92° 30'$; $a:b:c = 3.319:1:4.809$; $Z = 2$. d's: 3.45(4), 2.40(4), 2.17(10), 1.559(6), 1.402(6).

Physical Properties.   H 5–6. G 3.7–4.7. *Luster* submetallic. *Color* black. *Streak* brownish-black. Opaque.

Composition.   Small amounts of Mg, Ca, Ni, Co, and Cu may be present.

Diagnostic Features.   Distinguished from the other manganese oxides by its greater hardness and botryoidal form, and from limonite by its black streak. Infusible. With sodium carbonate gives an opaque bluish-green bead.

Occurrence.   Psilomelane is a secondary mineral; it occurs usually with pyrolusite, and its origin and associations are similar to those of that mineral.

Use.   An ore of manganese. (See pyrolusite, page 287).

Name.   Derived from two Greek words meaning *smooth* and *black*, in allusion to its appearance.

Diaspore—$HAlO_2$

Crystallography.   Orthorhombic; $2/m2/m2/m$. Usually in thin crystals, tabular parallel to {010}; sometimes elongated on [001]. Bladed, foliated massive, disseminated.

Angles: (010) $\wedge$ (110) = 64° 53'.

Pbnm; $a = 4.41$, $b = 9.40$, $c = 2.84$ Å; $a:b:c = 0.469:1:0.302$; $Z = 4$. d's: 3.98(10), 2.31(8), 2.12(7), 2.07(7), 1.629(8).

Physical Properties.   *Cleavage* {010} perfect. H $6\frac{1}{2}$–7. G 3.35–3.45. *Luster* vitreous except on cleavage face, where it is pearly. *Color* white, gray, yellowish, greenish. Transparent to translucent. *Optics:* (+), $\alpha = 1.702$, $\beta = 1.722$, $\gamma = 1.750$; 2V = 85°; $X = c$, $Y = b$; $r < v$.

Composition. Al_2O_3 85, H_2O 15 per cent. Diaspore corresponds to the general formula type ABO_2 and differs from boehmite, $AlO(OH)$, in not having (OH) groups. The hydrogen acts as a cation in 2 coordination with oxygen.

Diagnostic Features.   Characterized by its good cleavage, its bladed habit, and its high hardness. Infusible. Insoluble. Decrepitates and gives water when heated in the closed tube. Ignited with cobalt nitrate turns blue (aluminum).

Occurrence. Diaspore is commonly associated with corundum in emery rock, in dolomite, and in chlorite schist. In a fine-grained massive form it is a major constituent of much bauxite.

Notable localities are the Ural Mountains; Schemnitz, Czechoslovakia; Campolungo in Switzerland; and the island of Naxos, Greece. In the United States it is found in Chester County, Pennsylvania; at Chester, Massachusetts; with alunite, forming rock masses at Mt. Robinson, Rosita Hills, Colorado. It is found abundantly in the bauxite and aluminous clays of Akransas, Missouri, and elsewhere in the United States.

Use. As a refractory.

Name. Derived from a Greek word meaning *to scatter*, in allusion to its decrepitation when heated.

Similar Species. *Boehmite*, AlO(OH), and *gibbsite*, Al(OH)$_3$, are found in disseminated particles as constituents of bauxite.

GOETHITE—HFeO$_2$

Formerly the name *limonite* was given to most of the hydrated ferric oxide here described, and the name *goethite* was reserved for material showing definite evidence of crystallinity. It is now known that most of the amorphous-appearing limonite is in fact goethite. However, the name is retained for the truly amorphous substance and is also used as a field term to refer to natural hydrous iron oxides of uncertain identity.

Crystallography. Orthorhombic; $2/m2/m2/m$. Rarely in distinct prismatic, vertically striated crystals (Fig. 327). Often flattened parallel to {010}. In acicular crystals. Also massive, reniform, stalactitic in radiating fibrous aggregates. Foliated. The so-called bog ore is generally loose and porous.

Angles: $b(010) \wedge m(110) = 65° 20'$, $b(010) \wedge y(120) = 47° 26'$, $e(021) \wedge e'(0\bar{2}1) = 62° 30'$.

Pbnm; $a = 4.65$, $b = 10.02$, $c = 3.04$ Å; $a:b:c = 0.464:1:0.303$; $Z = 4$. *d's:* 4.21(10), 2.69(8), 2.44(7), 2.18(4), 1.719(5).

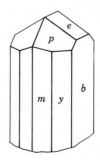

FIG. 327. Goethite.

Physical Properties. *Cleavage* {010} perfect. **H** 5–5½. **G** 4.37; may be as low as 3.3 for impure material. *Luster* adamantine to dull silky in certain fine scaly or fibrous varieties. *Color* yellowish-brown to dark brown. *Streak* yellowish-brown. Subtranslucent.

Composition. Fe 62.9, O 27.0, H_2O 10.1 per cent. Hydrogen acts as a cation in 2 coordination with oxygen, and thus goethite differs from lepidocrocite, FeO(OH), in not having OH groups. Manganese is often present in amounts up to 5 per cent. The massive varieties often contain adsorbed or capillary water.

Diagnostic Features. Distinguished from hematite by its streak. Fusible with difficulty (5–5½). Becomes magnetic in the reducing flame and gives water in the closed tube.

Occurrence. Goethite is one of the commonest minerals and is typically formed under oxidizing conditions as a weathering product of iron-bearing minerals. It also forms as a direct inorganic or biogenic precipitate from water and is widespread as a deposit in bogs and springs. Goethite forms the gossan or "iron hat" over metalliferous veins. Large quantities of goethite have been found as residual lateritic mantles resulting from the weathering of serpentine.

Goethite in some localities constitutes an important ore of iron. It is the principal constituent of the valuable minette ores of Alsace-Lorraine. Other notable European localities are: Eiserfeld in Westphalia; Příbram, Bohemia; and Cornwall, England. Large deposits of iron-rich laterites composed essentially of goethite are found in the Mayari and Moa districts of Cuba.

In the United States goethite is common in the Lake Superior hematite deposits, and has been obtained in fine specimens at Negaunee, near Marquette, Michigan. Goethite is found in iron-bearing limestones along the Appalachian Mountains, from western Massachusetts as far south as Alabama. Such deposits are particularly important in Alabama, Georgia, Virginia, and Tennessee. Finely crystallized material occurs with smoky quartz and microcline in Colorado at Fluorissant and in the Pikes Peak region.

Use. An ore of iron.

Name. In honor of Goethe, the German poet.

Similar Species. *Turgite*, $2Fe_2O_3 \cdot H_2O$, is frequently associated with goethite but distinguished from it by the red streak. *Lepidocrocite*, FeO(OH), is a platy mineral associated with goethite.

BAUXITE[3]

Crystallography. A mixture. Pisolitic, in round concretionary grains; also massive, earthy, claylike.

---

[3] Although bauxite is not a mineral species, it is described here because of its importance as the ore of aluminum.

Physical Properties. H 1–3. G 2–2.55. *Luster* dull to earthy. *Color* white, gray, yellow, red. Translucent.

Composition. A mixture of <u>hydrous aluminum oxides</u> in varying proportions. Some bauxites approach closely the composition of *gibbsite*, $Al(OH)_3$, but most are a mixture, and usually contain iron. As a result, bauxite is not a mineral and should be used only as a rock name. The principal constituents of the rock bauxite are *gibbsite*; *boehmite*, $AlO(OH)$; and *diaspore*, $HAlO_2$, any one of which may be dominant. *Cliachite* is the name proposed for the very fine-grained amorphous constituent of bauxite.

Diagnostic Features. Can usually be recognized by its pisolitic character. Infusible. Insoluble. Assumes a blue color when moistened with cobalt nitrate and then ignited (aluminum). Gives water in the closed tube.

Occurrence. Bauxite is of supergene origin, commonly produced under subtropical to tropical climatic conditions by prolonged weathering and leaching of silica from aluminum-bearing rocks. Also may be derived from the weathering of clay-bearing limestones. It has apparently originated as a colloidal precipitate. It may occur in place as a direct derivative of the original rock, or it may have been transported and deposited in a sedimentary formation. In the tropics deposits known as *laterites*, consisting largely of hydrous aluminum and ferric oxides, are found in the residual soils. These vary widely in composition and purity but many are valuable as sources of aluminum and iron.

Bauxite occurs over a large area in the south of France, an important district being at Baux, near Arles, France. The principal world producers are Surinam, Jamaica, and Guiana. Other major producing countries are Indonesia, U.S.S.R., and Hungary. In the United States, the chief deposits are found in Arkansas, Georgia, and Alabama. In Arkansas bauxite has formed by the alteration of a nepheline syenite.

Use. The ore of aluminum. Eighty-five per cent of the bauxite produced is consumed as aluminum ore. Because of its low density and great strength, aluminum has been adapted to many uses. Sheets, tubes, and castings of aluminum are used in automobiles, airplanes, and railway cars, where light weight is desirable. It is manufactured into cooking utensils, household appliances, and furniture. Aluminum is replacing copper to some extent in electrical transmission lines. Aluminum is alloyed with copper, magnesium, zinc, nickel, silicon, silver, and tin. Other uses are in paint, aluminum foil, and numerous salts.

The second largest use of bauxite is in the manufacture of Al_2O_3, which is used as an abrasive. It is also manufactured into aluminous refractories. Synthetic alumina is also used as the principal ingredient in heat-resistant porcelain such as spark plugs.

Name. From its occurrence at Baux, France.

HALIDES

The chemical class of halides is characterized by the dominance of the electronegative halogen ions, Cl^-, Br^-, F^-, and I^-. These ions are large, feebly charged, and easily polarized. When they combine with relatively large, weakly polarized, cations of low valence, both cations and anions behave as almost perfectly spherical bodies. The packing of these spherical units leads to structures of the highest possible symmetry. The AX type in halite (Fig. 328) and sylvite, and the AX_2 type in fluorite (Fig. 329) are isometric hexoctahedral.

Because the weak electrostatic charges are spread over the entire surface of the nearly spherical ions, the halides are the most perfect examples of the pure ionic-bonding mechanism. The isometric halides all have relatively low hardness and moderate to high melting points and are poor conductors of heat and electricity in the solid state. Such conduction of electricity as takes place does so by electrolysis, that is, by transport of charges by ions rather than by electrons. As the temperature increases and ions are liberated by thermal disorder, electrical conductivity increases rapidly, becoming excellent in the molten state. Advantage is taken of this conductivity of halide melts in the commercial methods for the preparation of sodium and chlorine by electrolysis of molten sodium chloride

FIG. 328. Halite, NaCl, packing model. Na white, Cl gray.

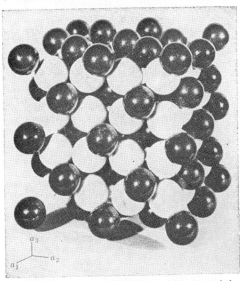

FIG. 329. Fluorite, CaF_2, packing model. F (white) is in 8 coordination about Ca (black). Because of requirements of electrical neutrality only half the possible Ca sites are filled and Ca is in 4 coordination about F. Compare with Fig. 315 (rutile). UO_2 also has the fluorite structure.

in the Downs cell, and in the Hall process for the electrolytic preparation of aluminum using molten cryolite. These properties are those conferred by the ionic bond.

When the halogen ions are combined with smaller and more strongly polarizing cations than those of the alkali metals, structures of lower symmetry result, and the bond has somewhat more covalent properties. In such structures, water and hydroxyl commonly enter as essential constituents, as in atacamite and carnallite.

Halides

Halite	NaCl	Fluorite	CaF_2
Sylvite	KCl	Atacamite	$Cu_2Cl(OH)_3$
Cerargyrite	AgCl	Carnallite	$KMgCl_3 \cdot 6H_2O$
Cryolite	Na_3AlF_6		

HALITE—NaCl

Crystallography. Isometric; $4/m\bar{3}2/m$. Habit cubic; other forms very rare. Some crystals hopper-shaped (Fig. 330). Found in crystals or granular crystalline masses showing cubic cleavage, known as *rock salt*. Also massive, granular to compact.

The crystal structure of halite was the first structure to be determined by x-rays (Fig. 331), and it typifies a large number of AX compounds with a radius ration between 0.41 and 0.73. It is also the classical example of a compound with ionic bond.

$Fm3m$; $a = 5.640$ Å, $Z = 4$. $d's$: 2.82(10), 1.99(4), 1.628(2), 1.261(2), 0.892(1).

Physical Properties. *Cleavage* {001} perfect. **H** $2\frac{1}{2}$. **G** 2.16. *Luster* transparent to translucent. *Color* colorless or white, or when impure may have shades of yellow, red, blue, purple. Salty taste. Diathermanous. *Refractive index* = 1.544.

Composition. Na 39.3, Cl 60.7 per cent. Commonly contains impurities, such as calcium and magnesium sulfates and calcium and magnesium chlorides.

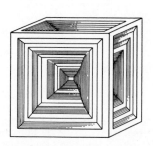

FIG. 330. Hopper-shaped halite crystal.

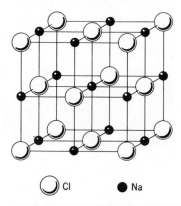

Cl　●Na

FIG. 331. Structure of NaCl.

Diagnostic Features.   Characterized by its cubic cleavage and taste, and distinguished from sylvite by its yellow flame color and by less bitter taste. Fusible at $1\frac{1}{2}$.

Occurrence.   Halite is a common mineral, occurring often in extensive beds and irregular masses, precipitated by evaporation with gypsum, sylvite, anhydrite, calcite, clay, and sand. Halite is dissolved in the waters of salt springs, salt lakes, and the ocean. It is a major salt in playa deposits of inclosed basins.

The deposits of salt have been formed by the gradual evaporation and ultimate drying up of inclosed bodies of salt water. The salt beds formed in this way may have subsequently been covered by other sedimentary deposits and gradually buried beneath the rock strata formed from them. Salt beds range between a few feet to over 200 feet in thickness and it is estimated that some are buried beneath as much as 35,000 feet of overlying strata.

Extensive bedded deposits of salt are widely distributed throughout the world and are mined in many countries. Important production comes from China, U.S.S.R., Great Britain, West Germany, Canada, and Italy.

The United States is the largest world producer and salt is recovered on a commercial scale, in some fifteen states, either from rock-salt deposits or by evaporation of saline waters. Thick beds of rock salt are found in New York State which extend through Ontario, Canada into Michigan. Salt is recovered from these beds at many localities. Notable deposits are also found in Ohio, Kansas, New Mexico, and in Canada in Nova Scotia and Saskatchewan. Salt is obtained by the evaporation of sea waters in California and Texas and from the waters of the Great Salt Lake in Utah.

Salt is also produced from *salt domes*, nearly vertical pipelike masses of salt that appear to have punched their way upward to the surface from an underlying salt bed. Anhydrite, gypsum, and native sulfur are commonly associated with salt domes. Geophysical prospecting for frequently associated petroleum has located several hundred salt domes along the Gulf coast of Louisiana and Texas and far out into the Gulf itself. Salt domes are also found in Germany, Roumania, Spain, and Iran. In Iran, in an area of complete aridity, salt that has punched its way to the surface is not dissolved but moves down slope as a salt glacier.

Use.   Halite finds its greatest use in the chemical industry, where it is the source of sodium and chlorine for the manufacture of hydrochloric acid and a large number of sodium compounds.

Salt is used extensively in the natural state in tanning hides, in fertilizers, in stock feeds, and as a weeed killer. In addition to its familiar functions in the home, salt enters into the preparation of foods of many kinds, such as the preservation of butter, cheese, fish, and meat.

Name.   Halite comes from the Greek word meaning *salt*.

SYLVITE—KCl

Crystallography. Isometric; $4/m\bar{3}2m$. Cube and octahedron frequently in combination. Usually in granular crystalline masses showing cubic cleavage; compact.

$Fm3m$; $a = 6.293$ Å. $Z = 4$. $d's$: 3.15(10), 2.22(6), 1.816(2), 1.407(2), 1.282(4).

Physical Properties. *Cleavage* $\{001\}$ perfect. **H** 2. **G** 1.99. Transparent when pure. *Color* colorless or white; also shades of blue, yellow, or red from impurities. Readily soluble in water. Salty taste but more bitter than halite. *Optics:* refractive index 1.490.

Composition. K 52.4, Cl 47.6 per cent. May contain admixed sodium chloride. Sylvite has the sodium chloride structure, but because of the difference in the ionic radii of Na (0.97 Å) and K (1.33 Å) there is little solid solution.

Diagnostic Features. Fusible at $1\frac{1}{2}$. Distinguished from halite by the violet flame color of potassium and its more bitter taste. Readily soluble in water; a solution made acid with nitric acid gives, with silver nitrate, a heavy precipitate of silver chloride.

Occurrence. Sylvite has the same origin, mode of occurrence, and associations as halite (page 309) but is much rarer. It remains in the mother liquor after precipitation of halite and is one of the last salts to be precipitated.

It is found in quantity and frequently well crystallized, associated with the salt deposits at Stassfurt, Germany; from Kalusz in Galicia. In the United States it is found in large amount in the Permian salt deposits near Carlsbad, New Mexico, and in western Texas. More recently, deposits have been located in Utah. The most important world reserves are in Saskatchewan, Canada where extensive bedded deposits have been found at depths greater than 3000 feet.

Use. The chief source of potassium compounds, which are principally used as fertilizers.

Name. Potassium chloride is the *sal digestivus Sylvii* of early chemistry, whence the name for the species.

Other potassium salts. Other potassium minerals commonly associated with sylvite and found in Germany and Texas in sufficient amount to make them valuable as sources of potassium salts are carnallite, $KMgCl_3 \cdot 6H_2O$ (see page 315); *kainite*, $MgSO_4 \cdot KCl \cdot 3H_2O$; *polyhalite*, $K_2SO_4 \cdot MgSO_4 \cdot 2CaSO_4 \cdot 2H_2O$.

CERARGYRITE—AgCl

Crystallography. Isometric; $4/m\bar{3}2/m$. Habit cubic but crystals rare. Usually massive, resembling wax; often in plates and crusts.

$Fm3m$; $a = 5.55$ Å; $Z = 4$. $d's$: 3.20(5), 2.80(10), 1.97(5), 1.67(2), 1.61(2).

Physical Properties. H 2–3. G 5.5±. Sectile, can be cut with a knife; hornlike appearance, hence the name *horn silver*. Transparent to translucent. *Color* pearl-gray to colorless. Rapidly darkens to violet-brown on exposure to light. *Optics:* refractive index 2.07.

Composition. Ag 75.3, Cl 24.7 per cent. A complete solid-solution series exists between AgCl and *bromyrite*, AgBr. Small amounts of iodine may be present in substitution for chlorine or bromine. Some specimens contain mercury.

Diagnostic Features. Distinguished chiefly by its horny or waxlike appearance and its sectility. Fusible at 1. Before the blowpipe on charcoal it gives a globule of silver.

Occurrence. Cerargyrite is an important supergene ore of silver found in the upper, enriched zone of silver deposits. It is found associated with native silver, cerussite, and secondary minerals in general.

Notable amounts have been found at Broken Hill, Australia; and in Peru, Chile, Bolivia, and Mexico. In the United States cerargyrite was an important mineral in the mines at Leadville and elsewhere in Colorado, at the Comstock Lode in Nevada, and in crystals at the Poorman's Lode in Idaho.

Use. A silver ore.

Name. Cerargyrite is derived from two Greek words meaning *horn* and *silver*, in allusion to its hornlike appearance and characteristics.

Similar Species. Other closely related minerals that are less common but formed under similar conditions, are *bromyrite*, AgBr and *iodobromite*, Ag(Cl, Br, I), isostructural with cerargyrite; and *iodyrite*, AgI, which is hexagonal.

CRYOLITE—Na_3AlF_6

Crystallography. Monoclinic; $2/m$. Prominent forms are {001} and {110}. Crystals rare, usually cubic in aspect, and in parallel groupings growing out of massive material. Usually massive.

Angles: $(110) \wedge (1\bar{1}0) = 88° 02'$, $(001) \wedge (101) = 55° 03'$.

$P2_1/n$; $a = 5.47$, $b = 5.62$, $c = 7.82$ Å, $\beta = 90° 11'$; $a:b:c = 0.973:1:1391$; $Z = 2$. d's: 4.47(2), 3.87(2), 2.75(7), 2.33(4), 1.939(10).

Physical Properties. Parting on {110} and {001} produces cubical forms. H $2\frac{1}{2}$. G 2.95–3.0. *Luster* vitreous to greasy. *Color* colorless to snow-white. Transparent to translucent. *Optics:* (+), $\alpha = 1.338$, $\beta = 1.338$, $\gamma = 1.339$; $2V = 43°$; $X = b$, $Z \wedge c = -44°$, $r < v$. The low refractive index, near that of water, gives the mineral the appearance of watery snow or paraffin, and causes the powdered mineral to almost disappear when immersed in water.

Composition. Na 32.8, Al 12.8, F 54.4 per cent.

Diagnostic Features. Characterized by pseudocubic parting, white color, and peculiar luster; and for the Greenland cryolite, the association of siderite, galena, and chalcopyrite. Fusible at $1\frac{1}{2}$ with strong yellow sodium flame.

Occurrence. The only important deposit of cryolite is at Ivigtut, on the west coast of Greenland. Here, in a large mass in granite, it is associated with siderite, galena, sphalerite, and chalcopyrite; and less commonly quartz, wolframite, fluorite, cassiterite, molybdenite, arsenopyrite, columbite. It is found at Miask, U.S.S.R., and in the United States, at the foot of Pikes Peak, Colorado.

Use. Cryolite is used for the manufacture of sodium salts, of certain kinds of glass and porcelain, and as a flux for cleansing metal surfaces. It was early used as a source of aluminum. When bauxite became the ore of aluminum, cryolite was used as a flux in the electrolytic process. Today most of the sodium aluminum fluoride used in the aluminum industry is produced artificially.

Name. Name is derived from two Greek words meaning *frost* and *stone*, in allusion to its icy appearance.

FLUORITE—CaF$_2$

Crystallography. Isometric; $4/m\overline{3}2/m$. Usually in cubes, often as penetration twins on $\{111\}$ (Fig. 333a). Other forms are rare, but examples of all the forms of the hexoctahedral class have been observed; the tetrahexahedron (Fig. 333b) and hexoctahedron (Fig. 333c) are characteristic. Usually in crystals or in cleavable masses. Also massive; coarse or fine granular; columnar.

$Fm3m; a = 5.46$ Å$; Z = 4. d's:$ 3.15(9), 1.931(10), 1.647(4), 1.366(1), 1.115(2).

Physical Properties. *Cleavage* $\{111\}$ perfect. **H** 4. **G** 3.18. Transparent to translucent. *Luster* vitreous. *Color* varies widely; most commonly light green, yellow, bluish-green, or purple; also colorless, white, rose, blue, brown. The color in some fluorite results from the presence of a hydrocarbon. A single crystal may show bands of varying colors; the massive variety is also often banded in color. The phenomenon of fluorescence (see page 139) is shown by some varieties of fluorite and hence receives its name. *Optics:* refractive index 1.433.

Composition. Ca 51.3, F 48.7 per cent. The rare earths, particularly Y and Ce may substitute for Ca. The fluorite structure is shown in Fig. 332.

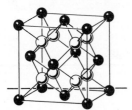

FIG. 332. Structure of fluorite.

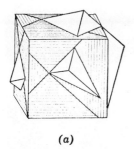 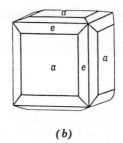 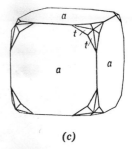

(*a*)   (*b*)   (*c*)

FIG. 333. Fluorite. (*a*) Penetration twin. (*b*) Cube and tetrahexahedron. (*c*) Cube and hexoctahedron.

Diagnostic Features. Determined usually by its cubic crystals and octahedral cleavage; also vitreous luster and usually fine coloring, and by the fact that it can be scratched with a knife. Fusible at 3 giving a reddish flame (calcium).

Occurrence. Fluorite is a common and widely distributed mineral. Usually found in hydrothermal veins in which it may be the chief mineral or as a gangue mineral with metallic ores, especially those of lead and silver. Common in dolomites and limestone and has been observed also as a minor accessory mineral in various igneous rocks and pegmatites. Associated with many different minerals, as calcite, dolomite, gypsum, celestite, barite, quartz, galena, sphalerite, cassiterite, topaz, tourmaline, and apatite.

Fluorite is found in quantity in England, chiefly from Cumberland, Derbyshire, and Durham; the first two localities are famous for their magnificent crystallized specimens (Fig. 334). Found commonly in the mines of Saxony. Fine specimens come from Switzerland, the Tyrol, Bohemia, and Norway. The large producers of commercial fluorite (fluorspar), aside from the United States, are Mexico, Canada, Germany, U.S.S.R., England, and Italy. The most important deposits in the United States are in southern Illinois near Rosiclare and Cave-in-Rock, and in the adjacent part of Kentucky. At Rosiclare fluorite which is without crystal form occurs in limestone, in fissure veins which in places are 40 feet in width. Twenty miles away at Cave-in-Rock the fluorite is in coarsely crystalline aggregates lining flat-lying open spaces. Fluorite is also mined in Colorado, New Mexico, Montana, and Utah.

Use. Fluorite is used mainly as a flux in the making of steel, in the manufacture of opalescent glass, in enameling cooking utensils, for the preparation of hydrofluoric acid. Formerly used extensively as an ornamental material and for carving vases and dishes. Small amounts of fluorite are used for lenses and prisms in various optical systems, but most of the optical material is now made synthetically.

Name. From the Latin *fluere*, meaning to flow, since it melts more easily than other minerals with which it was, in the form of cut stones, confused.

FIG. 334. Fluorite crystals coated with quartz, Northumberland, England.

Atacamite—$Cu_2Cl(OH)_3$

Crystallography. Orthorhombic; $2/m2/m2/m$. Commonly occurs in slender prismatic crystals with vertical striations. Also tabular parallel to the {010}. Usually in confused crystalline aggregates; fibrous; granular.

Pnam; $a = 6.02$, $b = 9.15$, $c = 6.85$ Å; $a:b:c = 0.658:1:0.749$; $Z = 4$. *d's:* 5.40(10), 5.00(10), 2.82(10), 2.75(10), 2.26(10).

Physical Properties. *Cleavage* {010} perfect. **H** $3-3\frac{1}{2}$. **G** 3.75–3.77. *Luster* adamantine to vitreous. *Color* various shades of green. Transparent to translucent. *Optics:* $(-)$, $\alpha = 1.831$, $\beta = 1.861$, $\gamma = 1.880$; 2V = 75°; $r < v$. X = b, Y = a.

Composition. Cu 14.88, CuO 55.87, Cl 16.60, H_2O 12.65 per cent.

Diagnostic Features. Characterized by its green color and granular crystalline aggregates. Distinguished from malachite by its lack of effervescence in acids, and from brochantite and antlerite by its azure-blue copper chloride flame without the use of HCl. Fuses at 3–4 and on charcoal with sodium carbonate gives a copper globule. Gives acid water in the closed tube.

Occurrence. Atacamite is a comparatively rare copper mineral. Found originally as sand in the province of Atacama in Chile. Occurs in arid regions as a supergene mineral in the oxidized zone of copper deposits. It is associated with other secondary minerals in various localities in Chile, Bolivia, Mexico, and in some of the copper districts of South Australia. In the United States occurs sparingly in the copper districts of Arizona.

Use. A minor ore of copper.

Name. From the province of Atacama, Chile.

Carnallite—KMgCl$_3$·6H$_2$O

Crystallography. Orthorhombic; $2/m2/m2/m$. Crystals rare. Usually massive, granular.

Pban; $a = 9.56$, $b = 16.05$, $c = 22.56$ Å; $a:b:c = 0.593:1:1.341$; $Z = 12$. $d's$: 4.65(5), 3.77(5), 3.56(5), 3.30(10), 2.92(7).

Physical Properties. H 1. G 1.6. *Luster* nonmetallic, shining, greasy. *Color* milk-white, often reddish, due to included hematite. Transparent to translucent. *Taste* bitter. Deliquescent. *Optics:* (+), $\alpha = 1.467$, $\beta = 1.475$, $\gamma = 1.494$; $2V = 70°$; $X = c$, $Y = b$, $r < v$.

Composition. KCl 26.81, MgCl$_2$ 34.19, H$_2$O 39.0 per cent. Bromine may substitute for chlorine in small amounts.

Diagnostic Features. Carnallite is distinguished from associated salts by lack of cleavage and its deliquescent nature. Fusible at 1 to 1$\frac{1}{2}$ with violet flame. Gives much water in the closed tube. Easily and completely soluble in water; on addition of nitric acid and silver nitrate gives a white precipitate of silver chloride.

Occurrence. Carnallite is found associated with halite, sylvite, etc., in the salt deposits at Stassfurt, Germany. It is also found to a lesser extent in the potash deposits of western Texas and eastern New Mexico.

Use. A source of potassium and magnesium.

Name. In honor of Rudolph von Carnall (1804–1874), Prussian mining engineer.

CARBONATES

When carbon unites with oxygen, it has a strong tendency to link to two oxygen atoms by sharing two of its four valence electrons with each and to form a stable chemical unit, a molecule of carbon dioxide. In nature, carbon also joins with oxygen to form the carbonate ion, $CO_3^=$. The ratio of the radii of carbon and oxygen requires that three oxygen ions be coordinated by each carbon ion. Since oxygen has a charge of (-2) and carbon $(+4)$, the carbon-oxygen bond has a strength equal to 1$\frac{1}{3}$ units of charge. This is greater than one-half the total charge of the oxygen ion, and, thus, each oxygen must be bonded to its coordinating carbon more strongly than it can possibly be bonded to any other ion in the structure. Further, oxygens are not shared between carbonate groups, and the carbon-oxygen triangles must be regarded as separate units of the structure. These flattened trefoil-shaped carbonate groups are the basic building units of all carbonate minerals and are largely responsible for the properties peculiar to the group (Fig. 336a).

Although the bond between the central carbon and its coordinated oxygens in the carbonate radical is strong, it is not as strong as the covalent bond in carbon dioxide. In the presence of the hydrogen ion, the carbonate radical becomes unstable and breaks down to yield carbon dioxide and water. This instability is the cause of the familiar "fizz" tests with acids, the widely used test for carbonates.

The important anhydrous carbonates fall into three isostructural groups: the *calcite group*, the *dolomite group*, and the *aragonite group*. Aside from the minerals of these groups, the basic copper carbonates, azurite and malachite, are the only important carbonates.

Carbonates

CALCITE GROUP		DOLOMITE GROUP	
Calcite	$CaCO_3$	Dolomite	$CaMg(CO_3)_2$
Magnesite	$MgCO_3$	Ankerite	$CaFe(CO_3)_2$
Siderite	$FeCO_3$	ARAGONITE GROUP	
Rhodochrosite	$MnCO_3$	Aragonite	$CaCO_3$
Smithsonite	$ZnCO_3$	Witherite	$BaCO_3$
		Strontianite	$SrCO_3$
		Cerussite	$PbCO_3$
	Malachite	$Cu_2CO_3(OH)_2$	
	Azurite	$Cu_3(CO_3)_2(OH)_2$	

Calcite Group

When divalent carbonate groups are combined with divalent cations, such that radius ratio considerations dictate 6 coordination, structures of simple geometry result. In this structure, which we call the *calcite type* (Fig. 335), layers of metal cations and carbonate anions alternate. As described by Bragg, calcite may be thought of as having a distorted sodium chloride structure in which Na ions are replaced by calcium and chlorine ions by carbonate groups. We picture this structure oriented with a 3-fold axis vertical and then compressed along this axis so that the faces make angles of 74° 55′ with each other instead of the 90° in the cube. The vertical axis is now a unique 3-fold axis and perpendicular to the alternating layers of calcium and carbonate ions. Each calcium ion is coordinated to six oxygen ions, and each oxygen ion is coordinated to two calcium ions as well as to the carbon ion at the center of the carbonate groups. The cleavage

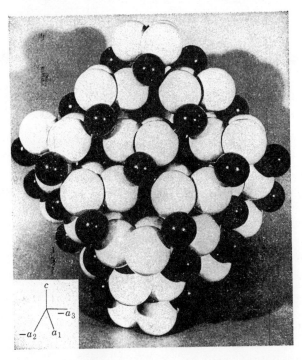

FIG. 335. Calcite, CaCO₃, packing model. Ca dark; O white; C at the center of the CO₃ triangle is not seen. Note that horizontal layers of Ca ions alternate with horizontal layers of CO₃ = ions.

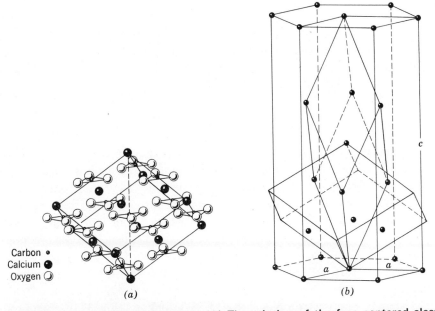

Carbon •
Calcium ●
Oxygen ◑

(a)

(b)

FIG. 336. (a) Structure of calcite. (b) The relation of the face-centered cleavage rhombohedron to the steep structural rhombohedral unit.

characteristic of the calcite group, like the cleavage of halite, is parallel to the mostly widely spaced planes of maximum atomic population, but because of the reduction in symmetry, cleavage is rhombohedral rather than cubic.

Although a very strong and partly covalent bonding is present within the carbonate ion, the bonds uniting it with the metal ions are simple ionic, and the properties of the individual members of the calcite group are largely conferred by the metal ions. Thus, the specific gravity of most of the members of the group is proportional to the atomic weight of the cation. The exception is magnesium, which is so much smaller than the rest that the closer packing permitted more than compensates for its low atomic weight. Hence its carbonate, magnesite, is denser than the carbonate of the heavier but much larger calcium ion.

Because all the members of the calcite group are isostructural, substitution of metal cations is possible with the limits set by their relative sizes. Thus, ferrous iron (R_A 0.74 Å), divalent manganese (R_A 0.80 Å), and magnesium (R_A 0.66 Å) substitute for each other producing substances intermediate between the pure compounds, siderite, rhodochrosite, and magnesite; and whose physical properties vary in proportion to the amounts of the three ions. The substitution of these ions for calcium in calcite is not as complete nor as perfectly random because of the large size of the calcium ion (R_A 0.99 Å).

Traditionally the cleavage rhombohedron of calcite has been considered the unit form with indices $\{10\bar{1}1\}$ and the axial ratio expressed accordingly. In the morphological descriptions and indexing of forms of the calcite and dolomite group minerals that follow, this convention has been preserved. However, structural determinations have shown that the simplest unit cell is a steep rhombohedron containing $2(CaCO_3)$ whereas the hexagonal cell contains $6(CaCO_3)$. (Fig. 336b). Therefore, the structural axial ratios differ from the morphological.

CALCITE—CaCO₃

Crystallography. Hexagonal—R; $\bar{3}2/m$. Crystals are extremely varied in habit and often highly complex. Over 300 different forms have been described (Fig. 337). Three important habits: (1) prismatic, in long or short prisms, in which the prism faces are prominent, with base or rhombohedral terminations; (2) rhombohedral, in which rhombohedral forms predominate, [the unit (cleavage) form r] is not common; (3) scalenohedral, in which scalenohedrons predominate, often with prism faces and rhombohedral truncations. The most common scalenohedron is $\{21\bar{3}1\}$. All possible combinations and variations of these types are found.

Twinning with the twin plane, $\{01\bar{1}2\}$, very common (Fig. 338); often produces twinning lamellae which may, as in crystalline limestones, be of secondary origin.

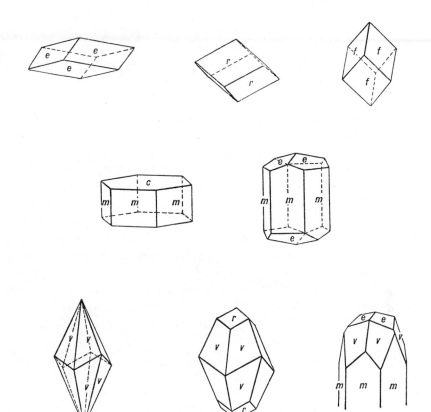

FIG. 337. Calcite crystals. Forms: $c\{0001\}$, $m\{10\bar{1}0\}$, $e\{01\bar{1}2\}$, $r\{10\bar{1}1\}$, $f\{02\bar{2}1\}$, $v\{21\bar{3}1\}$.

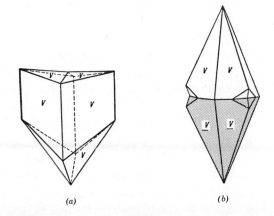

          (a)   (b)

FIG. 338. Twinned calcite crystals. Twin planes: (a) $\{01\bar{1}2\}$. (b) $\{0001\}$.

This twinning may be produced artificially (see page 100). Twins with $\{0001\}$, the twin plane common. Calcite is usually in crystals or in coarse to fine grained aggregates. Also fine grained to compact, earthy, and stalactitic.

Angles: $r(10\bar{1}1) \wedge r'(\bar{1}101) = 74° 55'$, $c(0001) \wedge r(10\bar{1}1) = 44° 37'$, $e(01\bar{1}2) \wedge e'(\bar{1}012) = 45° 3'$, $c(0001) \wedge e(01\bar{1}2) = 26° 15'$, $v(21\bar{3}1) \wedge v'(\bar{2}3\bar{1}1) = 75° 22'$, $c(0001) \wedge M(40\bar{4}1) = 75° 47'$.

$R3c$. Morphology $a:c = 1:0.855$. Structure $a = 4.99$, $c = 17.06$ Å; $a:c = 1:3.419$; $Z = 6$. *d's:* 3.04(10), 2.29(2), 2.10(2), 1.913(2), 1.875(2).

Physical Properties. *Cleavage* $\{10\bar{1}1\}$ perfect, (cleavage angle = 74° 55'). Parting along twin lamellae on $\{01\bar{1}2\}$. **H** 3 on cleavage, $2\frac{1}{2}$ on base. **G** 2.71. *Luster* vitreous to earthy. *Color* usually white or colorless, but may be variously tinted, gray, red, green, blue, yellow; also, when impure, brown to black. Transparent to translucent. The chemically pure and optically clear, colorless variety is known as *Iceland spar* because of its occurrence in Iceland. *Optics:* $(-)$; $\omega = 1.658$, $\varepsilon = 1.486$.

Composition. CaO 56.0, CO_2 44.0 per cent. Mn^2, Fe^2, and Zn may substitute for Ca, and a complete series extends to rhodochrosite and partial series toward siderite and smithsonite. Mg substitutes for Ca in only small amounts; when much Mg is present, the tendency is to form dolomite, $CaMg(CO_3)_2$.

Diagnostic Features. Infusible. Fragments effervesce readily in cold dilute hydrochloric acid. Characterized by its hardness (3), rhombohedral cleavage, light color, vitreous luster. Distinguished from dolomite by the fact that coarse fragments of calcite effervesce freely in cold hydrochloric acid, and distinguished from aragonite by lower specific gravity and rhombohedral cleavage.

Occurrence. *As a rock mineral:* Calcite is one of the most common and wide spread minerals. It occurs in extensive sedimentary rock masses in which it is the predominant mineral; and in some limestones, it is the only mineral present. Crystalline metamorphosed limestones are *marbles*. Chalk is a fine-grained pulverulent deposit of calcium carbonate. Calcite is an important constituent of calcareous marls and calcareous sandstones. The limestone rocks have, in great part, been formed by the deposition on a sea bottom of great thicknesses of calcareous material in the form of shells and skeletons of sea animals. A smaller proportion of these rocks have been formed directly by precipitation of calcium carbonate.

As cave deposits, etc.: Waters carrying calcium carbonate in solution and evaporating in limestone caves often deposit calcite as stalactites, stalagmites, and incrustations. Such deposits, usually semitranslucent and of light yellow colors, are often beautiful and spectacular. Both hot and cold calcareous spring waters may form cellular deposits of calcite known as *travertine*, or *tufa*, around their mouths. The deposit at Mammoth Hot Springs, Yellowstone Park, is more spectacular than most, but of similar origin. *Onyx marble* is banded calcite and/or aragonite used for decorative purposes. Since much of this material comes from Lower California, Mexico, it is also called Mexican onyx.

Siliceous calcites: Calcite crystals may inclose considerable amounts of quartz sand (up to 60 per cent) and form what are known as sandstone crystals. Such occurrences are found at Fontainebleau, France (Fontainebleau limestone), and in the Bad Lands, South Dakota.

Calcite occurs as a secondary mineral in igneous rocks as a product of decomposition of lime silicates and is found lining the amygdaloidal cavities in lavas. It is the cementing material in some light-colored sandstones and is also a common mineral in hydrothermal veins associated with sulfide ores.

It is impossible to specify all the important districts for the occurrence of calcite in its various forms. Some of the more notable localities in which finely crystallized calcite is found are as follows: Andreasberg in the Harz Mountains; various places in Saxony; in Cumberland, Derbyshire, Durham, Cornwall, and Lancashire, England; Iceland; and Guanajuato, Mexico. In the United States at Joplin, Missouri; Lake Superior copper district; and Rossie, New York.

Use. The most important use for calcite is for the manufacture of cements and lime for mortars. Limestone is the chief raw material, which when heated to about 900°C, loses CO_2 and is converted into quicklime, CaO. This, when mixed with water, forms calcium hydrate (slaked lime), swells, gives off much heat, and hardens or, as commonly termed, "sets." Quicklime when mixed with sand forms common mortar.

The greatest consumption of limestone is in the manufacture of cements. The type known as Portland cement is most widely produced. It is composed of about 75 per cent calcium carbonate (limestone) with the remainder essentially silica and alumina. Small amounts of magnesium carbonate and iron oxide are also present. In some limestones, known as *cement rocks*, the correct proportions of silica and alumina are present as impurities. In others these oxides are contributed by clay or shale mixed with the limestone before "burning." When water is mixed with cement, hydrous calcium silicates and calcium aluminates are formed.

Limestone is a raw material for the chemical industry, and finely crushed is used as a fertilizer, for whiting and whitewash. Great quantities are quarried each year as a flux for smelting various metallic ores, as an aggregate in concrete and as road metal. A fine grained limestone is used in lithography.

Calcite in several forms is used in the building industry. Limestone and marble as dimension stone are used both for construction purposes and decorative exterior facings. Polished slabs of travertine and Mexican onyx are commonly used as ornamental stone for interiors. Indiana is the chief source of building limestone in the United States and the most important marble quarries are in Vermont, New York, Georgia, and Tennessee.

Iceland spar is valuable for various optical instruments; its best known use is in the form of the Nicol prism to produce polarized light.

Name. From the Latin word, *calx*, meaning burnt lime.

MAGNESITE—MgCO$_3$

Crystallography. Hexagonal—R; $\bar{3}2/m$. Crystals rhombohedral, $\{10\bar{1}1\}$, but rare. Usually cryptocrystalline in white, compact, earthy masses; less frequently in cleavable granular masses, coarse to fine.

Angles: $(10\bar{1}1) \wedge (\bar{1}101) = 72° 33'$.

$R\bar{3}c$. Morphology $a:c = 1:0.810$. Structure $a = 4.63$, $c = 15.02$ Å; $a:c = 1:3.244$, $Z = 6$. *d's:* 2.74(10), 2.50(2), 2.10(4), 1.939(1), 1.700(3).

Physical Properties. *Cleavage* $\{10\bar{1}1\}$ perfect. **H** $3\frac{1}{2}$–5. **G** 3.0–3.2. *Luster* vitreous. *Color* white, gray, yellow, brown. Transparent to translucent. *Optics:* $(-)$; $\omega = 1.700$, $\varepsilon = 1.509$.

Composition. MgO 47.8, CO$_2$ 52.2 per cent. Ferrous iron substitutes for magnesium, and a complete series extends to siderite. Small amounts of calcium and manganese may be present.

Diagnostic Features. Cleavable varieties are distinguished from dolomite only by the higher specific gravity and absence of abundant calcium. The white massive variety resembles chert and is distinguished from it by inferior hardness. Infusible. Scarcely acted upon by cold, but dissolves with effervescence in hot, hydrochloric acid.

Occurrence. Magnesite commonly occurs in veins and irregular masses derived from the alteration of serpentine through the action of waters containing carbonic acid. Such magnesites are compact cryptocrystalline and often contain opaline silica. Beds of crystalline cleavable magnesite are (1) of metamorphic origin associated with talc schists, chlorite schists, and mica schists, and (2) of sedimentary origin, formed as a replacement of calcite rocks by magnesium containing solutions, dolomite being formed as an intermediate product.

Notable deposits of the sedimentary type of magnesite are in Manchuria; at Satka in the Ural Mountains; and at Styria, Austria. The most famous deposit of the cryptocrystalline type is on the Island of Euboea, Greece.

In the United States the compact variety is found in irregular masses in serpentine in the Coast Range, California. The sedimentary type is mined at Chewelah in Stevens County, Washington, and in the Paradise Range, Nye County, Nevada. There are numerous minor localities in the eastern United States in which magnesite is associated with serpentine, talc, or dolomite rocks.

Use. Dead-burned magnesite, MgO, that is, magnesite that has been calcined at a high temperature and contains less than 1 per cent CO$_2$, is used in manufacturing bricks for furnace linings. Magnesite is the source of magnesia for industrial chemicals. It has also been used as an ore of metallic magnesium, but at present the entire production of magnesium comes from brines and sea water.

Name. Magnesite is named in allusion to the composition.

SIDERITE—FeCO$_3$

Crystallography. Hexagonal—R; $\bar{3}2/m$. Crystals usually unit rhombohedrons, frequently with curved faces. In globular concretions. Usually cleavable granular. May be botryoidal, compact, and earthy.

Angles: $(10\bar{1}1) \wedge (\bar{1}101) = 73°\ 00'$.

$R\bar{3}c$. Morphology $a:c = 1:0.819$. Structure $a = 4.72$, $c = 15.45$ Å; $a:c = 1:3.275$; $Z = 6$. *d's:* 3.59(6), 2.79(10), 2.13(6), 1.963(6), 1.73(8).

Physical Properties. *Cleavage* $\{10\bar{1}1\}$ perfect. **H** $3\frac{1}{2}$–4. **G** 3.96 for pure FeCO$_3$, but decreases with presence of divalent manganese and magnesium. *Luster* vitreous. *Color* usually light to dark brown. Transparent to translucent. *Optics:* $(-)$; $\omega = 1.875$, $\varepsilon = 1.633$.

Composition. FeO 62.1, CO$_2$ 37.9 per cent. Fe 48.2 per cent. Mn^2 and Mg substitute for Fe^2 and complete series extend to rhodochrosite and magnesite. Calcium may be present in small amounts.

Diagnostic Features. Distinguished from other carbonates by its color and high specific gravity, and from sphalerite by its rhombohedral cleavage. Difficulty fusible ($4\frac{1}{2}$–5). Becomes strongly magnetic on heating. Soluble in hot hydrochloric acid with effervescence.

Alteration. Pseudomorphs of limonite after siderite are common.

Occurrence. Siderite is frequently found as *clay ironstone*, impure by admixture with clay materials, in concretions with concentric layers. As *black-band ore* it is found, contaminated by carbonaceous material, in extensive stratified formations lying in shales and commonly associated with coal measures. These ores have been mined extensively in Great Britain in the past, but at present are mined only in North Staffordshire and Scotland. Clay ironstone is also abundant in the coal measures of western Pennsylvania and eastern Ohio, but it is not used to any great extent as an ore. Siderite is also formed by the replacement action of ferrous solutions upon limestones, and if such deposits are extensive they may be of economic value. The most notable occurrence of this type is in Styria, Austria, where siderite is mined on a large scale. Siderite, in its crystallized form, is a common vein mineral associated with various metallic ores containing silver minerals, pyrite, chalcopyrite, tetrahedrite, and galena. When siderite predominates in such veins it may be mined, as in southern Westphalia, Germany.

Use. An ore of iron. Important in Great Britain and Austria, but unimportant elsewhere.

Name. From the Greek word meaning *iron*. The name *spherosiderite* of the concretionary variety was shortened to siderite to apply to the entire species.

Chalybite, used by some mineralogists, was derived from the Chalybes, ancient iron workers, who lived on the Black Sea.

RHODOCHROSITE—$MnCO_3$

Crystallography. Hexagonal—R; $\bar{3}2/m$. Only rarely in crystals of the unit rhombohedron; frequently with curved faces. Usually cleavable massive; granular to compact.

Angles: $(10\bar{1}1) \wedge (\bar{1}101) = 73° 02'$.

$R\bar{3}c$. Morphology $a:c = 1:0.818$. Structure $a = 4.78$, $c = 15.67$ Å; $a:c = 3.278$; $Z = 6$. *d's:* 3.66(4), 2.84(10), 2.17(3), 1.770(3), 1.763(3).

Physical Properties. *Cleavage* $\{10\bar{1}1\}$ perfect. **H** $3\frac{1}{2}$–4. **G** 3.5–3.7. *Luster* vitreous. *Color* usually some shade of rose-red; may be light pink to dark brown. *Streak* white. Transparent to translucent. *Optics:* $(-)$; $\omega = 1.816$, $\varepsilon = 1.597$.

Composition. MnO 61.7, CO_2 38.3 per cent. Fe^2 and Ca substitute for Mn and a complete series extends to siderite and a partial series to calcite. Only limited amounts of magnesium and zinc substitute for manganese.

Diagnostic Features. Characterized by its pink color and rhombohedral cleavage; the hardness (4) distinguishes it from rhodonite ($MnSiO_3$) with hardness of 6. Infusible. Soluble in hot hydrochloric acid with effervescence. Gives blue-green color to sodium carbonate bead. When heated on charcoal it turns black but it is nonmagnetic.

Occurrence. Rhodochrosite is a comparatively rare mineral, occurring in hydrothermal veins with ores of silver, lead, and copper, and with other manganese minerals. Found in the silver mines of Rumania and Saxony. Beautiful banded rhodochrosite is mined for ornamental and decorative purposes at Capillitas, Catamarca, Argentina. In the United States it is found at Branchville, Connecticut; Franklin, New Jersey; and at Butte, Montana it is mined as a manganese ore. In good crystals at Alicante, Lake County; Alma Park County; and elsewhere in Colorado.

Use. A minor ore of manganese. Small amounts used for ornamental purposes.

Name. Derived from two Greek words meaning *rose* and *color*, in allusion to its rose-pink color.

SMITHSONITE—$ZnCO_3$

Crystallography. Hexagonal—R; $\bar{3}2/m$. Rarely in small rhombohedral or scalenohedral crystals. Usually reniform, botryoidal, or stalactitic, and in crystalline incrustations or in honeycombed masses known as *dry-bone ore*. Also granular to earthy.

Angles: $(10\bar{1}1) \wedge (\bar{1}101) = 72° 20'$.

$R\bar{3}c$. Morphology $a:c = 1:0.806$. Structure $a = 4.66$, $c = 1.50$ Å; $a:c = 1:3.224$, $Z = 6$. $d's$: 2.75(10), 3.55(5), 2.33(3), 1.946(3), 1.703(4).

Physical Properties. *Cleavage* $\{10\bar{1}1\}$ perfect. **H** $4-4\frac{1}{2}$. **G** 4.30–4.45. *Luster* vitreous. *Color* usually dirty brown. May be colorless, white, green, blue, or pink. The yellow variety contains cadmium and is known as *turkey-fat ore*. *Streak* white. Translucent. *Optics:* $(-)$; $\omega = 1.850$, $\varepsilon = 1.623$.

Composition. ZnO 64.8, CO_2 35.2 per cent. Fe^2 and Mn^2 may substitute in part for Zn; also, less commonly, Ca, Mg, Cd, Cu, Co, and Pb.

Diagnostic Features. Infusible. Soluble in cold hydrochloric acid with effervescence. When heated before the blowpipe it gives bluish-green streaks in the flame, due to the burning of the volatilized zinc, and in the reducing flame on charcoal it gives a nonvolatile coating of zinc oxide, yellow when hot, white when cold. The coating moistened with cobalt nitrate and again heated turns green. Distinguished by its effervescence in acids, its tests for zinc, its hardness, and its high specific gravity.

Occurrence. Smithsonite is a zinc ore of supergene origin, usually found with zinc deposits in limestone rocks. Associated with sphalerite, galena, hemimorphite, cerussite, calcite, and limonite. Often found in pseudomorphs after calcite. Smithsonite is found in places in translucent green or greenish-blue material which is used for ornamental purposes. Laurium, Greece, is noted for this ornamental smithsonite; and Sardinia for yellow stalactites with concentric banding. Fine crystallized specimens have come from the Broken Hill mine, Zambia, and from Tsumeb, South-West Africa. In the United States smithsonite occurs as an ore in the zinc deposits of Leadville, Colorado; Missouri; Arkansas; Wisconsin; and Virgina. Fine greenish-blue material has been found at Kelly, New Mexico.

Use. An ore of zinc. A minor use is for ornamental purposes.

Name. Named in honor of James Smithson (1754–1829), who founded the Smithsonian Institution at Washington. English mineralogists formerly called the mineral calamine.

Similar Species. *Hydrozincite,* $Zn_5(OH)_6(CO_3)_2$, occurs as a secondary mineral in zinc deposits.

Dolomite Group

The minerals of the dolomite group are similar in structure to those of the calcite group with layers of CO_3^{2-} groups alternating with layers of cations. However, in dolomite the cation layers are alternately calcium and magnesium. This ordered structure results from the large difference in ionic radii (33 per cent) of calcium and magnesium and their limited solid solution. With the nonequivalence of the calcium and magnesium layers, the 2-fold symmetry axes of calcite do not exist and the symmetry is reduced to that of the rhombohedral class, $\bar{3}$. In addition to dolomite, the dolomite group includes the minerals ankerite (Ca and Fe) and kutnahorite (Ca and Mn).

DOLOMITE—CaMg(CO₃)₂

Crystallography. Hexagonal—R; $\bar{3}$. Crystals are usually the unit rhombohedron (Fig. 339a), more rarely a steep rhombohedron and base (Fig. 339c). Faces are often curved, some so acutely as to form "saddle-shaped" crystals (Fig. 339b). Other forms rare. In coarse, granular, cleavable masses to fine grained and compact. Twinning on {0001} common; lamellar twinning on {02$\bar{2}$1}.

Angles: $r(10\bar{1}1) \wedge r'(\bar{1}101) = 73° \ 16'$, $c(0001) \wedge M(40\bar{4}1) = 75° \ 25'$.

$R\bar{3}$. Morphology $a:c = 1:0.832$. Structure $a = 4.84$, $c = 15.95$ Å; $a:c = 1:3.294$; $Z = 3$. d's: 2.88(10), 2.19(4), 2.01(3), 1.800(1), 1.780(1).

Physical Properties. *Cleavage* {10$\bar{1}$1} perfect. **H** $3\frac{1}{2}$–4. **G** 2.85. *Luster* vitreous; pearly in some varieties, *pearl spar*. *Color* usually some shade of pink, flesh color; may be colorless, white, gray, green, brown, or black. Transparent to translucent. *Optics:* $(-)$; $\omega = 1.681$, $\varepsilon = 1.500$.

Composition. CaO 30.4, MgO 21.7, CO₂ 47.9 per cent. In ordinary dolomite the proportion of CaCO₃ to MgCO₃ is 1:1. However, calcium may substitute for magnesium up to about Ca:Mg = 1:5 in the magnesium positions, and magnesium may substitute for calcium up to about Mg:Ca = 1:20 in the calcium positions. Thus in dolomite the ratio of calcium to magnesium ranges between 58:42 and $47\frac{1}{2}:52\frac{1}{2}$. Probably a complete series extends to *ankerite* with the substitution of Fe² for Mg. Small amounts of Mn², CO², and Zn may substitute for magnesium and small amounts of Pb for Ca.

Diagnostic Features. Infusible. In cold dilute HCl large fragments are slowly attacked but are soluble with effervescence in hot HCl; the powdered mineral is readily soluble in cold acid. Crystallized variety told by its curved rhombohedral crystals and usually by its flesh-pink color. The massive rock variety is distinguished from limestone by the less vigorous reaction with hydrochloric acid.

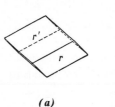

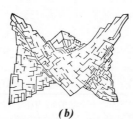

(a)   *(b)*   *(c)*

FIG. 339. Dolomite.

Occurrence. Dolomite is found in many parts of the world chiefly as extensive sedimentary strata, and the crystallized equivalent, dolomitic marble. Dolomite as a rock mass is thought to be secondary in origin, formed from ordinary limestone by the replacement of calcium by magnesium. The replacement may be only partial and thus most dolomite rocks are mixtures of dolomite and calcite. The mineral occurs also as a hydrothermal vein mineral, chiefly in the lead and zinc veins that traverse limestone, associated with fluorite, calcite, barite, and siderite.

Found in large rock strata in the dolomite region of southern Tyrol; in crystals from the Binnenthal, Switzerland; Traversella in Piedmont; northern England; and Guanajuato, Mexico. In the United States it is found as masses of sedimentary rock in many of the middle-western states, and in crystals in the Joplin, Missouri, district.

Use. As a building and ornamental stone. For the manufacture of certain cements. For the manufacture of magnesia used in the preparation of refractory linings of the converters in the basic steel process. Dolomite is a potential ore of metallic magnesium.

Name. In honor of the French chemist, Dolomieu (1750–1801).

Aragonite Group

When the carbonate ion combines with large divalent ions, the radius ratios do not permit stable 6 coordination, and orthorhombic structures result. This is the *aragonite* structure type (Fig. 340). As in calcite, the planar carbonate ions are in planes perpendicular to the *c* axis but with different coordination. In aragonite each calcium ion is coordinated to nine oxygen ions and each oxygen ion coordinated to three calcium ions. The cations have an arrangement in the structure approximating hexagonal closest-packing which gives rise to marked pseudo-hexagonal symmetry. This is reflected in both the crystal angles and in the pseudo-hexagonal twinning which is characteristic of all members of the group.

Solid solution within the aragonite group is somewhat more limited than in the calcite group, and it is interesting to note that calcium and barium, respectively the smallest and largest ions in the group, form a double salt closely analogous to dolomite. The differences in the physical properties of the minerals of the aragonite group are conferred largely by the cations. Thus the specific gravity is roughly proportional to the atomic weight of the metal ion (see page 131). The structure assumed by a carbonate is largely dependent on the radius of the cation, with the smaller ions entering into the rhombohedral structure and the larger ions entering into the orthorhombic structure. $CaCO_3$ is near the critical radius ratio of 0.73 between the two types and hence crystallizes with rhombohedral structure as calcite and with orthorhombic structure as aragonite.

FIG. 340. Aragonite, $CaCO_3$ packing model. Ca dark; O white; C at the center of the CO_3 triangle is not seen. Each CO_3 group lies between six calcium ions.

ARAGONITE—$CaCO_3$

Crystallography. Orthorhombic; $2/m2/m2/m$. Three habits of crystallization are common. (1) Acicular pyramidal; consisting of a vertical prism terminated by a combination of a very steep dipyramid and first-order prism (Fig. 341a). Usually in radiating groups of large to very small crystals. (2) Tabular; consisting of prominent $\{010\}$ modified by $\{110\}$ and a low prism, $k\{011\}$ (Fig. 341b); often twinned on $\{110\}$ (Fig. 341c). (3) In pseudohexagonal twins (Fig. 341d) showing a hexagonal-like prism terminated by a basal plane. It is formed by an intergrowth

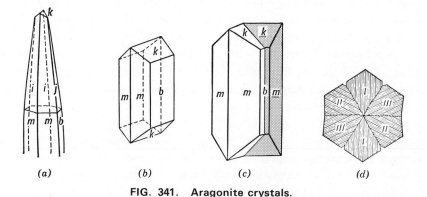

(a)	(b)	(c)	(d)

FIG. 341. Aragonite crystals.

of three individuals twinned on {110} with {001} planes in common. The cyclic twins are distinguished from true hexagonal forms by noting that the basal surface is striated in three different directions and that, because the prism angle of the simple crystals is not exactly 60°, the composite prism faces for the twin will often show slight reentrant angles. Also found in reniform, columnar, and stalactitic aggregates.

Angles: $m(110) \wedge m'(1\overline{1}0) = 63° 48'$, $b(010) \wedge m(110) = 58° 06'$, $k(011) \wedge k'(0\overline{1}1) = 71° 33'$, $b(010) \wedge k(011) = 54° 13'$.

Pmcn; $a = 4.95$, $b = 7.96$, $c = 5.73$ Å; $a:b:c = 0.622:1:0.720$. $Z = 4$. *d's:* 3.04(9), 2.71(6), 2.36(7), 1.975(10), 1.880(8).

Physical Properties. *Cleavage* {010} distinct, {110} poor. *Luster* vitreous. *Color* colorless, white, pale yellow, and variously tinted. Transparent to translucent. H $3\frac{1}{2}$–4. G 2.95 (harder and higher specific gravity than calcite). *Optics:* $(-)$; $\alpha = 1.530$, $\beta = 1.680$, $\gamma = 1.685$; $2V = 18°$; $X = c$, $Y = a$. $r < v$ weak.

Composition. $CaCO_3$, polymorphous with calcite. Sr, Pb, and more rarely Zn may substitute for Ca. Aragonite is unstable relative to calcite at ordinary temperatures and pressures. On heating in air aragonite begins to invert to calcite at 400°C. In contact with water or solutions containing dissolved $CaCO_3$, the inversion may take place at room temperature.

Diagnostic Features. Infusible; decrepitates on heating. Effervesces in cold HCl. Distinguished from calcite by its higher specific gravity and lack of rhombohedral cleavage. Cleavage fragments of columnar calcite are terminated by a cross cleavage which is lacking in aragonite. Distinguished from witherite and strontianite by being infusible, of lower specific gravity, and in lacking a distinctive flame color.

Alteration. Paramorphs of calcite after aragonite are common. Calcium carbonate secreted by molluscs as aragonite is usually changed to calcite on the outside of the shell.

Occurrence. Aragonite is less stable than calcite and much less common. It is formed under a narrow range of physiochemical conditions represented by low temperature, near surface deposits. Experiments have shown that carbonated waters containing calcium more often deposit aragonite when they are hot and calcite when they are cold. The pearly layer of many shells and the pearl itself is aragonite. Aragonite is deposited by hot springs; found associated with beds of gypsum and deposits of iron ore where it may occur in forms resembling coral, and is called *flos ferri*. It is found as fibrous crusts on serpentine and in amygdaloidal cavities in basalt.

Notable localities for the various crystalline types are as follows: Pseudo-hexagonal twin crystals are found in Aragon, Spain; Bastennes, in the south of France; and at Girgenti, Sicily associated with native sulfur. Tabular type crystals are found near Bilin, Bohemia. The acicular type is found at Alston Moor and

Cleator Moor, Cumberland, England. Flos ferri is found in the Styrian iron mines. Some *onyx marble* from Lower California, Mexico, is aragonite. In the United States pseudohexagonal twins are found at Lake Arthur, New Mexico. Flos ferri occurs in the Organ Mountains, New Mexico, and at Bisbee, Arizona.

Name. From Aragon, Spain, where the pseudohexagonal twins were first recognized.

WITHERITE—BaCO$_3$

Crystallography. Orthorhombic; $2/m2/m2/m$. Crystals always twinned on {110} forming pseudohexagonal dipyramids by the intergrowth of three individuals (Fig. 342). Crystals often deeply striated horizontally and by a series of reentrant angles have the appearance of one pyramid capping another. Also botryoidal to globular; columnar or granular.

Angles: $b(010) \wedge m(110) = 58°\ 54'$; $i(021) \wedge i'(02\bar{1}) = 111°\ 12'$.

Pmcn; $a = 5.26$, $b = 8.85$, $c = 6.55$ Å; $a:b:c = 0.594:1:0.740$. $Z = 4$. *d's:* 3.72(10), 2.63(6), 2.14(5), 2.03(5), 1.94(5).

Physical Properties. *Cleavage* {010} distinct, {110} poor. **H** $3\frac{1}{2}$. **G** 4.3. *Luster* vitreous. *Color* colorless, white, gray. Translucent. *Optics:* $(-)$, $\alpha = 1.529$, $\beta = 1.676$, $\gamma = 1.677$; $2V = 16°$; $X = c$, $Y = b$; $r > v$ weak.

Composition. BaO 77.7, CO$_2$ 22.3 per cent. Small amounts of strontium and calcium may substitute for barium.

Diagnostic Features. Fusible at $2\frac{1}{2}$–3, giving a yellowish-green flame (barium). Soluble in cold HCl with effervescence. All solutions, even the very dilute, give precipitate of barium sulfate with sulfuric acid (difference from calcium and strontium). Witherite is characterized by high specific gravity. It is distinguished from barite by its effervescence in acid and from strontianite by the flame test.

Occurrence. Witherite is a comparatively rare mineral, most frequently found in veins associated with galena. Found in England in fine crystals near Hexham in

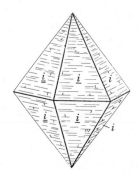

FIG. 342.   Witherite.

Northumberland and Alston Moor in Cumberland. Occurs at Leogang in Salzburg. In the United States found near Lexington, Kentucky, and in a large vein with barite at El Portal, Yosemite Park, California. Also at Thunder Bay, Lake Superior, Ontario.

Use. A minor source of barium.

Name. In honor of D. W. Withering (1741–1799), who discovered and first analyzed the mineral.

STRONTIANITE—$SrCO_3$

Crystallography. Orthorhombic; $2/m2/m2/m$. Crystals usually acicular, radiating like type (1) under aragonite. Twinning on {110} frequent, giving pseudohexagonal aspect. Also columnar, fibrous, and granular.

Angles: $b(010) \wedge m(110) = 58° 40'$, $(001) \wedge (111) = 54° 18'$, $(010) \wedge (021) = 34° 38'$, $(011) \wedge (0\bar{1}1) = 71° 48'$.

Pmcn; $a = 5.13$, $b = 8.42$, $c = 6.09$ Å; $a:b:c = 0.609:1:0.723$; $Z = 4$. *d's:* 3.47(10), 2.42(5), 2.02(7), 1.876(5), 1.794(7).

Physical Properties. *Cleavage* {110} good. **H** $3\frac{1}{2}$–4. **G** 3.7. *Luster* vitreous. *Color* white, gray, yellow, green. Transparent to translucent. *Optics:* $(-)$, $\alpha = 1.520$, $\beta = 1.667$, $\gamma = 1.669$; $2V = 7°$; $X = c$, $Y = b$. $r < v$ weak.

Composition. SrO 70.2, CO_2 29.8 per cent. Calcium may be present in substitution for strontium.

Diagnostic Features. Infusible, but on intense ignition swells and throws out fine branches and gives a crimson flame (strontium). Effervesces in HCl and the medium dilute solution will give precipitate of strontium sulfate on addition of a few drops of sulfuric acid. Characterized by high specific gravity and effervescence in hydrochloric acid. Can be distinguished from witherite and aragonite by flame test; from celestite by poorer cleavage and effervescence in acid.

Occurrence. Strontianite is a low-temperature hydrothermal mineral associated with barite, celestite, and calcite in veins in limestone or marl, and less frequently in igneous rocks and as a gangue mineral in sulfide veins. Occurs in commercial deposits in Westphalia, Germany; Spain; Mexico; and England. In the United States found in geodes and veins with calcite at Schoharie, New York; in the Strontium Hills north of Barstow, California; and near La Conner, Washington.

Use. Source of strontium. Strontium has no great commercial application; used in fireworks, red flares, military rockets, in the separation of sugar from molasses, and in various strontium compounds.

Name. From Strontian in Argyllshire, Scotland, where it was originally found.

CERUSSITE—PbCO$_3$

Crystallography. Orthohombic; $2/m2/m2/m$. Crystals of varied habit common and show many forms. Often tabular on {010} (Fig. 343a). May form reticulated groups with the plates crossing each other at 60° angles; frequently in pseudohexagonal twins with deep reentrant angles in the vertical zone (Fig. 343b). Also in granular crystalline aggregates; fibrous; compact; earthy.

Angles: $b(010) \wedge m(110) = 58° 37'$, $b(010) \wedge i(021) = 34° 40'$, $m(110) \wedge p(111) = 35° 46'$.

Pmcn; $a = 5.15$, $b = 8.47$, $c = 6.11$ Å; $a:b:c = 0.608:1:0.721$; $Z = 4$. *d's:* 3.59(10), 3.50(4), 3.07(2), 2.49(3), 2.08(3).

Physical Properties. *Cleavage* {110} good, {021} fair. **H** 3–3$\frac{1}{2}$. **G** 6.55. *Luster* adamantine. *Color* colorless, white, or gray. Transparent to subtranslucent. *Optics:* $(-)$; $\alpha = 1.804$, $\beta = 2.077$, $\gamma = 2.079$; $2V = 9°$; $X = c$, $Y = b$; $r > v$.

Composition. PbO 83.5, CO$_2$ 16.5 per cent.

Diagnostic Features. Fusible at 1$\frac{1}{2}$. With sodium carbonate on charcoal gives globule of lead and yellow to white coating of lead oxide. Soluble in warm dilute nitric acid with effervescence.

Recognized by its high specific gravity, white color, and adamantine luster. Crystal form and effervescence in nitric acid serve to distinguish it from anglesite.

Occurrence. Cerussite is an important and widely distributed supergene lead ore formed by the action of carbonated waters on galena. Associated with the primary minerals galena and sphalerite, and various secondary minerals such as anglesite, pyromorphite, smithsonite, and limonite.

Notable localities for its occurrence are Ems in Nassau; Mies, Bohemia; Nerchinsk, Siberia; on the island of Sardinia; in Tunis; at Tsumeb, South-West Africa; and Broken Hill, New South Wales. In the United States found at Phoenixville, Pennsylvania; Leadville, Colorado; various districts in Arizona; from the Organ Mountains, New Mexico; and in the Coeur d'Alene district in Idaho.

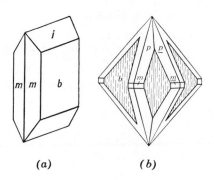

(a)   (b)

FIG. 343.   Cerussite.

Use. An important ore of lead.

Name. From the Latin word meaning *white lead.*

Similar Species. *Phosgenite*, $Pb_2Cl_2CO_3$, (tetragonal) is a rare anhydrous carbonate.

MALACHITE—$Cu_2CO_3(OH)_2$

Crystallography. Monoclinic; $2/m$. Crystals slender prismatic but seldom distinct. Crystals may be pseudomorphous after azurite. Usually in radiating fibers forming botryoidal or stalactitic masses. Often granular or earthy.

$P2_1/a;$ $a = 9.48$, $b = 12.03$, $c = 3.21$ Å, $\beta = 98° 44'$; $a:b:c = 0.788:1:0.267$; $Z = 4$. d's: 6.00(6), 5.06(8), 3.69(9), 2.86(10), 2.52(6).

Physical Properties. *Cleavage* $\{\overline{2}01\}$ perfect but rarely seen. **H** $3\frac{1}{2}$–4. **G** 3.9–4.03. *Luster* adamantine to vitreous in crystals; often silky in fibrous varieties; dull in earthy type. *Color* bright green. *Streak* pale green. Translucent. *Optics:* (−), $\alpha = 1.655$, $\beta = 1.875$, $\gamma = 1.909$; $2V = 43°$; $Y = b$, $Z \wedge c = 24°$; pleochroism: X colorless, Y yellow green, Z deep green.

Composition. CuO 71.9, CO_2 19.9, H_2O 8.2 per cent. Cu 57.4 per cent.

Diagnostic Features. Fusible at 3, giving a green flame. With fluxes on charcoal gives copper globule. Soluble in HCl with effervescence, yielding a green solution. Much water in the closed tube. Recognized by its bright green color

FIG. 344. Malachite, cut and polished.

and botryoidal forms, and distinguished from other green copper minerals by its effervescence in acid.

Occurrence. Malachite is a widely distributed supergene copper mineral found in the oxidized portions of copper veins associated with azurite, cuprite, native copper, iron oxides. It usually occurs in copper veins that lie in limestone.

Notable localities for its occurrence are at Nizhne Tagil in the Ural Mountains; at Chessy, near Lyons, France, associated with azurite; from Tsumeb, South-West Africa; Katanga, Congo; and Broken Hill, New South Wales. In the United States, formerly an important copper ore in the southwestern copper districts; at Bisbee, Morenci, and other localities in Arizona; and in New Mexico.

Use. An ore of copper. Has been used to some extent, particularly in Russia, as an ornamental material for vases, veneer for table tops, etc. (see Fig. 344.)

Name. Derived from the Greek word for *mallows*, in allusion to its green color.

AZURITE—$Cu_3(CO_3)_2(OH)_2$

Crystallography. Monoclinic; $2/m$. Habit varied (Fig. 345); crystals frequently complex and malformed. Also in radiating spherical groups.

Angles. $c(001) \wedge a(100) = 87° 35'$, $m(110) \wedge a(100) = 40° 33'$, $h(111) \wedge m(110) = 19° 58'$, $c(001) \wedge l(013) = 30° 30'$.

$P2_1c$; $a = 4.97$, $b = 5.84$, $c = 10.29$ Å, $\beta = 92° 24'$. $a:b:c = 0.851:1:1.762$. $Z = 2$. $d's$: 5.15(7), 3.66(4), 3.53(10), 2.52(6), 2.34(3).

Physical Properties. *Cleavage* $\{011\}$ perfect, $\{100\}$ fair. **H** $3\frac{1}{2}$–4. **G** 3.77. *Luster* vitreous. *Color* intense azure-blue. Transparent to translucent. *Optics:* (+), $\alpha = 1.730$, $\beta = 1.758$, $\gamma = 1.838$; $2V = 67°$; $X = b$, $Z \wedge c = -13°$; pleochroic in blue $Z > Y > X$; $r > v$.

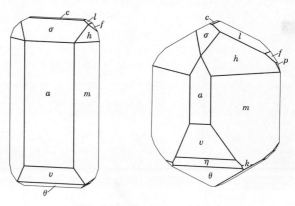

FIG. 345. Azurite crystals.

Composition. CuO 69.2, CO_2 25.6, H_2O 5.2 per cent. Cu 55.3 per cent.

Diagnostic Features. Characterized chiefly by its azure-blue color and effervescence in HCl. Gives the same tests as malachite (page 333).

Alteration. Pseudomorphs of malachite after azurite commonly observed; less common after cuprite.

Occurrence. Azurite is less common than malachite but has the same origin and associations. It is found in fine crystals at Chessy, near Lyons, France; Tsumeb, South-West Africa; in Rumania; Laurium, Greece; Siberia; and Broken Hill, New South Wales. In the United States at Copper Queen mine, Bisbee, and Morenci, Arizona. Widely distributed with copper ores.

Use. A minor ore of copper.

Name. Named in allusion to its color.

Rare Hydrous Carbonates. *Aurichalcite*, a basic carbonate of zinc and copper, pale green to blue in orthorhombic acicular crystals. *Gaylussite*, $Na_2Ca(CO_3)_2 \cdot 5H_2O$, monoclinic and *trona*, $Na_3H(CO_3)_2 \cdot 2H_2O$, monoclinic, are both found in saline-lake deposits.

NITRATES

Five-valent nitrogen forms flattened trefoil-shaped ionic groups with oxygen, much like the carbonate group. These triangles are the monovalent nitrate radical, NO_3^- and are the dominant building block of the nitrates. Like carbon in the carbonate group, the highly charged and highly polarizing nitrogen ion binds its three coordinated oxygens into a close-knit group in which the strength of the oxygen-nitrogen bond ($1\frac{2}{3}$) is greater than any other possible bond in the crystal. Because of the greater strength of this N–O bond, nitrates are less readily decomposed by acids than carbonates.

When these nitrogen-oxygen triangles combine in one-to-one proportions with monovalent cations whose radii permits 6 coordination, structures analogous to those of the calcite group result. Thus, soda niter, $NaNO_3$, and calcite are isostructural with the same crystallography and cleavage and are homeomorphs. Because of the lesser charge, however, soda niter is softer than calcite and melts at a lower temperature and, because of the lower atomic weight of sodium, has a lower specific gravity. Niter, KNO_3, is similarly a precise structural analogue of aragonite, except for the monovalence of the ions.

It is particularly significant that soda niter, like calcite, has an orthorhombic polymorph isostructural with niter, thus making the analogy complete. The calcite structure is often alluded to as the $NaNO_3$ structure, and the aragonite structure as the KNO_3 structure type.

Nitrates

Soda Niter $NaNO_3$ Niter KNO_3

Soda Niter—NaNO$_3$

Crystallography. Hexagonal—R; $\bar{3}2/m$. Rhombohedral crystals are rare. Usually massive, as an incrustation or in beds. Soda niter is homeomorphous with calcite and has similar crystal constants, cleavage, and optical properties.

$R\bar{3}c$. Morphology $a:c = 1:0.828$. Structure $a = 5.07$, $c = 16.81$ Å; $a:c = 1:3.316$; $Z = 6$. $d's$: 3.03(10), 2.13(6), 1.89(6), 1.65(3), 1.46(3).

Physical Properties. *Cleavage* perfect $\{10\bar{1}1\}$. **H** 1–2. **G** 2.29. Luster vitreous. Colorless or white, also reddish-brown, gray, yellow. Transparent to translucent. Taste cooling. Deliquescent. *Optics:* $(-)$; $\omega = 1.587$, $\varepsilon = 1.336$.

Composition. Na$_2$O 36.5, N$_2$O$_5$ 63.5 per cent.

Diagnostic Features. Fusible at 1, giving a strong yellow sodium flame. Easily and completely soluble in water. Distinguished by its cooling taste and its deliquescence.

Occurrence. Because of its solubility in water, soda niter is found only in arid and desert regions. Found in large quantities in the provinces of Tarapaca and Antofagasta, northern Chile, and the neighboring parts of Bolivia. Occurs over immense areas as a salt (caliche) bed interstratified with sand, beds of common salt, gypsum, etc. In the United States it has been noted in Humboldt County, Nevada, and in San Bernardino County, California.

Use. In Chile it is quarried, purified, and used as a source of nitrates. Soda niter now competes with nitrogen "fixed" from the air. Nitrates are used in explosives and fertilizer.

Name. From the composition, the sodium analogue of *niter*, KNO$_3$.

Niter—KNO$_3$

Saltpeter

Crystallography. Orthorhombic; $2/m2/m2/m$. Usually as thin encrustations or as silky acicular crystals. Twinning on $\{110\}$ is common, giving pseudo-hexagonal groupings analogous to aragonite.

Pcmn; $a = 5.43$, $b = 9.19$, $c = 6.46$; $a:b:c = 0.591:1:0.703$. $Z = 4$. $d's$: 3.77(10), 3.03(6), 2.66(5), 2.19(5), 1.96(3).

Physical Properties. Cleavage perfect $\{011\}$. **H** 2. **G** 2.09–2.14. *Luster* vitreous. *Color* white. Translucent. *Optics:* $(-)$; $\alpha = 1.332$, $\beta = 1.504$, $\gamma = 1.504$; $2V = 7°$; $X = c$, $Y = a$; $r < v$.

Composition. K$_2$O 46.5, N$_2$O$_5$ 53.5 per cent.

Diagnostic Features. Fusible at 1, giving violet flame (potassium). Easily soluble in water. Characterized by its cooling taste; distinguished from soda niter by the potassium test and by being nondeliquescent.

Occurrence. Niter is found as delicate crusts, as an efflorescence, on surfaces of earth, walls, rocks, etc. Found as a constituent of certain soils. Also in the loose soil of limestone caves. Not as common as soda niter, but produced from soils in Spain, Italy, Egypt, Arabia, Persia, and India.

Use. Used as a source of nitrogen compounds.

BORATES

The borates are of interest to the mineralogist because they, like the silicates, are capable of forming polymerized anionic groups, having the form of chains, sheets, or isolated multiple groups. This is possible because the small, trivalent boron ion coordinates three oxygens in its most stable configuration. Since the charge on the central cation is $+3$, and there are three closest oxygen neighbors, the strength of the B–O bond must be one unit, equal to exactly one-half the binding energy of the oxygen ion. This permits a single oxygen to be shared between two boron ions linking the BO_3 triangles into a larger unit.

Substances are known in which the fundamental structure of the crystal is the BO_2^- chain, of unlimited length, formed by sharing of oxygens in each boron triangle, each of the shared oxygens uniting two adjacent triangles. But most of the common borates seem to be built about interrupted sheets of BO_3 triangles in which all three oxygens are shared, the sheets being separated by layers of water molecules and joined together by sodium or calcium ions.

Although it is possible to prepare a three-dimensional boxwork made up of BO_3 triangles only and having the composition B_2O_3, such a configuration has a very low stability and disorders readily, yielding a glass. Because of the tendency to form somewhat disordered networks of BO_3 triangles, boron is regarded as a "network-former" in glass manufacture and is used in the preparation of special glasses of light weight and high transparency to energetic radiation.

Approximately 100 borate minerals are known but only the four most abundant are considered here.

Borates

Kernite	$Na_2B_4O_7 \cdot 4H_2O$	Ulexite	$NaCaB_5O_9 \cdot 8H_2O$
Borax	$Na_2B_4O_7 \cdot 10H_2O$	Colemanite	$Ca_2B_5O_{11} \cdot 5H_2O$

KERNITE—$Na_2B_4O_7 \cdot 4H_2O$

Crystallography. Monoclinic; $2/m$. Rarely in crystals; usually in coarse cleavable aggregates.

$P2/a$; $a = 15.68$, $b = 9.09$, $c = 7.02$ Å; $\beta = 108° 52'$; $a:b:c = 1.725:1:0.772$; $Z = 4$. d's: $7.41(10)$, $6.63(8)$, $3.70(3)$, $3.25(2)$, $2.88(2)$.

Physical Properties. *Cleavage* {001} and {100} perfect. Cleavage fragments are thus elongated parallel to the *b* crystallographic axis. The angle between the cleavages: (001) ∧ (100) = 71° 8'. **H** 3. **G** 1.95. *Luster* vitreous to pearly. *Color* colorless to white; colorless specimens become a chalky white on long exposure to the air owing to a surface film of tincalconite. *Optics:* (−); $\alpha = 1.454, \beta = 1.472,$ $\gamma = 1.488$, 2V = 80°; $Z = b, X \wedge c = 39°$.

Composition. Na_2O 22.7, B_2O_3 51.0, H_2O 26.3 per cent.

Diagnostic Features. Characterized by the long splintery cleavage fragments and low specific gravity. Before the blowpipe it swells and then fuses (at $1\frac{1}{2}$) to a clear glass. Slowly soluble in cold water.

Occurrence. The original locality and principal occurrence of kernite is on the Mohave Desert at Boron, California. Here, associated with borax and colemanite in a bedded series of tertiary clays, it is present by the million tons. This deposit of sodium borates is 4 miles long, 1 mile wide, and 100 feet thick, and lies from 350 to 800 feet beneath the surface. The kernite is believed to have formed from borax by recrystallization caused by increased temperature and pressure. Kernite is also found associated with borax at Tincalayu, Argentina.

Use. A source of borax and boron compounds.

Name. From Kern County, California where the mineral is found.

BORAX—$Na_2B_4O_7 \cdot 10H_2O$

Crystallography. Monoclinic; $2/m$. Commonly in prismatic crystals (Fig. 346). Also as massive cellular material or encrustations.

Angles: $c(001) \wedge a(100) = 73° 25'$, $b(010) \wedge m(110) = 43° 30'$, $c(001) \wedge o(\bar{1}12) = 40° 31'$, $c(001) \wedge z(\bar{1}11) = 64° 8'$.

$C2/c$; $a = 11.86, b = 10.67, c = 12.20$ Å; $\beta = 106°$; $a:b:c = 1.111:1:1.143$; $Z = 4$. *d's:* 5.7(2), 4.86(5), 3.96(4), 2.84(5), 2.57(10).

Physical Properties. *Cleavage* {100} perfect. **H** $2–2\frac{1}{2}$. **G** $1.7\pm$. *Luster* vitreous. *Color* colorless or white. Translucent. Sweetish-alkaline taste. Clear crystals effloresce and turn white with the formation of *tincalconite*, $Na_2B_4O_7 \cdot 5H_2O$.

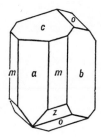

FIG. 346. Borax.

Optics: $(-)$; $\alpha = 1.447, \beta = 1.469, \gamma = 1.472$; $2V = 40°$. $X = b, Z \wedge c = -56°$. $r > v$.

Composition. Na_2O 16.2, B_2O_3 36.6, H_2O 47.2 per cent.

Diagnostic Features. Characterized by its crystals and tests for boron. Fusible at $1-1\frac{1}{2}$ with much swelling and gives strong yellow flame (sodium). Fused with boron flux gives the bright green flame of boron. Readily soluble in water. Much water in the closed tube.

Occurrence. Borax is the most widespread of the borate minerals. It is formed on the evaporation of inclosed lakes and as an efflorescence on the surface in arid regions. Deposits in Tibet furnished the first borax, under the name of *tincal*, to reach western civilization. In the United States it was first found in Lake County, California, and later in the desert region of southeastern California, in Death Valley, Inyo County, and in San Bernardino County. Associated with kernite, it is mined from the bedded deposits at Boron, California, and it is obtained commercially from the brines of Searles Lake at Trona, California.

Associated with other borates, borax is mined in the Andes Mountains in contiguous parts of Argentina, Bolivia, and Chile. Minerals commonly associated with borax are halite, gypsum, ulexite, colemanite, and various rare borates.

Use. Although boron is obtained from several minerals, it is usually converted to borax, the principal commercial product. There are many uses of borax. It is used for washing and cleansing; as an antiseptic and preservative; in medicine; as a solvent for metallic oxides in soldering and welding; and as a flux in various smelting and laboratory operations. Elemental boron is used as a deoxidizer and alloy in nonferrous metals; in rectifiers and control tubes; and a neutron absorber in shields for atomic reactors. Boron is used in rocket fuels and as an additive in motor fuel. Boron carbide, harder than corundum, is used as an abrasive.

Name. Borax comes from an Arabic name for the substance.

ULEXITE—$NaCaB_5O_9 \cdot 8H_2O$

Crystallography. Triclinic; $\bar{1}$. Usually in rounded masses of loose texture, consisting of fine fibers which are acicular or capillary crystals, "cotton-balls."

$P\bar{1}$; $a = 8.73$, $b = 12.75$, $c = 6.70$ Å; $\alpha = 90° 16'$, $\beta = 109° 8'$, $\gamma = 105° 7'$; $a:b:c = 1.104:1:0.526$; $Z = 2$. d's: 12.3(10), 7.89(7), 6.06(7), 4.19(7), 2.67(7).

Physical Properties. *Cleavage* {010} perfect. H $2\frac{1}{2}$; the aggregate has an apparent hardness of 1. G 1.96. *Luster* silky. *Color* white. Tasteless. *Optics:* $(+)$; $\alpha = 1.491, \beta = 1.504, \gamma = 1.520$; $2V = 73°$.

Composition. Na_2O 7.7, CaO 13.8, B_2O_3 43.0, H_2O 35.5 per cent.

Diagnostic Features. The soft "cotton-balls" with silky luster are characteristic of ulexite. Fusible at 1, with intumescence to a clear blebby glass, coloring the flame deep yellow. When moistened with sulfuric acid it gives momentarily the green flame of boron. Gives much water in closed tube.

Occurrence. Ulexite crystallizes in arid regions from brines which have concentrated in inclosed basins, as in playa lakes. Usually associated with borax. It occurs abundantly in the dry basins of northern Chile and in Argentina. In the United States it has been found abundantly in certain of the inclosed basins of Nevada and California and with colemanite in bedded Tertiary deposits.

Use. A source of borax.

Name. In honor of the German chemist, G. L. Ulex, who discovered the mineral.

COLEMANITE—$Ca_2B_6O_{11}\cdot5H_2O$

Crystallography. Monoclinic; $2/m$. Commonly in short prismatic crystals, highly modified. Cleavable massive to granular and compact. (See Fig. 347.)

Angles: $b(010) \wedge m(110) = 53°\ 54'$, $c(001) \wedge a(100) = 69°\ 53'$, $c(001) \wedge h(\bar{2}01) = 68°\ 25'$, $b(010) \wedge d(\bar{1}21) = 57°\ 56'$.

$P2_1/a$; $a = 8.74$, $b = 11.26$, $c = 6.10$ Å, $\beta = 110°\ 7'$; $a:b:c = 0.776:1:0.542$. $Z = 4$. d's: 5.64(5), 4.00(4), 3.85(5), 3.13(10), 2.55(5).

Physical Properties. *Cleavage* {010} perfect. **H** $4–4\frac{1}{2}$. **G** 2.42. *Luster* vitreous. *Color* colorless to white. Transparent to translucent. *Optics:* $(+)$; $\alpha = 1.586$, $\beta = 1.592$, $\gamma = 1.614$; $2V = 55°$; $X = b$, $Z \wedge c = 84°$. $r > v$.

Composition. CaO 27.2, B_2O_3 50.9, H_2O 21.9 per cent.

Diagnostic Features. Fusible at $1\frac{1}{2}$. Before the blowpipe it exfoliates, crumbles, and gives a green flame (boron). Water in the closed tube. Characterized by one direction of highly perfect cleavage and exfoliation on heating.

Occurrence. Colemanite deposits are interstratified with lake bed deposits of Tertiary age. Ulexite and borax are usually associated, and the colemanite is believed to have originated by their alteration. Found in California in Los Angeles, Ventura, San Bernardino, and Inyo counties, and in Nevada in the Muddy Mountains and White Basin, Clark County. Also occurs extensively in Argentina.

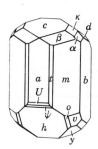

FIG. 347. Colemenite.

Use. A source of borax which, at the time of the discovery of kernite, yielded over half of the world's supply.

Name. In honor of William T. Coleman, merchant of San Francisco, who marketed the product of the colemanite mines.

Similar Species. Other borates, locally abundant are: *boracite*, $Mg_3B_7O_{13}Cl$; *hydroboracite*, $CaMgB_6O_{11}\cdot6H_2O$ and *inyoite*, $Ca_2B_6O_{11}\cdot13H_2O$.

SULFATES AND CHROMATES

We have seen that in the sulfide minerals sulfur is important as the large, divalent sulfide anion. This ion results from the filling by captured electrons of the two vacancies in the outer, or valence, electron shell. The six electrons normally present in this shell may be lost, giving rise to a small, highly charged and highly polarizing positive ion. (Ion. rad. = 0.30 Å.) The ratio of the radius of this hexavalent sulfur ion to that of oxygen ($R_A:R_X = 0.214$) indicates that 4 coordination will be stable. The sulfur to oxygen bond in such an ionic group is very strong and covalent in its properties and produces tightly bound groups that are not capable of sharing oxygens. These $SO_4^=$ groups, the sulfate radical of chemistry, are the fundamental structural units of the sulfate minerals.

The most important and common of the anhydrous sulfates are members of the barite group, with large divalent cations coordinated with the sulfate ion. The rather simple structure leads to orthorhombic symmetry, with perfect {001} and {210} cleavage. The calcium sulfate, anhydrite, because of the smaller size of the calcium ion, has a slightly different structure and has three pinacoidal cleavages. The physical properties are in general conferred by the dominant cation. Specific gravity, for example, is directly proportional to atomic weight.

Of the hydrous sulfates, gypsum is most important and abundant. As is suggested by the highly perfect {010} cleavage, the structure is sheetlike, consisting of layers of calcium and sulfate ions separated by water molecules. Loss of these water molecules allows collapse of the structure into the anhydrite configuration with a large decrease in specific volume and loss of the perfect cleavage.

A large number of minerals belong to this class but only a few of them are common. The class can be divided into (1) the anhydrous sulfates and (2) the hydrous and basic sulfates.

Anhydrous Sulfates		*Hydrous and Basic Sulfates*	
BARITE GROUP		Gypsum	$CaSO_4\cdot2H_2O$
Barite	$BaSO_4$	Chalcanthite	$CuSO_4\cdot5H_2O$
Celestite	$SrSO_4$	Epsomite	$MgSO_4\cdot7H_2O$
Anglesite	$PbSO_4$	Antlerite	$Cu_3(OH)_4SO_4$
Anhydrite	$CaSO_4$	Alunite	$KAl_3(OH)_6(SO_4)_2$
Crocoite	$PbCrO_4$		

Barite Group

The sulfates of barium, strontium, and lead form an isostructural group. They crystallize in the orthorhombic system with closely related crystal constants and similar habits. The members of the group are: *barite, celestite,* and *anglesite.*

BARITE—BaSO$_4$

Crystallography. Orthorhombic; $2/m2/m2/m$. Crystals usually tabular on {001}; often diamond-shaped. Both first- and second-order prisms are usually present, either beveling the corners of the diamond-shaped crystals or the edges of the tables and forming rectangular prismatic crystals elongated parallel to either *a* or *b* (Fig. 348). Crystals may be very complex. Frequently in divergent groups of tabular crystals forming "crested barite" or "barite roses." Also coarsely laminated; granular, earthy.

Angles: $m(210) \wedge m'(\overline{2}10) = 101° 19'$, $c(001) \wedge o(011) = 52° 43'$, $c(001) \wedge d(101) = 38° 52'$.

Pnma; $a = 8.87$, $b = 5.45$, $c = 7.14$ Å; $a:b:c = 1.627:1:1.310$; $Z = 4$. *d's:* 3.90(6), 3.44(10), 3.32(7), 3.10(10), 2.12(8).

Physical Properties. *Cleavage* {001} perfect, {210} less perfect. **H** $3-3\frac{1}{2}$. **G** 4.5 (heavy for a nonmetallic mineral). *Luster* vitreous; on some specimens pearly on base. *Color* colorless, white, and light shades of blue, yellow, red. Transparent to translucent. *Optics:* (+); $\alpha = 1.636$, $\beta = 1.637$, $\gamma = 1.648$; 2V = 37°. $X = c$, $Y = b$. $r > v$.

Composition. BaO 65.7, SO$_3$ 34.3 per cent. Strontium substitutes for barium and a complete solid solution probably extends to celestite, but most material is near one end or the other of the series. A small amount of lead may substitute for barium.

Diagnostic Features. Recognized by its high specific gravity and characteristic cleavage and crystals. Fusible at 4, giving yellowish-green barium flame. Fused with reducing mixture gives a residue which, when moistened, produces a dark stain of silver sulfide on a clean silver surface.

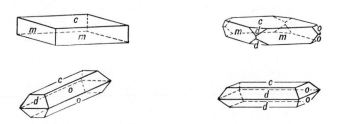

FIG. 348. Barite crystals.

Occurrence. Barite is a common mineral of wide distribution. It occurs usually as a gangue mineral in hydrothermal veins, associated with ores of silver, lead, copper, cobalt, manganese, and antimony. It is found in veins in limestone with calcite, or as residual masses in clay overlying limestone. Also in sandstone with copper ores. In places acts as a cement in sandstone. Deposited occasionally as a sinter by waters from hot springs.

Notable localities for the occurrence of barite crystals are in Westmoreland, Cornwall, Cumberland, and Derbyshire, England; Felsobanya and other localities, Rumania; Saxony; and Bohemia. In the United States at Cheshire, Connecticut; Dekalb, New York; and Fort Wallace, New Mexico. Massive barite, occurring usually as veins, nests, and irregular bodies in limestones, has been quarried in the United States in Georgia, Tennessee, Missouri, and Arkansas. At El Portal, California, at the entrance to Yosemite Park, barite is found in a vein with witherite.

Use. More than 80 per cent of the barite produced is used in oil and gas-well drilling. Barite is the chief source of barium for chemicals. A major use of barium is in *lithopone*, a combination of barium sulfide and zinc sulfate which combine to form an intimate mixture of zinc sulfide and barium sulfate. Lithopone is used in the paint industry and to a lesser extent in floor coverings and textiles. Precipitated barium sulfate, "blanc fixe," is employed as a filler in paper and cloth, in cosmetics, as a paint pigment, and for barium meals in medical radiology.

Name. From the Greek word meaning *heavy*, in allusion to its high specific gravity.

CELESTITE—SrSO$_4$

Crystallography. Orthorhombic; $2/m2/m2/m$. Crystals closely resemble those of barite. Commonly tabular parallel to {001} or prismatic parallel to a or b with prominent development of the first- and second-order prisms. Crystals elongated parallel to a are frequently terminated by nearly equal development of faces of $d\{101\}$ and $m\{210\}$ (Fig. 349). Also radiating fibrous; granular.

Angles: $b(010) \wedge m(210) = 52° 1'$, $c(001) \wedge o(011) = 52° 3'$, $c(001) \wedge d(101) = 39° 24'$, $c(001) \wedge l(102) = 22° 19'$.

Pnma; $a = 8.38$, $b = 5.37$, $c = 6.85$ Å; $a:b:c = 1.561:1:1.276$; $Z = 4$. *d's:* 3.30(10), 3.18(6), 2.97(10), 2.73(6), 2.04(6).

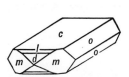
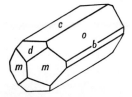

FIG. 349. Celestite crystals.

Physical Properties. *Cleavage* {001} perfect, {210} good. **H** $3-3\frac{1}{2}$. **G** 3.95–3.97. *Luster* vitreous to pearly. *Color* colorless, white, often faintly blue or red. Transparent to translucent.

Optics: (+); $\alpha = 1.622$, $\beta = 1.624$, $\gamma = 1.631$; $2V = 50°$. $X = c$, $Y = b$. $r < v$.

Composition. SrO 56.4, SO_3 43.6 per cent. Barium substitutes for strontium and a complete solid-solution series probably extends to barite.

Diagnostic Features. Closely resembles barite but is differentiated by lower specific gravity and lithium flame test. Fuses at $3\frac{1}{2}$–4 and colors the flame crimson (strontium). Usually decrepitates when touched with the blowpipe flame. Fused with sodium carbonate gives a residue which, when moistened, produces on a clean silver surface a dark stain of silver sulfide.

Occurrence. Celestite is found usually disseminated through limestone or sandstone, or in nests and lining cavities in such rocks. Associated with calcite, dolomite, gypsum, halite, sulfur, fluorite. Also found as a gangue mineral in lead veins.

Notable localities for its occurrence are with the sulfur deposits of Sicily; at Bex, Switzerland; Yate, Gloucestershire, England; and Herrengrund, Slovakia. Found in the United States at Clay Center, Ohio; Put-in-Bay, Lake Erie; Mineral County, West Virginia; Lampasas, Texas; and Inyo County, California.

Use. Used in the preparation of strontium nitrate for fireworks and tracer bullets, and other strontium salts used in the refining of beet sugar.

Name. Derived from *caelestis*, in allusion to the faint blue color of the first specimens described.

ANGLESITE—PbSO₄

Crystallography. Orthorhombic; $2/m2/m2/m$. Frequently in crystals with habit often similar to that of barite but more varied. Crystals may be prismatic parallel to any one of the crystal axes and frequently show many forms, with a complex development (Fig. 350). Also massive, granular to compact. Frequently earthy, in concentric layers which may have an unaltered core of galena.

Angles: $b(010) \wedge m(210) = 51° 52'$, $c(001) \wedge d(101) = 39° 24'$, $c(001) \wedge o(011) = 52° 13'$, $c(001) \wedge z(211) = 64° 25'$.

Pnma; $a = 8.47$, $b = 5.39$, $c = 6.94$ Å; $a:b:c = 1.571:1:1.288$. $Z = 4$. *d's:* 4.26(9), 3.81(6), 3.33(9), 3.22(7), 3.00(10).

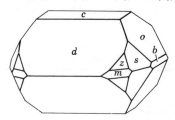

FIG. 350. Anglesite.

Physical Properties. *Cleavage* {001} good, {210} imperfect. Fracture conchoidal. **H** 3.0. **G** 6.2–6.4 (unusually high). *Luster* adamantine when crystalline, dull when earthy. *Color* colorless, white, gray, pale shades of yellow. May be colored dark gray by impurities. Transparent to translucent. *Optics:* (+); $\alpha = 1.877$, $\beta = 1.883$, $\gamma = 1.894$; $2V = 75°$; $X = c$, $Y = b$. $r < v$.

Composition. PbO 73.6, SO_3 26.4 per cent.

Diagnostic Features. Recognized by its high specific gravity, its adamantine luster, and frequently by its association with galena. Distinguished from cerussite in that it will not effervesce in nitric acid. Fusible at $1\frac{1}{2}$. On charcoal with sodium carbonate reduced to a lead globule with yellow to white coating of lead oxide; the residue, when moistened, produces on clean silver surface a dark stain of silver sulfide.

Occurrence. Anglesite is a common supergene mineral found in the oxidized portions of lead deposits. It is formed through the oxidation of galena, either directly, as is shown by concentric layers of anglesite surrounding a core of galena, or by an intermediate solution and subsequent recrystallization. Anglesite is commonly associated with galena, cerussite, sphalerite, smithsonite, hemimorphite, and iron oxides.

Notable localities for its occurrence are Monte Poni, Sardinia; Island of Anglesey, Wales; Derbyshire; and Leadhills, Scotland. From Sidi-Amor-ben-Salem, Tunis; Tsumeb, South-West Africa; Broken Hill, New South Wales; and Dundas, Tasmania. It is found in crystals imbedded in sulfur at Los Lamentos, Chihuahua, Mexico. Occurs in the United States at Phoenixville, Pennsylvania; Tintic district, Utah; and Coeur d'Alene district, Idaho.

Use. A minor ore of lead.

Name. Named from the original locality on Island of Anglesey.

ANHYDRITE—$CaSO_4$

Crystallography. Orthorhombic. $2/m2/m2/m$. Crystals rare; when observed are thick tabular on {010}, {100} or {001}, also prismatic parallel to b. Usually massive or in crystalline masses resembling an isometric mineral with cubic cleavage. Also fibrous, granular, massive. (See Fig. 351.)

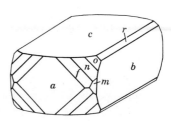

FIG. 351. Anhydrite.

Angles: $b(010) \wedge m(110) = 45° 1'$, $c(001) \wedge r(011) = 41° 45'$, $c(001) \wedge o(111) = 51° 38'$, $c(001) \wedge n(211) = 63° 24'$.

Bmmb; $a = 6.96$, $b = 6.97$, $c = 6.22$ Å; $a:b:c = 0.999:1:0.892$; $Z = 4$. *d's:* 3.50(10), 2.85(3), 2.33(2), 2.08(2), 1.869(2).

Physical Properties. *Cleavage* {010} perfect, {100} nearly perfect, {001} good. **H** $3-3\frac{1}{2}$. **G** 2.89–2.98. *Luster* vitreous to pearly on cleavage. *Color* colorless to bluish or violet. Also may be white or tinged with rose, brown, or red. *Optics:* (+); $\alpha = 1.570$, $\beta = 1.575$, $\gamma = 1.614$; $2V = 44°$; $X = b$, $Y = a$; $r < v$.

Composition. CaO 41.2, SO_3 58.8 per cent.

Diagnostic Features. Fusible at 3. When moistened with hydrochloric acid and ignited it gives orange-red flame of calcium. When fused with reducing mixture it gives a residue that, when moistened with water, darkens silver. Anhydrite is characterized by its three cleavages at right angles. It is distinguished from calcite by its higher specific gravity and from gypsum by its greater hardness. Some massive varieties are difficult to recognize, and one should test for the sulfate radical.

Alteration. Anhydrite by hydration changes to gypsum with an increase in volume, and in places large masses have been thus altered.

Occurrence. Anhydrite occurs in much the same manner as gypsum and is often associated with that mineral but is not nearly as common. Found in beds associated with salt deposits in the cap rock of salt domes, and in limestone rocks. Found in some amygdaloidal cavities in basalt.

Notable foreign localities are: Wieliczka, Poland; Aussee, Styria; Stassfurt, Prussia; Berchtesgaden, Bavaria; Hall near Innsbruck, Tyrol; and Bex, Switzerland. In the United States found in Lockport, New York; West Paterson, New Jersey; Nashville, Tennessee; New Mexico; and Texas. Found in large beds in Nova Scotia.

Use. Ground anhydrite is used as a soil conditioner and to a minor extent as a setting retarded in Portland cement. In Great Britain and Germany it is used as a source of sulfur for the production of sulfuric acid.

Name. Anhydrite is from the Greek meaning *without water*, in contrast to the more common hydrous calcium sulfate, gypsum.

Crocoite—$PbCrO_4$

Crystallography. Monoclinic; $2/m$. Commonly in slender prismatic crystals, vertically striated and columnar aggregates. Also granular.

$P2_1/n;$ $a = 7.11$, $b = 7.41$, $c = 6.81$ Å, $\beta = 102° 33'$; $a:b:c = 0.960:1:0.919$; $Z = 4$. *d's:* 4.96(3), 2.85(3), 2.33(2), 2.08(2), 1.869(2).

Physical Properties. *Cleavage* {110} imperfect. **H** $2\frac{1}{2}$–3. **G** 5.9–6.1. *Luster* adamantine. *Color* bright hyacinth-red. Streak orange-yellow. Translucent.

Composition. PbO 68.9, CrO_3 31.1 per cent.

Diagnostic Features. Crocoite is characterized by its color, high luster, and high specific gravity. It may be confused with wulfenite, lead molybdate, but can be distinguished from it by its redder color, lower specific gravity, and crystal form. Fusible at $1\frac{1}{2}$. Fused with sodium carbonate on charcoal gives a lead globule. Gives a green (chromium) borax bead in the oxidizing flame.

Occurrence. Crocoite is a rare mineral found in the oxidized zones of lead deposits in those regions where lead veins have traversed rocks containing chromite. Associated with pyromorphite, cerussite, and wulfenite. Notable localities are: Dundas, Tasmania; Beresovsk near Sverdlovsk, Ural Mountains; and Rezbanya, Rumania. In the United States is found in small quantities in the Vulture district, Arizona.

Use. Not abundant enough to be of commercial value, but of historic interest, since the element chromium was first discovered in crocoite.

Name. From the Greek meaning *saffron*, in allusion to the color.

GYPSUM—CaSO$_4$·2H$_2$O

Crystallography. Monoclinic; $2/m$. Crystals are of simple habit (Fig. 352); tabular on $\{010\}$; diamond-shaped, with edges beveled by $\{120\}$ and $\{\bar{1}11\}$. Other forms rare. Twinning common, on $\{100\}$ (Fig. 353), often resulting in swallowtail twins.

Angles: $b(010) \wedge f(120) = 55^\circ_\circ\ 45'$, $b(010) \wedge l(\bar{1}11) = 71^\circ\ 54'$, $b(010) \wedge n(011) = 69^\circ\ 20'$.

$A2/n;\ a = 5.68,\ b = 15.18,\ c = 6.29$ Å, $\beta = 113^\circ\ 50'$; $a:b:c = 0.374:1:0.414$; $Z = 4$. d's: $7.56(10),\ 4.27(5),\ 3.06(6),\ 3.71(9),\ 3.30(6)$.

Physical Properties. *Cleavage* $\{010\}$ perfect yielding thin folia; $\{100\}$, with conchoidal surface; $\{011\}$, with fibrous fracture. **H** 2. **G** 2.32. *Luster* usually vitreous; also pearly and silky. *Color* colorless, white, gray; various shades of yellow, red, brown, from impurities. Transparent to translucent.

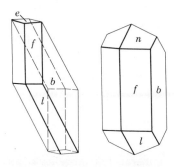

FIG. 352. Gypsum crystals.

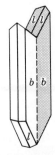

FIG. 353. Gypsum twin.

Satin spar is a fibrous gypsum with silky luster. *Alabaster* is the fine-grained massive variety. *Selenite* is a variety that yields broad colorless and transparent cleavage folia.

Optics: (+); $\alpha = 1.520$, $\beta = 1.523$, $\gamma = 1.530$; $2V = 58°$; $Y = b$, $X \wedge c = -37°$; $r > v$.

Composition. CaO 32.6, SO_3 46.5, H_2O 20.9 per cent.

Diagnostic Features. Characterized by its softness and its three unequal cleavages. Fusible at 3. Soluble in hot dilute hydrochloric acid, and solution with barium chloride gives white precipitate of barium sulfate. In the closed tube turns white and yields much water. Its solubility in acid and the presence of much water distinguish it from anhydrite.

Occurrence. Gypsum is a common mineral widely distributed in sedimentary rocks, often as thick beds. It frequently occurs interstratified with limestones and shales and is usually found as a layer underlying beds of rock salt, having been deposited there as one of the first minerals to crystallize on the evaporation of salt waters. May recrystallize in veins, forming *satin spar*. Occurs also as lenticular bodies or scattered crystals in clays and shales. Frequently formed by the alteration of anhydrite and under these circumstances may show folding because of increased volume. Found in volcanic regions, especially where limestones have been acted upon by sulfur vapors. Also common as a gangue mineral in metallic veins. Associated with many different minerals, the more common being halite, anhydrite, dolomite, calcite, sulfur, pyrite, and quartz.

Gypsum is the most common sulfate, and extensive deposits are found in many localities throughout the world. The principal world producers are the United States, Canada, France, England, and the U.S.S.R. In the United States commercial deposits are found in many states, but the chief producers are located in New York, Michigan, Iowa, Texas, and California. Gypsum is found in large deposits in Arizona and New Mexico in the form of wind-blown sand.

Use. Gypsum is used chiefly for the production of plaster of Paris. In the manufacture of this material, the gypsum is ground and then heated at 190–200°C until about 75 per cent of the water has been driven off. This plaster, when mixed with water, slowly absorbs the water, crystallizes, and thus hardens or "sets." Plaster of Paris is used extensively for "staff," the material from which temporary exposition buildings are built, for gypsum lath, wallboard, and for molds and casts of all kinds. Gypsum is employed in making adamant plaster for interior use. Serves as a soil conditioner "land plaster," for fertilizer. Uncalcined gypsum is used as a retarder in Portland cement. Satin spar and alabaster are cut and polished for various ornamental purposes but are restricted in their uses because of their softness.

Name. From the Greek name for the mineral, but more especially for the calcined mineral.

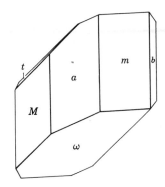

FIG. 354. Chalcanthite.

Chalcanthite—CuSO$_4$·5H$_2$O

Crystallography. Triclinic; $\bar{1}$. Found in crystals commonly $\bar{1}$ prismatic on [001] or tabular parallel to $\{\bar{1}\bar{1}1\}$. Also massive, stalactitic, and reniform; may have fibrous appearance. (See Fig. 354.)

Angles: $b(010) \wedge m(110) = 69° 49'$, $b(010) \wedge a(100) = 100° 55'$, $a(100) \wedge \omega(\bar{1}\bar{1}1) = 120° 28'$, $m(110) \wedge M(1\bar{1}0) = 57° 10'$.

$P\bar{1}$; $a = 6.12$, $b = 10.69$, $c = 5.96$ Å; $\alpha = 97° 35'$, $\beta = 107° 10'$; $\gamma = 77° 33'$; $a:b:c = 0.572:1:0.555$; $Z = 2$. *d's:* 5.48(6), 4.73(10), 3.99(6), 3.71(9), 3.30(6).

Physical Properties. *Cleavage* $\{1\bar{1}0\}$ imperfect. **H** $2\frac{1}{2}$. **G** 2.12–2.30. *Luster* vitreous. *Color* deep azure-blue. Transparent to translucent. Taste metallic. *Optics:* (−); $\alpha = 1.516$, $\beta = 1.539$, $\gamma = 1.546$, $2V = 56°$; $r < v$.

Composition. CuO 31.8, SO$_3$ 32.1, H$_2$O 36.1 per cent.

Diagnostic Features. Infusible. Gives copper globule when fused with sodium carbonate on charcoal. Soluble in water. Turns white and gives much water in the closed tube. Characterized by its blue color, metallic taste, and solubility in water.

Occurrence. Chalcanthite is rare, found only in arid regions as a supergene mineral, occurring near the surface in copper veins, and derived from the original copper sulfides by oxidation. Often deposited on iron from the waters in copper mines.

Chalcanthite is found abundantly at Chuquicamata and other arid localities in Chile, where it has served as an important ore mineral. Other occurrences are unimportant commercially.

Use. A minor ore of copper. The artificial *blue vitriol* is used in calico printing, in galvanic cells, as an insecticide, and for industrial purposes.

Name. From two Greek words meaning *brass* and *flower*.

Epsomite—MgSO$_4$·7H$_2$O

Crystallography. Orthorhombic; 222. Rarely in crystals. Usually in botryoidal masses and delicately fibrous crusts.

$P222$; $a = 11.96$, $b = 12.05$, $c = 6.88$ Å; $a:b:c = 0.993:1:0.571$; $Z = 4$. *d's:* 5.99(2), 5.35(3), 4.21(10), 2.68(2), 2.66(2).

Physical Properties. *Cleavage* {010} perfect. **H** $2-2\frac{1}{2}$. **G** 1.68. *Luster* vitreous to earthy. *Color* colorless to white. Transparent to translucent. Taste very bitter. *Optics:* $(-)$; $\alpha = 1.433$, $\beta = 1.455$, $\gamma = 1.461$; $2V = 52°$; $X = a$, $Y = c$; $r < v$.

Composition. MgO 16.3, SO_3 32.5, H_2O 51.2 per cent. Ni, Fe, Mn, Cu, and Co may substitute for Mg.

Diagnostic Features. In the closed tube it gives much acid water, and liquefies in its water of crystallization. Soluble in water. Characterized by its mode of occurrence in delicate fibrous and capillary aggregates, its easy solubility in water, and its bitter taste.

Occurrence. Epsomite is usually deposited as an efflorescence on the rocks in mine workings and on the walls of caves. More rarely it is found in lake deposits; associated with other soluble salts as at Stassfurt, Germany. In the United States it is found on the floors of limestone caves in Kentucky, Tennessee, and Indiana, and in abandoned mines in California and Colorado. Found in lake deposits in Stevens County, Washington.

Use. The mineral epsomite has little use in itself. It is manufactured as epsom salt from other magnesium minerals.

Name. Epsomite was named from the original locality of Epsom, England.

Similar Species. Melanterite, $FeSO_4 \cdot 7H_2O$ is a secondary mineral found as an efflorescence on mine timbers.

Antlerite—$Cu_3(OH)_4SO_4$

Crystallography. Orthorhombic; $2/m2/m2/m$. Crystals usually tabular or {010} (Fig. 355). May be in slender prismatic crystals vertically striated. Also in parallel aggregates, reniform, massive.

Angles: $m(110) \wedge m(1\bar{1}0) = 111° 3'$, $b(010) \wedge o(011) = 63° 19'$, $b(010) \wedge f(130) = 25° 54'$, $m(110) \wedge r(111) = 48° 23'$.

Pnam; $a = 8.24$, $b = 11.99$, $c = 6.03$ Å; $a:b:c = 0.687:1:0.503$; $Z = 4$. *d's:* 4.86(10), 3.60(8), 3.40(3), 2.68(8), 2.57(8).

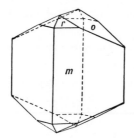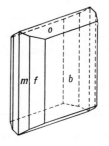

FIG. 355. **Antlerite crystals.**

Physical Properties.   *Cleavage* {010} perfect. **H** $3\frac{1}{2}$–4. **G** 3.9±. *Luster* vitreous. *Color* emerald to blackish-green. *Streak* pale green. Transparent to translucent. *Optics:* (+); $\alpha = 1.726$, $\beta = 1.738$, $\gamma = 1.789$; $2V = 53°$; $X = b$, $Y = a$; pleochroism X yellow-green, Y blue-green, Z green.

Composition.   CuO 67.3, SO_3 22.5, H_2O 10.2 per cent.

Diagnostic Features.   Fusible at $3\frac{1}{2}$. Yields a copper globule when fused with sodium carbonate on charcoal. Hydrochloric acid solution with barium chloride gives a white precipitate of barium sulfate. Yields water in the closed tube, and at a high temperature, yields sulfuric acid. Antlerite is characterized by its green color, {010} cleavage, and associations. Lack of effervescence in hydrochloric acid distinguishes it from malachite. It is commonly associated with atacamite and brochantite and one must use optical properties to distinguish it from these minerals.

Occurrence.   Antlerite is found in the oxidized portions of copper veins, especially in arid regions. It was formerly considered a rare mineral, but in 1925 it was recognized as the chief ore mineral at Chuquicamata, Chile, the world's largest copper mine. It may form directly as a secondary mineral on chalcocite, or the copper may go into solution and later be deposited as antlerite, filling cracks. In the United States it is found at Bisbee, Arizona, and near Black Mountain, Nevada. Also found at Kennecott, Alaska.

Use.   An ore of copper.

Name.   From the Antler mine, Arizona, from which locality it was originally described.

Similar Species.   *Brochantite,* $Cu_4(OH)_6SO_4$, is similar in all its properties to antlerite, but, although more widespread, it is nowhere abundant. Until 1925 it was considered to be the chief ore mineral at Chuquicamata, Chile.

Alunite—$KAl_3(OH)_6(SO_4)_2$

Crystallography.   Hexagonal—R; $3m$. Crystals usually a combination of positive and negative trigonal pyramids resembling rhombohedrons with nearly cubic angles (90° 50′). May be tabular on {0001}. Commonly massive or disseminated.

$R3m$; $a = 6.97$, $c = 17.38$ Å; $a:c = 1:2.494$; $Z = 3$. d's: 4.94(5), 2.98(10), 2.29(5), 1.89(6), 1.74(5).

Physical Properties.   *Cleavage* {0001} imperfect. **H** 4. **G** 2.6–2.8. *Color* white, gray, or reddish. Transparent to translucent. *Optics:* (+); $\omega = 1.572$, $\varepsilon = 1.592$.

Composition. K_2O 11.4, Al_2O_3 37.0, SO_3 38.6, H_2O 13.0 per cent. Sodium may replace potassium at least up to Na:K = 7:4. When sodium exceeds potassium the mineral is called natroalunite.

Diagnostic Features.   Infusible and decrepitates before the blowpipe, giving a potassium flame. Heated with cobalt nitrate solution turns a fine blue color. In the closed tube gives acid water. Soluble in sulfuric acid. Alunite is usually massive and difficult to distinguish from rocks such as limestone and dolomite, and other massive minerals as anhydrite and granular magnesite. A positive test for acid water will serve to distinguish alunite from similar-appearing minerals.

Occurrence. Alunite also called *alumstone*, is usually formed by sulfuric acid solutions acting on rocks rich in potash feldspar, and in some places large masses have thus been formed. Found in smaller amounts about volcanic fumeroles. In the United States it is found at Red Mountain in the San Juan district, Colorado; Goldfield, Nevada; and Marysvale, Utah.

Use. In the production of alum. At Marysvale, Utah, alunite has been mined and treated in such a way as to recover potassium and aluminum.

Name. From the Latin meaning *alum*.

Similar Species. *Jarosite*, $KFe_3(OH)_6(SO_4)_2$, the iron analogue of alunite, is a secondary mineral found as crusts and coatings on ferruginous ores.

TUNGSTATES AND MOLYBDATES

The six-valent ions of tungsten and molybdenum (Ion. rad. of both $= 0.62$ Å) are considerably larger than those of six-valent sulfur and five-valent phosphorus. Hence, when these ions enter into anisodesmic ionic groups with oxygen, the four coordinated oxygen ions do not occupy the apices of regular tetrahedra, as is the case in the sulfates and phosphates, but form a somewhat flattened grouping of square outline. Although W 184 has a much greater atomic weight than Mo 96, both belong to the same family of the periodic table and, because of the lanthanide contraction, have the same ionic radii. As a result, each may freely substitute for the other as the coordinating cation in the warped tetrahedral oxygen groupings. In nature, however, the operation of the processes of geochemical differentiation often separate these elements, perhaps because of their very different atomic weights, and it is not uncommon to find primary tungstates almost wholly free of molybdenum, and vice versa. In secondary minerals, the two elements are more commonly associated in solid-solution relations with each other.

The minerals of this chemical class fall in the main into two isostructural groups. The wolframite group consists of fairly small divalent cations such as iron, manganese, magnesium, nickel, and cobalt in 6 coordination with tungstate. Complete solid solution between ferrous iron and divalent manganese is observed in minerals of this group.

The scheelite group contains compounds of larger divalent ions such as calcium and lead in 8 coordination with tungstate and molybdate ions. Tungsten and molybdenum may substitute for each other forming partial series between *scheelite*, $CaWO_4$, and *powellite*, $CaMoO_4$; and *stolzite*, $PbWO_4$, and *wulfenite*, $PbMoO_4$.

The substitution of calcium and lead for one another forms partial series between scheelite and stolzite and between powellite and wulfenite.

Tungstates and Molybdates

Wolframite	$(Fe, Mn)WO_4$
Scheelite	$CaWO_4$
Wulfenite	$PbMoO_4$

WOLFRAMITE—$(Fe,Mn)WO_4$

Crystallography. Monoclinic; $2/m$. Crystals commonly tabular parallel to $\{100\}$ (Fig. 356) giving bladed habit, with faces striated parallel to c. In bladed, lamellar, or columnar aggregates. Massive granular.

Angles: $m(110) \wedge m'(1\bar{1}0) = 79° 5'$, $f(011) \wedge f'(0\bar{1}1) = 81° 51'$, $a(100 \wedge t(102) = 61° 57'$, $a(100 \wedge y(\bar{1}02) = 117° 20'$.

$P2/c$; $a = 4.79$, $b = 5.74$, $c = 4.99$ Å; $\beta = 90° 28'$; $a:b:c = 0.835:1:0.869$; $Z = 2$. d's: 4.76(5), 3.74(5), 2.95(10), 2.48(6), 1.716(5). Data for wolframite; cell parameters slightly less for $FeWO_4$ and greater for $MnWO_4$.

Physical Properties. *Cleavage* $\{010\}$ perfect. **H** $4–4\frac{1}{2}$. **G** 7.0–7.5, higher with higher iron content. *Luster* submetallic to resinous. *Color* black in *ferberite* to brown in *huebnerite*. *Streak* from nearly black to brown.

Composition. Ferrous iron and divalent manganese substitute for each other in all proportions and a complete solid-solution series exists between *ferberite*, $FeWO_4$, and *huebnerite*, $MnWO_4$. The percentage of WO_3 is 76.3 in ferberite and 76.6 in huebnerite.

Diagnostic Features. The dark color, one direction of perfect cleavage, and high specific gravity serve to distinguish wolframite from other minerals. Fusible at 3–4 to a magnetic globule. Insoluble in acids. Fused with sodium carbonate, dissolves in hydrochloric acid; tin added and solution boiled gives a blue color (tungsten). In the oxidizing flame with sodium carbonate gives bluish-green bead (manganese).

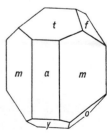

FIG. 356. Wolframite.

Occurrence. Wolframite is a comparatively rare mineral found usually in pegmatites and high-temperature quartz veins associated with granites. More rarely in sulfide veins. Minerals commonly associated include cassiterite, scheelite, bismuth, quartz, pyrite, galena, sphalerite, and arsenopyrite. In some veins wolframite may be the only metallic mineral present.

Found in fine crystals from Schlaggenwald and Zinnwald, Bohemia, and in the various tin districts of Saxony and Cornwall. Important producing countries are China, the U.S.S.R., Korea, Bolivia, and Australia. Wolframite occurs in the United States in the Black Hills, South Dakota. Ferberite has been mined extensively in Boulder County, Colorado. Huebnerite is found near Silverton, Colorado; Mammoth district, Nevada; and Black Hills, South Dakota.

Use. Chief ore of tungsten. Tungsten is used as hardening metal in the manufacture of high-speed tool steel, valves, springs, chisels, files, etc. Its high melting point (3410°C) requires a special chemical process for reduction of the metal which is produced in the form of a powder. By powder-metallurgy, pure metal products such as lamp filaments are fabricated. A large use for tungsten is in the manufacture of carbides, harder than any natural abrasives (other than diamond), and are used for cutting tools, rock bits, and hard facings. Sodium tungstate is used in fireproofing cloth and as a mordant in dyeing.

Name. Wolframite is derived from an old word of German origin.

SCHEELITE—CaWO$_4$

Crystallography. Tetragonal; $4/m$. Crystals usually simple dipyramids of the second order (Fig. 357). The first-order dipyramid, (112), closely resembles the octahedron in angles. Also massive granular.

Angles: $p(101) \wedge p'(\bar{1}01) = 130° 33'$, $\beta(013) \wedge \beta'(0\bar{1}3) = 71° 48'$.

$I4/a$; $a = 5.25$, $c = 11.40$ Å; $a:c = 1:2.171$; $Z = 4$. *d's:* 4.77(7), 3.11(10), 1.94(8), 1.596(9), 1.558(7).

Physical Properties. *Cleavage* {101}, distinct. **H** $4\frac{1}{2}$–5. **G** 5.9–6.1 (unusually high for a nonmetallic mineral). *Luster* vitreous to adamantine. *Color* white, yellow, green, brown. Translucent; some specimens transparent. Most scheelite will fluoresce. *Optics:* (+); $\omega = 1.920$, $\varepsilon = 1.934$.

FIG. 357. Scheelite.

Composition. CaO 19.4, WO_3 80.6 per cent. Mo can substitute for W and a partial series extends to *powellite*, $CaMoO_4$.

Diagnostic Features. Difficulty fusible (5). Decomposed by boiling in hydrochloric acid, leaving a yellow residue of tungstic oxide which, when tin is added to the solution and boiling continued, turns first blue and then brown. Recognized by its high specific gravity and crystal form. The test for tungsten may be necessary for identification.

Occurrence. Scheelite is found in granite pegmatites, contact metamorphic deposits, and high-temperature hydrothermal veins associated with granitic rocks. Associated with cassiterite, topaz, fluorite, apatite, molybdenite, and wolframite. Occurs with tin in deposits of Bohemia, Saxony, and Cornwall; and in quantity in New South Wales and Queensland. In the United States schellite is mined near Mill City and Mina, Nevada; near Atolia, San Bernardino County, California; and in lesser amounts in Arizona, Utah, and Colorado.

Use. An ore of tungsten. Wolframite furnishes most of the world's supply of tungsten, but scheelite is more important in the United States.

Name. After K. W. Scheele (1742–1786), the discoverer of tungsten.

WULFENITE—$PbMoO_4$

Crystallography. Tetragonal; 4. Crystals usually square, tabular in habit with prominent {001}. Some crystals very thin. More rarely pyramidal. Some crystals may be twinned on {001} giving them a dipyramidal habit (Fig. 358).

Angles: $c(001) \wedge n\{101\} = 65° 52'$, $c(001) \wedge s(103) = 36° 38'$.

$a = 5.42$, $c = 12.10$ Å; $a:c = 2.233$; $Z = 4$. *d's:* 3.24(10), 3.03(2), 2.72(2), 2.02(3), 1.653(3).

Physical Properties. *Cleavage* {011} distinct. **H** 3. **G** 6.8±. *Luster* vitreous to adamantine. *Color* yellow, orange, red, gray, white. *Streak* white. Transparent to subtranslucent.

Composition. PbO 60.8, MoO_3 39.2 per cent. Ca may substitute for Pb indicating at least a partial series to powellite, $Ca(Mo,W)O_4$.

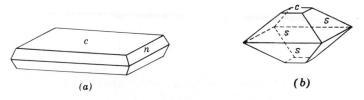

(a) (b)

FIG. 358. Wulfenite crystals.

Diagnostic Features. Wulfenite is characterized by its square tabular crystals, orange to yellow color, high luster, and association with other lead minerals. Distinguished from crocoite by test for molybdenum. Fusible at 2. Gives a lead globule when fused with sodium carbonate on charcoal. With salt of phosphorus in the reducing flame it gives green bead; in the oxidizing flame, yellowish-green when hot to almost colorless when cold.

Occurrence. Wulfenite is found in the oxidized portion of lead veins with other secondary lead minerals, especially cerussite, vanadinite, and pyromorphite. Found in the United States at Phoenixville, Pennsylvania, and in a number of places in Utah, Nevada, Arizona, and New Mexico. Found in beautiful crystals at Red Cloud, Arizona.

Use. A minor source of molybdenum. Molybdenite is the chief ore.

Name. After X. F. Wulfen, Austrian mineralogist.

PHOSPHATES, ARSENATES, AND VANADATES

Five-valent phosphorus is only slightly larger than six-valent sulfur and, hence, like sulfur, forms a tetrahedral ionic group with oxygen. This group, the PO_4^{-3} tetrahedron, is, like the sulfate tetrahedron, a separate radical that cannot share oxygen or form polymerized groups. All phosphates are built about this phosphate ion as the fundamental building unit. Similar units, having the same configuration of the oxygens, the same kind and intensity of the binding forces, are built about the five-valent arsenic and vanadium ions. Phosphorus, arsenic, and vanadium

Phosphates, Arsenates, and Vanadates

Triphylite—	$Li(Fe,Mn)PO_4$
Lithiophilite	$Li(Mn,Fe)PO_4$
Monazite	$(Ce,La,Y,Th)PO_4$
APATITE GROUP	
Apatite	$Ca_5(F,Cl,OH)(PO_4)_3$
Pyromorphite	$Pb_5Cl(PO_4)_3$
Vanadinite	$Pb_5Cl(VO_4)_3$
Vivianite	$Fe_3(PO_4)_2 \cdot 8H_2O$
Erythrite	$Co_3(AsO_4)_2 \cdot 8H_2O$
Amblygonite	$LiAlFPO_4$
Lazulite—	$(Mg,Fe)Al_2(OH)_2(PO_4)_2$
Scorzalite	$(Fe,Mg)Al_2(OH)_2(PO_4)_2$
Wavellite	$Al_3(OH)_3(PO_4)_2 \cdot 5H_2O$
Turquoise	$CuAl_6(PO_4)_4(OH)_8 \cdot 2H_2O$
Autunite	$Ca(UO_2)_2(PO_4)_2 \cdot 10\text{-}12H_2O$
Carnotite	$K_2(UO_2)_2(VO_4)_2 \cdot 3H_2O$

may substitute for each other as the central coordinating ion in the tetrahedral oxygen group.

This freedom of substitution of phosphorus, arsenic, and vanadium is best shown in the pyromorphite series of the apatite group. Pyromorphite, mimetite, and vanadinite are isostructural, and all gradations of composition between the pure compounds may exist.

Apatite, the most important and abundant phosphate, displays solid solution with respect to anions both by substitution of chlorine and hydroxyl for the common fluorine and more rarely by substitution of carbonate groups for phosphate. Manganese, strontium, and other cations may substitute for calcium. This complex ionic substitution is typical of the phosphates, which have involved chemical relationships and generally rather complicated structures.

This mineral class, composed mostly of phosphates, is very large, but most of its members are so rare that they need not be mentioned here. Of the minerals listed, apatite is the only one that can be considered common.

Triphylite—$Li(Fe,Mn)PO_4$;   Lithiophilite—$Li(Mn,Fe)PO_4$

Crystallography.   Orthorhombic; $2/m2/m2/m$. Crystals rare. Commonly in cleavable masses. Also compact.

Pmcn. For triphylite: $a = 6.01, b = 4.86, c = 10.36$ Å; $a:b:c = 1.284:1:2.214$; $Z = 4$. *d's:* 4.29(8), 3.51(9), 3.03(9), 2.54(10), 1.75(5).

Physical Properties.   *Cleavage* {001} nearly perfect, {010} imperfect. **H** $4\frac{1}{2}$–5. **G** 3.42–3.56 increasing with Fe content. *Luster* vitreous to reinous. *Color* bluish-gray in triphylite to salmon-pink or clove-brown in lithiophilite. May be stained black by manganese oxide. Translucent. *Optics:* (+); $\alpha = 1.669$–1.694, $\beta = 1.673$–1.695, $\gamma = 1.682$–1.700; $2V = 0°$–55°, $X = c$, $Y = a$, $Z = b$. Indices increase with Fe content.

Composition.   By mutual substitution of Fe^2 and Mn^2 a complete series extends between essentially pure and members.

Diagnostic Features.   Fusible at $2\frac{1}{2}$, giving red lithium flame. Triphylite becomes magnetic on heating in the reducing flame. Some manganese is usually present, and therefore it gives a manganese bead test. Characterized by two cleavages at right angles, resinous luster, and association.

Occurrence.   Triphylite and lithiophilite are pegmatite minerals associated with other phosphates, spodumene, and beryl. Notable localities are in Bavaria and Finland. Lithiophilite is found at several localities in Argentina. In the United States triphylite is found at Huntington, Massachusetts; Peru, Maine; Grafton, North Grafton, and Newport, New Hampshire; and the Black Hills, South Dakota. Lithiophilite is found at Branchville and Portland, Connecticut.

Name.   Triphylite, from the Greek words meaning *three* and *a tribe*, because it contains the three bases iron, lithium, and manganese.

Monazite—$(Ce,La,Y,Th)PO_4$

Crystallography. Monoclinic; 2/m. Crystals rare and usually small, often flattened on {100}, or elongated on b. Usually in granular masses, frequently as sand.

$P2_1/n;$ $a = 6.79,$ $b = 7.01,$ $c = 6.46$ Å; $\beta = 103° 38';$ $a:b:c = 0.969;$ $Z = 4.$ $d's:$ 4.17(3), 3.15(3), 3.30(5), 3.09(10), 2.87(7).

Physical Properties. *Cleavage* {100} poor. Parting {001}. **H** $5-5\frac{1}{2}$. **G** 4.6–5.4. *Luster* resinous. *Color* yellowish to reddish-brown. Translucent. *Optics:* (+); $\alpha = 1.785-1.800,$ $\beta = 1.787-1.801,$ $\gamma = 1.840-1.850;$ $2V = 10°-20°;$ $X = b,$ $Z \wedge c = 2°-6°.$

Composition. A phosphate of the rare-earth metals, essentially $(Ce,La,Y,Th)PO_4$. Th is present between a few per cent and 20 per cent ThO_2. Si is usually present and has been ascribed to *thorite*, $ThSiO_4$.

Diagnostic Features. Infusible. Insoluble in hydrochloric acid. Decomposed by heating with concentrated sulfuric acid; solution after dilution with water and filtering gives, with ammonium oxalate, a precipitate of the oxalates of the rare earths. Large specimens may be distinguished from zircon by crystal form and inferior hardness, and from sphene by crystal form and higher specific gravity. On doubtful specimens it is usually well to make the chemical phosphate test.

Occurrence. Monazite is a comparatively rare mineral occurring as an accessory mineral in granites, gneisses, aplites, and pegmatites, and as rolled grains in the sands derived from the decomposition of such rocks. It is concentrated in sands because of its resistance to chemical attack and its high specific gravity, and is thus associated with other resistant and heavy minerals such as magnetite, ilmenite, rutile, and zircon.

The bulk of the world's supply of monazite comes from beach sands in Brazil, India, and Australia. A dikelike body of massive granular monazite is mined near VanRhynsdorp, Cape Province, South Africa. Found in the United States in North Carolina, both in gneiss and in the stream sands; and in the beach sands of Florida.

Use. Monazite is the chief source of thorium oxide, which it contains in amounts varying between 1 and 20 per cent; commercial monazite usually contains between 3 and 9 per cent. Thorium oxide is used in the manufacture of mantles for incandescent gas lights.

Thorium is a radioactive element and is receiving considerable attention as a source of atomic energy. The natural isotope of thorium, Th-232, can be converted by neutron bombardment, first to Th-233, and then to U-233, a fissionable isotope.

Name. The name *monazite* is derived from a Greek word meaning *to be solitary*, in allusion to the rarity of the mineral.

Apatite Group

APATITE—$Ca_5(F,Cl,OH)(PO_4)_3$

Crystallography. Hexagonal; 6/m. Commonly occurs in crystals of long prismatic habit; some shortly prismatic or tabular. Usually terminated by prominent pyramid of first order and frequently a basal plane. Some crystals show faces of a hexagonal dipyramid (μ, Fig. 359c) which reveals the true symmetry. Also in massive granular to compact masses.

Angles: $m(10\bar{1}0) \wedge x(10\bar{1}1) = 49° 41'$, $c(0001) \wedge s(11\bar{2}1) = 55° 46'$.

$P6_3/m$; $a = 9.39$, $c = 6.89$ Å; $a:c = 1:0.734$; $Z = 2$. *d's:* 2.80(10), 2.77(4), 2.70(6), 1.84(6), 1.745(3).

Physical Properties. *Cleavage* {0001} poor. **H** 5 (can just be scratched by a knife). **G** 3.15–3.20. *Luster* vitreous to subresinous. *Color* usually some shade of green or brown; also blue, violet, colorless. Transparent to translucent. *Optics:* $(-)$; $\omega = 1.633$, $\varepsilon = 1.630$ (fluorapatite).

Composition. $Ca_5F(PO_4)_3$, *fluorapatite* is most common; more rarely $Ca_5Cl(PO_4)_3$, *chlorapatite*, and $Ca_5(OH)(PO_4)_3$, *hydroxylapatite*. F, Cl, and OH can substitute for each other, giving complete series. CO_3 may substitute for PO_4 giving *carbonate-apatite*. Mn can substitute in part for Ca.

Collophane. The name collophane has been given to the massive, cryptocrystalline types of apatite that constitute the bulk of phosphate rock and fossil bone. X-ray study shows that collophane is essentially apatite and does not warrant designation as a separate species. In its physical appearance collophane is usually dense and massive with a concretionary or colloform structure. It is usually impure and contains small amounts of calcium carbonate.

Diagnostic Features. Difficulty fusible ($5–5\frac{1}{2}$). Soluble in acids and gives the phosphate test with ammonium molybdate. Apatite is usually recognized by its crystals, color, and hardness. Distinguished from beryl by the prominent pyramidal terminations of its crystals and by its being softer than a knife blade.

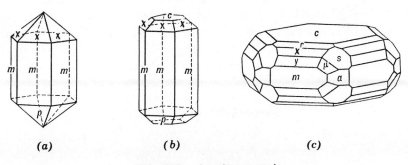

(a)   *(b)*   *(c)*

FIG. 359. Apatite crystals.

Occurrence. Apatite is widely disseminated as an accessory constituent in all classes of rocks—igneous, sedimentary, and metamorphic. It is also found in pegmatites and other veins, probably of hydrothermal origin. Found in titaniferous magnetite bodies. Occasionally concentrated into large deposits or veins associated with alkalic rocks.

Apatite occurs in large amounts along the southern coast of Norway, between Lagnesund and Arendal, where it is found in veins and pockets associated with gabbro. It is distributed through the magnetite iron ore at Kiruna, Sweden and Tahawas, New York. In Ontario and Quebec, Canada, large apatite crystals in crystalline limestone were formerly mined. Unusually fine crystals have come from Renfrew County, Ontario. The world's largest deposit of apatite is located on the Kola Peninsula, near Kirovsk, U.S.S.R. Here apatite in granular aggregates intimately associated with nepheline and sphene, is found in a great lens between two types of alkalic rocks.

Finely crystallized apatite occurs at various localities in the Tyrol; in Switzerland; and Jumilla, Spain. In the United States at Auburn, Maine; St. Lawrence County, New York; Alexander County, North Carolina; and San Diego County, California.

The variety collophane is an important constituent of the rock *phosphorite* or *phosphate rock*. Bone is calcium phosphate, and large bodies of phosphorite are derived from the accumulation of animal remains as well as chemical precipitation from sea water. Commercial deposits of phosphorite are found in northern France, Belgium, Spain, and especially in northern Africa in Tunisia, Algeria, and Morocco. In the United States, high-grade phosphate deposits are found in Tennessee and in Wyoming and Idaho. Deposits of "pebble" phosphate are found at intervals all along the Atlantic coast from North Carolina to Florida. The most productive deposits in the United States are in Florida.

Use. Crystallized apatite has been used extensively as a source of phosphate for fertilizer but today only the deposits on the Kola Peninsula are of importance and phosphorite deposits supply most of the phosphate for fertilizer. The calcium phosphate is treated with sulfuric acid and changed to superphosphate to render it more soluble in the dilute acids that exist in the soil. Transparent varieties of apatite of fine color are occasionally used for gems. The mineral is too soft, however, to allow its extensive use for this purpose.

Name. From the Greek word *to deceive*, since the gem varieties were confused with other minerals.

Pyromorphite—$Pb_5Cl(PO_4)_3$

Crystallography. Hexagonal; $6/m$. Crystals usually prismatic with basal plane (Fig. 360). Rarely shows pyramid truncations. Often in rounded barrel-shaped forms. Sometimes cavernous, the crystals being hollow prisms. Also in parallel groups. Frequently globular, reniform, fibrous, and granular.

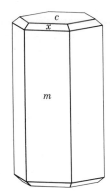

FIG. 360. Pyromorphite.

Angles: $c(0001) \wedge x(10\bar{1}1) = 40° 22'$.

$P6_3m$; $a = 9.97$, $c = 7.32$ Å; $a:c = 1:0.734$; $Z = 2$. *d's:* 4.31(6), 4.09(9), 2.95(10), 2.05(8), 1.94(7).

Physical Properties. **H** $3\frac{1}{2}$–4. **G** 7.04. *Luster* resinous to adamantine. *Color* usually various shades of green, brown, yellow; rarely orange-yellow, gray, white. Subtransparent to translucent. *Optics:* $(-)$, $\omega = 2.058$, $\varepsilon = 2.048$.

Composition. $Pb_3Cl(PO_4)_3$. PbO 82.2, Cl 2.6, P_2O_5 15.7 per cent. AsO_4 substitutes for PO_4 and a complete series extends to *mimetite*. Calcium may substitute in part for lead.

Diagnostic Features. Fusible at 2. Gives a lead globule with sodium carbonate. When fused alone on charcoal gives a globule which on cooling appears to show crystal forms. Gives phosphate test with ammonium molybdate. Pyromorphite is characterized by its crystal form, high luster, and high specific gravity.

Occurrence. Pyromorphite is a supergene mineral found in the oxidized portions of lead veins, associated with other oxidized lead and zinc minerals.

Notable localities for its occurrence are the lead mines of Poullaouen and Huelgoat, Brittany; at Ems in Nassau; at Zschopau, Saxony; Příbram, Bohemia; Beresovsk, Ural Mountains; in Cumberland, England and at Leadhills, Scotland. In the United States found at Phoenixville, Pennsylvania; Davidson County, North Carolina; and Idaho.

Use. A subordinate ore of lead.

Name. Derived from two Greek words meaning *fire* and *form*, in allusion to the apparent crystalline form it assumes on cooling from fusion.

Similar Species. Mimetite, $Pb_5Cl(AsO_4)_3$ is similar to pyromorphite in appearance, occurrence and most of its physical and chemical properties.

Vanadinite—$Pb_5Cl(VO_4)_3$

Crystallography. Hexagonal; $6/m$. Most commonly occurs in prismatic crystals $\{10\bar{1}0\}$ and $\{0001\}$. May have small pyramidal faces, rarely the hexagonal dipyramid. In rounded crystals; in some cases cavernous. Also in globular forms. As incrustations.

$P6_3/m$; $a = 10.33$, $c = 7.35$ Å; $a:c = 1:0.711$; $Z = 2$. d's: 4.47(3), 4.22(4), 3.38(6), 3.07(9). 2.99(10).

Physical Properties. **H** 3. **G** 6.9. *Luster* resinous to adamantine. *Color* ruby-red, orange-red, brown, yellow. Transparent to translucent. *Optics:* $(-)$; $\omega = 2.25–2.42$, $\varepsilon = 2.20–2.35$. Indices lowered by substitution of As or P for V.

Composition. PbO 78.7, Cl 2.5, V_2O_5 19.4 per cent. PO_4 and AsO_4 may substitute in small amounts for VO_4. In the variety *endlichite*, intermediate between vanadinite and mimetite, the proportion of V_2O_5 to As_2O_5 is nearly 1:1. Small amounts of Ca, Zn, and Cu substitute for Pb.

Diagnostic Features. Characterized by crystal form, high luster, and high specific gravity; distinguished from pyromorphite and mimetite by color. Fusible at $1\frac{1}{2}$. Gives globule of lead when fused with sodium carbonate on charcoal. Gives an amber color in the oxidizing flame to salt of phosphorus bead (vanadium). Dilute nitric acid solution gives with silver nitrate a white precipitate of silver chloride.

Occurrence. Vanadinite is a rare secondary mineral found in the oxidized portion of lead veins associated with other secondary lead minerals. Found in fine crystals near Oudjda, Morocco, and Grootfontein, South-West Africa. In the United States it occurs in various districts in Arizona and New Mexico.

Use. Source of vanadium and minor ore of lead. Vanadium is obtained chiefly from other ores, such as *patronite*, a substance of indefinite composition formerly thought to be the sulfide; the vanadate *carnotite;* and a vanadium mica, *roscoelite*. Vanadium is used chiefly as a steel-hardening metal. Metavanadic acid, HVO_3, is a yellow pigment, known as vanadium bronze. Vanadium oxide is a mordant in dyeing.

Vivianite—$Fe_3(PO_4)_2 \cdot 8H_2O$

Crystallography. Monoclinic; $2/m$. Prismatic crystals, vertically striated; often in radiating groups. Also nodular and earthy. (See Fig. 361.)

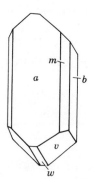

FIG. 361.   Vivianite.

Angles: $b(010) \wedge m(110) = 54°$, $b(010) \wedge v(\bar{2}21) = 60° \ 13'$, $a(100) \wedge \omega(\bar{2}01) = 125° \ 20'$.

$C2/m$; $a = 10.08$, $b = 13.43$, $c = 4.70$ Å; $\beta = 104° \ 30'$. $a:b:c = 0.750:1:0.351$; $Z = 2$. $d's$: 6.80(10), 4.91(4), 3.20(5), 2.97(7), 2.71(7).

Physical Properties. *Cleavage* {010} perfect. **H** $1\frac{1}{2}$–2. **G** 2.58–2.68. *Luster* vitreous; pearly on cleavage face. *Color* colorless when unaltered; blue to green when altered. Transparent when fresh, becoming translucent on exposure. *Optics:* (+); $\alpha = 1.579$, $\beta = 1.602$, $\gamma = 1.629$; 2V = 83°. $X = b$, $Z \wedge c = 29°$. $r < v$. Pleochroism X blue, Y and Z yellow-green.

Composition. FeO 43.0, P_2O_5 28.3, H_2O 28.7 per cent. The darkening of vivianite results from the partial oxidation of Fe^2 to Fe^3.

Diagnostic Features. Usually altered, and in this state characterized by blue to green color. Cleavage lamellae flexible. Fusible at 2 to a magnetic globule. Yields water in the closed tube and gives phosphate test with ammonium molybdate solution.

Occurrence. Vivianite is a rare mineral of secondary origin, associated with pyrrhotite and pyrite in copper and tin veins and forms as a weathering product from primary iron-manganese phosphates in pegmatites. Found also in beds of clay; may be associated with limonite; often in cavities of fossils.

Name. In honor of the eighteenth-century English mineralogist J. G. Vivian, the discoverer of the mineral.

Erythrite—$Co_3(AsO_4)_2 \cdot 8H_2O$

Crystallography. Monoclinic; $2/m$. Crystals prismatic and vertically striated. Usually as crusts in globular and reniform shapes. Also pulverulent and earthy.

$C2/m$; $a = 10.20$, $b = 13.37$, $c = 4.74$ Å; $\beta = 105°$. $a:b:c = 0.763:1:0.355$; $Z = 2$. $d's$: 6.65(10), 3.34(1), 3.22(1), 2.70(1), 2.32(1).

Physical Properties. *Cleavage* {010} perfect. **H** $1\frac{1}{2}$–$2\frac{1}{2}$. **G** 3.06. *Luster* adamantine to vitreous, pearly on cleavage. *Color* crimson to pink. Translucent. *Optics:* (−); $\alpha = 1.626$, $\beta = 1.661$, $\gamma = 1.699$; 2V = 90° ±; $X = b$, $Z \wedge c = 31°$; $r > v$. Pleochroism X pink, Y violet, Z red.

Composition. CoO 37.5, As_2O_5 38.4, H_2O 24.1 per cent. Ni substitutes for Co to form a complete series to annabergite, $Ni_3(AsO_4)_2 \cdot 8H_2O$. *Annabergite*, or *nickel bloom*, is light green in color.

Diagnostic Features. The association of erythrite with other cobalt minerals and its pink color are usually sufficient to distinguish it from all other minerals. Fusible at 2 to a gray bead. When heated on charcoal gives arsenical odor. Imparts a deep blue to the borax bead (cobalt). Soluble in hydrochloric acid, giving a red solution.

Occurrence. Erythrite is a rare secondary mineral. In pink crusts known as *cobalt bloom* it occurs as an alteration product of cobalt arsenides. It is rarely present in large amounts and usually forms as crusts or fine aggregates filling cracks. Notable localities are at Schneeberg, Saxony, and Cobalt, Ontario.

Use. Although erythrite has no economic importance, it is used by the prospector as a guide to other cobalt minerals and associated native silver.

Name. From the Greek word meaning *red*.

AMBLYGONITE—LiAlFPO$_4$

Crystallography. Triclinic; $\bar{1}$. Usually occurs in coarse, cleavable masses. Crystals are rare, equant, and usually rough when large. Frequently twinned on $\{\bar{1}\bar{1}1\}$.

Angles: $(001) \wedge (100) = 90° 13'$, $(110) \wedge (100) = 44° 51'$, $(100) \wedge (0\bar{1}1) = 75° 22'$.

$P\bar{1}$; $a = 5.19$, $b = 7.12$, $c = 5.04$ Å; $\alpha = 112° 02'$, $\beta = 97° 50'$, $\gamma = 68° 8'$; $a:b:c = 0.729:1:0.708$; $Z = 2$. $d's$: 4.64(10), 3.15(10), 2.93(10), 2.39(5), 2.11(4).

Physical Properties. *Cleavage* $\{100\}$ perfect, $\{110\}$ good, $\{0\bar{1}1\}$ distinct. **H** 6. **G** 3.0–3.1. *Luster* vitreous, pearly on $\{100\}$ cleavage. *Color* white to pale green or blue, rarely yellow. Translucent. *Optics:* usually $(-)$; $\alpha = 1.58$–1.60, $\beta = 1.59$–1.62, $\gamma = 1.60$–1.63; $2V = 50$–$90°$; $r > v$. Indices increase with increase in (OH).

Composition. Li$_2$O 10.1, Al$_2$O$_3$ 34.4, F 12.9, P$_2$O$_5$ 47.9 per cent. Na substitutes for Li; (OH) also substitutes for F and probably forms a complete series. When OH > F the mineral is *montebrasite*.

Diagnostic Features. Fusible at 2 with intumescence, giving a red flame (lithium). Insoluble in acids. Gives test for phosphate with ammonium molybdate solution. Cleavage fragments may be confused with feldspar, but are distinguished by cleavage angles, easy fusibility and flame test.

Occurrence. Amblygonite is a rare mineral found in granite permatite with spodumene, tourmaline, lepidolite, and apatite. Found at Montebras, France. In the United States it occurs at Hebron, Paris, Auburn, and Peru, Maine; Pala, California; and Black Hills, South Dakota.

Use. A source of lithium.

Name. From the two Greek words meaning *blunt* and *angle*, in allusion to the angle between the cleavages.

Lazulite—(Mg,Fe)Al$_2$(OH)$_2$(PO$_4$)$_2$;
Scorzalite—(Fe,Mg)Al$_2$(OH)$_2$(PO$_4$)$_2$

Crystallography. Monoclinic; $2/m$. Crystals showing steep fourth-order prisms rare. Usually massive, granular to compact.

$P2_1/c$; $a = 7.12$, $b = 7.26$, $c = 7.24$ Å; $\beta = 118° 55'$; $a:b:c = 0.981:1:0.997$; $Z = 2$. $d's$: 6.15(8), 3.23(8), 3.20(7), 3.14(10), 3.07(10).

Physical Properties. *Cleavage* {110} indistinct. **H** 5–5$\frac{1}{2}$. **G** 3.0–3.1. *Luster* vitreous. *Color* azure-blue. Translucent. *Optics:* (−); $\alpha = 1.640$–1.639, $\beta = 1.626$–1.670, $\gamma = 1.637$–1.680, 2V = 60°; $Y = b$, $X \wedge c = 10°$; $r < v$. Absorption $X < Y < Z$. Indices increase with increasing Fe^2 content.

Composition. Probably a complete series exists from lazulite to scorzalite with the substitution of Fe^2 for Mg.

Diagnostic Features. Infusible. Before the blowpipe it swells, loses its color, and falls to pieces. In the closed tube it whitens and yields water. Insoluble. Gives phosphate test with ammonium molybdate. If crystals are lacking, lazulite is difficult to distinguish from other blue minerals without optical, chemical, or blowpipe tests.

Occurrence. The members of the lazulite-scorzalite series are rare minerals found in high-grade quartz-rich metamorphic rock and in pegmatites. They are usually associated with kyanite, andalusite, corundum, rutile, sillimanite, and garnet. Notable localities are Salzburg, Austria; Krieglach, Styria; and Horrsjoberg, Sweden. In the United States they are found with corundum on Crowder's Mountain, Gaston County, North Carolina; with rutile on Graves Mountain, Lincoln County, Georgia; and with andalusite in the White Mountains, Inyo County, California.

Use. A minor gem stone.

Name. Lazulite derived from an Arabic word meaning *heaven*, in allusion to the color of the mineral. Scorzalite after E. P. Scorza, Brazilian mineralogist.

Wavellite—$Al_3(OH)_3(PO_4)_2 \cdot 5H_2O$

Crystallography. Orthorhombic; 2/m2/m2/m. Crystals rare. Usually in radiating spherulitic and globular aggregates.

Pcmn; $a = 9.62$, $b = 17.34$, $c = 6.99$ Å; $a:b:c = 0.555:1:0.403$; $Z = 4$. *d's:* 8.39(10), 5.64(6), 3.44(8), 3.20(8), 2.56(8).

Physical Properties. *Cleavage* {110} and {101} good. **H** 3$\frac{1}{2}$–4. **G** 2.36. *Luster* vitreous. *Color* white, yellow, green, and brown. Translucent. *Optics:* (+); $\alpha = 1.525$, $\beta = 1.535$, $\gamma = 1.550$; 2V = 70°. $X = b$, $Y = a$; $r > v$.

Composition. Al_2O_3 38.0, P_2O_5 35.2, H_2O 26.8 per cent. F may substitute for OH.

Diagnostic Features. Almost invariably in radiating globular aggregates. Infusible, but on heating swells and splits into fine particles. Insoluble. Yields much water in the closed tube. When moistened with cobalt nitrate and then ignited it assumes a blue color (aluminum). Gives phosphate test.

Occurrence. Wavellite is a secondary mineral found in small amounts in crevices in aluminous, low-grade metamorphic rocks and in limonite and phosphorite deposits. Although it occurs in many localities, it only rarely is found in quantity. It is abundant in the tin veins of Llallagua, Bolivia. In the United States wavellite occurs in a number of localities in Pennsylvania and near Avant, Arkansas.

Name. After Dr. William Wavel, who discovered the mineral.

Turquoise—$CuAl_6(PO_4)_4(OH)_8 \cdot 4H_2O$

Crystallography. Triclinic; $\bar{1}$. Rarely in minute crystals, usually crypto-crystalline. Massive compact, reniform, stalactitic. In thin seams, incrustations, and disseminated grains.

$P\bar{1}$; $a = 7.48$, $b = 9.95$, $c = 7.69$; $\alpha = 111°\,39'$, $\beta = 115°\,23'$, $\gamma = 69°\,26'$; $a:b:c = 0.752:1:0.773$; $Z = 1$. ds: 6.17(7), 4.80(6), 3.68(10), 3.44(7), 3.28(7).

Physical Properties. *Cleavage* {001} perfect, {010} good (rarely seen). **H** 6. **G** 2.6–2.8. *Luster* waxlike. *Color* blue, bluish-green, green. Transmits light on thin edges. *Optics:* (+); $\alpha = 1.61$, $\beta = 1.62$, $\gamma = 1.65$; $2V = 40°$; $r < v$ strong.

Composition. Fe^3 substitutes for Al and a complete series exists between turquoise and *chalcosiderite* in which $Fe^3 > Al$.

Diagnostic Features. Turquoise can be easily recognized by its color. It is harder than chrysocolla, the only common mineral which it resembles. Infusible. When moistened with hydrochloric acid and heated gives blue copper chloride flame. Soluble in hydrochloric acid after ignition and solution gives test for phosphate. In the closed tube it turns dark and gives water.

Occurrence. Turquoise is a secondary mineral usually found in the form of small veins and stringers traversing more or less decomposed volcanic rocks in arid regions. The famous Persian deposits are found in trachyte near Nishapur in the province of Khorasan. In the United States it is found in a much altered trachytic rock in the Los Cerillos Mountains, near Santa Fe, and elsewhere in New Mexico. Turquoise has also been found in Arizona, Nevada, and California. Small crystals have been found in Virginia.

Use. As a gem stone. It is always cut in round or oval forms. Much turquoise is cut which is veined with the various gangue materials, and such stones are sold under the name of *turquoise matrix*.

Name. Turquoise is French and means *Turkish*, the original stones having come into Europe from the Persian locality through Turkey.

Similar Species. *Variscite*, $Al(PO_4) \cdot 2H_2O$, is a massive, bluish-green mineral somewhat resembling turquoise. It has been found in nodules in a large deposit at Fairfield, Utah.

Autunite—$Ca(UO_2)_2(PO_4)_2 \cdot 10\text{-}12H_2O$

Crystallography. Tetragonal; $4/m2/m2/m$. Crystals tabular on {001}; sub-parallel growths are common; also foliated and scaly aggregates.

$I4/mmm$; $a = 7.00$, $c = 20.67$ Å; $a:c = 1:2.953$; $Z = 2$. $d's:$ 10.33(10), 4.96(8), 3.59(7), 3.49(7), 3.33(7).

Physical Properties. *Cleavage* {001} perfect. **H** $2-2\frac{1}{2}$. **G** 3.1–3.2. *Luster* vitreous, pearly on {001}. *Color* lemon yellow to pale green. *Streak* yellow. In ultraviolet light fluoresces strongly yellow-green. *Optics:* $(-)$; $\omega = 1.577$, $\varepsilon = 1.553$. Pleochroism E pale yellow, O dark yellow.

Composition. A hydrated phosphate of calcium and uranium, $Ca(UO_2)_2$-$(PO_4)_2 \cdot 10-12H_2O$. Small amounts of Ba and Mg may substitute for Ca. On slight heating antunite passes reversibly to *meta-autunite I* with $6\frac{1}{2}-2\frac{1}{2}H_2O$; on heating to about 80°C autunite passes irreversibly to *meta-autunite II* with $0-6H_2O$.

Diagnostic Features. Autunite is characterized by yellow-green tetragonal plates and strong fluorescence in ultraviolet light. Fusible at 2–3. Soluble in acids. A soda bead with dissolved torbernite fluoresces in ultraviolet light.

Occurrence. Autunite is a secondary mineral found chiefly in the zone of oxidation and weathering derived from the alteration of uraninite or other uranium minerals. Notable localities are near Autun, France; Sabugal and Vizeu, Portugal; Johanngeorgenstadt district and Falkenstein, Germany; Cornwall, England; and Katanga district of the Congo. In the United States autunite is found in many permatites, notably at the Ruggles mine, Grafton Center, New Hampshire; Black Hills, South Dakota; and Spruce Pine, Mitchell County, North Carolina. The finest specimens have come from the Daybreak mine near Spokane, Washington.

Use. An ore of uranium (see uraninite, page 291).

Name. From Autun, France.

Similar Species. *Torbernite*, $Cu(UO_2)_2(PO_4)_2 \cdot 8-12H_2O$, is isostructural with autunite and with similar properties, but there is no evidence of a solid-solution series. Color green, nonfluorescent. Associated with autunite.

Carnotite—$K_2(UO_2)_2(VO_4)_2 \cdot 3H_2O$

Crystallography. Monoclinic; $2/m$ only rarely in imperfect microscopic crystals flattened on {001} or elongated on b. Usually found as a powder or as loosely coherent aggregates; disseminated.

$P2_1a$; $a = 10.47$, $b = 8.41$, $c = 6.91$ Å; $\beta = 103° 40'$. $a:b:c = 1.245:1:0.822$; $Z = 2$. $d's:$ 6.56(10), 4.25(3), 3.53(5), 3.25(3), 3.12(7).

Physical Properties. *Cleavage* {001} perfect. Hardness unknown, but soft. **G** 4.7–5. *Luster* dull or earthy. *Color* bright yellow to greenish yellow. *Optics:* $(-)$; $\alpha = 1.75$, $\beta = 1.93$, $\gamma = 1.95$; $2V = 40° \pm$, $Y = b$, $X = c$. $r < v$. Indices increase with loss of water.

Composition. The water content varies with humidity at ordinary temperatures; the $3H_2O$ is for fully hydrated material. Small amounts of Ca, Ba, Mg, Fe, and Na have been reported.

Diagnostic Features. Carnotite is characterized by its yellow color, its pulverant nature, and its occurrence. Unlike many secondary uranium minerals, carnotite will not fluoresce in ultraviolet light but a soda bead with dissolved carnotite will fluoresce (test for uranium). Infusible. Soluble in acids.

Occurrence. Carnotite is of secondary origin, and its formation is usually ascribed to the action of meteoric waters on preexisting uranium and vanadium minerals. It has a strong pigmenting power and when present in a sandstone in amounts even less than 1 per cent will color the rock yellow. It is found principally in the plateau region of southwestern Colorado and in adjoining districts of Utah where it occurs disseminated in a cross-bedded sandstone. Concentrations of relatively pure carnotite are found around petrified tree trunks.

Use. Carnotite is an ore of vanadium and, in the United States, a principal ore of uranium.

Name. After Marie-Adolphe Carnot (1839–1920), French mining engineer and chemist.

Similar Species. *Tyuyamunite*, $Ca(UO_2)_2(VO_4)_2 \cdot 3H_2O$, is the calcium analogue of carnotite and similar in physical properties except for a slightly more greenish color and yellow-green fluorescence. It is found in almost all carnotite deposits. Named from Tyuya Muyum, southeastern Turkistan, U.S.S.R., where it is mined as a uranium ore.

SILICATES

The silicate mineral class is of greater importance than any other, for about 25 per cent of the known minerals and nearly 40 per cent of the common ones are silicates. With a few minor exceptions all the igneous rock-forming minerals are silicates, and they thus constitute well over 90 per cent of the earth's crust.

Of every 100 atoms in the crust of the earth, more than 60 are oxygen, over 20 silicon, and 6 to 7 aluminum. Iron, calcium, magnesium, sodium, and potassium each accounts for about two more atoms. With the possible exception of titanium, all other elements are volumetrically insignificant in the architecture of the earth's crust. Since our approach to the nature of minerals in this book is structural rather than stoichiometric, it is entirely proper that we should think of the constituents of the crust in terms of the space they occupy rather than their percentages by weight. If we do so, we are led to picture the crust as a boxwork of oxygen ions bound together by the small, highly charged silicon and aluminum ions. The interstices of this more or less continuous oxygen-silicon-aluminum network are occupied by ions of magnesium, iron, calcium, sodium, and potassium. This startling simplification of the composition of the earth's crust results from consideration of *atomic proportions* rather than weight per cents, the form in which the composition of rocks and minerals is usually stated.

The dominant minerals of the crust are thus shown to be the silicates and oxides, whose properties depend upon the chemical and physical conditions of origin. Of the different assemblages of silicate minerals, which characterize igneous, sedimentary, and metamorphic rocks, ore veins, pegmatites, weathered rocks, and soils, each tells something of the environment in which it was formed. If the rocks are the pages in the book of geologic history, the minerals are the characters in which the book is printed and only with an understanding of them and their structures does the document become readable.

We have a further deep and compelling reason to study the silicates. The soil from which our food is ultimately drawn is made up in large part of silicates. The brick, stone, concrete, and glass used in the construction of our buildings are either silicates or largely derived from silicates. Even with the coming of the space age, we need not fear obsolescence of our studies of the silicates, but rather an enlargement of their scope, since we now know that the moon and all the planets of our solar system have rocky crusts made of silicates and oxides much like those of Earth.

The radius ratio of the silicon ion (Rad. $= 0.42$ Å) to that of the oxygen ion (Rad. $= 1.40$ Å) is 0.300. This radius ratio indicates that 4-fold coordination will be the stable state of silicon-oxygen groupings. The fundamental unit on which the structure of all silicates is based consists of four oxygen ions at the apices of a regular tetrahedron surrounding and coordinated by the four-valent silicon ion (Fig. 362). The powerful bond which unites the oxygen and silicon ions is literally the cement that holds the earth's crust together. This bond may be estimated by use of Pauling's (page 186) electronegativity concept as 50 per cent ionic 50 per cent covalent. That is, although the bond arises in part from the attraction of oppositely charged ionic units, it also involves sharing of electrons and interpenetration of the electronic superstructures of the ions involved. The bond is strongly localized in the vicinity of these shared electrons.

Although electron sharing is present in the silicon-oxygen bond, the total bonding energy of the silicon ion is still distributed equally among its four closest oxygen neighbors. Hence, the strength of any single silicon-oxygen bond is equal to just one-half the total bonding energy available in the oxygen ion. Each oxygen

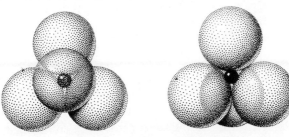

FIG. 362. SiO_4 tetrahedra.

ion has, therefore, the potentiality of bonding to another silicon ion and entering into another tetrahedral grouping, thus uniting the tetrahedral groups through the shared oxygen. This sharing may involve one, two, three, or all four of the oxygen ions in the tetrahedron, giving rise to a diversity of structural configurations. In no case, however, are three or even two oxygens shared between two adjacent tetrahedra in nature. Such sharing would place two highly charged positive silicon ions close together and the repulsion between them renders the structure unstable. Sharing of one oxygen between any two adjacent tetrahedra may, if all four oxygens are so shared, give rise to structures with a very high degree of connectivity, such as the quartz structure. We may call this linking of tetrahedra by sharing of oxygens *polymerization*, to borrow a term from organic chemistry, and the capacity for polymerization is the origin of the great variety of silicate structures.

There is a simple and highly significant relation between the conditions of origin of silicate minerals and the degree of polymerization. All other things being equal, *the higher the temperature of formation, the lower the degree of polymerization and vice versa.* This relation is subject to the disturbing effect of a host of external factors, chief among which are pressure and chemical concentration. Within a single geological body such as a mass of crystallizing igneous rock, the generalization seems to be supported by observation. It has long been noted that the silicate minerals in igneous rocks display a fairly regular and predictable sequence of crystallization, beginning with olivine and progressing through pyroxene to amphibole and thence to micas. This sequence is in the order of increasing polymerization of the silicate tetrahedra.

Next to oxygen and silicon, the most important constituent of the crust is aluminum. Aluminum has an ionic radius of 0.51 Å. Thus the radius ratio $Al:O = 0.364$, corresponding to normal 4 coordination number with oxygen. However, the radius ratio is sufficiently close to the upper limit for 4 coordination so that 6 coordination is also possible. It is this capacity for playing a double role in silicate minerals that gives aluminum its outstanding significance in the crystal chemistry of the silicates. When aluminum coordinates four oxygens arranged at the apices of a regular tetrahedron, the resultant grouping occupies the same space as a silicon-oxygen tetrahedron and may link with silicon tetrahedra in polymerized groupings. On the other hand, aluminum in 6 coordination may serve to link the tetrahedral groupings together through simple ionic bonds, much weaker than those which unite the ions in the tetrahedra. It is thus possible to have aluminum in silicate structures both in the tetrahedral sites, substituting for silicon, and in the octahedral sites with 6 coordination, involved in solid-solution relations with magnesium and divalent and trivalent iron.

Magnesium, divalent iron, trivalent iron, divalent manganese, aluminum, and tetravalent titanium all tend to occur in silicate structures in 6 coordination with respect to oxygen. Although divalent, trivalent, and tetravalent ions are included here, all have about the same space requirements and about the same radius ratio relations with oxygen and, hence, tend to occupy the same type of atomic site. Solid-solution relations between ions of such diverse valence introduces a problem of compensation to maintain electrical neutrality. Thus, if a tetravalent cation substitutes for a trivalent, such as titanium for ferric iron, gaining one positive charge, then somewhere in the crystal another substitution must be made in which a positive charge is lost or a negative charge gained.

The larger and more weakly charged cations, calcium, and sodium, of ionic radii 0.99 Å and 0.97 Å respectively, generally enter sites having 8 coordination with respect to oxygen. It is again obvious that substitution of divalent calcium for monovalent sodium creates a problem of electrical unbalance that must be solved by concomitant substitution elsewhere in the structure. If, for instance, every time Al^3 substitutes for Si^4 in a tetrahedral site, resulting in a loss of one positive charge, Ca^2 substitutes for Na^+ in 8 coordination, electrical balance will be maintained. It is by just this mechanism that electrical neutrality is preserved in the sodium-calcium feldspars, the scapolite group and other groups in which sodium and calcium substitute freely for each other.

Coordination of Important Elements in Silicates

	Coordination Number	Ion	Ionic Radius (Å)	$R_A : R_O$
Z	4	Si^4	0.42	0.300
	4	Al^3	0.51	0.364
Y	6	Al^3	0.51	0.364
	6	Fe^3	0.64	0.457
	6	Mg^2	0.66	0.471
	6	Ti^4	0.68	0.486
	6	Fe^2	0.74	0.529
	6	Mn^2	0.80	0.571
X	8	Na^+	0.97	0.693
	8	Ca^2	0.99	0.707
X	8–12	K^+	1.33	0.950
	8–12	Ba^2	1.34	0.957
	8–12	Rb^+	1.47	1.050

The largest ions common in silicate structures are those of potassium, rubidium, barium, and the rarer alkalis and alkali earths. These ions generally do not enter readily into sodium-calcium sites and are found in high-coordination-number sites of unique type. Hence, solid-solution relations between these ions and the common ions are limited and are generally confined to high-temperature crystallization, where solid solution is favored.

Silicate Classification[a]

Class	Arrangement of SiO_4 Tetrahedra	Ratio Si:O	Mineral Example
Nesosilicates	Isolated	1:4	Olivine $(Mg,Fe)_2SiO_4$
Sorosilicates	Double	2:7	Hemimorphite $Zn_4(Si_2O_7)(OH)_2 \cdot H_2O$
Cyclosilicates	Rings	1:3	Beryl, $Be_3Al_2(Si_6O_{18})$
Inosilicates	Chains (single)	1:3	Enstatite, $Mg_2(Si_2O_6)$
	Chains (double)	4:11	Tremolite, $Ca_2Mg_5(Si_8O_{22})(OH)_2$
Phyllosilicates	Sheets	2:5	Talc, $Mg_3(Si_4O_{10})(OH)_2$
Tectosilicates	Frameworks	1:2	Quartz, SiO_2

[a] The names of the silicate classes are those proposed by H. Strunz.[5] The prefixes are from the Greek: *neso*, island; *soro*, group; *cyclo*, ring; *ino*, chain or thread; *phyllo*, sheet; *tecto*, framework.

Ionic substitution is generally common and extensive between elements whose symbols lie between a pair of horizontal lines in the table, page 371, but it is rare and difficult between elements separated by a horizontal line. The usual role played by the commomest elements permits us to write a general formula for all silicates:

$$X_m Y_n (Z_p O_q) W_r$$

where X represents large, weakly charged ions in 8-fold or higher coordination with oxygen; Y represents medium sized, two to four valent ions in 6 coordination; Z represents small, highly charged ions in tetrahedral coordination; O is oxygen; and W represents additional anionic groups such as (OH) or anions such as Cl^-, F^-, etc. The ratio $p:q$ depends on the degree of polymerization of the silicate framework, and the other subscript variables, m, n, and r, depend on the condition of electrical neutrality. Any common silicate may be expressed by suitable substitution in this general formula.

Depending on the degree of polymerization and the extent of oxygen-sharing between tetrahedra, the silicate framework may consist of separate tetrahedra, separate multiple tetrahedral groups, chains, double chains, sheets or three-dimensional boxworks. The silicate framework governs the ratio of $p:q$ in the general formula, the stoichiometric proportion of the oxides, and to a large extent the physical properties and chemical stability of the mineral. Hence, it is appropriate that this criterion should be used as a basis for the classification of the silicates.

Up to the 1930's, the analyses of silicates were interpreted and their formulas generally written in terms of a number of hypothetical oxyacids of silicon. Thus, olivine, Mg_2SiO_4, was termed an "orthosilicate" and considered to be a salt of orthosilicic acid H_4SiO_4; enstatite, $MgSiO_3$, was called a "metasilicate" and considered to be a salt of metasilicic acid H_2SiO_3. In some simple cases this theory of the silicates worked fairly well, and some of these acids could be prepared. We now know that, because of the peculiar nature of the hydrogen bond, such acids have no significance for the silicates. It is little wonder that the chemistry of the silicates was in a chaotic morass of uncertainty. Into this chaos the structural determinations of the Braggs brought order. It is this scheme, summarized in 1937 by Berman[4] and revised and elaborated on by Strunz[5] (1966), that is followed in this book.

Nesosilicates

In the nesosilicates the SiO_4 tetrahedra are isolated (Fig. 362) and bound to each other only by ionic bonds though interstitial cations. Their structures depend chiefly on the size and charge of the interstitial cations.

Nesosilicates

PHENACITE GROUP
| Phenacite | $Be_2(SiO_4)$ |
| Willemite | $Zn_2(SiO_4)$ |

OLIVINE GROUP
| Forsterite | $Mg_2(SiO_4)$ |
| Fayalite | $Fe_2(SiO_4)$ |

GARNET GROUP $A_3B_2(SiO_4)_3$
| Pyrope | Spessartite | Andradite |
| Almandite | Grossularite | Uvarovite |

ZIRCON GROUP
| Zircon | $Zr(SiO_4)$ |

Al_2SiO_5 GROUP[a]
Andalusite	$Al^{[6]}Al^{[5]}O(SiO_4)$
Sillimanite	$Al^{[6]}Al^{[4]}O(SiO_4)$
Kyanite	$Al^{[6]}Al^{[6]}O(SiO_4)$
Topaz	$Al_2(SiO_4)(F,OH)_2$
Staurolite	$Fe_2Al_9O_6(SiO_4)_4(O,OH)_2$

CHONDRODITE GROUP
Chondrodite	$Mg_5(SiO_4)_2(OH,F)_2$
Datolite	$CaB(SiO_4)(OH)$
Sphene	$CaTiO(SiO_4)$

[a] Numbers in square brackets equal the coordination number of the aluminum ion.

[4] Harry Berman, *Constitution and classification of the natural silicates.* Am. Min. 22, pp. 342–408, 1937.

[5] Hugo Strunz, *Mineralogische Tabellen,* 4th ed. Akademische Verlagsgesellschaft, Leipzig, 1966.

Phenacite—Be$_2$(SiO$_4$)

Crystallography. Hexagonal—R; $\bar{3}$. Crystals usually flat rhombohedral or short prismatic. Often with complex development. Frequently twinned on $\{10\bar{1}0\}$.
$R\bar{3}$; $a = 12.45$, $c = 8.23$ Å; $a:c = 1:0.661$; $Z = 18$. d's: 3.58(6), 2.51(8), 2.35(6), 2.18(8), 1.258(10).

Physical Properties. *Cleavage* $\{11\bar{2}0\}$ imperfect. **H** $7\frac{1}{2}$–8. **G** 2.97–3.00. *Luster* vitreous. *Color* colorless, white. Transparent to translucent. *Optics:* (+); $\omega = 1.654$, $\varepsilon = 1.670$.

Composition. BeO 45.6, SiO$_2$ 54.4 per cent.

Diagnostic Features. Characterized by its crystal form and great hardness. Infusible. Fused with sodium carbonate yields a white enamel.

Occurrence. Phenacite is a rare pegmatite mineral associated with topaz, chrysoberyl, beryl, and apatite. Fine crystals are found at the emerald mines in the Ural Mountains, U.S.S.R., and in Minas Geraes, Brazil. In the United States found at Mount Antero, Colorado.

Use. Occasionally cut as a gem stone.

Name. From the Greek meaning *a deceiver*, in allusion to its resemblance to quartz.

WILLEMITE—Zn$_2$(SiO$_4$)

Crystallography. Hexagonal—R; $\bar{3}$. In hexagonal prisms with rhombohedral terminations. Usually massive to granular. Rarely in crystals.
$R\bar{3}$; $a = 13.96$, $c = 9.34$ Å; $a:c = 0.669$; $Z = 18$. d's: 2.84(8), 2.63(9), 2.32(8), 1.849(8), 1.423(10).

Physical Properties. *Cleavage* $\{0001\}$ good. **H** $5\frac{1}{2}$. **G** 3.9–4.2. *Luster* vitreous to resinous. *Color* yellow-green, flesh-red, and brown; white when pure. Transparent to translucent. Most willemite from Franklin, New Jersey, fluoresces. *Optics:* (+); $\omega = 1.691$, $\varepsilon = 1.719$.

Composition. ZnO 73.0, SiO$_2$ 27.0 per cent. Manganese often replaces a considerable part of the zinc (manganiferous variety called *troostite*); iron may also be present in small amount.

Diagnostic Features. Willemite from Franklin, New Jersey, can usually be recognized by its association with franklinite and zincite. Pure willemite infusible. When heated on charcoal with cobalt nitrate assay turns blue. Troostite will give reddish-violet color to the borax bead in the oxidizing flame (manganese). Distinguished from hemimorphite by absence of water.

Occurrence. Willemite is found in crystalline limestone and may be the result of metamorphism of earlier hemimorphite or smithsonite. It is also found sparingly as a secondary mineral in the oxidized zone of zinc deposits.

Found at Altenberg, near Moresnet, Belgium; Algeria; French Congo; Zambia; South-West Africa; and Greenland. The most important locality is in the United States at Franklin, New Jersey, where willemite occurs associated with franklinite and zincite and as grains imbedded in calcite. It has also been found at the Merritt mine, New Mexico, and Tiger, Arizona.

Use. A valuable zinc ore at Franklin, New Jersey.

Name. In honor of the King of the Netherlands, William I.

Olivine Group

The simplest structures of nesosilicates are found in those minerals with only one type of cation site. If the cation is divalent, a compound of formula type A_2SiO_4, with olivine the chief representative, results when Mg and Fe^2 enter into a compound of this type. SiO_4 tetrahedra are arranged so that each A ion coordinates six oxygens. This, the olivine structure, can be thought of as a regular stacking of alternating tetrahedra and octahedra, with apices of tetrahedra pointing alternately up and down. The octahedral sites are randomly occupied by Mg or Fe^2 giving rise to the solid solution between Mg_2SiO_4 and Fe_2SiO_4. Mn^2 may also enter the octahedral sites resulting in solid solution between Fe_2SiO_4 and Mn_2SiO_4, *tephroite*.

OLIVINE—$(Mg,Fe)_2(SiO_4)$

Crystallography. Orthorhombic; $2/m2/m2/m$. Crystals are usually a combination of the three prisms, the three pinacoids, and a dipyramid. Often flattened parallel to either $\{100\}$ or $\{010\}$. Usually appear as imbedded grains or in granular masses. (See Fig. 363.)

Angles: $b(010) \wedge m(110) = 55° 1'$, $b(010) \wedge s(120) = 32° 58'$, $k(021) \wedge k'(0\bar{2}1) = 99° 6'$, $d(101) \wedge d'(\bar{1}01) = 103° 6'$, $c(001) \wedge e(111) = 54° 11'$.

Pmcn; $Z = 4$. Mg_2SiO_4: $a = 4.76$, $b = 10.20$, $c = 5.98$; $a:b:c = 0.467:1:0.586$. Fe_2SiO_4: $a = 4.82$, $b = 10.48$, $c = 6.11$; $a:b:c = 0.460:1:0.583$. d's common olivine: 2.49(10), 2.41(8), 2.24(7), 1.734(8), 1.498(7).

Physical Properties. *Fracture* conchoidal. **H** $6\frac{1}{2}$–7. **G** 3.27–4.37, increasing with increase in iron content. (See Fig. 364.) *Luster* vitreous. *Color* olive to

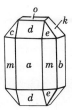

FIG. 363. Olivine.

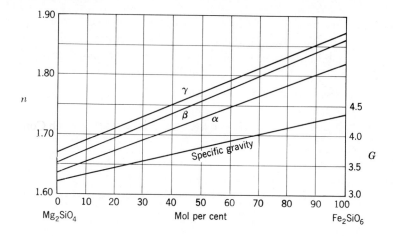

FIG. 364. Olivine. Variation of refractive indices and specific gravity with composition.

grayish-green, brown. Transparent to translucent. *Optics:* forsterite $(+)$, others $(-)$. For $Mg_{1.6}Fe_{0.4}SiO_4$: $\alpha = 1.674$, $\beta = 1.692$, $\gamma = 1.712$, $2V = 87°$; $X = b$, $Z = a$.

Composition. A complete solid-solution series exists, grading from *forsterite*, $Mg_2(SiO_4)$ to *fayalite*, $Fe_2(SiO_4)$. The more common olivines are richer in magnesium than in iron.

Diagnostic Features. Distinguished usually by its glassy luster, conchoidal fracture, green color, and granular nature. Infusible. Rather slowly soluble in hydrochloric acid and yields gelatinous silica upon evaporation.

Occurrence. Olivine is a rather common rock-forming mineral, varying in amount from that of an accessory to that of a main constituent. It is found principally in the dark-colored igneous rocks such as gabbro, peridotite, and basalt. The rock, *dunite*, is made up almost wholly of olivine. Found also as glassy grains in meteorites. Occasionally in crystalline dolomitic limestones. Associated often with pyroxene, calcic plagioclase feldspar, magnetite, corundum, chromite, and serpentine.

The transparent gem variety is known as *peridot*. It was used as a gem in ancient times in the East, but the exact locality for the stones is not known. At present peridot is found in Burma, and in rounded grains associated with pyrope garnet in the surface gravels of Arizona and New Mexico, but the best quality material comes from St. John's Island in the Red Sea. Crystals of olivine are found in the lavas of Vesuvius. Larger crystals, altered to serpentine, come from Snarum, Norway. Olivine occurs in granular masses in volcanic bombs in the Eifel district, Germany, and in Arizona. Dunite rocks are found at Dun Mountain, New Zealand, and with the corundum deposits of North Carolina.

Alteration. Very readily altered to *serpentine* and less commonly to *iddingsite*. Magnesite and iron oxides may form at the same time as a result of the alteration.

Use. As the clear green variety, peridot, it has some use as a gem.

Name. Olivine derives its name from the usual olive-green color. *Chrysolite* is a synonym for olivine. Peridot is an old name for the species.

Similar Species. Other rarer members of the olivine group are *monticellite*, $CaMgSiO_4$; *tephroite*, Mn_2SiO_4; *larsenite*, $PbZnSiO_4$.

GARNET GROUP—$A_3B_2(SiO_4)_3$

The garnet group includes a series of subspecies that crystallize in the hexoctahedral class of the isometric system and are similar in crystal habit.

This group, with its striking uniformity of morphology, great diversity of chemical composition, and strict dependence of physical properties upon composition, presents the finest example of an isostructural group. In the structure isolated SiO_4 tetrahedra are united by oxygen-cation-oxygen bonds through two structurally distinct cation sites. One of these sites, *A*, is occupied by rather large divalent ions, the other, *B*, by smaller trivalent ions leading to the formula $A_3B_2(SiO_4)_3$. The structural arrangement is such that the atomic population of the {100} and {111} families of planes is much depleted. As a result, the cube and octahedron, common on most isometric hexoctahedral crystals, are rarely found on garnets. Within the framework of this structure type, there is ready and substantially complete interchange of magnesium, ferrous iron, and divalent manganese in the A cation sites. Calcium substitutes less readily for the foregoing ions in the *A* sites. In the *B* sites there is limited substitution with respect to aluminum, ferric iron, and chromium.

Because of size considerations in the filling of the A cation sites, we may expect a fairly well-defined division of the garnets into those with calcium and those with the easily interchangeable divalent ions. Likewise, because of the limited substitution of the B ions, we may expect a separation of garnets into aluminum, ferric iron, and chromium bearing. These two trends are both well marked and each has given rise to a mode of classification. The first, proposed by Winchell,[6] on the basis of the *A* ion, divides the garnets into two groups:

Pyralspite		*Ugrandite*	
Pyrope	MgAl	Uvarovite	CaCr
Almandite	FeAl	Grossularite	CaAl
Spessartine	MnAl	Andradite	CaFe

[6] N. H. Winchell and A. N. Winchell, *Elements of Optical Mineralogy*. John Wiley and Sons, New York, 1927.

This classification serves as an excellent mnemonic aid for the names and formulas. The second grouping, on the basis of the B ion, yields three unequal groups:

Aluminum Garnets	*Ferri-garnets*	*Chrome-garnets*
Pyrope	Andradite	Uvarovite
Almandite		
Spessartine		
Grossularite		

$(OH)_4$ may substitute to a limited extent for SiO_4 groups in the *hydro-garnets*, such as hydrogrossularite, and titanium may enter the B sites concomitant with replacement of calcium by sodium in the A sites, producing the black *melanite*.

Crystallography. Isometric; $4/m\bar{3}2/m$. Common forms are dodecahedron d (Fig. 365a) and trapezohedron, n (Fig. 365b), often in combination (Figs. 365c, d). Hexoctahedrons are observed occasionally (Fig. 365e). Other forms are rare. Usually distinctly crystallized; also appear in rounded grains; massive granular, coarse or fine.

Ia3d; cell edge (see table page 379); $Z = 8$. d's for pyrope: 2.89(8), 2.58(9), 1.598(9), 1.542(10), 1.070(8).

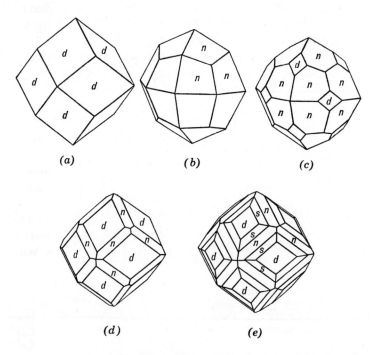

(a)   *(b)*   *(c)*

(d)   *(e)*

FIG. 365.   **Garnet crystals.**

Physical Properties. H $6\frac{1}{2}$–$7\frac{1}{2}$. G 3.5–4.3, varying with the composition (see table). *Luster* vitreous to resinous. *Color* varying with composition; most commonly red, also brown, yellow, white, green, black. Streak white. Transparent to translucent.

Composition. The garnets are silicates which conform to the general formula $A_3B_2(SiO_4)_3$. *A* may be Ca, Mg, Fe^2, Mn^2; *B* may be Al, Fe^3, Ti, Cr. The formulas of the chief subspecies are given, with the specific gravity refractive indices and cell edges for the pure compounds. There is usually extensive ionic substitution and rarely are the pure subspecies found.

Pyrope. Precious garnet in part. Ca and Fe^2 usually present. Color deep red to nearly black. Often transparent and then used as a gem. Name derived from Greek, meaning *firelike*. *Rhodolite* is the name given to a pale rose-red or purple garnet, corresponding in composition to two parts of pyrope and one of almandite.

Subspecies	Composition	G	n	Cell Edge
Pyrope	$Mg_3Al_2(SiO_4)_3$	3.58	1.714	11.46 Å
Almandite	$Fe_3Al_2(SiO_4)_3$	4.32	1.830	11.53
Spessartite	$Mn_3Al_2(SiO_4)_3$	4.19	1.800	11.62
Grossularite	$Ca_3Al_2(SiO_4)_3$	3.59	1.734	11.85
Andradite	$Ca_3Fe_2(SiO_4)_3$	3.86	1.887	12.05
Uvarovite	$Ca_3Cr_2(SiO_4)_3$	3.80	1.868	11.97

Almandite. Precious garnet in part, common garnet in part. Fe^3 may replace Al, and Mg, Fe^2. Color fine deep red, transparent in precious garnet; brownish-red, translucent in common garnet. Name derived from Alabanda, where in ancient times garnets were cut and polished.

Spessartite. Fe^2 usually replaces some Mn^2 and Fe^3 some Al. Color brownish to red.

Grossularite (Essonite, Cinnamon Stone). Often contains Fe^2 replacing Ca and Fe^3 replacing Al. Color, white, green, yellow, cinnamon-brown, pale red. Name derived from the botanical name for gooseberry, in allusion to the light green color of the original grossularite.

Andradite. Common garnet in part. Al may replace Fe^3; Fe^2, Mn^2, and Mg may replace Ca. Color various shades of yellow, green, brown to black. *Demantoid* is a green variety with a brilliant luster, used as a gem. Named after the Portuguese mineralogist, d'Andrada.

Uvarovite. Calcium-chromium garnet. Color emerald-green. Named after Count Uvarov.

Diagnostic Features. Garnets are usually recognized by their characteristic isometric crystals, their hardness, and their color. Specific gravity, refractive index, and unit-cell dimension taken together serve to distinguish members of the group. With the exception of uvarovite, all garnets fuse at $3-3\frac{1}{2}$; uvarovite is almost infusible. The iron garnets, almandite and andradite, fuse to magnetic globules. Spessartite when fused with sodium carbonate gives a bluish-green bead (manganese). Uvarovite gives a green color to salt of phosphorus bead (chromium).

Occurrence. Garnet is a common and widely distributed mineral, occurring as an accessory constituent of metamorphic rocks and some igneous rocks. Its most characteristic occurrence is in mica schists, hornblende schists, and gneisses. Found in pegmatite dikes, more rarely in granitic rocks. Grossularite is found chiefly as a product of contact or regional metamorphism in crystalline limestones. Almandite is especially characteristic of the mica schists. Pyrope is often found in peridotite rocks and the serpentines derived from them. Spessartite occurs in the igneous rock, *rhyolite*. *Melanite*, a black variety of andradite, occurs mostly in certain igneous rocks. Uvarovite is found in serpentine associated with chromite. Garnet frequently occurs as rounded grains in stream and beach sands.

Almandite, of gem quality, is found in northern India, Ceylon, and Brazil. Fine crystals, although for the most part too opaque for cutting, are found in a mica schist on the Stikine River, Alaska. Pyrope of gem quality is found associated with clear grains of olivine (peridot) in the surface sands near Fort Defiance, close to the Utah-Arizona line. A locality near Meronitz, Bohemia, is famous for pyrope gems. Grossularite is used only a little in jewelry, but essonite or cinnamon stones of good size and color are found in Ceylon.

Alteration. Garnet often alters to other minerals, particularly talc, serpentine, and chlorite.

Use. Chiefly as a rather inexpensive gem stone. A green andradite, known as *demantoid*, comes from the Ural Mountains, U.S.S.R., and yields fine gems known as *Uralian emeralds*. At Gore Mountain, New York large crystals of almandite in an amphibolite are mined. The unusual angular fractures and high hardness of these garnets makes them desirable for a variety of abrasive purposes including garnet paper.

Name. *Garnet* is derived from the Latin *granatus*, meaning like a grain.

ZIRCON—$Zr(SiO_4)$

Crystallography. Tetragonal; $4/m2/m2/m$. Crystals usually show a simple combination of $a\{010\}$ and $p\{011\}$, but $m\{110\}$ and a ditetragonal dipyramid also observed (Fig. 366). Usually in crystals; also in irregular grains.

Angles: $a(100) \wedge p(101) = 47°\ 50'$, $p(101) \wedge p'(\bar{1}01) = 64°\ 20'$, $p(101) \wedge p'(011) = 56°\ 41'$.

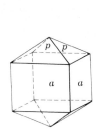 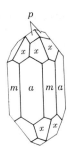 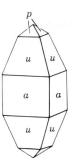

FIG. 366. Zircon crystals.

I4/amd; $a = 6.60$, c 5.98 Å; $a:c = 1:0.906$; $Z = 4$. *d's:* 4.41(7), 3.29(10), 2.52(8), 1.710(9).

Physical Properties. *Cleavage* {010} poor. **H** $7\frac{1}{2}$. **G** 4.68. *Luster* adamantine. *Color* commonly some shade of brown; also colorless, gray, green, red. *Streak* uncolored. Usually translucent; in some cases transparent. *Optics:* (+), $\omega = 1.923–1.960$, $\varepsilon = 1.968–2.015$.

Composition. For $Zr(SiO_4)$, ZrO_2 67.2, SiO_2 32.8 per cent. Zircon always contains some hafnium. Although the amount is usually small (1 to 4 per cent), analyses have reported over 20 per cent HfO_2. Zircon may be metamict due to structural damage from the presence of thorium and uranium.

Diagnostic Features. Usually recognized by its characteristic crystals, color, luster, hardness, and high specific gravity. Infusible; insoluble.

Occurrence. Zircon is a common and widely distributed accessory mineral in all types of igneous rocks. It is especially frequent in the more silicic types such as granite, granodiorite, syenite, and monzonite, and very common in nepheline syenite. Found also commonly in crystalline limestone, in gneiss, schist, etc. Found frequently as rounded grains in stream and beach sands, often with gold. Zircon has been produced from beach sands in Australia, Brazil, and Florida.

Gem zircons are found in the stream sands at Matura, Ceylon, and in the gold gravels in the Ural Mountains and Australia. In large crystals from Madagascar. Found in the nepheline syenites of Norway. Found in the United States at Litchfield, Maine, and in Orange and St. Lawrence counties, New York; in considerable quantity in the sands of Henderson and Buncombe counties, North Carolina. Large crystals have been found in Renfrew County, Ontario, Canada.

Use. When transparent it serves as a gem stone. It is colorless in some specimens, but more often of a brownish or red-orange color. Blue is not a natural color for zircon but is produced by heat treatment. The colorless, yellowish, or smoky stones are called *jargon*, because although resembling the diamond they have little value; and thence the name *zircon*. Serves as the source of zirconium oxide, which is one of the most refractory substances known. Platinum, which fuses at 1755°C, can be melted in crucibles of zirconium oxide.

Overshadowing its other uses since 1945 is the use of zircon as the source of metallic zirconium. Pure zirconium metal is used in the construction of nuclear reactors. Its low neutron-absorption cross section, coupled with retention of strength at high temperatures and excellent corrosion resistance, makes it a most desirable metal for this purpose.

Similar Species. *Thorite*, $Th(SiO_4)$, is like zircon in form and structure; usually hydrated and reddish-brown to black; radioactive.

Aluminum Silicate Group

Andalusite, sillimanite, and kyanite, the three polymorphs of Al_2SiO_5, have rather complex chain or fiberlike structures. In all of them one of the two aluminum ions is consistently coordinated to six oxygens. In sillimanite the other aluminum is in 4 coordination producing a true chain structure of alternating silicon and aluminum tetrahedra, much resembling the pyroxene structure. It is significant that, of the three polymorphs, sillimanite has the most fibrous habit. In andalusite the second aluminum is in 5 coordination with oxygen; and in kyanite both aluminums are in 6 coordination with oxygen. As a result, andalusite and kyanite have a more columnar or bladed habit.

All three minerals are characteristic of metamorphic rocks. The stability relations between the three polymorphs have been determined experimentally as shown in Fig. 367. However, all three may, occur together because of sluggish transformations.

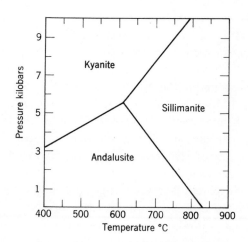

FIG. 367. Al_2SiO_5 phase diagram.

ANDALUSITE—AlAlO(SiO$_4$)

Crystallography. Orthorhombic; $2/m2/m2/m$. Usually occurs in coarse, nearly square prisms terminated by {001}.

Angles: {110} $\wedge$ {1$\bar{1}$0} $= 89°$ 12′.

Pnmm; $a = 7.78$, $b = 7.92$, $c = 5.57$ Å; $a:b:c = 0.982:1:0.703$; $Z = 4$. *d's:* 4.53(10), 3.96(8), 2.76(9), 2.17(10), 1.46(10).

Physical Properties. H 7$\frac{1}{2}$. G 3.16–3.20. *Luster* vitreous. *Color* flesh-red, reddish-brown, olive-green. The variety *chiastolite* has dark-colored carbonaceous inclusions arranged in a regular manner forming a cruciform design (Fig. 368). Transparent to translucent. *Optics:* (−), $\alpha = 1.632$, $\beta = 1.638$, $\gamma = 1.643$; 2V $= 85°$; $X = c$, $Z = a$. In some crystals strong pleochroism: X red Y and Z green to colorless; $r > v$.

Composition. Al$_2$O$_3$ 63.2, SiO$_2$ 36.8 per cent.

Diagnostic Features. Characterized by the nearly square prism and hardness. Chiastolite is readily recognized by the symmetrically arranged inclusions. Infusible. Insoluble. When fine powder is made into a paste with cobalt nitrate and intensely ignited it turns blue (aluminum).

Alteration. Pseudomorphs of fine-grained muscovite (sericite) after andalusite are common.

Occurrence. Andalusite is formed usually by the metamorphism of aluminous shales as the result of both regional and contact metamorphism.

Notable localities are in Andalusia, Spain; the Austrian Tyrol; in water-worn pebbles from Minas Geraes, Brazil. Crystals of chiastolite are found at Bimbowrie, South Australia. In the United States found in the White Mountains near Laws, California; at Standish, Maine; and Delaware County, Pennsylvania. Chiastolite is found at Westford, Lancaster, and Sterling, Massachusetts.

Use. Andalusite has been mined in large quantities in California for use in the manufacture of spark plugs and other porcelains of a highly refractory nature. When clear and transparent it may serve as a gem stone.

Name. From Andalusia, a province of Spain.

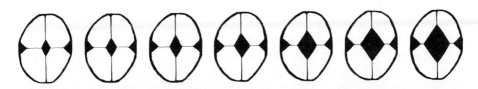

FIG. 368. Successive cross sections through a chiastolite crystal.

SILLIMANITE—AlAlO(SiO₄)

Crystallography. Orthorhombic; $2/m2/m2/m$. Occurs in long slender crystals without distinct terminations; often in parallel groups; frequently fibrous and called *fibrolite*.

Pbnm; $a = 7.44$, $b = 7.60$, $c = 5.75$ Å; $a:b:c = 0.979:1:0.757$; $Z = 4$. *d's:* 3.32(10), 2.49(7), 2.16(8), 1.677(7), 1.579(7).

Physical Properties. *Cleavage* {010} perfect. **H** 6–7. **G** 3.23. *Luster* vitreous. *Color* brown, pale green, white. Transparent to translucent. *Optics:* (+), $\alpha = 1.657$, $\beta = 1.658$, $\gamma = 1.677$; $2V = 20°$, $X = b$, $Z = c$; $r > v$.

Diagnostic Features. Characterized by slender crystals with one direction of cleavage. Infusible. Insoluble. The finely ground mineral turns blue when heated with cobalt nitrate solution.

Occurrence. Sillimanite is a comparatively rare mineral, found as a constituent of gneiss and schist in the highest grade of metamorphic rocks; rarely a contact metamorphic mineral. Often occurs with corundum.

Notable localities for its occurrence are Maldau, Bohemia; Fassa, Austrian Tyrol; Bodenmais, Bavaria; Freiberg, Saxony; and waterworn masses in diamantiferous sands of Minas Gerais, Brazil. In the United States found at Worcester, Massachusetts; at Norwich and Willimantic, Connecticut; and New Hampshire.

Name. In honor of Benjamin Silliman (1779–1864), professor of chemistry at Yale University.

KYANITE—AlAlO(SiO₄)

Crystallography. Triclinic; $\bar{1}$. Usually in long, tabular crystals, rarely terminated. In bladed aggregates.

P$\bar{1}$; $a = 7.10$, $b = 7.74$, $c = 5.57$ Å; $\alpha = 90° 6'$, $\beta = 101° 2'$, $\gamma = 105° 45'$; $a:b:c = 0.899:1:0.709$; $Z = 4$. *d's:* 3.18(10), 2.52(4), 2.35(4), 1.93(5), 1.37(8).

Physical Properties. *Cleavage* {100} perfect. **H** 5 parallel to length of crystals, 7 at right angles to this direction. **G** 3.55–3.66. *Luster* vitreous to pearly. *Color* usually blue, often of darker shade toward the center of the crystal. Also, in some cases, white, gray, or green. Color may be in irregular streaks and patches. *Optics:* (−); $\alpha = 1.712$, $\beta = 1.720$, $\gamma = 1.728$; $2V = 82°$; $r > v$.

Diagnostic Features. Characterized by its bladed crystals, good cleavage, blue color, and different hardness in different directions. Infusible. Insoluble. A fragment moistened with cobalt nitrate solution and ignited assumes a blue color.

Occurrence. Kyanite is an accessory mineral in gneiss and mica schist, often associated with garnet, staurolite, and corundum. Crystals of exceptional quality are found at St. Gothard, Switzerland (Fig. 369); in the Austrian Tyrol; and at

FIG. 369. Bladed kyanite crystals and prismatic andalusite (dark) in mica schist, St. Gothard, Switzerland.

Pontivy and Morbihan, France. Commercial deposits are located in India, Kenya and in the United States in North Carolina and Georgia.

Use. Kyanite is used as is andalusite in the manufacture of spark plugs and other highly refractory porcelains.

Name. Derived from a Greek word meaning *blue*.

Similar Species. *Mullite*, $Al_6Si_2O_{13}$, is rare as a mineral but common in artificial melts; it forms when kyanite, andalusite, and sillimanite are heated to high temperatures. *Dumortierite*, $Al_7O_3(BO_3)(SiO_4)_3$, has been used in the manufacture of high-grade porcelain.

TOPAZ—$Al_2(SiO_4)(F,OH)_2$

Crystallography. Orthorhombic, $2/m2/m2/m$. Commonly in prismatic crystals terminated by dipyramids, first- and second-order prisms and basal pinacoid (Fig. 370). Vertical prism faces frequently striated. Usually in crystals but also in crystalline masses; granular, coarse or fine.

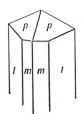
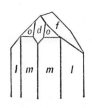
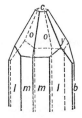
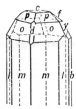

FIG. 370. Topaz crystals.

Angles: $m(110) \wedge m'(1\bar{1}0) = 55° 43'$, $l(120) \wedge l'(1\bar{2}0) = 93° 08'$, $c(001) \wedge f(011) = 43° 39'$, $c(001) \wedge o(111) = 63° 54'$.

Pbnm; $a = 4.65$, $b = 8.80$, $c = 8.40$ Å; $a:b:c = 0.528:1:0.955$; $Z = 4$. *d's:* 3.20(9), 2.96(10), 2.07(9), 1.65(9), 1.403(10).

Physical Properties. *Cleavage* {001} perfect. **H** 8. **G** 3.4–3.6. *Luster* vitreous. *Color* colorless, yellow, pink, wine-yellow, bluish, greenish. Transparent to translucent. *Optics:* (+); $\alpha = 1.606–1.629$, $\beta = 1.609–1.631$, $\gamma = 1.616–1.638$; $2V = 48°–68°$; $X = a$, $Y = b$; $r > v$.

Diagnostic Features. Recognized chiefly by its crystals, basal cleavage, hardness (8), and high specific gravity. Infusible. Insoluble. With cobalt nitrate solution it gives test for aluminum.

Occurrence. Topaz is a mineral formed through the agency of fluorine-bearing vapors given off during the last stages of the solidification of igneous rocks. Found in cavities in rhyolite lavas and granite; a characteristic mineral in pegmatites, especially in those carrying tin. Associated with tourmaline, cassiterite, apatite, and fluorite; also with beryl, quartz, mica, and feldspar. Found in some localities as rolled pebbles in stream sands.

Notable localities for its occurrence are in the U.S.S.R. in the Nerchinsk district in Siberia in large wine-yellow crystals, and in Mursinsk, Ural Mountains, in pale blue crystals; in Saxony from various tin localities; Omi and Mino provinces, Japan; and San Luis Potosí, Mexico. Minas Gerais, Brazil has long been the principal source of yellow gem-quality topaz. In the 1940's several well-formed crystals of colorless topaz were found there; the largest weighs 596 pounds. In the United States found at Pikes Peak, near Florissant and Nathrop, Colorado; Thomas Range, Utah; Streeter, Texas; San Diego County, California; Stoneham and Topsham, Maine; Amelia, Virginia; and Jefferson, South Carolina.

Use. As a gem stone. Frequently sold as "precious topaz" to distinguish it from citrine quartz, commonly called topaz. The color of the stones varies, being colorless, wine-yellow, golden brown, pale blue, and pink. The pink color is usually artificial, being produced by gently heating the dark yellow stones.

Name. Derived from Topazion, the name of an island in the Red Sea, but originally probably applied to some other species.

STAUROLITE—$Fe_2Al_9O_6(SiO_4)_4(O,OH)_2$

Crystallography. Monoclinic; $2/m$ (pseudoorthorhombic). Prismatic, crystals with common forms {110}, {010}, {001}, and {101} (Fig. 371a). Cruciform twins very common, of two types: (1) with twin plane {031} in which the two individuals cross at nearly 90° (Fig. 371b); (2) with twin plane {231} in which they cross at nearly 60° (Fig. 371c). In some cases both types are combined in one twin group. Usually in crystals; rarely massive.

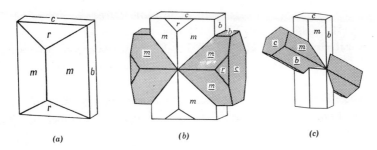

FIG. 371. Staurolite.

Angles: $m(110) \wedge m'(1\bar{1}0) = 50°\ 40'$, $c(001) \wedge r(201) = 55°\ 16'$.

$C2/m$; $a = 7.83$, $b = 16.62$, $c = 5.65$ Å, $\beta = 90°$; $a:b:c = 0.471:1:0.340$; $Z = 4$.

d's: 3.01(8), 2.38(10), 1.974(9), 1.516(5), 1.396(10).

Physical Properties. H 7–$7\frac{1}{2}$. G 3.65–3.75. *Luster* resinous to vitreous when fresh; dull to earthy when altered or impure. *Color* red-brown to brownish-black. Translucent. *Optics:* (+), $\alpha = 1.739$–1.747, $\beta = 1.745$–1.753, $\gamma = 1.752$–1.761; $2V = 82°$–88°; $X = b$ (colorless), $Y = a$ (pale yellow), $Z = c$ (deep yellow); $r > v$.

Composition. FeO 16.7, Al_2O_3 53.3, SiO_2 27.9, H_2O 2.0 per cent for pure material but often impure. Fe^3 may substitute for Al, and Mg for Fe^2.

Diagnostic Features. Recognized by its characteristic crystals and twins. Distinguished from andalusite by its obtuse prism. Infusible. Insoluble. On intense ignition in the closed tube it yields a little water.

Occurrence. Staurolite is formed during metamorphism of aluminum-rich rocks and is found in schists and gneisses. Often associated with garnet, kyanite, and tourmaline. May grow on kyanite in parallel orientation.

Notable localities are Monte Campione, Switzerland; Goldenstein, Moravia; Aschaffenburg, Bavaria; and in large twin crystals in Brittany and Scotland. In the United States found at Windham, Maine: Franconia and Lisbon, New Hampshire; Chesterfield, Massachusetts; also in North Carolina, Georgia, Tennessee, Virginia, New Mexico, and Montana.

Use. Occasionally a transparent stone from Brazil is cut as a gem. In North Carolina the right angle twins are sold as amulets under the name "fairy stone," but most of the crosses offered for sale are imitations carved from a fine-grained rock and dyed.

Name. Derived from a Greek word meaning *cross*, in allusion to its cruciform twins.

Chondrodite—$Mg_5(SiO_4)_2(F,OH)_2$

Crystallography. Monoclinic; $2/m$. Crystals are frequently complex with many forms. Usually in isolated grains. Also massive.

$P2_1/c$; $a = 7.89$, $b = 4.74$, $c = 10.29$, $\beta = 109°$; $a:b:c = 1.665:1:2.171$; $Z = 2$. $d's$: 3.02(5), 2.76(4), 2.51(5), 2.26(10), 1.740(7).

Physical Properties. H $6-6\frac{1}{2}$. G 3.1–3.2. *Luster* vitreous to resinous. *Color* light yellow to red. Translucent. *Optics*: $(+)$; $\alpha = 1.592-1.615$, $\beta = 1.602-1.627$, $\gamma = 1.621-1.646$; $2V = 71°-85°$; $Z = b$, $X \wedge c = 25°$; $r > v$.

Composition. Hydroxyl replaces part of the fluorine, and iron often takes the place of magnesium. Chondrodite is the most common member of the *humite group*. The species of the group are:

	Chemical Composition	β index
Norbergite	$Mg_3(SiO_4)_1(F, OH)_2$	1.573
Chondrodite	$Mg_5(SiO_4)_2(F, OH)_2$	1.614
Humite	$Mg_7(SiO_4)_3(F, OH)_2$	1.636
Clinohumite	$Mg_9(SiO_4)_4(F, OH)_2$	1.642

Diagnostic Features. Characterized by its light yellow to red color and its mineral associations in crystalline limestone. The members of the *humite group* cannot be distinguished from one another without optical tests. Infusible. Yields water in the closed tube.

Occurrence. Chondrodite occurs most commonly in metamorphosed dolomitic limestones. The mineral association including phlogopite, spinel, pyrrhotite, and graphite is highly characteristic. Noteworthy localities of chondrodite are Monte Somma, Italy; Paragas, Finland; and Kafveltorp, Sweden. In the United States is found abundantly at the Tilly Foster magnetite deposit near Brewster, New York.

Name. Chondrodite is from the Greek meaning *a grain;* alluding to its occurrence as isolated grains. Humite is named in honor of Sir Abraham Hume.

DATOLITE—$CaB(SiO_4)(OH)$

Crystallography. Monoclinic; $2/m$. Crystals usually nearly equidimensional in the three axial directions and often complex in development (Fig. 372). Usually in crystals. Also coarse to fine granular. Compact and massive, resembling unglazed porcelain.

Angles: $m(110) \wedge m'(1\bar{1}0) = 64° 47'$, $n(111) \wedge n'(1\bar{1}1) = 59° 5'$, $f(011) \wedge f'(0\bar{1}1) = 103° 23'$, $e(\bar{1}12) \wedge e'(\bar{1}\bar{1}2) = 48° 20'$.

$P2_1/a$; $a = 4.83$, $b = 7.64$, $c = 9.66$, $\beta = 90° 9'$; $a:b:c = 0.632:1:1.264$; $Z = 4$. $d's$: 3.76(5), 3.40(3), 3.11(10), 2.86(7), 2.19(6).

Physical Properties. H $5-5\frac{1}{2}$. G 2.8–3.0. *Luster* vitreous. *Color*, white, often with faint greenish tinge. Transparent to translucent. *Optics*: $(-)$; $\alpha = 1.624$, $\beta = 1.652$, $\gamma = 1.668$; $2V = 74°$; $Y = b$, $Z = c$. $r > v$.

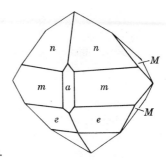

FIG. 372.   Datolite.

Composition.   CaO 35.0, B_2O_3 21.8, SiO_2 37.6, H_2O 5.6 per cent.

Diagnostic Features.   Characterized by its glassy luster, pale green color, and its crystals with many faces. Fuses at $2-2\frac{1}{2}$ to a clear glass and colors the flame green (boron). It is thus distinguished from quartz.

Occurrence.   Datolite is a secondary mineral found usually in cavities in basalt lavas and similar rocks. Associated with zeolites, prehnite, apophyllite, and calcite. Notable foreign localities are Andreasberg Harz Mountains; in Italy near Bologna; from Seiser Alpe and Theiso, Trentino; and Arendal, Norway. In the United States it occurs in the trap rocks of Massachusetts, Connecticut, and New Jersey, particularly at Westfield, Massachusetts, and Bergen Hill, New Jersey. Found associated with the copper deposits of Lake Superior.

Name.   Derived from a Greek word meaning *to divide*, in allusion to the granular character of a massive variety.

SPHENE (*titanite*)—CaTiO(SiO$_4$)

Crystallography.   Monoclinic; $2/m$. Wedge-shaped crystals common resulting from a combination of {001}, {110}, and {111} (Fig. 373). May be lamellar or massive.

Angles: $m(110) \wedge m(1\bar{1}0) = 66° 29'$, $p(111) \wedge p'(1\bar{1}1)$ 43° 29', $c(001) \wedge p(111) = 38° 16$.

$C2/c$, $a = 6.56$, $b = 8.72$, $c = 7.44$ Å, $\beta = 119° 43'$; $a:b:c = 0.752:1:0.853$; $Z = 4$. d's: 3.23(10), 2.99(9), 2.60(9), 2.27(3), 2.06(4).

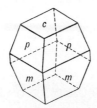
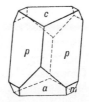

FIG. 373.   Sphene crystals.

Physical Properties. *Cleavage* {110} distinct. Parting on {100}, may be present. **H** 5–5½. **G** 3.4–3.55. *Luster* resinous to adamantine. *Color* gray, brown, green, yellow, black. Transparent to translucent. *Optics:* $(-)$; $\alpha = 1.900$, $\beta = 1.907$, $\gamma = 2.034$; $2V = 27°$; $Y = b$, $Z \wedge c = 51°$; $r > v$.

Composition. CaO 28.6, TiO_2 40.8, SiO_2 30.6 per cent. Small amounts of rare earths, Fe, Al, Mn, Mg, and Zr may be present.

Diagnostic Features. Characterized by its wedge-shaped crystals and high luster. Hardness is less than that of staurolite and greater than that of sphalerite. Fusible at 4.

Occurrence. Sphene in small crystals is a common accessory mineral in granites, granodiorites, diorites, syenites, and nepheline syenites. Found in crystals of considerable size in the metamorphic rocks, gneiss, chlorite schist, and crystalline limestone. Also found with iron ores, pyroxene, amphibole, scapolite, zircon, apatite, feldspar, and quartz.

The most notable occurrence of sphene is on the Kola Peninsula, U.S.S.R., where it is associated with apatite and nepheline in nepheline syenite. It is mined there as a source of titanium. It is found in crystals at Tavetsch, Binnental, and St. Gothard, Switzerland; Zillertal, Tyrol; Ala, Piedmont; Vesuvius; and Arendal, Norway. In the United States in Diana, Rossie, Fine, Pitcairn, Edenville, and Brewster, New York; and Riverside, California. Also in various places in Ontario and Quebec, Canada.

Use. As a source of titanium for use as a paint pigment.

Name. *Sphene* comes from a Greek word meaning *wedge*, in allusion to the wedge-shaped crystals.

Similar Species. *Benitoite* is a blue, hexagonal, barium titanium silicate associated with *neptunite* in San Benito, California. *Astrophyllite, aenigmatite, lamprophyllite, ramsayite, fersmannite* are rare titanium-bearing silicates found associated with alkalic rocks.

Sorosilicates

The sorosilicates are characterized by isolated double tetrahedral groups formed by two SiO_4 tetrahedra sharing a single apical oxygen, (Fig. 374). The resulting ratio of silicon to oxygen is 2:7. The most important sorosilicates are idocrase and the members of the *epidote group.*

Sorosilicates

Hemimorphite	$Zn_4(Si_2O_7)(OH)_2 \cdot H_2O$
Lawsonite	$CaAl_2(Si_2O_7)(OH)_2 \cdot H_2O$
EPIDOTE GROUP	
Clinozoisite	$Ca_2Al_3O(SiO_4)(Si_2O_7)(OH)$
Epidote	$Ca_2(Al,Fe)Al_2O(SiO_4)(Si_2O_7)(OH)$
Allanite	$X_2Y_3O(SiO_4)(Si_2O_7)(OH)$
Idocrase	$Ca_{10}(Mg,Fe)_2Al_4(SiO_4)_5(Si_2O_7)_2(OH)_4$
Prehnite	$Ca_2Al_2(Si_3O_{10})(OH)_2$

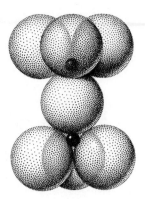

FIG. 374. Si$_2$O$_7$ group.

HEMIMORPHITE—Zn$_4$(Si$_2$O$_7$)(OH)$_2$·H$_2$O

Crystallography. Orthorhombic; *mm*2. Crystals usually tabular parallel to {010}. They show prism faces and are terminated above usually by a combination of domes and pedion, and below by a pyramid (Fig. 375). Crystals often divergent, giving rounded groups with slight reentrant notches between the individual crystals, forming knuckle or coxcomb masses. Also mammillary, stalactitic, massive, and granular.

Angles: b(010) ∧ m(110) = 51° 55′, c(001) ∧ i(031) = 55° 06′, c(001) ∧ t(301) = 61° 20′.

*Imm*2, a = 8.38, b = 10.72, c = 5.12 Å; $a:b:c$ = 0.782:1:0.478. Z = 2. d's: 6.60(9), 5.36(6), 3.30(8), 3.10(10), 2.56(5).

Physical Properties. *Cleavage* {110} perfect. **H** 4$\frac{1}{2}$–5. **G** 3.4–3.5. *Luster* vitreous. *Color* white, in some cases with faint bluish or greenish shade; also yellow to brown. Transparent to translucent. Strongly pyroelectric and piezoelectric. *Optics:* (+); α = 1.614, β = 1.617, γ = 1.636; 2V = 46°; X = b, Z = c; r > v.

Composition. ZnO 67.5, SiO$_2$ 25.0, H$_2$O 7.5 per cent. Small amounts of aluminum and iron may be present.

Diagnostic Features. Characterized by the grouping of crystals. Resembles prehnite but has a higher specific gravity. Fusible with difficulty at 5. When fused

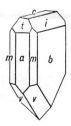

FIG. 375. Hemimorphite.

on charcoal with sodium carbonate it gives a nonvolatile coating of zinc oxide (yellow when hot, white when cold). When fused with cobalt nitrate on charcoal it turns blue. Gives water in the closed tube.

Occurrence. Hemimorphite is a secondary mineral found in the oxidized portion of zinc deposits, associated with smithsonite, sphalerite, cerussite, anglesite, and galena.

Notable localities for its occurrence are at Moresnet, Belgium; Aixla-Chapelle, Germany; Carinthia; Rumania; Sardinia; Cumberland and Derbyshire, England; Algeria; and Chihuahua, Mexico. In the United States it is found at Sterling Hill, Ogdensburg, New Jersey; Friedensville, Pennsylvania; Wythe County, Virginia; with the zinc deposits of southwestern Missouri; Leadville, Colorado; Organ Mountains, New Mexico; and Elkhorn Mountains, Montana.

Use. An ore of zinc.

Name. From the hemimorphic character of the crystals. The mineral was formerly called calamine.

Lawsonite—$CaAl_2(Si_2O_7)(OH)_2 \cdot H_2O$

Crystallography. Orthorhombic; 222. Usually in tabular or prismatic crystals. Frequently twinned polysynthetically on {110}.

$C222_1$; $a = 8.79$, $b = 5.84$, $c = 13.12$ Å; $a:b:c = 1.505:1:2.247$; $Z = 4$. d's: 4.17(5), 3.65(6), 2.72(10), 2.62(7), 2.13(7).

Physical Properties. *Cleavage* {010} and {110} good. **H** 8. **G** 3.09. *Color* colorless, pale blue to bluish-gray. *Luster* vitreous to greasy. Translucent. *Optics:* (+); $\alpha = 1.665$, $\beta = 1.674$, $\gamma = 1.684$; 2V = 84°; $X = a$, $Z = c$; $r > v$.

Composition. The composition of lawsonite is the same as that of anorthite plus water, but has a more closely packed structure resulting in a higher hardness and specific gravity.

Diagnostic Features. Lawsonite is characterized by its high hardness. It fuses to a blebby glass but once fused is not again fusible. On intense ignition in the closed tube it yields water.

Occurrence. Lawsonite is found in gneisses and schists in well-formed grains as well as in veins in metamorphic rocks. The type locality is on the Tiburon Peninsula, San Francisco Bay, California. Lawsonite is also found in schists in France and New Caledonia.

Name. In honor of Professor Andrew Lawson of the University of California.

Similar Species. *Ilvaite*, $CaFe^2Fe^3O(Si_2O_7)(OH)$, is related to lawsonite with a similar although not identical structure. A combination of ferrous iron and ferric iron in ilvaite seems to be roughly structurally equivalent to aluminum in lawsonite; $Fe^2Fe^3(OH)$ for $Al_2(OH)_2$.

Epidote Group

In the minerals of the epidote group both isolated SiO_4 tetrahedra and Si_2O_7 groups occur. In the rather complicated structure there are two kinds of cation sites: one, the X site, is usually occupied by rather large, weakly charged ions such as Ca and Na; two, the Y site, is occupied by smaller, more highly charged ions including Al, Fe^3, Mn^3, and more rarely Mn^2. The general formula may thus be written: $X_2Y_3(SiO_4)(Si_2O_7)(OH)$.

All members of the group are isostructural and form monoclinic crystals characteristically elongate on *b*. Orthorhombic zoisite has a structure which may be derived from that of its monoclinic polymorph, clinozoisite, by a simple twinlike doubling of the cell along the *a* axis. As a result of the generally consistent structure, the crystal chemistry of this group is chiefly concerned with the kind and degree of solid-solution relations. X may further be constituted in part by divalent manganese, lead, or strontium; and Y may be in part chromium. Most common epidote contains little manganese and almost none of the rarer ions listed and has a composition that may be represented fairly well by a simple ratio of aluminum to ferric iron.

Idocrase has a composition closely analogous to that of epidote and a similar structure containing both isolated SiO_4 and Si_2O_7 groups. Magnesium and ferrous iron occur in variable proportions, replacing each other freely.

Ions in Epidote Group

	X	Y
Clinozoisite	Ca	Al
Epidote	Ca	Al, Fe^3
Piedmontite	Ca	Al, Fe^3, Mn^3
Allanite	Ca, Ce, La, Na	Al, Fe^3, Be, Mg, Mn^3

CLINOZOISITE—$Ca_2Al_3O(SiO_4)(Si_2O_7)(OH)$

EPIDOTE—$Ca_2(Al, Fe)Al_2O(SiO_4)(Si_2O_7)(OH)$

Crystallography. Monoclinic; $2/m$. Crystals are usually elongated parallel to *b* with a prominent development of the faces of the [010] zone, giving them a prismatic aspect. Striated parallel to *b* (Fig. 376). Twinning on {100} common. Usually coarse to fine granular; also fibrous.

Angles: $c(001) \wedge a(100) = 64° 36'$, $n(\overline{1}11) \wedge n'(11\overline{1}) = 70° 29'$, $a(100) \wedge m(110) = 55° 0'$, $c(001) \wedge o(011) = 58° 28'$.

$P2_1/m$; $a = 8.98$, $b = 5.64$, $c = 10.22$ Å, $\beta = 115° 24'$, $a:b:c = 1.592:1:1.812$; $Z = 2$. *d's:* 5.02(4), 2.90(10), 2.86(6), 2.53(6), 2.40(7).

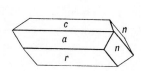
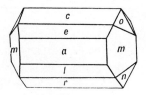

FIG. 376. Epidote crystals.

Physical Properties. *Cleavage* {001} perfect and {100} imperfect. **H** 6–7. **G** 3.35–3.45. *Luster* vitreous. *Color* epidote: pistachio-green to yellowish-green to black; clinozoisite: pale green to gray. Transparent to translucent. *Optics.* Refractive indices and birefringence increase with iron content. Clinozoisite: $(+)$: $\alpha = 1.670$–1.715, $\beta = 1.674$–1.725, $\gamma = 1.690$–1.734; $2V = 14°$–$90°$; $Y = b$, $X \wedge c -2°$ to $-7°$; $r < v$. Epidote: $(-)$, $\alpha = 1.715$–1.751, $\beta = 1.725$–1.784, $\gamma = 1.734$–1.797; $2V = 64°$–$90°$; $Y = b$, $X \wedge c -1$ to $-5°$; $r > v$. Absorption $Y > Z > X$. Transparent crystals may show strong absorption in ordinary light.

Composition. A complete solid solution extends from clinozoisite (Al:Fe = 3:0) to epidote (Al:Fe = 2:1).

Diagnostic Features. Fuses at 3–4 with intumescence to a black slag; once formed, this slag is not again fusible. On intense ignition in the closed tube yields a little water. Epidote is characterized by its peculiar green color and one perfect cleavage. The slag formed on fusing which is again infusible is diagnostic.

Occurrence. Epidote and clinozoisite occur in metamorphic rocks, alteration products of such minerals as feldspar, pyroxene, amphibole, and biotite. Often associated with chlorite. Epidote is also frequently formed during metamorphism of an impure limestone, and is especially characteristic of contact metamorphic deposits in limestone.

Epidote is a widespread mineral. Notable localities for its occurrence in fine crystals are Knappenwand, Untersulzbachthal, Salzburg, Austria; Bourg d'Oisans, Isere, France; and the Ala Valley and Traversella, Piedmont. In the United States found at Haddam, Connecticut; Riverside, California; and on Prince of Wales Island, Alaska (Fig. 377).

Name. Epidote from the Greek meaning *increase*, since the base of the vertical prism has one side longer than the other.

Similar Species. *Piedmontite*, similar in crystal structure and chemical composition to epidote but contains trivalent manganese; occurs in crystalline schists and with manganese ores.

Zoisite is the orthorhombic polymorph of clinozoisite. It is similar in appearance and occurrence to clinozoisite but is less common. In 1967 gem quality zoisite crystals were found in Tanzania.

FIG. 377. Epidote with quartz crystals, Prince of Wales Island, Alaska.

Allanite

Crystallography. Monoclinic; $2/m$. Habit of crystals similar to epidote. Commonly massive and in imbedded grains.

$P2_1/m$; $a = 8.98$, $b = 5.75$, $c = 10.23$ Å, $\beta = 115° 0'$; $a:b:c = 1.562:1:1.779$; $Z = 2$. $d's$: 3.57(6), 2.94(10), 2.74(8), 2.65(6), 2.14(4).

Physical Properties. H $5\frac{1}{2}$–6. G 3.5–4.2. Luster submetallic to pitchy and resinous. *Color* brown to pitch-black. Often coated with a yellow-brown alteration product. Subtranslucent, will transmit light on thin edges. Slightly radioactive. *Optics:* $(-)$, usually, with $2V = 40°$–$90°$; when $(+)$, $2V = 60°$–$90°$, $\alpha = 1.690$–1.791, $\beta = 1.700$–1.815, $\gamma = 1.706$–1.828; $Y = b$, $X \wedge c$ $1°$–$40°$. Metamict allanites are isotropic with $n = 1.54$–1.72.

Composition. A silicate of variable composition, $X_2Y_3O(SiO_4)(Si_2O_7)(OH)$ where $X = $ Ca, Ce, La, Th, Na; and $Y = $ Al, Fe, Mn, Be, Mg.

Diagnostic Features. Characterized by its black color, pitchy luster, and association with granitic rocks. Fuses at $2\frac{1}{2}$ with intumescence to a black magnetic glass.

Occurrence. Allanite occurs as a minor accessory constituent in many igneous rocks, such as granite, syenite, diorite, and pegmatites. Frequently associated with epidote. Has been noted in limestone as a contact mineral.

Notable localities are at Miask, Ural Mountains, U.S.S.R.; Greenland; Falun, Ytterby, and Sheppsholm, Sweden; and Madagascar. In the United States allanite is found at Moriah, Monroe, and Edenville, New York; Franklin, New Jersey; Amelia Court House, Virginia; and Barringer Hill, Texas.

Name. In honor of Thomas Allan, who first observed the mineral. *Orthite* is sometimes used as a synonym.

IDOCRASE—$Ca_{10}(Mg,Fe)_2Al_4(SiO_4)_5(Si_2O_7)_2(OH)_4$

Crystallography. Tetragonal; $4/m2/m2/m$. Crystals prismatic, often vertically striated. Common forms are $\{110\}$, $\{010\}$, and $\{001\}$ (Fig. 378). Frequently found in crystals, but striated columnar aggregates more common. Also granular, massive.

Angles: $c(001) \wedge p(011) = 37° 08'$, $(001) \wedge (112) = 28° 11'$.

P4nnc; $a = 15.66$, $c = 11.85$ Å; $a:c = 1:0.757$; $Z = 4$. *d's:* 2.95(4), 2.75(10), 2.59(8), 2.45(5), 1.621(6).

Physical Properties. *Cleavage* $\{010\}$ poor. **H** $6\frac{1}{2}$. **G** 3.35–3.45. Luster vitreous to resinous. *Color* usually green or brown; also yellow, blue, red. Subtransparent to translucent. *Streak* white. *Optics:* $(-)$; $\omega = 1.703–1.752$, $\varepsilon = 1.700–1.746$.

Composition. There is some substitution of Na for Ca; Mn^2 for Mg, Fe^3, and Ti for Al, and F for (OH). B and Be have been reported in some varieties.

Diagnostic Features. Brown tetragonal prisms and striated columnar masses are characteristic of idocrase. Fuses at 3 with intumescence to a greenish or brownish glass. Gives water in the closed tube.

Occurrence. Idocrase is usually formed as the result of contact metamorphism of impure limestones. Associated with other contact minerals, such as grossularite and andradite garnet, wollastonite, and diopside. Originally discovered in the ancient ejections of Vesuvius and in the dolomitic blocks of Monte Somma. Thus *vesuvianite* is sometimes used as a synonym.

Important localities are Zermatt, Switzerland; Ala, Piedmont; Monzoni, Trentino; Vesuvius; Christiansand, Norway; Achmatovsk, Ural Mountains, and River Vilui, Siberia, U.S.S.R.; and Morelos and Chiapas, Mexico. In the United States it is found at Auburn and Sanford, Maine; near Amity, New York; and Franklin, New Jersey. Found at many contact metamorphic deposits in western United States. A compact green variety resembling jade found in Siskiyou, Fresno, and Tulare counties, California, is known as *californite*. In Quebec, Canada, found at Litchfield, Pontiac County; and at Templeton, Ottawa County.

Use. The green variety californite is used to a small extent as a semiprecious stone which is sometimes sold as jade.

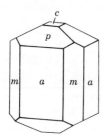

FIG. 378. Idocrase.

Name. From two Greek words meaning *form* and a *mixture* because the crystal forms present appear to be a combination of those found on different minerals.

PREHNITE—$Ca_2Al_2(Si_3O_{10})(OH)_2$

Crystallography. Orthorhombic; $2mm$. Distinct crystals are rare, commonly tabular parallel to {001}. Usually reniform, stalactitic, and in rounded groups of tabular crystals.

$P2cm$; $a = 4.65$, $b = 5.48$, $c = 18.49$ Å; $a:b:c = 0.849:1:3.374$; $Z = 2$. d's: 3.48(9), 3.28(6), 3.08(10), 2.55(10), 1.77(7).

Physical Properties. H $6-6\frac{1}{2}$. G 2.8–2.95. *Luster* vitreous. *Color* usually light green, passing into white. Translucent. *Optics:* (+); $\alpha = 1.616$, $\beta = 1.626$, $\gamma = 1.649$; 2V = 66°; $X = a$, $Z = c$; $r > v$.

Composition. CaO 27.1, Al_2O_3 24.8, SiO_2 43.7, H_2O 4.4 per cent. Some iron may replace aluminum.

Diagnostic Features. Characterized by its green color and crystalline aggregates forming reniform surfaces. Resembles hemimorphite but is of lower specific gravity and fuses easily. ($2\frac{1}{2}$). Ease of fusion also distinguishes it from quartz and beryl. Yields water in the closed tube.

Occurrence. Prehnite occurs as a secondary mineral lining cavities in basalt and related rocks. Associated with zeolites, datolite, pectolite, and calcite. In the United States it occurs at Farmington, Connecticut; Paterson and Bergen Hill, New Jersey; Westfield, Massachusetts; and Lake Superior copper district. Found in good crystals at Coopersburg, Pennsylvania.

Name. In honor of Colonel Prehn, who brought the mineral from the Cape of Good Hope.

Cyclosilicates

The cyclosilicates are built about rings of linked SiO_4 tetrahedra having a ratio of $Si:O = 1:3$. Three possible closed cyclic configurations of this kind may exist as shown in Figs. 379 to 381. The simplest is the Si_3O_9 ring, represented among minerals only by the rare titano-silicate *benitoite*, $BaTiSi_3O_9$. The Si_4O_{12} ring occurs together with BO_3 triangles and (OH) groups in the complex structure of the triclinic mineral *axinite*. The Si_6O_{18} ring, however, is the basic framework of the structures of the common and important minerals, beryl and tourmaline.

The hexagonal Si_6O_{18} rings are arranged, in the structure of beryl, in planar sheets parallel to {0001}. These sheets are so firmly bonded by the small beryllium and aluminum ions with their high surface density of charge and high polarizing power that only poor cleavage results. The very simple crystal morphology of

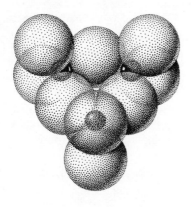

FIG. 379. Si_3O_9 ring.

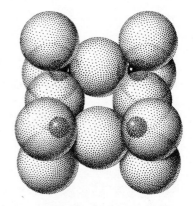

FIG. 380. Si_4O_{12} ring.

beryl is an outward expression of the simple, firmly bonded architecture of the crystals. Beryllium is in 4 coordination with oxygen, whereas the slightly larger aluminum is in 6 coordination. The silicon-oxygen rings are so arranged as to be nonpolar; that is, a symmetry plane can be pictured as passing through the tetrahedra in the plane of the ring (Fig. 381). The bonds extending from oxygens to beryllium and aluminum ions have the same total strength viewed in either direction along the c axis. Rings are positioned one above the other in the basal sheets so that the central holes correspond, forming prominent channels parallel to the c axis. In these channels a wide variety of ions, neutral atoms and molecules may be trapped. Hydroxyl, fluorine, atomic helium, water, and ions of rubidium, cesium, sodium, and potassium are housed in beryl in this way. To compensate for the presence of these elements, lithium may substitute for aluminum and

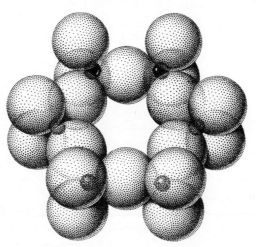

FIG. 381. Si_6O_{18} ring.

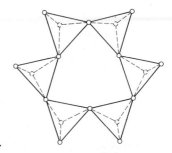

FIG. 382. Si_6O_{18} ring in tourmaline.

aluminum for beryllium. Cordierite has a structure similar to beryl but forms orthorhombic pseudohexagonal crystals in which some of the aluminum enters tetrahedrally coordinated silicon sites in the rings and some is in 6 coordination.

Tourmaline, whose structure long remained a mystery because of its complexity, has now been shown to be built about Si_6O_{18} rings. In tourmaline, however, the rings are polar; that is, the net strength of bonds to one face of the ring is not the same as the strength of bonds extending to the other, looking first in one direction, then in the other along the c axis (Fig. 382). This polarity of the fundamental structural unit leads to the well-known polar character of the tourmaline crystal. In the structure of tourmaline, in addition to the Si_6O_{18} rings, there are independent BO_3 triangles and (OH) groups. All these structural units are bound together by ionic bonds through X- and Y-type cations. X-type ions may be sodium or calcium, and Y-type ions may be magnesium, ferrous iron, aluminum, ferric iron, divalent manganese, and lithium, much as in other silicates discussed earlier. The varieties are determined by the relative proportions of different X and Y ions, and ionic substitution follows the usual pattern, with extensive mutual substitution of magnesium, ferrous iron, and divalent manganese in Y sites, and of sodium and calcium in the X sites accompanied by concomitant coupled substitution to maintain electrical neutrality.

Cyclosilicates

Axinite	$Ca_2(Fe,Mn)Al_2(BO_3)(Si_4O_{12})(OH)$
BERYL GROUP	
Beryl	$Be_3Al_2(Si_6O_{18})$
Cordierite	$(Mg, Fe)_2Al_3(AlSi_5O_{18})$
Tourmaline	$XY_3Al_6(BO_3)_3(Si_6O_{18})(OH)_4$
Chrysocolla	Amorphous

Axinite—$Ca_2(Fe,Mn)Al_2(BO_3)(Si_4O_{12})(OH)$

Crystallography. Triclinic; 1. Axinite is the only common mineral crystallizing in this class. Crystals usually thin with sharp edges but varied in habit (Fig. 383). Frequently in crystals and crystalline aggregates; also massive, lamellar to granular.

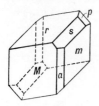

FIG. 383.   Axinite.

*P*1; *a* = 7.15, *b* = 9.16, *c* = 8.96 Å, *α* = 88° 04′, *β* = 81° 36′, *γ* = 77° 42′; *a:b:c* = 0.779:1:0.978. *Z* = 2. *d's:* 6.30(7), 3.46(8), 3.28(6), 3.16(9), 2.81(10).

Physical Properties. *Cleavage* {100} distinct. **H** 6½–7. **G** 3.27–3.35. *Luster* vitreous. *Color* clove-brown, violet, gray, green, yellow. Transparent to translucent. Pyroelectric. *Optics:* (−); *α* = 1.674–1.693, *β* = 1.681–1.701, *γ* = 1.684–1.704; 2V = 63–80°; *r* < *v*.

Composition. A considerable range exists in composition with varying amounts of Ca, Mn, and Fe. Some Mg may be present.

Diagnostic Features. Characterized by the triclinic crystals with very acute angles. Fusible at 2½–3 with intumescence. When mixed with boron flux and heated on platinum wire it gives a green flame (boron). Gives water in the closed tube.

Occurrence. Axinite occurs in cavities in granite, and in the contact zones surrounding granitic intrusions. Notable localities for its occurrence are Bourg d'Oisans, Isere, France; various points in Switzerland; St. Just, Cornwall; and Obira, Japan. In the United States at Luning, Nevada, and a yellow manganous variety at Franklin, New Jersey.

Name. Derived from a Greek word meaning *axe*, in allusion to the wedgelike shape of the crystals.

BERYL—$Be_3Al_2(Si_6O_{18})$

Crystallography. Hexagonal; 6/*m*2/*m*2/*m*. Strong prismatic habit. Frequently vertically striated and grooved. Cesium beryl frequently flattened on {0001}. Forms usually present consist only of {10$\bar{1}$0} and {0001} (Fig. 384a). Dihexagonal forms are rare (Fig. 384b). Crystals frequently of considerable size with rough faces. At Albany, Maine, a tapering crystal 27 feet long weighed over 25 tons.

Angles: *c*(0001) ∧ *p*(10$\bar{1}$2) = 29° 57′, *c*(0001) ∧ *s*(11$\bar{2}$2) = 44° 56′.

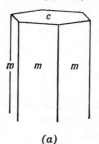

(a)

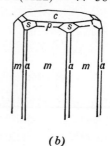

(b)

FIG. 384.   Beryl crystals.

P6mcc; $a = 9.23$, $c = 9.19$ Å; $a:c = 1:0.996$; $Z = 2$. *d's:* 7.98(9), 4.60(5), 3.99(5), 3.25(10), 2.87(10).

Physical Properties. *Cleavage* {0001} imperfect. **H** $7\frac{1}{2}$–8. **G** 2.75–2.8. *Luster* vitreous. *Color* commonly bluish-green or light yellow, may be deep emerald-green, gold-yellow, pink, white, or colorless. Transparent to translucent. Frequently the larger, coarser crystals show a mottled appearance due to the alteration of clear transparent spots with cloudy portions.

Color serves as the basis for several variety names of gem beryl. *Aquamarine* is the pale greenish-blue transparent variety. *Morganite*, or *rose beryl*, is pale pink to deep rose. *Emerald* is the deep green transparent beryl. *Golden beryl* is a clear golden-yellow variety. *Optics:* $(-)$; $\omega = 1.566$–1.608, $\varepsilon = 1.562$–1.600.

Composition. BeO 14.0, Al_2O_3 19.0, SiO_2 67.0 are the theoretical percentages of the oxides in the formula. However, the presence of alkalis (see page 398) and lithium may reduce considerably the percentage of BeO.

Diagnostic Features. Recognized usually by its hexagonal crystal form and color. Distinguished from apatite by greater hardness and from quartz by higher specific gravity. Fuses with difficulty at 5–$5\frac{1}{2}$ to an enamel. Insoluble in acids.

Occurrence. Beryl, although containing the rare element beryllium, is rather common and widely distributed. It occurs usually in granitic rocks, or in pegmatites. It is also found in mica schists and associated with tin ores. Emeralds of gem quality occur in a dark bituminous limestone at Muso, Colombia; and in a mica schist associated with phenacite, chrysoberyl, and rutile near Sverdlovsk, U.S.S.R. Rather pale emeralds have been found in small amount in Alexander County, North Carolina, associated with the green variety of spodumene, hiddenite. Beryl of the lighter aquamarine color is much more common and is found in gem quality in many countries.

For many years Brazil has led in production not only of gem beryls other than emerald but also in the production of common beryl used as beryllium ore. India, the U.S.S.R., and the United States are also major producers of common beryl. In the United States gem beryls, chiefly aquamarine, have been found in various places in Maine, New Hampshire, Massachusetts, Connecticut, North Carolina, and Colorado. Rose-colored beryl has been found in San Diego County, California, associated with pink tourmaline and the pink spodumene, kunzite.

Use. Used as a gem stone of various colors. The emerald ranks as the most valuable of stones and may have a much greater value than the diamond. Beryl is also the major source of beryllium, a light metal similar to aluminum in many of its properties. One of its chief uses is as an alloy with copper. One and one-half per cent of beryllium in copper greatly increases the hardness, tensile strength, and fatigue resistance.

Name. The name beryl is of ancient origin, derived from the Greek word which was applied to green gem stones.

Similar Species. *Euclase,* BeAl(SiO$_4$)(OH), and *gadolinite,* YFe″Be$_2$(SiO$_4$)$_2$O$_2$, are rare beryllium silicates.

Cordierite—(Mg, Fe)$_2$Al$_3$(AlSi$_5$O$_{18}$)

Crystallography. Orthorhombic; $2/m2/m2/m$. Crystals are usually short prismatic, pseudohexagonal twins twinned on {110}. Also is found as imbedded grains and massive.

Cccm; a = 17.13, b = 9.80, c = 9.35 Å; $a:b:c$ = 1.748:1:0.954; Z = 4. *d's:* 8.54(10), 4.06(8), 3.43(8), 3.13(7), 3.03(8).

Physical Properties. *Cleavage* {010} poor. **H** 7–7$\frac{1}{2}$. **G** 2.60–2.66. *Luster* vitreous. *Color* different shades of blue. Transparent to translucent. *Optics:* usually (−), may be (+). Indices increasing with Fe content. α = 1.522–1.558, β = 1.532–1.568, γ = 1.527–1.573. 2V = 0°–90°. X = c, Y = a; $r < v$. Pleochroism: cordierite is sometimes called *dichroite* because of pleochroism. Fe rich varieties: X colorless, Y and Z violet.

Composition. Manganese may replace part of the magnesium, and ferric iron part of the aluminum. Water may also be present.

Diagnostic Features. Cordierite resembles quartz and is distinguished from it with difficulty. Unlike quartz, it is fusible on thin edges. Distinguished from corundum by lower hardness. Pleochroism is characteristic if observed.

Alteration. Commonly altered to some form of mica, chlorite, or talc and is then various shades of grayish-green.

Occurrence. Cordierite is found as an accessory mineral in granite, gneiss, schists, and in contact metamorphic zones. Notable localities for its occurrence are Bavaria, Finland, Greenland, and Madagascar. Gem material has come from Ceylon. In the United States it is found chiefly in Connecticut and New Hampshire.

Use. Transparent cordierite of good color has been used as a gem known by jewelers as *saphir d'eau* or *dichroite.*

Name. After the French geologist P. L. A. Cordier (1777–1861). Iolite is sometimes used as a synonym.

TOURMALINE

Crystallography. Hexagonal—R; $3m$. Usually in prismatic crystals with a prominent trigonal prism and subordinate second-order hexagonal prism vertically striated. The prism faces may round into each other giving the crystals a cross section like a spherical triangle. When doubly terminated, crystals usually show different forms at the opposite ends of the vertical axis (Fig. 385). May be massive compact; also coarse to fine columnar, either radiating or parallel.

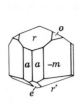  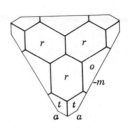

FIG. 385. Tourmaline.

Angles: $r(10\bar{1}1) \wedge r'(\bar{1}101) = 46° 52'$, $-m(01\bar{1}0) \wedge o(02\bar{2}1) = 44° 3'$, $r(10\bar{1}1) \wedge$
$m(10\bar{1}0) = 62° 40'$.

$R3m$; $a = 15.95$, $c = 7.24$ Å; $a;c = 1:0.448$; $Z = 3$. *d's:* 4.24(7), 4.00(7), 3.51(7), 2.98(9), 2.58(10).

Physical Properties. H $7-7\frac{1}{2}$. G 3.0–3.25. Luster vitreous to resinous. *Color* varied, depending upon the composition. *Fracture* conchoidal.

Black, iron-bearing tourmaline (*shortlite*) is most common; brown tourmaline (dravite) contains magnesium. The rarer lithium-bearing varieties (elbaite) are light colored in fine shades of green (verdelite), yellow, red-pink (rubellite), and blue (indicolite). Rarely white or colorless *achroite*. A single crystal may show several different colors arranged either in concentric bands about the center of the crystal or in layers transverse to the length. Strongly pyroelectric and piezo-electric. *Optics:* $(-)$; $\omega = 1.635–1.675$, $\varepsilon = 1.610–1.650$. Some varieties strongly pleochroic, $O > E$ (see page 146).

Composition. A complex silicate of boron and aluminum, whose composition can be expressed by the general formula $XY_3Al_6(BO_3)_3(Si_6O_{18})(OH)_4$. X is usually Na but may be replaced by Ca, and $Y = Fe^2$, Mg, Li, Al.

Diagnostic Features. Usually recognized by the characteristic rounded triangular cross section of the crystals and conchoidal fracture. Distinguished from hornblende by absence of cleavage. Fusibility varies with composition: magnesium varieties fusible at 3; iron varieties fusible with difficulty; lithium varieties infusible. Fused with boron flux gives momentary green flame of boron.

Occurrence. The most common and characteristic occurrence of tourmaline is in granite pegmatites and in the rocks immediately surrounding them. It is found also as an accessory mineral in igneous rocks and in metamorphic rocks such as gneisses and schists. Most pegmatitic tourmaline is black and is associated with the common pegmatite minerals, microcline, albite, quartz, and muscovite. Pegmatites are also the home of the light colored lithium-bearing tourmalines which are frequently associated with lepidolite, beryl, apatite, fluorite, and rarer

FIG. 386. Tourmaline crystals with quartz and cleavelandite, Pala, California.

minerals (see Fig. 386). The brown magnesium-rich tourmaline is found in crystalline limestones.

Famous localities for the occurrence of the gem tourmalines are the Island of Elba; the state of Minas Gerais, Brazil; Ural Mountains near Sverdlovsk; and Madagascar. In the United States found at Paris and Auburn, Maine; Chesterfield, Massachusetts; Haddam Neck, Connecticut; and Mesa Grande, Pala, Rincon, and Ramona in San Diego County, California. Brown crystals are found near Gouverneur, New York, and fine black crystals at Pierrepont, New York.

Use. Tourmaline forms one of the most beautiful of the semiprecious gem stones. The color of the stones varies, the principal shades being olive-green, pink to red, and blue. Sometimes a stone is so cut as to show different colors in different parts. The green colored stones are usually known by the mineral name, tourmaline, or as *Brazilian emeralds*. The red or pink stones are known as *rubellite*, and the rarer dark blue stones are called *indicolite*.

Because of its strong piezoelectric property, tourmaline is used in the manufacture of pressure gauges to measure transient blast pressures.

Name. *Tourmaline* comes from *turamali*, a name given to the early gems from Ceylon.

CHRYSOCOLLA

Crystallography. Amorphous and thus in the strict sense cannot be considered a mineral. Massive compact; in some cases earthy. Individual specimens inhomogeneous.

Physical Properties. Fracture conchoidal. **H** 2–4. **G** 2.0–2.4. *Luster* vitreous to earthy. *Color* green to greenish-blue; brown to black when impure. Refractive index 1.40±.

Composition. Chrysocolla is a hydrogel or gelatinous precipitate and shows a wide range in composition. Chemical analyses report: CuO 32.4–42.2, SiO_2 37.9–42.5, H_2O 12.2–18.8 per cent. In addition Al_2O_3 and Fe_2O_3 are usually present in small amounts.

Diagnostic Features. Characterized by its green or blue color and conchoidal fracture. Distinguished from turquoise by inferior hardness. Infusible. Gives a copper globule when fused with sodium carbonate on charcoal. In the closed tube darkens and gives water.

Occurrence. Chrysocolla forms in the oxidized zones of copper deposits associated with malachite, azurite, cuprite, native copper, etc. It is commonly found in the oxidized portions of porphyry copper ores. Found in the copper districts of Arizona and New Mexico and at Chuquicamata, Chile,

Use. A minor ore of copper.

Name. Chrysocolla, derived from two Greek words meaning *gold* and *glue*, which was the name of a similar-appearing material used to solder gold.

Similar Species. *Dioptase*, $Cu_6(Si_6O_{18})\cdot6H_2O$, is a rhombohedral hydrous copper silicate occurring in well-defined green rhombohedral crystals.

Inosilicates

SiO_4 tetrahedra may link into chains by sharing oxygens (Fig. 387). Such simple chains may be joined side by side by further sharing of oxygens in alternate tetrahedra to form bands or double chains (Fig. 388). In the simple chain structure, two of the four oxygens in each SiO_4 tetrahedron are shared giving a ratio of $Si:O = 1:3$. In the band structure half of the tetrahedra share three oxygens, the other half share two oxygens yielding a ratio of $Si:O = 4:11$.

Included in the inosilicates are two important rock-forming groups of minerals: the pyroxenes as single chain members and the amphiboles as double chain members. The structure of both may be visualized as parallel silicon-oxygen chains extending indefinitely in the c direction and bound together through cations, chiefly Ca, Mg, and Fe. The most characteristic property of both pyroxenes and amphiboles is their {110} cleavage. When the chains are viewed end on, that is, in the direction of c (Figs. 389 and 390), it can be seen how they account

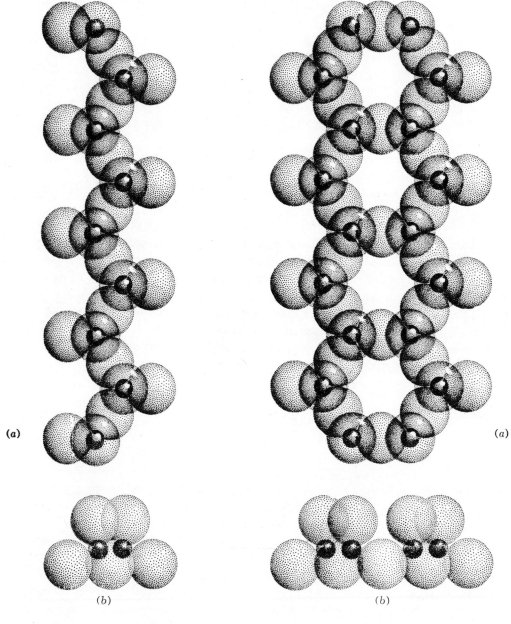

(a)

(b)

(a)

(b)

FIG. 387.   SiO₃. (a) Chain.
(b) Cross section.

FIG. 388.   Si₄O₁₁. (a) Chain. (b) Cross section.

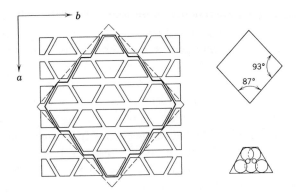

FIG. 389. Section of pyroxene structure at right angles to *c* axis showing how it controls cleavage (modified from Bragg).

for the cleavages: in pyroxenes $(110) \wedge (1\bar{1}0) \approx 93°$; in amphiboles $(110) \wedge (1\bar{1}0) \approx 56°$. The cleavage angles are the only important, easily determined distinguishing criteria.

Many similarities exist between the two groups in crystallographic, physical and chemical properties. Although most pyroxenes and amphiboles are monoclinic both groups have orthorhombic members. Also in both the repeat distance along the chains, that is, the *c* dimension of the unit cell, is approximately 5.3 Å. The *a* cell dimensions are also analogous but the *b* dimension of amphiboles, because of the double chain, is roughly twice that of the corresponding pyroxenes (Fig. 387 and 388).

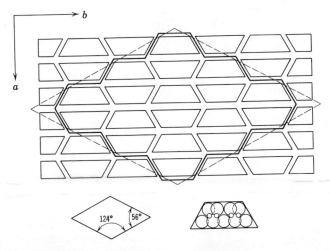

FIG. 390. Section of amphibole structure at right angles to cleavage showing how it controls cleavage (modified from Bragg).

The same cations are present in both groups; but the amphiboles are characterized by the presence of (OH), which is lacking in pyroxenes. Although the color, luster and hardness of analogous species are similar, the (OH) in amphiboles gives them, in general, slightly lower specific gravity and refractive indices than their pyroxene counterparts. Furthermore, the crystals have somewhat different habits. Pyroxenes commonly occur in stout prisms whereas amphiboles tend to form more elongated crystals, often acicular.

Pyroxenes crystallize at higher temperatures than their amphibole analogous and hence are generally formed early in a cooling igneous magma. If water is present in the melt, the early-formed pyroxene may react with the residual liquid at lower temperatures to form amphibole.

Pyroxene Group

The simple chain structure of the pyroxenes can be represented by the general formula: $XY(Si_2O_6)$. The parallel silicon-oxygen chains are bound together by ionic bonds through the X and Y cations. The X cations are large and weakly charged, generally Na or Ca in 8 coordination with oxygen. The Y cations are smaller, leading to 6 coordination with oxygen, and may be Mg, Fe^2, Fe^3, Al,

Inosilicates

PYROXENE GROUP
  Enstatite Series
    Enstatite $Mg_2(Si_2O_6)$
    Hypersthene $(Mg, Fe)_2(Si_2O_6)$
  Diopside Series
    Diopside $CaMg(Si_2O_6)$
    Hedenbergite $CaFe(Si_2O_6)$
    Augite $XY(Z_2O_6)$
  Spodumene Series
    Spodumene $LiAl(Si_2O_6)$
    Jadeite $NaAl(Si_2O_6)$
    Aegirine $NaFe(Si_2O_6)$

PYROXENOID GROUP
  Wollastonite $Ca(SiO_3)$
  Pectolite $Ca_2NaH(SiO_3)_3$
  Rhodonite $Mn(SiO_3)$

AMPHIBOLE GROUP
  Anthophyllite $(Mg,Fe)_7(Si_8O_{22})(OH)_2$
  Tremolite Series
    Tremolite $Ca_2Mg_5(Si_8O_{22})(OH)_2$
    Actinolite $Ca_2(Mg,Fe)_5(Si_8O_{22})(OH)_2$
  Hornblende Series $X_{2-3}Y_{5-7}Z_8O_{22}(O, OH, F)_2$
  Riebeckite Series
    Glaucophane $Na_2Mg_3Al_2(Si_8O_{22})(OH)_2$
    Riebeckite $Na_2Fe_3^2Fe_3^2(Si_8O_{22})(OH)_2$

Ions in Common Pyroxenes and Amphiboles

X	Y	Pyroxene	Amphibole
Mg	Mg	Enstatite Clinoenstatite	Anthophyllite Kupfferite
Mg, Fe	Mg, Fe	Bronzite, hypersthene Clinohypersthene	Anthophyllite Cummingtonite
Ca	Mg	Diopside	Tremolite
Ca	Fe	Hedenbergite	Actinolite
Ca	Mn	Johannssenite	
Na	Al	Jadeite	Glaucophane
Na	Fe^3	Aegirine	Riebeckite Arfvedsonite
Li	Al	Spodumene	
Ca, Na	Mg, Fe, Mn, Al, Fe^3, Ti	Augite	Hornblende

Mn^2 or Mn^3, or even Li or Ti. The introduction of an ion of greater or lesser charge may be compensated for by a simultaneous substitution, as Al for Si in the tetrahedral sites.

Pyroxenes are, in general, monoclinic when X and Y sites are occupied by large and small ions respectively. If, however, both X and Y sites are occupied by small ions, an orthorhombic symmetry results. This symmetry is produced by a twinlike reflection across (100) accompanied by a doubling of the a cell dimension. The occupation of both X and Y sites by larger ions may result in a triclinic lattice, as is the case in rhodonite, $Mn(SiO_3)$, and wollastonite, $Ca(SiO_3)$. All pyroxenes display the prominent near-90° {110} cleavage parallel to the SiO_3 chains (Fig. 391) and generally have prominent parting parallel to {100} and {001}.

In accord with the principles set forth above, we find that when both X and Y sites are occupied by Mg or Fe^2, as in enstatite-hypersthene, orthorhombic symmetry results. Mg and Fe^2 are free to substitute for each other in all proportions with random distribution resulting in a linear change of properties with composition. Under certain conditions of pressure and temperature, these compounds may form monoclinic polymorphs (clinoenstatite-clinohypersthene) with half the a-cell dimension of the orthorhombic modification.

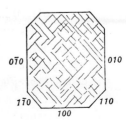

FIG. 391. Pyroxene cleavage.

When the X sites are occupied by Ca in 8 coordination and Y sites by Mg, Fe^2, or Mn^2 in 6 coordination, a member of the *diopside series* results. Within this series there can be substantially complete mutual substitution of Mg, Fe^2, and Mn^2, resulting in minor but nearly linear changes in cell dimensions and properties.

Occupation of the X sites by a moderate to large-sized monovalent alkali metal ion and the Y sites by a small trivalent cation, results in a member of the *spodumene series:* spodumene, jadeite, or aegirine. Solid solution, not only within this series but also between this series and the diopside series, is possible and gives rise to many varieties and variety names.

The common pyroxenes of the igneous and metamorphic rocks contain both Ca and Na in the X sites, Mg, Fe^2, Al, Fe^3, and some Ti^4 in the Y sites, as well as some Al substituting for Si in the tetrahedral sites. The proportions of these constituents vary depending on the environment of formation. The name "common augite" is used to cover pyroxenes of variable and complex composition of this kind.

ENSTATITE—$Mg_2(Si_2O_6)$;   HYPERSTHENE—$(Mg,Fe)_2(Si_2O_6)$

Crystallography. Orthorhombic; $2/m2/m2/m$. Prismatic habit, crystals rare. Usually massive, fibrous, or lamellar.

Angles: $(210) \wedge (2\bar{1}0) = 91° 44'$.

Pbca; $a = 18.22$, $b = 8.81$, $c = 5.21$ Å; $a:b:c = 2.068:1:0.590$; $Z = 8$. *d's:* 3.17(10), 2.94(4), 2.87(9), 2.53(4), 2.49(5).

Physical Properties. *Cleavage* {210} good. Because of the doubling of the a dimension in orthorhombic pyroxenes, the cleavage form is {210} rather than {110} as in monoclinic pyroxenes. Frequently good parting on {100}, less common on {001}. **H** $5\frac{1}{2}$–6. **G** 3.2–3.6 increasing with Fe content. *Luster* vitreous to pearly on cleavage surfaces; *bronzite* has submetallic, bronzelike luster. *Color* grayish, yellowish, or greenish-white to olive-green and brown. Translucent. *Optics:* enstatite (+); bronzite and hypersthene (−). $\alpha = 1.650$–1.715, $\beta = 1.653$–1.728, $\gamma = 1.658$–1.731 (enstatite-hypersthene). Indices increase with iron content; in orthoferrosilite $\beta = 1.785$ (see Fig. 392).

Composition. Ferrous iron may substitute for magnesium in all proportions up to nearly 90 per cent $FeSiO_3$. However, in the more common orthorhombic pyroxenes the ratio of Fe:Mg rarely exceeds 1:1. Pure enstatite, $Mg_2(Si_2O_6)$ contains MgO 40.0, SiO_2 60 per cent. If the amount of FeO lies between 5 and 13 per cent, the variety is called *bronzite;* if between 13 and 20 per cent, the mineral is called *hypersthene*. Aluminum may also be present. Pure $Fe_2(Si_2O_6)$, *orthoferrosilite*, does not occur in nature, but the varieties containing more iron than does hypersthene, are known.

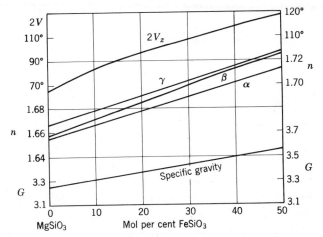

FIG. 392. Enstatite-hypersthene. Variation of refractive index, 2V, and specific gravity with composition.

Diagnostic Features. Usually recognized by its color, cleavage, and unusual luster. Varieties high in iron are black and difficult to distinguish from augite without optical tests. Almost infusible; thin edges will become slightly rounded. Fuses more easily with increasing amounts of iron.

Occurrence. The orthorhombic pyroxenes are common in pyroxenites, peridotites, gabbros, norites, and basalts, and are thus widespread minerals. Enstatite is a common mineral in both metallic and stony meteorites.

Enstatite is found in the United States at the Tilly Foster Mine, Brewster, New York, and at Edwards, St. Lawrence County, New York; at Texas, Pennsylvania; Bare Hills, near Baltimore, Maryland; and Webster, North Carolina. Hypersthene occurs in New York in the norites of the Cortland region, on the Hudson River, and in the Adirondack region.

Name. Enstatite is named from the Greek word meaning *opponent* because of its refractory nature. Hypersthene is named from two Greek words meaning *very* and *strong* because its hardness is greater than that of hornblende.

Similar Species. *Clinoenstatite* and *clinohypersthene* are monoclinic, dimorphs of $Mg_2(Si_2O_6)$ and $(Mg, Fe)_2(Si_2O_6)$ respectively.

DIOPSIDE—HEDENBERGITE—AUGITE

Diopside, $CaMgSi_2O_6$ and hedenbergite $CaFeSi_2O_6$, form a complete solid solution series with physical and optical properties varying linearly with composition. Augite, essentially $Ca(Mg,Fe,Al)(Si,Al)_2O_6$ can be considered an intermediate member in which Al substitutes for both Mg and Si. Although the crystal constants vary slightly from one member to another, a single description suffices for all.

Crystallography. Monoclinic; $2/m$. In prismatic crystals showing square or eight-sided cross section. Also granular massive, columnar, and lamellar. Frequently twinned polysynthetically on $\{001\}$; less commonly twinned on $\{100\}$ (see Fig. 393).

Angles: $m(110) \wedge m'(1\bar{1}0) = 92° 50'$, $c(001) \wedge p(111) = 33° 50'$, $s(\bar{1}11) \wedge s'(\bar{1}\bar{1}1) = 59° 11'$, $c(001) \wedge a(100) = 74° 10'$.

$C2/c$; $a = 9.73$, $b = 8.91$, $c = 5.25$ Å; $\beta = 105° 50'$; $a:b:c = 1.092:1:0.589$; $Z = 4$. d's: 3.23(8), 2.98(10), 2.94(7), 2.53(4), 1.748(4).

Physical Properties. *Cleavage* $\{110\}$ imperfect. Frequently parting on $\{001\}$, and less commonly on $\{100\}$ in the variety *diallage*. **H** 5–6. **G** 3.2–3.3. *Luster* vitreous. *Color* white to light green in diopside; deepens with increase of iron. Augite is black. Transparent to translucent. *Optics:* $(+)$; $\alpha = 1.66$–1.75, $\beta = 1.67$–1.73, $\gamma = 1.69$–1.75. $2V = 55°$–$65°$; $Y = b$, $Z \wedge c = 39°$–$48°$; $r > v$. Darker members show pleochroism: X pale green, Y yellow-green, Z dark green.

Composition. In the diopside-hedenbergite series Mg and Fe^2 substitute for each other in all proportion. In augite, in addition to varying Mg and Fe^2, Al substitutes for both Mg and Si and Na, Mn, Fe^3, and Ti may be present.

Diagnostic Features. Characterized by crystal form and imperfect prismatic cleavage at 87° and 93°. Fusible at 4 to a green glass. Insoluble in acids. With fluxes gives tests for calcium, magnesium, and iron.

Occurrence. Diopside is characteristically found as a contact metamorphic mineral in crystalline limestones and dolomites. In such deposits it is associated with tremolite, scapolite, idocrase, garnet, and sphene. It and hedenbergite are also found in regionally metamorphosed rocks. The variety diallage is frequently found in gabbros, peridotites, and serpentines.

Fine crystals have been found in the Ural Mountains, U.S.S.R.; Austrian Tyrol; Binnenthal, Switzerland; and Piedmont, Italy. At Nordmark Sweden, fine crystals

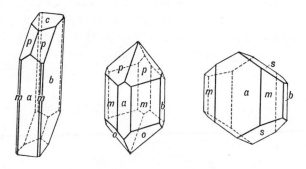

FIG. 393. Augite crystals.

range between diopside and hedenbergite. In the United States they are found at Canaan, Connecticut; and DeKalb Junction and Gouverneur, New York.

Augite is the most common pyroxene and an important rock-forming mineral. It is found chiefly in the dark-colored igneous rocks, such as basaltic lavas and intrusives, andesites, gabbros and peridotites. Some notable localities, particularly for fine crystals of augite are: the lavas of Vesuvius; at Val di Fassa, Trentino, Italy; and Bilin, Bohemia.

Use. Transparent varieties of diopside have been cut and used as gem stones.

Name. Diopside is from two Greek words meaning *double* and *appearance*, since the vertical prism zone can apparently be oriented in two ways.

Augite comes from a Greek word meaning *luster*. The name pyroxene, *stranger to fire*, is a misnomer and was given to the mineral because it was thought that it did not occur in igneous rocks.

SPODUMENE—$LiAl(Si_2O_6)$

Crystallography. Monoclinic; $2/m$. Crystals prismatic, frequently flattened on $\{100\}$. Deeply striated vertically. Crystals usually coarse with roughened faces; some very large. Occurs also in cleavable masses. Twinning on $\{100\}$ common.

$C2/c$; $a = 9.52$, $b = 8.32$, $c = 5.25$ Å; $\beta = 110° 28'$, $a:b:c = 1.144:1:0.631$; $Z = 4$. d's: 4.38(5), 4.21(6), 2.93(10), 2.80(8), 2.45(6).

Physical Properties. Cleavage $\{110\}$ perfect at angles of 87° and 93°. Usually a well-developed parting on $\{100\}$. **H** $6\frac{1}{2}$–7. **G** 3.15–3.20. *Luster* vitreous. *Color* white, gray, pink, yellow, green. Transparent to translucent. *Optics:* $(+)$; $\alpha = 1.660$, $\beta = 1.666$, $\gamma = 1.676$; $2V = 58°$; $Y = b$, $Z \wedge c = 24°$. $r < v$. Absorption: $X > Y > Z$.

The clear lilac-colored variety is called *kunzite*, and the clear emerald-green variety *hiddenite*.

Composition. Li_2O 8.0, Al_2O_3 27.4, SiO_2 64.6 per cent. A small amount of sodium usually substitutes for lithium.

Diagnostic Features. Fusible at $3\frac{1}{2}$, throwing out fine branches and then fusing to a clear glass. Gives crimson flame (lithium). Insoluble. Characterized by its prismatic cleavage and $\{100\}$ parting. The angle formed by one cleavage direction and the $\{100\}$ parting resembles the cleavage angle of tremolite. A careful angular measurement or a lithium flame is necessary to distinguish them.

Alteration. Spodumene very easily alters to other species, becoming dull. The alteration products include clay minerals, albite, eucryptite, $LiAl(SiO_4)$, muscovite, microcline.

Occurrence. Spodumene is a comparatively rare mineral found almost exclusively in lithium-rich pegmatites. It occasionally forms very large crystals.

Occurs in Goshen, Chesterfield, Huntington, and Sterling, Massachusetts; Branchville, Connecticut; Newry, Maine; and Dixon, New Mexico, At the Etta mine, Black Hills, South Dakota, crystals measuring as much as 40 feet in length and weighing many tons have been found. Spodumene is mined as a source of lithium in the Black Hills, South Dakota; Kings Mountain district, North Carolina; and Gunnison County, Colorado. Hiddenite occurs with emerald beryl at Stony Point, Alexander County, North Carolina. Kunzite is found with pink beryl at Pala, San Diego County, California, and at various localities in Madagascar and Brazil.

Use. As a source of lithium. Lithium's largest use is in grease to which it is added to help them retain their lubricating properties over a wide range of temperatures. It is used in ceramics, storage batteries, air conditioning, and as a welding flux. The gem varieties of spodumene, hiddenite and kunzite furnish very beautiful gem stones but are limited in their occurrence.

Names. *Spodumene* comes from a Greek word meaning *ash colored. Hiddenite* is named for W. E. Hidden, *kunzite* for G. F. Kunz.

Similar Species. Eucryptite, $LiAl(SiO_4)$ is a common alteration product of spodumene.

Jadeite—$NaAl(Si_2O_6)$

Crystallography. Monoclinic; $2/m$. Rarely in isolated crystals. Usually fibrous in compact massive aggregates.

$C2/c$; $a = 9.50$, $b = 8.61$, $c = 5.24$ Å; $\beta = 107°\ 26'$; $a:b:c = 1.103:1:0.609$; $Z = 4$. d's: 3.27(3), 3.10(3), 2.92(8), 2.83(10), 2.42(9).

Physical Properties. *Cleavage* $\{110\}$ at angles of 87° and 93°. Extremely tough and difficult to break. **H** $6\frac{1}{2}$–7. **G** 3.3–3.5. *Color* apple-green to emerald-green, white. May be white with spots of green. *Luster* vitreous, pearly on cleavage surfaces. *Optics:* (+); $\alpha = 1.654$, $\beta = 1.659$, $\gamma = 1.667$; $2V = 70°$; $X = b$, $Z \wedge c = 34°$; $r > v$.

Composition. Na_2O 15.4, Al_2O_3 25.2, SiO_2 59.4 per cent. Contains some Fe^3, Ca, Mg.

Diagnostic Features. Fuses at $2\frac{1}{2}$ to a transparent blebby glass. Insoluble in acids. Characterized by its green color and tough aggregates of compact fibers. Distinguished from nephrite by its ease of fusion. On polished surfaces nephrite has an oily luster, jadeite is vitreous.

Occurrence. Jadeite occurs in large masses in serpentine apparently formed by the metamorphism of a nepheline-albite rock. It is found chiefly in eastern Asia in Upper Burma. It is also found in Tibet and southern China.

Use. Jadeite has long been highly prized in the Orient, especially in China, where it is worked into ornaments and utensils of great variety and beauty. It was also used by early man for various weapons and implements.

Name. The term *jade* includes both jadeite and the amphibole, nephrite.

Aegirine—NaFe(Si$_2$O$_6$)

Crystallography. Monoclinic; $2/m$. Crystals slender prismatic with steep terminations. Often in fibrous aggregates. Faces often imperfect.

$C2/c$; $a = 9.66$, $b = 8.79$, $c = 5.26$ Å; $\beta = 107°\ 20'$, $a:b:c = 1.099:1:0.598$; $Z = 4$. d's: 6.5(4), 4.43(4), 2.99(10), 2.91(4), 2.54(6).

Physical Properties. *Cleavage* {110} imperfect at angles of 87° and 93°. **H** 6–6$\frac{1}{2}$. **G** 3.40–3.55. *Luster* vitreous. *Color* brown or green. Translucent. *Optics* for aegirine: (−); $\alpha = 1.776$, $\beta = 1.819$, $\gamma = 1.836$; 2V $= 60°$; $Y = b$, $Z \wedge c = 8°$; $r > v$. Aegirine-augite: (+) with lower indices and pleochroism in greens and brown.

Composition. NaFe(Si$_2$O$_6$). The mineral with this exact composition is known as *acmite* but is rarely found. Some Ca substitutes for Na, some Mg and Al for Fe^3, and a complete series (aegirine-augite) extends to augite. Some vanadium may be present.

Diagnostic Features. Fusible at 3, giving yellow sodium flame. Fused globule slightly magnetic. With fluxes gives tests for iron. The slender prismatic crystals, brown to green color, and associations are characteristic. However, it is not easily distinguished without optical tests.

Occurrence. Aegirine is a comparatively rare rock-forming mineral found chiefly in rocks rich in sodium and poor in silica, such as nepheline syenite and phonolite. Associated with orthoclase, feldspathoids, augite, and soda-rich amphiboles. It occurs in the nepheline syenites and related rocks of Norway; southern Greenland; Kola Peninsula, the U.S.S.R. In the United States it is found in fine crystals at Magnet Cove, Arkansas. Found in Montana at Libby and in the Highwood and Bear Paw Mountains.

Name. From *Aegir*, the Icelandic god of the sea.

Pyroxenoid Group

There are a number of silicate minerals that have, as do the pyroxenes, a ratio of Si:O $= 1:3$ but do not have the pyroxene structure. These are known as the pyroxenoids, three of which—wollastonite, pectolite, and rhodonite—are considered here.

The single chains of linked SiO$_4$ tetrahedra that in pyroxenes extend indefinitely in the c direction are not present in the pyroxenoids. In wollastonite and pectolite there is a different type of chain structure which is parallel to the b axis. A chain is made up of an Si$_2$O$_7$ group, as in the sososilicates, alternating with and linked to a single SiO$_4$ tetrahedron. The chain structure of these minerals is reflected in their fibrous habit and splintery cleavages. The structure of rhodonite is similar with chains parallel to b. But in this case the chain is composed of 2 Si$_2$O$_7$ groups alternating with a single SiO$_4$ tetrahedron.

WOLLASTONITE—Ca(SiO₃)

Crystallography. Triclinic; $\bar{1}$. Rarely in tabular crystals, with either {001} or {100} prominent. Commonly massive, cleavable to fibrous; also compact.

Pseudowollastonite, Ca(SiO₃), is formed above 1200°C; it is triclinic, pseudo-hexagonal with different properties from wollastonite.

P$\bar{1}$; $a = 7.94$, $b = 7.32$, $c = 7.07$; $\alpha = 90° 2'$, $\beta = 95° 22'$, $\gamma = 103° 26'$; $a:b:c = 1.084:1:0.966$, $Z = 6$. *d's:* 3.83(8), 2.52(8), 3.31(8), 2.97(10), 2.47(6).

Physical Properties. *Cleavage* {100} perfect, {001} and {$\bar{1}$01} good giving splintery fragments elongated on b. **H** 5–5$\frac{1}{2}$. **G** 2.8–2.9. *Luster* vitreous, pearly on cleavage surfaces. May be silky when fibrous. *Color* colorless, white, or gray. Translucent. *Optics:* (−); $\alpha = 1.670$, $\beta = 1.632$, $\gamma = 1.634$; 2V = 40°; *Y* near *b*, $X \wedge c = 32°$.

Composition. CaO 48.3, SiO₂ 51.7 per cent.

Diagnostic Features. Fusible at 4 to a white, almost glassy globule. Decomposed by hydrochloric acid, with the separation of silica but without the formation of a jelly. Characterized by its two perfect cleavages at about 84°. It resembles tremolite but is distinguished from it by the cleavage angle and solubility in acid.

Occurrence. Wollastonite occurs chiefly as a contact metamorphic mineral in crystalline limestones. It is associated with calcite, diopside, andradite and grossularite garnet, tremolite, calcium feldspars, idocrase, and epidote.

In places it may be so plentiful as to constitute the chief mineral of the rock mass. Such wollastonite rocks are found in the Black Forest, in Brittany; Willsboro, New York; in California; and Mexico. Crystals of the mineral are found at Csiklova in Rumania; Harz Mountains, Germany; and Chiapas, Mexico. In the United States found in New York at Diana, Lewis County, and also in Orange and St. Lawrence counties. In California at Crestmore, Riverside County.

Use. Wollastonite is mined in those places where it constitutes a major portion of the rock mass and is used in the manufacture of tile.

Name. In honor of the English chemist, W. H. Wollaston (1766–1828).

Pectolite—Ca₂NaH(SiO₃)₃

Crystallography. Triclinic; $\bar{1}$. Crystals elongated parallel to the *b* axis. Usually in aggregates of acicular crystals. Frequently radiating, with fibrous appearance. In compact masses.

P$\bar{1}$; $a = 7.99$, $b = 7.04$, $c = 7.02$ Å; $\alpha = 90° 31'$, $\beta = 95° 11'$, $\gamma = 102° 28'$; $a:b:c = 1.135:1:0.997$; $Z = 2$. *d's:* 3.28(7), 3.08(9), 2.89(10), 2.31(7), 2.28(7).

Physical Properties. *Cleavage* {001} and {100} perfect. **H** 5. **G** 2.8±. *Luster* vitreous to silky. *Color* colorless, white, or gray. Transparent. *Optics:* (+); $\alpha = 1.595$, $\beta = 1.604$, $\gamma = 1.633$; 2V = 60°; $Z \approx b$, $X \wedge c = 19°$.

Composition. CaO 33.8, Na$_2$O 9.3, SiO$_2$ 54.2, H$_2$O 2.7 per cent. In some pectolite Mn substitutes for Ca.

Diagnostic Features. Fuses quietly at $2\frac{1}{2}$–3 to a glass coloring the flame yellow (sodium). Decomposed by hydrochloric acid, with the separation of silica but without the formation of a jelly. Water in the closed tube. Characterized by two directions of perfect cleavage yielding sharp acicular fragments that will puncture the skin if not handled carefully. Resembles wollastonite but gives test for water and sodium. Distinguished from similar-appearing zeolites by absence of aluminum.

Occurrence. Pectolite is a secondary mineral similar in its occurrence to the zeolites. Found lining cavities in basalt, associated with various zeolites, prehnite, calcite, etc. Found at Bergen Hill and West Paterson, New Jersey.

Name. From the Greek word meaning *compact*, in allusion to its habit.

RHODONITE—Mn(SiO$_3$)

Crystallography. Triclinic; $\bar{1}$. Crystallographically closely related to the pyroxenes. Crystals commonly tabular parallel to {001} (Fig. 394); often rough with rounded edges. Commonly massive, cleavable to compact; in imbedded grains.

Angles: $m(110) \wedge M(1\bar{1}0) = 92° 29'$, $a(100) \wedge c(001) = 72° 37'$, $c(001) \wedge n(\bar{2}\bar{2}1) = 73° 52'$. These angles are given for the traditional morphological orientation with the intersection of the cleavages defining c, as in the pyroxenes. The cell parameters refer to a more recent orientation.

$P\bar{1}$; $a = 7.79$, $b = 12.47$, $c = 6.75$ Å; $\alpha = 85° 10'$, $\beta = 94° 4'$, $\gamma = 111° 29'$; $a:b:c = 0.625:1:0.541$; $Z = 10$. $d's$: 3.08(6), 2.97(9), 2.94(10), 2.76(8), 2.60(8).

Physical Properties. *Cleavage* {110} and {1$\bar{1}$0} perfect. **H** $5\frac{1}{2}$–6. **G** 3.4–3.7. *Luster* vitreous. *Color* rose-red, pink, brown; frequently with black exterior of manganese oxide. Transparent to translucent. *Optics:* (+); $\alpha = 1.716$–1.733, $\beta = 1.720$–1.737, $\gamma = 1.728$–1.747; $2V = 60°$–$75°$, $r < v$.

Composition. MnO 54.1, SiO$_2$ 45.9 per cent. Ca and Fe substitute in part for Mn, and in the variety *fowlerite*, Zn substitutes for Mn.

Diagnostic Features. Fusible at 3 to a nearly black glass. Insoluble in hydrochloric acid. Gives a blue-green color to the sodium carbonate bead. Characterized

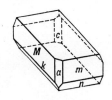

FIG. 394. Rhodonite.

by its pink color and near 90° cleavages. Distinguished from rhodochrosite by its greater hardness and insolubility.

Occurrence. Rhodonite is found at Långban, Sweden, with other manganese minerals and iron ore; in large masses near Sverdlovsk in the Ural Mountains, U.S.S.R.; and from Broken Hill, New South Wales. In the United States *fowlerite*, occurs in good-sized crystals in crystalline limestone with franklinite, willemite, zincite, etc., at Franklin, New Jersey.

Use. Some rhodonite is polished for use as an ornamental stone. This material is obtained chiefly from the Ural Mountains.

Name. Derived from the Greek word for *a rose*, in allusion to the color.

Similar Species. *Tephroite*, Mn_2SiO_4, is a red to gray mineral associated with rhodonite.

Amphibole Group

The *amphiboles*, constitute a mineral family closely paralleling the pyroxenes. All the members have as their basic structure the Si_4O_{11} double chain parallel to *c* defining the perfect {110} cleavage (Fig. 395). The chains are bound together, as in the pyroxenes, by ionic bonds through two types of cation sites, generally occupied by ions of *X* and *Y* type, as defined above. OH ions occupy the open spaces resulting from the joining of single chains side by side to form the double chain. A geometry of arrangement is such that the larger *X* cations are in 8 coordination, whereas the smaller *Y* cations are in 6 coordination. Consistent with the requirement of electrical neutrality a general formula for the amphiboles may be written as:

$$X_{0-7}Y_{7-14}Z_{16}O_{44}(OH)_4$$

There is considerable latitude in the interpretation of even so general a formula. *X* ions are generally Na and Ca with minor K. *Y* ions include Mg, Fe^2, and Fe^3, Al, Mn^2, Mn^3, and Ti as in the pyroxenes. Essentially complete ionic substitution may take place between Na and Ca and among Mg, Fe^2, and Mn^2. There is limited substitution between Fe^3 and Al and between Ti and other *Y*-type ions; and partial substitution of Al for Si in the *Z*-type sites within the double chains, as required for valence compensation and electrical neutrality. Partial substitution of F and O for OH in the hydroxyl sites is also common. In the common rock-forming amphibole hornblende, many if not all of the above substitution possibilities take place leading to a complex formula.

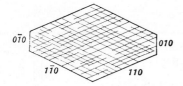

FIG. 395. Amphibole cleavage.

Some amphiboles, like some pyroxenes, are dimorphous, having orthorhombic and monoclinic members. The orthorhombic members have a structure which may be derived from the monoclinic structure by reflection across (100) and accordingly show doubling of the a cell dimension. This doubling of the unit cell occurs in those members in which both X- and Y-type sites are occupied by the smaller cations such as Mg and Fe^2, thus producing amphiboles analogous to the enstatite series of pyroxenes. These amphiboles are assigned to the *anthophyllite series*, and their monoclinic dimorphs to the *cummingtonite series*. The resemblance between the pyroxene and amphibole analogues can not be carried further, for it has been shown that the limits of ionic substitution are not identical in anthophyllite and cummingtonite, whereas complete solid solution is possible in both enstatite-ferrosilite and clinoenstatite-clinoferrosilite.

Anthophyllite—$(Mg,Fe)_7(Si_8O_{22})(OH)_2$

Crystallography. Orthorhombic; $2/m2/m2/m$. Rarely in distinct crystals. Commonly lamellar or fibrous.

Pnma; $a = 18.56$, $b = 18.08$, $c = 5.28$ Å; $a:b:c = 1.027:1:0.292$; $Z = 4$. *d's:* 8.26(6), 3.65(4), 3.24(6), 3.05(10), 2.84(4).

Physical Properties. *Cleavage* {210} perfect; (210) $\wedge$ $(2\bar{1}0) = 55°$. **H** $5\frac{1}{2}$–6. **G** 2.85–3.2. *Luster* vitreous. *Color* gray to various shades of green and brown. *Optics:* $(-)$; $\alpha = 1.60$–1.69, $\beta = 1.61$–1.71, $\gamma = 1.62$–1.72. 2V = 70°–100°. Absorption $Z > Y$ and X. Indices increase with Fe content.

Composition. *Gedrite* is a variety in which in addition of the substitution of Fe^2 for Mg, Al substitutes for both (Mg, Fe^2) and Si.

Diagnostic Features. Characterized by its clove-brown color but unless in crystals cannot be distinguished from other amphiboles without optical or x-ray tests. Fusible at 5 to a black magnetic enamel. Insoluble in acids. Yields water in the closed tube.

Occurrence. Anthophyllite is a comparatively rare mineral, occurring in the crystalline schists and thought to have been derived from the metamorphism of olivine. Occurs at Kongsberg, Norway; in many localities in southern Greenland. In the United States it is found at several localities in Pennsylvania and Montana and at Franklin, North Carolina.

Use. *Amosite*, an iron-rich anthophyllite, occurs in long, flexible fibers and is used as asbestos. It is mined for this purpose in South Africa.

Name. From the Latin *anthophyllum*, meaning *clove*, in allusion to the clove-brown color.

Similar Species. *Cummingtonite*, a monoclinic amphibole, has the same composition as anthophyllite. It is difficult to distinguish from anthophyllite but usually contains more iron and has a higher specific gravity.

TREMOLITE—ACTINOLITE—$Ca_2(Mg,Fe)_5(Si_8O_{22})(OH)_2$

Crystallography. Monoclinic; $2/m$. Crystals usually prismatic. The termination of the crystals is almost always formed by the two faces of a low first-order prism (similar to hornblende, see Fig. 396). Tremolite is often bladed and frequently in radiating columnar aggregate. In some cases in silky fibers. Coarse to fine granular, Compact.

$Cm/2$; $a = 9.84$, $b = 18.05$, $c = 5.28$; $\beta = 104° 42'$; $a:b:c = 0.545:1:0.293$; $Z = 2$. d's: 8.38(10), 3.27(8), 3.12(10), 2.81(5), 2.71(9).

Physical Properties. *Cleavage* {110} perfect; (110) $\wedge$ ($1\bar{1}0$) = 56° often yielding a splintery surface. **H** 5–6. **G** 3.0–3.3. *Luster* vitreous; often with silky sheen in the prism zone. *Color* varying from white to green in actinolite. The color deepens and the specific gravity increases with increase in iron content. Transparent to translucent. A felted aggregate of tremolite fibers goes under the name of *mountain leather* or *mountain cork*. A tough, compact variety which supplies much of the material known as *jade* is called *nephrite* (see Jadeite). *Optics:* ($-$); $\alpha = 1.56–1.63$, $\beta = 1.61–1.65$, $\gamma = 1.63–1.66$; $2V = 80°$; $Y = b$, $Z \wedge c = 15°$. Indices increase with iron content.

Composition. Tremolite usually contains some Fe^2 substituting for Mg. When Fe^2 is present in amounts greater than 2 per cent, the mineral is called *actinolite*.

Diagnostic Features. Characterized by slender prisms and good prismatic cleavage. Distinguished from pyroxenes by the cleavage angle and from hornblende by lighter color. Fusible at 4 to a clear glass.

Occurrence. Tremolite is most frequently found in metamorphosed dolomitic limestones. It is also found in talc schists. Actinolite commonly occurs in the crystalline schists, and is often the chief constituent of green-colored schists and greenstones.

Notable localities for crystals of tremolite are: Ticino, Switzerland; in the Tyrol; and in Piedmont, Italy. In the United States from Russell, Gouverneur, Amity, Pierrepont, DeKalb, and Edwards, New York. Crystals of actinolite are

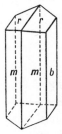
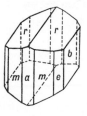

FIG. 396. Hornblende crystals.

found at Greiner, Zillerthal, Tyrol. Tremolite and actinolite frequently are fibrous and is the material to which the name *asbestos* was originally given. The asbestos forms have been found in the metamorphic rocks in various states along the Appalachian Mountains. The original jade was probably nephrite that came from the Khotan and Yarkand mountains of Chinese Turkestan. Nephrite has been found in New Zealand, Mexico, and Central America; and, in the United States, near Lander, Wyoming.

Use. The fibrous varieties are used as asbestos, but fibrous serpentine, chrysotile, is more widely used and usually a better grade. The compact variety, nephrite, is used as ornamental material.

Names. Tremolite is derived from the Tremola Valley near St. Gothard, Switzerland. Actinolite comes from two Greek words meaning *a ray* and *stone*, in allusion to its frequently somewhat radiated habit.

HORNBLENDE

Crystallography. Monoclinic; $2/m$. Crystals prismatic, usually terminated by $\{011\}$. The vertical prism zone shows, in addition to the prism faces, usually $\{010\}$, and more rarely $\{100\}$. May be columnar or fibrous; coarse to fine grained. (See Fig. 396.)

Angles: $m(110) \wedge m'(1\bar{1}0) = 55° 49'$, $r(011) \wedge r'(0\bar{1}1) = 31° 32'$, $b(010) \wedge e(130) = 32° 11'$, $c(001) \wedge a(100) = 74° 16'$.

$C2/m$; $a = 9.87$, $b = 18.01$, $c = 5.33$ Å; $\beta = 105° 44'$. $a:b:c = 0.548:1:0.296$; $Z = 2$. d's: 3.38(9), 3.29(7), 3.09(10), 2.70(10), 2.59(7).

Physical Properties. Cleavage $\{110\}$ perfect. **H** 5–6. **G** 3.0–3.4. *Luster* vitreous; fibrous; fibrous varieties often silky. *Color* various shades of dark green to black. Translucent; will transmit light on thin edges. *Optics:* $(-)$; $\alpha = 1.62–1.71$, $\beta = 1.62–1.71$, $\gamma = 1.63–1.73$; $2V = 30°–90°$; $Y = b$, $Z \wedge c = 15°–34°$. $r > v$ usually; may be $r < v$. Pleochroism X yellow-green, Y olive green, Z deep green.

Composition. What passes under the name of hornblende is in reality a complex series that varies with respect to the ratios of Ca:Na, Mg:Fe^2, Al:Fe^3, Al:Si, and OH:F. A formula for common hornblende is: $Ca_2Na(Mg, Fe^2)_4(Al, Fe^3, Ti)(Al, Si)_8O_{22}(O, OH)_2$.

Diagnostic Features. Crystal form and cleavage angle serve to distinguish hornblende from dark pyroxenes. Usually distinguished from other amphiboles by its dark color. Fusible at 4. Yields water in the closed tube.

Occurrence. Hornblende is an important and widely distributed rock-forming mineral, occurring in both igneous and metamorphic rocks; it is particularly characteristic of medium-grade metamorphic rocks. It characteristically alters from pyroxene both during the late magmatic stages of crystallization of igneous rocks and during metamorphism. Hornblende is common in syenites and diorites and is the chief constituent of the rock, *amphibolite.*

Name. From an old German word for any dark prismatic mineral occurring in ores but containing no recoverable metal.

Similar Species. Because of the variation in the hornblende series, many names have been proposed for members on the basis of chemical composition and physical and optical properties. The most common names on the basis of chemical composition are: *edenite, pargasite, hastingsite, kaersutite.*

Glaucophane—$Na_2Mg_3Al_2(Si_8O_{22})(OH)_2$;

Riebeckite—$Na_2Fe_3{}^2Fe_2{}^3(Si_8O_{22})(OH)_2$

Crystallography. Monoclinic; $2/m$. In slender acicular crystals; frequently aggregated; sometimes asbestiform.

$C2/m$; $a = 9.58$, $b = 17.80$, $c = 5.30$ Å; $\beta = 103° 48'$; $a:b:c = 0.538:1:0.298$; $Z = 2$. d's: 8.42(10), 4.52(5), 3.43(6), 3.09(8), 2.72(10).

Physical Properties. *Cleavage* {110} perfect. **H** 6. **G** 3.1–3.3. *Luster* vitreous. *Color* blue to lavender-blue to black, darker with increasing iron content. *Streak* white to light blue. Translucent. *Optics:* (−); $\alpha = 1.61–1.70$, $\beta = 1.62–1.71$, $\gamma = 1.63–1.72$; $2V = 40°–90°$; $Y = b$, $Z \wedge c = 8°$. Pleochroism in blue $X < Y < Z$.

Composition. In the series between the end-members the substitution of Fe^2 for Mg is limited but there may be complete substitution of Fe^3 for Al.

Diagnostic Features. Glaucophane and riebeckite are characterized by their fibrous habit and blue color. Fusible at 3.

Occurrence. Glaucophane is found only in metamorphic rocks such as schist, eclogite, and marble. It is a major constituent of the glaucophane schists of the California Coast Ranges. Riebeckite is present in some metamorphic rocks but is also found in sodium-rich igneous rocks and related pegmatites. It is a conspicuous mineral in the granite of Quincy, Massachusetts. The asbestiform variety of riebeckite, crocidolite, occurs in South Africa in a metamorphosed ironstone and is mined on a large scale. It is also mined in Australia and Brazil.

Use. Crocidolite makes up about 4 per cent of the total world production of asbestos. In many places in South Africa there has been a pseudomorphous replacement of crocidolite by quartz. The preservation of the fibrous texture makes it an attractive ornamental material with a chatoyancy. It is widely used for jewelry under the name of *tiger eye.*

Name. Glaucophane is from the two Greek words meaning *bluish* and *to appear.*

Similar Species. *Arfvedsonite*, $Na_3Mg_4Al(Si_8O_{22})(OH,F)$, is a deep green rock-forming amphibole similar to riebeckite in occurrence.

Phyllosilicates

As the derivation of the name of this important group implies (Greek: *phyllon*, leaf), all its many members have platy or flaky habit and one prominent cleavage. They are generally soft, of relatively low specific gravity and may show flexibility or even elasticity of the cleavage lamellae. All these characterizing peculiarities arise from the dominance in the structure of the indefinitely extended silicon-oxygen sheet. In this sheet, shown in Fig. 397 (packing model) and schematically in Fig. 398, three of the four oxygens in each SiO_4 tetrahedron are shared with neighboring tetrahedra, leading to a ratio of $Si:O = 2:5$. This structural unit is sometimes called the "siloxane sheet."

Most of the minerals of this class are hydroxyl bearing, and the structural peculiarities associated with the hydroxyl ion are of great importance in determining their properties. All the common phyllosilicates may be placed in one or the other of two clans, depending on the mode of coordination of the hydroxyl groups.

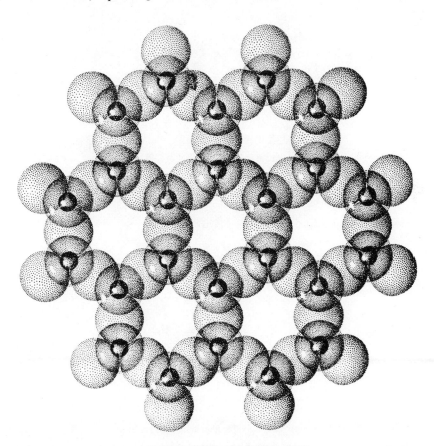

FIG. 397. Si_2O_5 **sheet structure.**

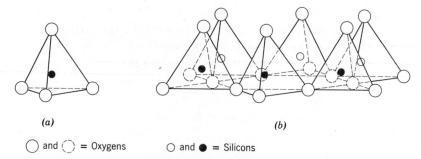

(a)   (b)

◯ and ◌ = Oxygens   ◯ and ● = Silicons

FIG. 398.   Diagrammatic sketch showing (a) single SiO$_4$ tetrahedron and (b) sheet structure of tetrahedra arranged in a hexagonal network. (From Ralph E. Grim, *Clay Mineralogy*, McGraw-Hill Book Co., 1953.)

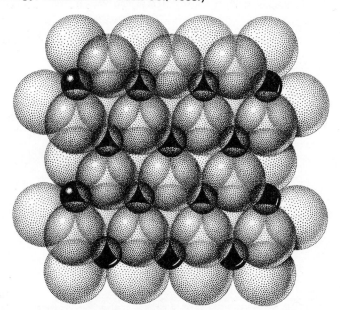

FIG. 399.   Brucite sheet. Large spheres hydroxyl, small spheres magnesium.

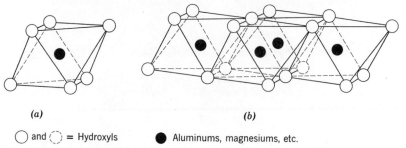

(a)   (b)

◯ and ◌ = Hydroxyls   ● Aluminums, magnesiums, etc.

FIG. 400.   Diagrammatic sketch showing (a) single octahedral unit and (b) sheet of octahedral units (after Grim).

424   Trioctahedral

One of these clans has as an integral part of its structure sheets of OH ions coordinated by magnesium ions, forming the *brucite*, $Mg_3(OH)_6$, structure. The brucite sheet (Fig. 399) consists of two layers of hydroxyl ions in hexagonal closest packing with magnesium ions occupying the interstices. The radius ratio of magnesium to hydroxyl is such that 6 coordination of hydroxyl about magnesium has maximum stability. Hence, the hydroxyl ions may be regarded as occupying the apices of a regular octahedron (Fig. 400a) with the magnesium ion at its center. The brucite sheet may then be visualized as made up of these octahedra tipped over and placed together so that certain of the (111) faces are coplanar (Fig. 400b). Figure 399 reveals that the magnesium ions in the resulting sheet form a pattern of interlocking hexagonal rings such that a magnesium ion falls at the center of each ring of six hydroxyl ions. Hydroxyl ions are shared between adjacent octahedra so that there are three magnesium ions for each octahedron of hydroxyl ions. This configuration is accordingly called the *trioctahedral* sheet and can accommodate divalent ions of such size that they may enter 6 coordination with hydroxyl.

The other clan is built about similar sheets of hydroxyl ions coordinated by aluminum ions in the *gibbsite*, $Al_2(OH)_6$, structure. Aluminum, like magnesium, forms a stable 6 coordination polyhedron with hydroxyl, but, because of the higher charge of the aluminum ion, only two-thirds as many aluminum ions may enter the sheet structure (Fig. 401). Hence, although the hydroxyl ions form double sheets with hexagonal closest packing, not all the interstices may be occupied. The

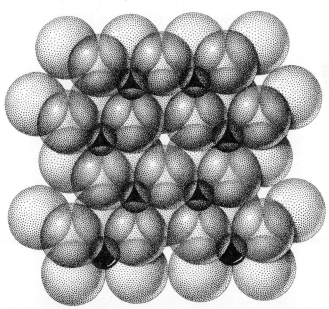

FIG. 401. Gibbsite sheet. Large spheres hydroxyl, small spheres aluminum. Note that aluminum occupies two-thirds of possible sites.

dioctahedral

aluminum ions form a pattern of hexagonal rings in which the site at the center of the ring is not occupied. This omission leads to a ratio of two aluminum ions to each octahedron of hydroxyl ions, and the configuration is called the *dioctahedral sheet*. This structure can accommodate only trivalent ions of a size appropriate to octahedral coordination with hydroxyl.

The structures of the common and important members of the phyllosilicate class can all be derived by combination of the Si_2O_5 sheet with either the gibbsite or the brucite sheet.

The simplest derivative structures formed in this way are those of the minerals of the kaolinite group and the antigorite, or serpentine, group. These structures are formed, respectively, by combination of a single gibbsite sheet with a siloxane sheet and of a single brucite sheet with a siloxane sheet. Both the brucite and gibbsite sheets are electrically neutral, and the minerals gibbsite and brucite are bound together only by the weak van der Waals' bond. The Si_2O_5 sheet, however, is not electrically neutral and cannot form a stable structure alone. If the apical oxygens, those not shared in the plane of the sheet, are permitted to enter the hydroxyl sites of the gibbsite or brucite sheet, the space requirements are fulfilled. Moreover, an electrically neutral structure results, since hydroxyl and oxygen have essentially the same ionic radius and the residual unit charge on the apical oxygen equals the hydroxyl charge. This structure, in the dioctahedral clan, is the kaolinite sheet (Figs. 402, 403); and in the trioctahedral clan, the antigorite sheet. From the first of these structural units the kaolinite group of clay minerals is built up, whereas from the second is built the group of the serpentines. Following the nomenclature of P.

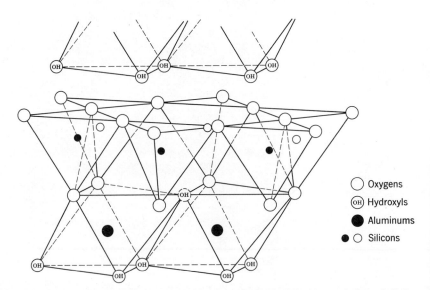

FIG. 402.   Diagrammatic sketch of the structure of the kaolinite layer (after Grim).

FIG. 403. Kaolinite, $Al_2Si_2O_5(OH)_4$, packing model. O white, (OH) gray, Al black; *c* axis vertical.

Niggli, we may call the siloxane sheet a "t" sheet, since the coordination is tetrahedral, and brucite or gibbsite sheets "o" sheets since the coordination is octahedral. Hence, their union builds a "$t{-}o$" sheet, and we may allude to the basic structural unit of either the kaolinite or serpentine groups as a "$t{-}o$" sheet. If we recall that the apical oxygens of a "t" sheet replace two hydroxyls of a gibbsite or brucite sheet, it is easy to remember that the formulas resulting in each case are

Dioctahedral

$$\overset{\text{``}o\text{''}}{Al_2(OH)_6} - (OH)_2 + \overset{\text{``}t\text{''}}{Si_2O_5} \rightarrow \overset{\text{``}t-o\text{''}}{Al_2(Si_2O_5)(OH)_4} \qquad \text{Kaolinite}$$

Trioctahedral

$$Mg_3(OH)_6 - (OH)_2 + Si_2O_5 \rightarrow Mg_3(Si_2O_5)(OH)_4 \qquad \text{Antigorite}$$

In exactly similar fashion, we may derive more complex members of the phyllosilicate family tree. Let us substitute the apical oxygens of an Si_2O_5 sheet for two out of the six hydroxyls of a gibbsite or brucite sheet on one side and then do the same on the other side of the same sheet. Now we have a triple sheet made up of an Si_2O_5 "t" sheet joined to an "o" sheet which in turn is joined to another inverted Si_2O_5 "t" sheet. Since each siloxane sheet replaces two out of the six hydroxyls per octahedron of the brucite or gibbsite sheet, two remain, and the formulas of the resulting triple sheets must be

Dioctahedral

$$\overset{\text{``}t\text{''}}{Si_2O_5} + \overset{\text{``}o\text{''}}{Al_2(OH)_6} - (OH)_4 + \overset{\text{``}t\text{''}}{Si_2O_5} \rightarrow \overset{\text{``}t-o-t\text{''}}{Al_2(Si_4O_{10})(OH)_2} \qquad \text{Pyrophyllite}$$

Trioctahedral

$$Si_2O_5 + Mg_3(OH)_6 - (OH)_4 + Si_2O_5 \rightarrow Mg_3(Si_4O_{10})(OH)_2 \qquad Talc$$

These structural units, which we call "t–o–t" sheets, are electrically neutral and form stable structures in which the sheets are joined only by the van der Waals' bond (Figs. 404 and 405). Since this is a very weak bond, we may expect such structures to have excellent cleavage, easy gliding, and a greasy feel. These expectations are borne out, in the minerals talc and pyrophyllite.

If we carry the process of evolution a step farther, we may substitute aluminum ions for some of the silicon in the tetrahedral sites in the Si_2O_5 sheets. Since

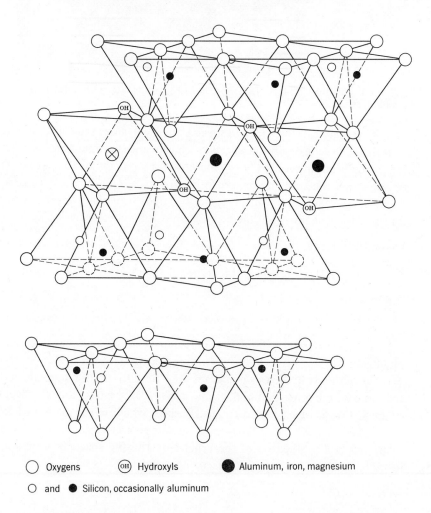

○ Oxygens	(OH) Hydroxyls	● Aluminum, iron, magnesium

○ and ● Silicon, occasionally aluminum

FIG. 404. Diagrammatic sketch of the "t–o–t" layer (after Grim).

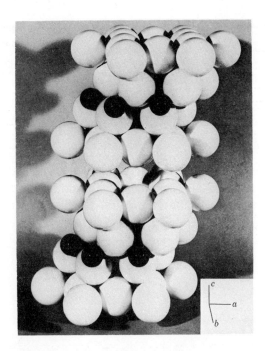

FIG. 405. Talc, $Mg_3Si_4O_{10}(OH)_2$, packing model. O white, Mg black. (OH) not visible. *c* axis vertical. The model shows the wide spacing between two closely packed *"t–o–t"* sheets.

aluminum is trivalent, whereas silicon is tetravalent, each substitution of this kind causes a free electrical charge to appear on the surface of the sheet. If aluminum substitutes for every fourth silicon, a charge of sufficient magnitude is produced to bind univalent cations in regular 12 coordination between the *"t–o–t"* sheets. By virtue of these sheet-cation-sheet bonds, the layers are more firmly held together, the ease of gliding is diminished, the hardness is increased, and slippery feel is lost. The resulting mineral structures are those of the true micas (Figs. 406 and 407). In the dioctahedral line of descent, the cation is commonly potassium (muscovite) or sodium (paragonite); in the trioctahedral branch of the family it is potassium in both phlogopite and biotite. It is easy to remember the formulas of the micas if it is recalled that *one* of the aluminums belongs in the tetrahedral sites and the formulas written accordingly. Thus,

Dioctahedral	$KAl_2(AlSi_3O_{10})(OH)_2$	Muscovite
Trioctahedral	$KMg_3(AlSi_3O_{10})(OH)_2$	Phlogopite

If half the silicon ions in tetrahedral sites in Si_2O_5 layers are replaced by aluminum, two charges per *"t–o–t"* sheet become available for binding an inter-layer cation. Such ions as calcium, magnesium, and ferrous iron may enter the structure of the mica held between the triple-sheet layers by ionic bonds. In these structures the interlayer binding is so strong that the quality of the cleavage is diminished, the hardness is increased, the flexibility of the layers is almost wholly lost, and the density is increased. The resulting minerals are the *brittle micas*, typified by

Dioctahedral $\quad$ $CaAl_2(Al_2Si_2O_{10})(OH)_2$ $\quad$ Margarite

Trioctahedral $\quad$ $CaMg_3(Al_2Si_2O_{10})(OH)_2$ $\quad$ Xanthophyllite

There is little solid solution between members of the dioctahedral and triocta-hedral series, although there may be extensive and substantially complete ionic substitution of ferrous iron for magnesium, of ferric iron for aluminum, of calcium for sodium in appropriate sites. Barium may replace potassium, chromium may substitute for aluminum, and fluorine may replace hydroxyl to a limited extent.

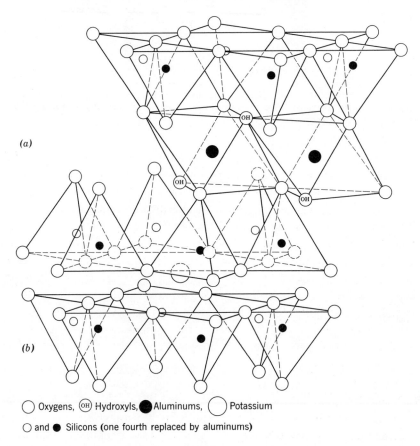

○ Oxygens, (OH) Hydroxyls, ● Aluminums, ◯ Potassium

○ and ● Silicons (one fourth replaced by aluminums)

FIG. 406. Diagrammatic sketch of the muscovite structure (after Grim).

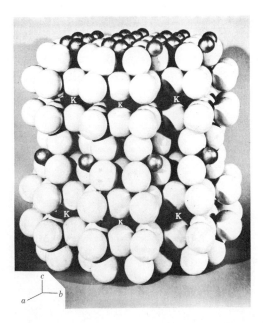

FIG. 407. Muscovite, $KAl_2(AlSi_3O_{10})(OH)_2$, packing model. Al in 6 coordination (metallic), Al in 4 coordination not seen. O is white, One complete *'t–o–t''* sheet constitutes the central part of the model, joined by large K ions to portions of adjacent *''t–o–t''* sheets above and below. Compare Fig. 405 (talc), Fig. 403 (kaolinite) and Fig. 409 (chlorite).

Manganese, titanium, and cesium are rarer constituents of some micas. The lithium micas are structurally distinct from muscovite and biotite because of the small lithium ion.

A number of branches may be added to our phyllosilicate family tree (Fig. 408). The important family of *chlorites* may be simply described as having the triple-layered talc structure interlayered with a simple brucite sheet (Fig. 409). This leads to the formula $Mg_3Si_4O_{10}(OH)_2 \cdot Mg_3(OH)_6$. However, in most chlorites aluminum and ferrous and ferric iron substitute for magnesium in octahedral sites both in the talc layers and in the brucite sheets, and aluminum substitutes for silicon in the tetrahedral sites. The formula may now be generalized as:

$$(Mg,Fe'',Fe''',Al)_3(Al,Si)_4O_{10}(OH)_2 \cdot (Mg,Fe'',Fe''',Al)_3(OH)_6$$

The various members of the group differ from one another in amounts of substitution and the manner in which the layers are arranged one over the other.

The important group of the *montmorillonites* may be derived from the pyrophyllite structure by the insertion of sheets of molecular water containing free cations between pyrophyllite "t–o–t" triple layers. The essentially uncharged pyrophyllite sheets may be expanded in this way to unusually large cell dimensions,

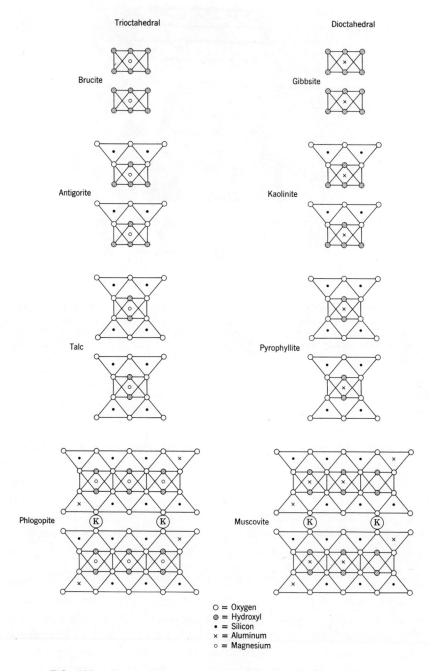

FIG. 408. Family tree of the phyllosilicates. (Schematic.)

FIG. 409.   Chlorite, $Mg_3Si_4O_{10}(OH)_2 \cdot Mg_3(OH)_6$, packing model. O white, (OH) large black, Mg small black; c axis vertical. The model shows "t–o–t" sheets alternating with "o" sheets. Fe^2, Fe^3, and Al may substitute for Mg and Al for Si.

and the minerals of this group accordingly show unrivalled capacity for swelling when wetted. The *vermiculites* are similarly derived from the talc structure by interlamination of sheets of molecular water.

If occasional random substitution of aluminum for silicon in the tetrahedral sites of pyrophyllite sheets takes place, there may not be enough of an aggregate charge on the triple layers to produce an ordered mica structure with all possible interlayer cation sites filled. Locally, however, occasional cation sites may be occupied, leading to properties intermediate between those of clays and micas. Introduction of some molecular water may further complicate this picture. Minerals of this type, intermediate between the montmorillonite clays and the true micas, are referred to the *illite* or *hydromica* group.

The great importance of the phyllosilicates lies in part in the fact that the products of rock weathering and hence the constituents of soils are mostly of this structural type. The release and retention of plant foods, the stockpiling of water in the soil from wet to dry seasons, and the accessibility of the soil to atmospheric gases and organisms depend in large part on the properties of the sheet silicates.

Geologically, the phyllosilicates are of great significance. The micas are the chief minerals of the schists and are widespread in igneous rocks. They form at

Phyllosilicates

Apophyllite	$KCa_4(Si_4O_{10})_2F\cdot8H_2O$
Kaolinite	$Al_4(Si_4O_{10})(OH)_8$
Serpentine	$Mg_6(Si_4O_{10})(OH)_8$
Pyrophyllite	$Al_2(Si_4O_{10})(OH)_2$
Talc	$Mg_3(Si_4O_{10})(OH)_2$
Muscovite	$KAl_2(AlSi_3O_{10})(OH)_2$
Phlogopite	$KMg_3(AlSi_3O_{10})(OH)_2$
Biotite	$K(Mg,Fe)_3(AlSi_3O_{10})(OH)_2$
Lepidolite	$K(Li,Al)(AlSi_3O_{10})(O,OH,F)_2$
Margarite	$CaAl_2(Al_2Si_2O_{10})(OH)_2$
Chlorite	$Mg_3(Si_4O_{10})(OH_2)\cdot Mg_3(OH)_6$

lower temperatures than amphiboles or pyroxenes and frequently are formed as replacements of earlier minerals as a result of hydrothermal alteration.

Apophyllite—$KCa_4(Si_4O_{10})_2F\cdot8H_2O$

Crystallography. Tetragonal; $4/m2/m2/m$. Usually in crystals showing a combination of {110}, {011}, and {001} (Fig. 410). Crystals may resemble a combination of cube and octahedron, but are shown to be tetragonal by difference in luster between faces of prism and base.

Angles: $c(001) \wedge p(101) = 60° 17'$, $c(001) \wedge e(114) = 23° 39'$.

$P4mnc$; $a = 9.02$, $c = 15.8$ Å; $a:c = 1:1.752$; $Z = 2$. *d's:* 4.52(2), 3.94(10), 3.57(1), 2.98(7), 2.48(3).

Physical Properties. *Cleavage* {001} perfect. **H** $4\frac{1}{2}$–5. **G** 2.3–2.4. *Luster* of base pearly, other faces vitreous. *Color* colorless, white, or grayish; may show pale shades of green, yellow, rose. Transparent to translucent. *Optics:* (+); $\omega = 1.537$, $\varepsilon = 1.535$. May be optically (−).

Composition. The structure of apophyllite differs from that of the ordinary phyllosilicates in that the sheets are composed of 4-membered rings of tetrahedra. The tetrahedra are linked together in such a way that the ratio of Si:O = 2:5 as in other phyllosilicates.

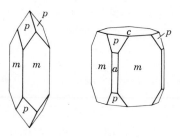

FIG. 410. Apophyllite.

Diagnostic Features. Recognized usually by its crystals, color, luster, and basal cleavage. Fusible at 2, with swelling, forms a white vesicular enamel. Colors the flame pale violet (potassium). Yields much water in the closed tube.

Occurrence. Apophyllite occurs as a secondary mineral lining cavities in basalt and related rocks, associated with zeolites, calcite, datolite, and pectolite.

It is found in fine crystals at Andreasberg, Harz Mountains; Aussig, Bohemia; on the Seiser Alpe in Trentino, Italy; near Bombay, India; Faeroe Islands; Iceland; Greenland; and Guanajuato, Mexico. In the United States at Bergen Hill and Paterson, New Jersey; and Lake Superior copper district. Found in fine crystals in Nova Scotia.

Name. Apophyllite is named from two Greek words meaning *from* and *a leaf*, because of its tendency to exfoliate when ignited.

Clay Minerals[7]

Clay is a rock term, and like most rocks it is made up of a number of different minerals in varying proportions. Clay also carries the implication of small particle size. Usually the term clay is used in reference to fine-grained, earthy material which becomes plastic when mixed with a small amount of water. With the use of x-ray techniques, clays have been shown to be made up dominantly of a group of crystalline substances known as the *clay minerals*. They are all essentially hydrous aluminum silicates. In some, magnesium or iron substitute in part for aluminum, and alkalies or alkaline earths may be present as essential constituents. Although a clay may be made up of a single clay mineral, there are usually several mixed with other minerals such as feldspar, quartz, carbonates, and micas.

KAOLINITE—$Al_4(Si_4O_{10})(OH)_8$

Crystallography. Triclinic; $\bar{1}$. In very minute, thin, rhombic or hexagonal-shaped plates. Usually in claylike masses, either compact or friable.

$P\bar{1}$; $a = 5.14$, $b = 8.93$, $c = 7.37$ Å; $\alpha = 91° 48'$; $\beta = 104° 30'$, $\gamma = 90°$; $a:b:c = 0.576:1:0.825$. $Z = 1$. d's: 7.15(10), 3.57(10), 2.55(8), 2.49(9), 2.33(10).

Physical Properties. *Cleavage* {001} perfect. **H** 2. **G** 2.6. *Luster* usually dull earthy; crystal plates pearly. *Color* white. Often variously colored by impurities. Usually unctuous and plastic. *Optics:* (−); $\alpha = 1.553–1.565$, $\beta = 1.559–1.569$, $\gamma = 1.560–1.570$; $2V = 24°–50°$. $r > v$.

Composition. Al_2O_3 39.5, SiO_2 46.5, H_2O 14.0 per cent.

Diagnostic Features. Infusible. Insoluble. Assumes a blue color when moistened with cobalt nitrate and ignited (aluminum). Gives water in the closed tube.

[7] For a discussion of the clay minerals, their mineralogy, chemistry, occurrence, and origin, see Ralph E. Grim, *Clay Mineralogy*, McGraw-Hill Book Co., New York.

Recognized usually by its claylike character, but without optical or x-ray tests it is impossible to distinguish from the other clay minerals of similar composition which collectively make up *kaolin*.

Occurrence. Kaolinite is a common mineral, the chief constituent of kaolin or clay. It is always a secondary mineral formed by weathering or hydrothermal alteration of aluminum silicates, particularly feldspar. It is found mixed with feldspar in rocks that are undergoing alteration; in places it forms entire deposits where such alteration has been carried to completion. As one of the common products of the decomposition of rocks it is found in soils and transported by water is deposited, mixed with quartz and other materials, in lakes, etc., in the form of beds of clay.

Use. Clay is one of the most important of the natural industrial substances. It is available in every country of the world and is commercially produced in nearly every state in the United States. Many and varied products are made from it which include common brick, paving brick, drain tile, sewer pipe. The commercial users of clay recognize many different kinds having slightly different properties, each of which is best suited for a particular purpose. High-grade clay which goes under the name of *china clay* or *kaolin* has many uses in addition to the manufacture of china and pottery. Its largest is as a filler in paper, but it is also used in the rubber industry and in the manufacture of refractories.

The chief value of clay for ceramic products lies in the fact that when wet it can be easily molded into any desired shape, and then, when it is heated, part of the combined water is driven off, producing a hard, durable substance.

Name. Kaolinite is derived from *kaolin*, which is a corruption of the Chinese *kauling*, meaning *high ridge*, the name of a hill near Jauchu Fa, where the material is obtained.

Similar Species. *Dickite* and *nacrite* are similar to kaolinite in structure and chemical composition but are less important constituents of clay deposits. *Anauxite* is also assigned to the kaolinite group but has a higher silicon to aluminum ratio than kaolinite. *Halloysite* has two forms: one with kaolinite composition, $Al_4(Si_4O_{10})(OH)_8$, the other with composition $Al_4(Si_4O_{10})(OH)_8 \cdot 4H_2O$. The second type dehydrates to the first with loss of interlayered water molecules.

The *montmorillonite* group (see page 431) comprises a number of clay minerals composed of "t–o–t" silicate layers of both dioctahedral and trioctahedral type. The outstanding characteristic of members of this group is their capacity of absorbing water molecules between the sheets producing marked expansion of the structure. The dioctahedral members are *montmorillonite*, *beidellite*, and *nontronite*; the trioctahedral members are *hectorite* and *saponite*.

Montmorillonite is the dominant clay mineral in *bentonite*, altered volcanic ash. Bentonite has the unusual property of expanding several times its original volume when placed in water. This property gives rise to interesting industrial uses.

Most important is as a drilling mud in which the montmorillonite is used to give the fluid a viscosity several times that of water. It is also used for stopping leakage in soil, rocks, and dams.

Illite is a general term for the micalike clay minerals. The illites differ from the micas in having less substitution of aluminum for silicon, in containing more water, and in having potassium partly replaced by calcium and magnesium. Illite is the chief constituent in many shales.

SERPENTINE—$Mg_6(Si_4O_{10})(OH)_8$

Crystallography. Monoclinic; $2/m$. Crystals, except as pseudomorphs, unknown. Serpentine occurs in two distinct forms: (1) a platy variety, *antigorite*, which conforms in its properties to those of the phyllosilicates, and (2) a fibrous variety, *chrysotile* (Fig. 411). It has been postulated that the fibers of chrysotile result from the layered silicate structure being curved to form cylindrical tubes. Orthorhombic polymorphs of both antigorite and chrysotile are known.

FIG. 411. Veins of chrysotile asbestos in serpentine, Thetford, Quebec.

Cm, *C2* or *C2/m*. For antigorite: $a = 5.30$, $b = 9.20$, $c = 7.46$ Å; $\beta = 91° 24'$; $Z = 1$. *d's*: 7.30(10), 3.63(8), 2.52(2), 2.42(2), 2.19(1). For chrysotile: $a = 5.34$, $b = 9.25$, $c = 14.65$; $Z = 2$.

Physical Properties. H 3–5, usually 4. G 2.5–2.6. *Luster* greasy, waxlike in the massive varieties, silky when fibrous. *Color* often variegated, showing mottling in lighter and darker shades of green. Translucent. *Optics:* $(-)$; Chrysotile: $\alpha = 1.532$–1.549 (across fiber), $\gamma = 1.545$–1.556 (parallel to fiber length). Antigorite: nearly isotropic, $n = 1.55$–1.56.

Composition. Close to $Mg_6(Si_4O_{10})(OH)_8$. MgO 43.0, SiO_2 44.1, H_2O 12.9 per cent. Fe and Ni may be present substituting for Mg.

Diagnostic Features. Infusible. Gives water in the closed tube. Recognized by its variegated green color and its greasy luster or by its fibrous nature. Distinguished from fibrous amphibole by the presence of a large amount of water.

Occurrence. Serpentine is a common mineral and widely distributed; usually as an alteration of magnesium silicates, especially olivine, pyroxene, and amphibole. Frequently associated with magnesite, chromite, and magnetite. Found in both igneous and metamorphic rocks, frequently in disseminated particles, in places in such quantity as to make up practically the entire rock mass. Serpentine, as a rock name, is applied to such rock masses made up mostly of the variety antigorite. Large deposits of the fibrous variety, chrysotile, are located in the province of Quebec, Canada; in the Ural Mountains, U.S.S.R.; and in Rhodesia. In the United States chrysotile is found in Vermont, New York, New Jersey, and in Arizona near Globe from the Sierra Ancha, and in the Grand Canyon.

Use. The variety *chrysotile* is the chief source of asbestos. The uses of asbestos depend upon its fibrous, flexible nature, which allows it to be made into felt and woven into cloth and other fabrics, and upon its incombustibility and slow conductivity of heat. Asbestos products, therefore, are used for fireproofing and as an insulation material against heat and electricity. Massive serpentine, which is translucent and of a light to dark green color, is often used as an ornamental stone and may be valuable as building material. Mixed with white marble and showing beautiful variegated coloring, it is called *verd antique* marble.

Name. The name refers to the green serpentlike cloudings of the massive variety.

Similar Species. *Garnierite*, $(Ni, Mg)_6(Si_4O_{10})(OH)_8$, is an apple green nickel-bearing serpentine formed as an alteration product of nickel-bearing peridotites. It is mined as a nickel ore in New Caledonia, U.S.S.R., and Australia. In the United States it is found at Riddle, Oregon.

Sepiolite (meerschaum), a hydrous magnesium silicate, is a claylike, secondary mineral associated with serpentine. It is used in the manufacture of meerschaum pipes.

Pyrophyllite—$Al_2(Si_4O_{10})(OH)_2$

Crystallography. Monoclinic; $2/m$. Not in distinct crystals. Foliated, in some cases in radiating lamellar aggregates. Also granular to compact. Identical with talc in appearance.

$C2/c$; $a = 5.15$, $b = 8.92$, $c = 18.59$ Å; $\beta = 99° 55'$; $a:b:c = 0.577:1:2.084$; $Z = 4$. d's: 9.21(6), 4.58(5), 4.40(2), 3.08(10), 2.44(2).

Physical Properties. *Cleavage* {001} perfect. Folia somewhat flexible but not elastic. **H** 1–2 (will make a mark on cloth). **G** 2.8. *Luster* pearly to greasy. *Color* white, apple-green, gray, brown. Translucent, will transmit light on thin edges. *Optics:* ($-$); $\alpha = 1.552$, $\beta = 1.588$, $\gamma = 1.600$; $2V = 57°$; $X \perp$ {001}; $r > v$.

Composition. Al_2O_3 28.3, SiO_2 66.7, H_2O 5.0 per cent.

Diagnostic Features. Characterized chiefly by its micaceous habit, cleavage, and greasy feel. When moistened with cobalt nitrate and ignited it assumes a blue color (aluminum); talc under the same conditions becomes pale violet. Infusible, but, on heating, radiated varieties exfoliate in a fanlike manner. At high temperature it yields water in the closed tube.

Occurrence. Pyrohyllite is a comparatively rare mineral. Found in metamorphic rocks; frequently with kyanite. Occurs in considerable amount in Guilford and Orange counties, North Carolina.

Use. Quarried in North Carolina and used for the same purposes as talc (see page 440), but does not command as high a price as the best grades of talc. A considerable part of the so-called *agalmatolite*, from which the Chinese carve small images, is this species.

Name. From the Greek meaning *fire* and *a leaf*, since it exfoliates on heating.

TALC—$Mg_3(Si_4O_{10})(OH)_2$

Crystallography. Monoclinic; $2/m$. Crystals rare. Usually tabular with rhombic or hexagonal outline. Foliated and in radiating foliated groups. When compact and massive known as *steatite* or *soapstone*.

$C2/c$; $a = 5.27$, $b = 6.19$, $c = 18.85$ Å; $\beta = 100° 0'$; $a:b:c = 0.578:1:2.067$; $Z = 4$. d's: 9.34(10), 4.66(9), 3.12(10), 2.48(7), 1.870(4).

Physical Properties. *Cleavage* {001} perfect. Thin folia somewhat flexible but not elastic. Sectile. **H** 1 (will make a mark on cloth). **G** 2.7–2.8. *Luster* pearly to greasy. *Color* apple-green, gray, white, or silver-white; in soapstone often dark gray or green. Translucent. Greasy feel. *Optics:* ($-$); $\alpha = 1.539$, $\beta = 1.589$, $\gamma = 1.589$; $2V = 6°–30°$. $Z = b$, $X \perp$ {001}; $r > v$.

Composition. MgO 31.7, SiO_2 63.5, H_2O 4.8 per cent.

Diagnostic Features. Fusible with difficulty (5). Unattacked by acids. Yields water in the closed tube when heated intensely. Characterized by its micaceous habit, cleavage, softness, and greasy feel. Distinguished from pyrophyllite by moistening a fragment with cobalt nitrate and heating intensely; talc will assume a pale violet color, pyrophyllite a blue color.

Occurrence. Talc is a secondary mineral formed by the alteration of magnesium silicates, such as olivine, pyroxenes, and amphiboles, and may be found as pseudomorphs after these minerals. Characteristically in low-grade metamorphic rocks, where, in massive form, *soapstone*, it may make up nearly the entire rock mass. It may also occur as a prominent constituent in the schistose rocks, as in talc schist.

In the United States many talc or soapstone quarries are located along the line of the Appalachian Mountains from Vermont to Georgia. The major producing states are California, North Carolina, Texas, and Georgia.

Use. As slabs of the rock soapstone, talc is used for laboratory table tops, electric switchboards, and sanitary appliances. Most of the talc and soapstone produced is used in powdered form as an ingredient in paint, ceramics, rubber, insecticides, roofing, paper, and foundry facings. The most familar use is in talcum powder.

Name. The name talc is of ancient and doubtful origin, probably derived from the Arabic, *talk*.

Mica Group

The micas crystallize in the monoclinic system but with crystallographic angle β close to 90°, so that their monoclinic symmetry is not clearly seen. The crystals are usually tabular with prominent basal planes and have either a diamond-shaped or hexagonal outline with angles of approximately 60° and 120°. Crystals, as a rule, therefore, appear to be either orthorhombic or hexagonal. They are characterized by a highly perfect {001} cleavage. A blow with a somewhat dull-pointed instrument on a cleavage plate develops in all the species a six-rayed *percussion figure* (Fig. 412), two lines of which are nearly parallel to the prism edges, and the third, which is most strongly developed, is parallel to the plane of symmetry {010}.

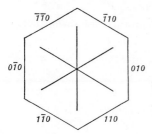

FIG. 412.   Percussion figure in mica.

There is limited ionic substitution between the different members. Two members of the group, however, frequently crystallize together in parallel position in the same crystal plate with the cleavage extending through both.

Each of the mica minerals has one or more polymorphs that depend on the stacking of the "t–o–t" layers. Most commonly the stacking repeats monoclinic cells with every layer or with every two layers and are designated respectively as 1M or 2M. The distinction between nM layered micas is made only by careful single crystal x-ray study.

MUSCOVITE—$KAl_2(AlSi_3O_{10})(OH)_2$

Crystallography. Monoclinic; $2/m$. Distinct crystals rare; usually tabular with prominent {001}. The presence of prism faces {110} at angles of nearly 60° gives some plates a diamond-shaped outline, making them simulate orthorhombic symmetry (Fig. 413). If {010} is also present, the crystals have a hexagonal appearance. The prism faces are roughened by horizontal striations and frequently taper. Foliated in large to small sheets; in scales which are in some cases aggregated into plumose or globular forms. Also cryptocrystalline and compact massive.

$C2/c$; $a = 5.19$, $b = 9.04$, $c = 20.08$ Å; $\beta = 95°\ 30'$; $a:b:c = 0.574:1:2.221$; $Z = 4$. d's: 9.95(10), 3.37(10), 2.66(8), 2.45(8), 2.18(8).

Physical Properties. *Cleavage* {001} perfect allowing the mineral to be split into very thin sheets; folia flexible and elastic. **H** $2-2\frac{1}{2}$. **G** 2.76–2.88. *Luster* vitreous to silky or pearly. *Color* colorless and transparent in thin sheets. In thicker blocks translucent, with light shades of yellow, brown, green, red. Some

FIG. 413. Muscovite. Diamond-shape crystals.

crystals allow more light to pass parallel to the cleavage than perpendicular to it. *Optics:* $(-)$; $\alpha = 1.560$–1.572, $\beta = 1.593$–1.611, $\gamma = 1.599$–1.615; $2V = 30°$–$47°$; $Z = Y$, $X \wedge c = 0$–$5°$; $r > v$.

Composition. Essentially $KAl_2(AlSi_3O_{10})(OH)_2$. Minor substitutions may be: Na, Rb, Cs for K; Mg, Fe^2, Fe^3, Li, Mn, Ti, Cr for Al; F for OH.

Diagnostic Features. Characterized by its highly perfect cleavage and light color. Distinguished from phlogopite by not being decomposed in sulfuric acid and from lepidolite by not giving a crimson flame. Gives water in the closed tube.

Occurrence. Muscovite is a widespread and common rockforming mineral. Characteristic of granites and granite pegmatites. In pegmatites, muscovite associated with quartz and feldspar may be in large crystals, called books, which in some localities are several feet across. It is also very common in metamorphic rocks, as gneiss and schist, forming the chief constituent in certain mica schists. In some schistose rocks it occurs as fibrous aggregates of minute scales with a silky luster. This variety, known as *sericite*, is usually the product of alteration of feldspar. Sericite also forms as an alteration of the wall rock of hydrothermal ore veins. Muscovite may form as an alteration of several other minerals, as topaz, kyanite, spodumene, andalusite, scapolite. As *illite* muscovite is a constituent of some shales. *Pinite* is a name given to the micaceous alteration product of various minerals, and corresponds in composition more or less closely to muscovite.

Several notable localities for muscovite are found in the Alps; also Mourne Mountains, Ireland; Cornwall, England; Norway; and Sweden. Large and important deposits occur in India. The most productive pegmatites in the United States are in New Hampshire, North Carolina, and in the Black Hills of South Dakota. Of less importance are the deposits in Colorado, Alabama, Maine, and Virginia. Crystals measuring 7 to 9 feet across have been mined in Mattawan Township, Ontario, Canada.

Use. Because of its high dielectric and heat-resisting properties, *sheet mica,* single cleavage plates, is used as an insulating material in the manufacture of electrical apparatus. The *isinglass* used in furnace and stone doors is sheet mica. Many small parts used for electrical insulation are built up of thin sheets of mica cemented together. They may thus be pressed into shape before the cement hardens. India is the largest supplier of mica used in this way. Ground mica is used in many ways: in the manufacture of wallpapers to give them a shiny luster; as a lubricant when mixed with oils; a filler; and as a fireproofing material.

Name. *Muscovite* was so called from the popular name of the mineral, *Muscovy-glass,* because of its use as a substitute for glass in Old Russia (Muscovy). *Mica* was probably derived from the Latin *micare,* meaning *to shine.*

Similar Species. *Paragonite,* $NaAl_2(AlSi_3O_{10})(OH)_2$, occurs with and is physically indistinguishable from muscovite. There is limited substitution of potassium for sodium in paragonite.

Phlogopite—$KMg_3(AlSi_3O_{10})(OH)_2$

Crystallography. Monoclinic; $2/m$. Usually in six-sided plates or in tapering prismatic crystals. Crystals frequently large and coarse. Found also in foliated masses.

$C2/m$; $a = 5.33, b = 9.23, c = 10.26$ Å; $\beta = 100°\ 12'$; $a:b:c = 0.577:1:1.112$; $Z = 2$. d's: 10.13(10), 3.54(4), 3.36(10), 3.28(4), 2.62(10).

Physical Properties. *Cleavage* {001} perfect. Folia flexible and elastic. **H** $2\frac{1}{2}$–3. **G** 2.86. *Luster* vitreous to pearly. *Color* yellowish-brown, green, white, often with copperlike reflections from the cleavage surface. Transparent in thin sheets. When viewed in transmitted light, some phlogopite shows asterism because of tiny oriented inclusions of rutile. *Optics:* $(-)$; $\alpha = 1.53$–1.59, $\beta = 1.56$–1.64, $\gamma = 1.56$–1.64; $2V = 0°$–$15°$; $Y = b$, $X \wedge c = 0°$–$5°$. $r < v$.

Composition. Small amount of Na and lesser amounts of Rb, Cs, and Ba may substitute for K. Fe^2 substitutes for Mg forming a series grading into biotite. F may substitute in part for OH.

Diagnostic Features. Characterized by its micaceous cleavage and yellowish-brown color. Distinguished from muscovite by its decomposition in sulfuric acid and from biotite by its lighter color. But it is impossible to draw a sharp distinction between biotite and phlogopite. Fusible at $4\frac{1}{2}$–5. Yields water in the closed tube.

Occurrence. Phlogopite is found in metamorphosed magnesium limestones, dolomites, and ultrabasic rocks. It is a common mineral in kimberlite. Notable localities are in Finland; Sweden; Campolungo, Switzerland; Ceylon; and Madagascar. In the United States found chiefly in Jefferson and St. Lawrence Counties, New York. Found abundantly in Canada in Ontario at North and South Burgess, and in various other localities in Ontario and Quebec.

Use. Same as for muscovite; chiefly as electrical insulator.

Name. Named from a Greek word meaning *firelike*, in allusion to its color.

BIOTITE—$K(Mg,Fe)_3(AlSi_3O_{10})(OH)_2$

Crystallography. Monoclinic; $2/m$. Rarely in tabular or short prismatic crystals with prominent {001} and pseudo-hexagonal outline. Usually in irregular foliated masses; often in disseminated scales or in scaly aggregates.

$C2/m$; $a = 5.31, b = 9.23, C = 10.18$ Å; $\beta = 99°\ 18'$; $a:b:c = 0.575:1:1.103$; $Z = 2$. d's: 10.1(10), 3.37(10), 2.66(8), 2.54(8), 2.18(8).

Physical Properties. *Cleavage* {001} perfect. Folia flexible and elastic. **H** $2\frac{1}{2}$–3. **G** 2.8–3.2. *Luster* splendent. *Color* usually dark green, brown to black. More rarely light yellow. Thin sheets usually have a smoky color (differing from the almost colorless muscovite). *Optics:* $(-)$; $\alpha = 1.57$–1.63, $\beta = 1.61$–1.70; $\gamma = 1.61$–1.70; $2V = 0°$–$25°$; $Y = b$, $Z \wedge a = 0°$–$9°$. $r < v$.

Composition. The composition is similar to phlogopite but with considerable substitution of Fe^2 for Mg. There is also substitution by Fe^3 and Al for Mg and by Al for Si. In addition Na, Ca, Ba, Rb, and Cs may substitute for K.

Diagnostic Features. Characterized by its micaceous cleavage and dark color. Fusible with difficulty at 5. Decomposed by boiling concentrated sulfuric acid, giving a milky solution. Gives water in the closed tube.

Occurrence. Biotite is an important and widely distributed rock-forming mineral, and occurs in igneous rocks varying from granite to gabbro. It is found in some pegmatites in large sheets. Found also in many felsite lavas and porphyries. Is also present in some gneisses and schists often associated with muscovite. Occurs in fine crystals in blocks included in the lavas of Vesuvius.

Name. In honor of the French physicist, J. B. Biot.

Similar Species. *Glauconite*, similar in composition to biotite, is an authigenic mineral found in green pellets in sedimentary rocks.

Vermiculite forms as an alteration of biotite and phlogopite. The structure is made up of mica sheets interlayered with water molecules. On heating, it loses water and expands into wormlike forms. Vermiculite is mined at Libby, Montana, and Macon, North Carolina. In the expanding form it is used extensively in heat and sound insulating.

LEPIDOLITE—$K(Li,Al)(AlSi_3O_{10})(O,OH,F)_2$

Crystallography. Monoclinic; $2/m$. Crystals usually in small plates or prisms with hexagonal outline. Commonly in coarse- to fine-grained scaly aggregates.

$C2/m$; $a = 5.21$, $b = 8.97$, $c = 20.16$ Å; $\beta = 100° 48'$; $a:b:c = 0.581:1:1.124$; $Z = 4$. d's: 10.0(6), 5.00(5), 4.50(5), 2.58(10), 1.989(8).

Physical Properties. *Cleavage* {001} perfect. **H** $2\frac{1}{2}$–4. **G** 2.8–2.9. *Luster* pearly. *Color* pink and lilac to grayish-white. Translucent. *Optics:* $(-)$; $\alpha = 1.53$–1.55, $\beta = 1.55$–1.59, $\gamma = 1.55$–1.59; $2V = 0°$–60°; $Y = b$, $Z \wedge a = 0°$–7°; $r > v$.

Composition. The composition of lepidolite varies depending chiefly on the relative amounts of Al and Li in octahedral coordination. Analyses show a range of Li_2 from 3.3–7 per cent. In addition Na, Rb, and Cs may substitute for K.

Diagnostic Features. Fusible at 2, giving a crimson flame (lithium). Insoluble in acids. Gives acid water in the closed tube. Characterized chiefly by its micaceous cleavage and usually by its lilac to pink color. Muscovite may be pink, or lepidolite white, and therefore a flame test should be made to distinguish them.

Occurrence. Lepidolite is a comparatively rare mineral, found in pegmatites, usually associated with other lithium-bearing minerals such as pink and green tourmaline, amblygonite, and spodumene. Often intergrown with muscovite in parallel position. Notable foreign localities for its occurrence are the Ural Mountains, U.S.S.R.; Isle of Elba; Rozna, Moravia; Bikita, Rhodesia; South West Africa and Madagascar. In the United States is found in several localities in Maine; near Middletown, Connecticut; Pala, California; Dixon, New Mexico; and Black Hills, South Dakato.

Use.   A source of lithium. Used in the manufacture of heat-resistant glass.

Name.   Derived from a Greek word meaning *scale*.

Margarite—$CaAl_2(Al_2Si_2O_{10})(OH)_2$

Crystallography.   Monoclinic; $2/m$. Seldom in distinct crystals. Usually in foliated aggregates with micaceous habit.

$C2/c$; $a = 5.13$, $b = 8.92$, $c = 19.50$ Å; $\beta = 100°\ 48'$; $a:b:c = 0.575:1:2.186$; $Z = 4$. d's: 4.40(8), 3.39(8), 3.20(9), 2.51(10), 2.42(8).

Physical Properties.   *Cleavage* $\{001\}$ perfect. **H** $3\frac{1}{2}$–5 (harder than the true micas). **G** 3.0–3.1. *Luster* vitreous to pearly. *Color* pink, white, and gray. Translucent. Folia somewhat brittle, Because of this brittleness margarite is known as a *brittle mica. Optics:* $(-)$; $\alpha = 1.632$–1.638, $\beta = 1.643$–1.648, $\gamma = 1.645$–1.650; $2V = 40°$–$67°$; $Z = b$, $Y \wedge a = 7°$.

Composition.   CaO 14.0, Al_2O_3 51.3, SiO_2 30.2, H_2O 4.5 per cent. A small amount of Na may replace Ca.

Diagnostic Features.   Characterized by its micaceous cleavage, brittleness, and association with corundum. Fuses at 4–$4\frac{1}{2}$, turning white. Slowly and incompletely decomposed by boiling hydrochloric acid.

Occurrence.   Margarite occurs usually with corundum and apparently as one of its alteration products. It is found in this way with the emery deposits of Asia Minor and on the islands Naxos and Nicaria, of the Grecian Archipelago. In the United States associated with emery at Chester, Massachusetts; Chester County, Pennsylvania; and with corundum deposits in North Carolina.

Name.   From the Greek meaning *pearl*.

Similar Species.   Several minerals similar to margarite in structure and physical and chemical properties are included in the group of *brittle micas*. Aside from margarite, the most important members of the group are *ottrelite* and *chloritoid*, found in metamorphosed sedimentary rocks.

Chlorite Group

A number of minerals are included in the chlorite group all of which have similar chemical, crystallographic, and physical properties. Without quantitative chemical analyses or careful study of the optical properties, it is extremely difficult to distinguish between the members. The following is a composite description of the principal members of the group: *clinochlore*, *penninite*, and *prochlorite*.

CHLORITE—$Mg_3(Si_4O_{10})(OH)_2 \cdot Mg_3(OH)_6$

Crystallography. Monoclinic; $2/m$, some chlorites are triclinic. In pseudo-hexagonal tabular crystals, with prominent $\{001\}$. Similar in habit to crystals of the mica group, but distinct crystals rare. Usually foliated massive or in aggregates of minute scales; also in finely disseminated particles.

Cell parameters vary with composition. For clinochlore: $a = 5.2$–5.3, $b = 9.2$–9.3, $c = 28.6$ Å; $\beta = 96°\ 50'$; $Z = 4$. d's: 3.53(10), 1.998(9), 1.564(9), 1.535(10), 1.393(10).

Physical Properties. *Cleavage* $\{001\}$ perfect. Folia flexible but not elastic. **H** 2–$2\frac{1}{2}$. **G** 2.6–3.3. *Luster* vitreous to pearly. *Color* green of various shades. Rarely yellow, white, rose-red. Transparent to translucent. *Optics:* Most $(+)$, some $(-)$; in all Bxa nearly $\perp \{001\}$. $\alpha = 1.57$–1.66, $\beta = 1.57$–1.67, $\gamma = 1.57$–1.67; $2V = 20°$–$60°$. Pleochroism in green $(+)\ X, Y > Z$; $(-)\ X < Y, Z$. The indices increase with increasing iron content.

Composition. The above formula represents a chlorite with "t–o–t" talc layers alternating with brucite layers. In most chlorites there is considerable deviation from this composition with Fe^2, Fe^3 and Al substituting for Mg in both the talc layers and brucite sheets, and Al substituting in tetrahedral sites for Si. (See page 431). The great range in composition is reflected in the variations in physical and optical properties.

Diagnostic Features. Characterized by its green color, micaceous habit and cleavage, and by the fact that the folia are not elastic. Decomposed by boiling concentrated sulfuric acid, giving a milky solution. Water in the closed tube at high temperature.

Occurrence. Chlorite is a common mineral, usually formed as an alteration of iron, magnesium silicates such as pyroxenes, amphiboles, biotite, garnet, idocrase. The alteration may result from low-grade metamorphism or hydro-thermal solutions. Some schists are composed almost entirely of chlorite. The green color of many igneous rocks is due to the chlorite to which the ferromagnesian silicates have altered; and the green color of many schists and slates is due to finely disseminated particles of the mineral.

Name. Chlorite is derived from a Greek word meaning *green*, in allusion to the common color of the mineral.

Tectosilicates

Nearly three-quarters of the rocky crust of the earth is made up of minerals built about a three-dimensional framework of linked SiO_4 tetrahedra. These minerals belong to the *tectosilicate* class in which all the oxygen ions in each SiO_4 tetrahedron are shared with neighboring tetrahedra. This results in a stable, strongly bonded structure in which the ratio of Si:O is $1:2$ (Fig. 414).

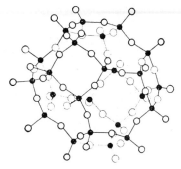

FIG. 414. Three-dimensional linkage of SiO_4 tetrahedra.

SiO_2 Group

In its simplest form the SiO_2 framework is electrically neutral and does not contain other structural units. There are, however, at least nine different ways in which the linked tetrahedra may share all oxygens and at the same time build a continuous, electrically neutral three-dimensional network. These modes of geometrical arrangement correspond to the nine known polymorphs of SiO_2,

Tectosilicates

SiO_2 GROUP	
Quartz	SiO_2
Tridymite	SiO_2
Cristobalite	SiO_2
Opal	$SiO_2 \cdot nH_2O$
FELDSPAR GROUP	
K-Feldspar Series	
Microcline	$K(AlSi_3O_8)$
Orthoclase	$K(AlSi_3O_8)$
Na-Ca Feldspar Series	
Albite	$Na(AlSi_3O_8)$
Anorthite	$Ca(Al_2Si_2O_8)$
Danburite	$Ca(B_2Si_2O_8)$
FELDSPATHOID GROUP	
Leucite	$K(AlSi_2O_6)$
Nepheline	$(Na,K)(AlSiO_4)$
Sodalite	$Na_8(AlSiO_4)_6Cl_2$
Lazurite	$(Na,Ca)_8(AlSiO_4)_6(SO_4,S,Cl)_2$
Petalite	$Li(AlSi_4O_{10})$
SCAPOLITE SERIES	
Marialite	$Na_4(AlSi_3O_8)_3(Cl,CO_3,SO_4)$
Meionite	$Ca_4(Al_2Si_2O_8)_3(Cl,CO_3,SO_4)$
ZEOLITE GROUP	
Analcime	$Na(AlSi_2O_6)\cdot H_2O$
Natrolite	$Na_2(Al_2Si_3O_{10})\cdot 2H_2O$
Chabazite	$Ca(Al_2Si_4O_{12})\cdot 6H_2O$
Heulandite	$Ca(Al_2Si_7O_{18})\cdot 6H_2O$
Stilbite	$Ca(Al_2Si_7O_{18})\cdot 7H_2O$

Polymorphs of SiO$_2$

Name	Crystal System	Specific Gravity	Refractive Index (mean)
Stishovite	Tetragonal	4.35	1.81
Coesite	Monoclinic	2.93	1.59
Quartz	Hexagonal	2.65	1.55
Keatite	Tetragonal	2.50	1.52
Cristobalite	Tetragonal	2.32	1.49
Tridymite	Hexagonal	2.26	1.47
Lechatelierite	Amorphous	2.20	1.46
Opal	Amorphous	2.0–2.2	1.44

one of which is known only as a synthetic substance. Each of these polymorphs has its characteristic morphology, cell dimensions, and lattice energy. Which polymorph is stable is determined chiefly by energy considerations; the higher-temperature forms with greater lattice energy possess the more expanded structures which are reflected in lower specific gravity and lower refractive index.

The principal SiO$_2$ polymorphs fall into three structural categories: *quartz*, with the lowest symmetry and most compact lattice; *tridymite*, with higher symmetry and more open structure; and *cristobalite*, with the highest symmetry and the most expanded lattice. Each of these structural types may be transformed into the other only by breaking silicon-oxygen bonds and rearranging the tetrahedra into a new pattern. The transformation of one type into another is accordingly a sluggish process, and all may exist metastably for long periods of time. The temperature of inversion varies widely, depending chiefly on the rate and direction of temperature change. Each structure type has, however, high- and low-temperature modifications which differ from each other only in the length or direction of the bonds joining the silicon and oxygen ions. (See Figs. 416 and 417.) Hence, these transformations take place quickly and reversibly at a fairly constant and sharply defined temperature of inversion and may be repeated over and over again without physical disintegration of the crystal. The nearly instant high → low inversions take place with the release of a fairly constant amount of energy. Very near the same temperature the low → high inversions take place with the absorption of energy.

High-Temperature Polymorph	Crystallizes as Stable Form at 1 atm. P Above	Inverts to the Low-Temperature Form at 1 atm. P
High-cristobalite	1470°	163–275°(?)
High-tridymite	870°	117–163°(?)
High-quartz	574°	573°

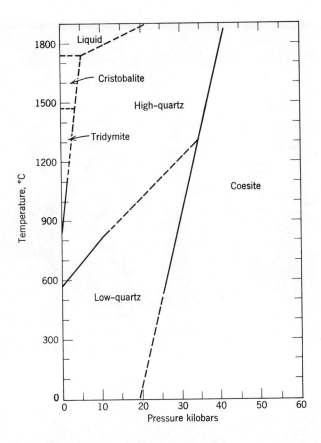

FIG. 415. Pressure-temperature diagram for SiO₂ polymorphs.

The low-temperature form of each type has lower symmetry than the higher-temperature form, but the high-low change of symmetry is less than in the transformation from one of the principal types to another. The effect of increased pressure is to raise all inversion temperatures and for any temperature to favor the crystallization of the polymorph most economical of space (Fig. 415).

The most dense of the silica polymorphs are coesite and stishovite. Coesite was synthesized in 1953 and stishovite in 1961 and it was not until later that they were found in nature. They have both been identified in very small amounts at Meteor Crater, Arizona. Their formation is attributed to the high pressure and high temperature resulting from the impact of a meteorite. Keatite has not been found in nature.

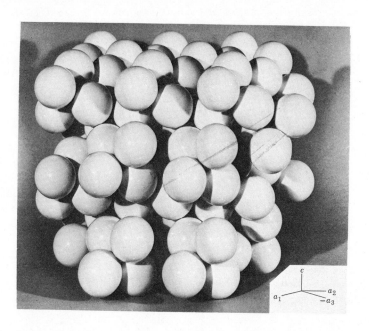

FIG. 416. Low-temperature quartz, SiO_2, packing model. The tetrahedral SiO_4 groups are arranged spirally about the *c*-axis with trigonal symmetry. Three groups form a unit cell.

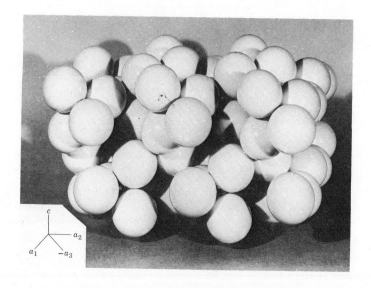

FIG. 417. High-temperature quartz, SiO_2, packing model. *c*-axis vertical. As in low-temperature quartz, the SiO_4 tetrahedra are arranged in three layers but are shifted in such a way as to convert the 3-fold axis to a 6-fold axis.

QUARTZ—SiO$_2$

Crystallography. Quartz, hexagonal—*R;* 32. High-quartz, hexagonal; 622. Crystals commonly prismatic, with prism faces horizontally striated. Terminated usually by a combination of positive and negative rhombohedrons, which often are so equally developed as to give the effect of a hexagonal dipyramid (Fig. 418a). In some crystals one rhombohedron predominates or occurs alone (Fig. 418b). The prism faces may be wanting, and the combination of the two rhombohedrons gives what appears to be a hexagonal dipyramid (a *quartzoid*) (Fig. 418c). Some crystals are malformed, but the recognition of the prism faces by their horizontal striations assists in the orientation. The trigonal trapezohedral faces x are occasionally observed and reveal the true symmetry. They occur at the upper right of alternate prism faces in right-hand quartz and to the upper left of alternate prism faces in left-hand quartz (Fig. 418d and e). The right- and left-hand trigonal trapezohedrons are enantiomorphous forms and reflect the arrangement of the SiO$_4$ tetrahedra, either in the form of a right- or left-hand screw. In the absence of x faces, the "hand" can be recognized by observing whether plane polarized light passing parallel to c is rotated to the left or right.

Crystals may be elongated in tapering and sharply pointed forms, and some appear twisted or bent. Equant crystals with apparent 6-fold symmetry (Fig. 418a, c) are characteristic of high-temperature quartz, but the same habit is found

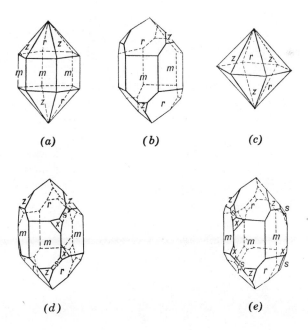

(a)   (b)   (c)

(d)   (e)

FIG. 418.   Quartz crystals.

in quartz that crystallized as the low-temperature form. Most quartz is twinned according to one or both of two laws (see page 100). These are Dauphiné, c the twin axis; and Brazil, $\{11\overline{2}0\}$ the twin plane. Both types are penetration twins and external evidence of them is rarely seen.

The size of crystals varies from individuals weighing several tons to finely crystalline coatings, forming "drusy" surfaces. Also common in massive forms of great variety. From coarse- to fine-grained crystalline to flintlike or crypto-crystalline, giving rise to many variety names (see below). May form in concretionary masses.

Angles: $m(10\overline{1}0) \wedge r(10\overline{1}1) = 38° 13'$, $r(10\overline{1}) \wedge z(01\overline{1}1) = 46° 16'$, $m(10\overline{1}0) \wedge s(11\overline{2}1) = 37° 58'$, $m(10\overline{1}0) \wedge x(51\overline{6}1) = 12° 1'$.

$C32$; $a = 4.91$, $c = 5.41$ Å; $a:c = 1:1.102$; $Z = 3$. *d's:* 4.26(8), 3.34(10), 1.818(6), 1.541(4), 1.081(5).

Physical Properties. H 7. G 2.65. Fracture conchoidal. *Luster* vitreous, in some specimens greasy, splendent. *Color* usually colorless or white, but frequently colored by impurities and may then be any color. Transparent to translucent. Strongly piezoelectric and pyroelectric. *Optics:* $(+)$; $\omega = 1.544$, $\varepsilon = 1.553$.

Composition. Si 46.7, O 53.3 per cent. Of all the minerals, quartz is most nearly a pure chemical compound with constant physical properties.

Diagnostic Features. Characterized by its glassy luster, conchoidal fracture, and crystal form. Distinguished from calcite by its high hardness, and from white varieties of beryl by its inferior hardness. Infusible. Insoluble except in hydrofluoric acid.

Varieties. A great many different forms of quartz exist, to which varietal names have been given. The more important varieties, with a brief description of each, follow.

Coarsely Crystalline Varieties

Rock Crystal. Colorless quartz, commonly in distinct crystals (Fig. 419).

Amethyst. Quartz colored various shades of violet, often in crystals. The color apparently results from small amounts of ferric iron.

Rose Quartz. Coarsely crystalline but usually without crystal form, color a rose-red or pink. Often fades on exposure to light. Small amounts of titanium appear to be the coloring agent.

Smoky Quartz; Cairngorm Stone. Frequently in crystals; smoky yellow to brown to almost black. Named *cairngorm* from the locality of Cairngorm in Scotland. Spectrographic analyses of smoky quartz show no dominant impurity and are similar to those of colorless quartz. The dark color is attributed to free silicon formed by exposure to radioactive material.

Citrine. Light yellow resembling topaz in color.

Milky Quartz. Milky white owing to minute fluid inclusions. Some specimens have a greasy luster.

FIG. 419.   Quartz crystals, Hot Springs, Arkansas.

Quartz may contain parallel fibrous inclusions which give the mineral a chatoy-ancy. When stones are cut *en cabochon* they are called *quartz cat's eye*. *Tiger's eye* is a yellow fibrous quartz pseudomorphic after the fibrous mineral crocidolite. It is also chatoyant.

With Inclusions.   Many other minerals occur as inclusions in quartz and thus give rise to variety names. *Rutilated quartz* has fine needles of rutile penetrating it. Tourmaline and other minerals are found in quartz in the same way. *Aventurine* is quartz including brilliant scales of hematite or mica. Liquids and gases may occur as inclusions; both liquid and gaseous carbon dioxide exist in some quartz.

Cryptocrystalline Varieties

Depending on their microstructure the cryptocrystalline varieties of quartz may be divided into two types: fibrous and granular. It is difficult to distinguish between them macroscopically.

A. Fibrous Varieties

Chalcedony is the general term applied to fibrous varieties. More specifically it is a brown to gray, translucent variety, with a waxy luster, often mammillary and in other imitative shapes. Chalcedony has been deposited from aqueous solutions and is frequently found lining or filling cavities in rocks. Color and banding give rise to the following varieties:

Carnelian, a red chalcedony grades into brown *sand*.

Chrysoprase is an apple-green chalcedony colored by nickel oxide.

Agate is a variety with alternating layers of chalcedony having different colors and porosity. The colors are usually in delicate, fine parallel bands which are commonly curved, in some specimens concentric (Fig. 420). Most agate used for commercial purposes is colored by artificial means. Some agates have the different colors not arranged in bands but irregularly distributed. *Moss agate* is a variety in which the variation in color is due to visible impurities, often manganese oxide in mosslike patterns.

Wood that has been petrified by replacement by clouded agate is known as *silicified* or *agatized wood*.

Onyx, like agate, is a layered chalcedony, with layers arranged in parallel planes. *Sardonyx* is an onyx with sard alternating with white or black layers.

Heliotrope or *bloodstone*, is a green chalcedony with small red spots of jasper in it.

FIG. 420.   Agate cut and polished, Brazil.

B. Granular Varieties

Flint and *chert* resemble each other and there is no sharp distinction between them. Dark siliceous nodules, usually found in chalk are called flint; whereas lighter colored bedded deposits are called chert.

Jasper is a granular cryptocrystalline quartz with dull luster usually colored red by included hematite.

Prase has a dull green color; otherwise it is similar to jasper, and occurs with it.

Occurrence. Quartz is a common and abundant mineral occurring in a great variety of geological environments. It is present in many igneous and metamorphic rocks and is a major constituent of granite pegmatites. It is the most common gangue mineral in hydrothermal metal-bearing veins and in many veins is essentially the only mineral present. In the form of flint and chert, quartz is deposited on the sea floor contemporaneously with the enclosing rock; or solutions carrying silica may replace limestone to form chert horizons. On the breakdown of quartz-bearing rocks, the quartz, because of its mechanical and chemical stability, persists as detrital grains to accumulate as sand. Sandstone and its metamorphic equivalent, quartzite, are composed essentially of quartz.

Rock crystal is found widely distributed, some of the more notable localities being: the Alps; Minas Gerais, Brazil; the island of Madagascar; and Japan. The best quartz crystals from the United States are found at Hot Springs, Arkansas, and Little Falls and Ellenville, New York. Important occurrences of amethyst are in the Ural Mountains, Czechoslovakia, Tyrol, Zambia, and Brazil. Found at Thunder Bay on the north shore of Lake Superior. In the United States found in Delaware and Chester counties, Pennsylvania; Oxford County, Maine; Black Hills, South Dakota; and Wyoming. Smoky quartz is found in large and fine crystals in Switzerland; and in the United States at Pikes Peak, Colorado; Alexander County, North Carolina; and Oxford County, Maine.

The chief source of agates at present is in southern Brazil and northern Uruguay. Most of these agates are cut at Idar-Oberstein, Germany, itself a famous agate locality. In the United States agate is found in numerous places, notably in Oregon and Wyoming. The chalk cliffs of Dover, England, are famous for the flint nodules that weather from them. Similar nodules are found on the French coast of the English channel and on islands off the coast of Denmark. Massive quartz, occurring in veins or with feldspar in pegmatite dikes, is mined in Connecticut, New York, Maryland, and Wisconsin for its various commercial uses.

Use. Quartz has many and varied uses. It is widely used as gem stones or ornamental material, as amethyst, rose quartz, cairngorm, tiger's-eye, aventurine, carnelian, agate, and onyx. As sand, quartz is used in mortar, in concrete, as a flux, as an abrasive, and in the manufacture of glass and silica brick. In powdered form it is used in porcelain, paints, sandpaper, scouring soaps, and as a wood filler. In the form of quartzite and sandstone it is used as a building stone and for paving purposes.

Quartz has many uses in scientific equipment. Because of its transparency in both the infrared and ultraviolet portions of the spectrum, quartz is made into lenses and prisms for optical instruments. The *optical activity* of quartz (the ability to rotate the plane of polarization of light) is utilized in the manufacture of an instrument to produce monochromatic light of differing wave lengths. Quartz wedges, cut from transparent crystals, are used as an accessory to the polarizing microscope. Because of its piezoelectric property, quartz has specialized uses. It is cut into small oriented plates and used as radio oscillators to permit both transmission and reception on a fixed frequency. This property also renders it useful in the measurement of instantaneous high pressures such as result from firing a gun or atomic explosion.

Name. The name quartz is a German word of ancient derivation.

Similar Species. *Lechatelierite*, SiO_2 is fused silica or silica glass. Found in fulgurites, tubes of fused sand formed by lightning, and in cavities in some lavas. Lechatelierite is also found at Meteor Crater, Arizona where sandstone has been fused by the heat generated by the impact of a meteorite.

Tridymite—SiO_2

Crystallography. Hexagonal; $6/m2/m2/m$ (high-tridymite). Orthorhombic crystal class (?) (low-tridymite). Crystals are small and commonly twinned and at room temperature are paramorphs after high-tridymite.

High-tridymite: $a = 5.04$, $c = 8.24$ Å; $a:b = 1:1.635$. $Z = 4$.

Low-tridymite: $a = 9.98$, $b = 17.26$, $c = 8.18$ Å; $a:b:c = 0.578:1:0.474$; $Z = 32$. *d's:* 4.30(10), 4.08(9), 3.81(9), 2.96(6), 2.47(6).

Physical Properties. H 7. G 2.26. *Luster* vitreous. *Color* colorless to white. Transparent to translucent. *Optics:* (+); $\alpha = 1.468–1.479$, $\beta = 1.470–1.480$, $\gamma = 1.475–1.483$; 2V = 40°–90°.

Diagnostic Features. It is impossible to identify tridymite by macroscopic means, but under the microscope its crystalline outline and refractive index distinguish it from the other silica minerals.

Occurrence. Tridymite is present on a large scale in certain siliceous volcanic rocks and for this reason may be considered an abundant mineral. Usually associated with cristobalite. It is found in large amounts in the lavas of the San Juan district of Colorado.

Name. From the Greek meaning *threefold*, in allusion to its common occurrence in trillings.

Cristobalite—SiO_2

Crystallography. High-cristobalite: Isometric; 23(?). Low-cristobalite: Tetragonal; 422. Crystals small octahedrons; the form is retained on inversion to low-cristobalite. Also in spherical aggregates.

$P2_13$. High-cristobalite: $a = 7.12$ Å; $Z = 8$. $P4_12_1(?)$ low-cristobalite: $a = 4.97$, $c = 6.93$ Å; $a:c = 1:1.394$; $Z = 4$. $d's:$ 4.03(10), 2.83(7), 2.48(8), 1.924(6), 1.876(7).

Physical Properties. **H** $6\frac{1}{2}$. **G** 2.32. *Luster* vitreous. Colorless. Translucent. *Optics:* $(+)$; $\omega = 1.484$, $\varepsilon = 1.487$.

Diagnostic Features. Infusible, but when heated to 200°C inverts to high-cristobalite and becomes nearly transparent; on cooling, again inverts and assumes its initial white, translucent appearance. The occurrence in small lava cavities in spherical aggregates and its behavior when heated are characteristic, but it cannot be determined with certainty without optical or x-ray determinations.

Occurrence. Cristobalite is present in many siliceous volcanic rocks, both as the lining of cavities and as an important constituent in the fine-grained ground mass. It is, therefore, an abundant mineral. Associated with tridymite in the lavas of the San Juan district, Colorado.

Name. From the Cerro San Cristobal near Pachuca, Mexico.

OPAL—$SiO_2 \cdot nH_2O$

Crystallography. Amorphous. Massive; often botryoidal, stalactitic. X-ray studies indicate that, although opal is essentially amorphous, it contains crystalline aggregates of cristobalite.

Physical Properties. *Fracture* conchoidal. **H** 5–6. **G** 2.0–2.25. *Luster* vitreous; often somewhat resinous. *Color* colorless, white, pale shades of yellow, red, brown, green, gray, and blue. The darker colors result from impurities. Often has a milky or "opalescent" effect and may show a fine play of colors. Transparent to translucent. Some opal, especially hyalite, shows a greenish-yellow fluorescence in ultraviolet light. *Optics:* Refractive index 1.44–1.46.

Composition. $SiO_2 \cdot nH_2O$. The water content, usually between 4 and 9 per cent, may be as high as 20 per cent. The specific gravity and refractive index decrease with increasing water content.

Diagnostic Features. Distinguished from cryptocrystalline varieties of quartz by lesser hardness and specific gravity and by the presence of water. Soluble in hot strongly alkaline solutions. Upon intense ignition gives water in the closed tube.

Varieties. *Precious opal* is characterized by a brilliant internal play of colors that may be red, orange, green, or blue. The body color is white, milky-blue, yellow, or black (*black opal*). Interference of light within the opal is the chief cause of the play of colors. *Fire opal* is a variety with intense orange to red reflections.

Common Opal. Milk-white, yellow, green, red, etc., without internal reflections.

Hyalite. Clear and colorless opal with a globular or botryoidal surface.

Geyserite or Siliceous Sinter. Opal deposited by hot springs and geysers. Found about the geysers in Yellowstone National Park.

Wood Opal. Fossil wood with opal as the petrifying material.

Diatomite. Fine-grained deposits, resembling chalk in appearance. Formed by sinking from near the surface and the accumulation on the sea floor of the siliceous tests of diatoms. Also known as *diatomaceous earth* or *infusorial earth.*

Occurrence. Opal may be deposited by hot springs at shallow depths by meteoric waters or low-temperature hypogene solutions. It is found lining and filling cavities in rocks and may replace wood buried in volcanic tuff. The largest accumulations of opal are as siliceous tests of silica-secreting organisms.

Precious opals are found at Caernowitza, Hungary; in Querétaro, Mexico; Queensland and New South Wales, Australia. Black opal has been found in the United States in Nevada and Idaho. Diatomite is mined in several western states, principally at Lompoc, California.

Use. As a gem, opal is usually cut in round shapes, *en cabochon.* Stones of large size and exceptional quality are very highly prized. Diatomite is used extensively as an abrasive, filler, filtration powder, and in insulation products.

Name. The name opal originated in the Sanskrit, *upala,* meaning stone or precious stone.

Feldspar Group

The feldspars, silicates of aluminum with potassium, sodium, and calcium and, rarely, barium, form the most important of mineral groups. They belong to both the monoclinic and the triclinic systems, but all crystals resemble each other closely in angles and habit. They all show good cleavages in two directions which make an angle of 90°, or close to 90°, with each other. Hardness is about 6, and specific gravity ranges from 2.55 to 2.76.

The Aluminosilicate Framework. In all tectosilicates other than the SiO_2 minerals, aluminum is present in 4 coordination forming aluminum-oxygen tetrahedra almost identical in size and shape to the silicon-oxygen tetrahedra. These AlO_4 tetrahedra link with SiO_4 tetrahedra by sharing oxygen ions to form a three-dimensional framework. However, because aluminum is trivalent, the AlO_4 tetrahedron has an aggregate charge of -5 rather than the -4 of SiO_4. This excess negative charge permits the introduction into the structure of one monovalent cation per AlO_4 tetrahedron. Introduction of a divalent cation calls for two AlO_4 tetrahedra, and so on. This introduction of aluminum in 4 coordination cannot be regarded as "solid solution" or "ionic proxying" of aluminum for silicon. In such minerals as orthoclase, $KAlSi_3O_8$, aluminum is an essential constituent, present in stoichiometric amount, and is not replaceable by silicon without breakdown of the structure. However, where a monovalent cation is replaced by a divalent cation, as, sodium by calcium in the plagioclase feldspars, the amount of aluminum in 4 coordination varies in proportion to the relative amounts of calcium and sodium so as to maintain electrical neutrality; the more calcium, the greater the amount of aluminum. Here the variation in amount of aluminum may be properly regarded as part of a process of coupled ionic substitution. In order to express such relationships quantitatively, a general formula for the plagioclase feldspars may be written as: $Na_{1-x}Ca_xAl(Si_{3-x}Al_x)O_8$, where x may have all the values between 0 and 1.

This formula shows that the number of ions of calcium that substitute for sodium is equaled by the number of aluminum ions substituting for silicon in 4 coordination. The total number of sodium and calcium ions must equal 1, and the total number of silicon and aluminum ions must equal 3. One tetrahedron out of every four must be AlO_4, even if all the cation sites are occupied by sodium, so one aluminum appears in the formula irrespective of the changes in the proportion of the other cations. If the writing of a formula containing parts of atoms seems unnatural, multiply all the quantities by 100. For instance, if the proportion of atoms of Na:Ca is such that twelve calcium are present out of every 100 Ca + Na, the composition may be written as:

$$Na_{88}Ca_{12}Al_{100}(Si_{288}Al_{12})O_{800} \qquad \text{or} \qquad Na_{0.88}Ca_{0.12}Al(Si_{2.88}Al_{0.12})O_8$$

Note that the aluminum atoms which balance the calcium atoms are written separately from those that are invariably present.

Structure. The feldspars are the most important of the alumino-silicate minerals that result from the partial substitution of aluminum for silicon in the tectosilicate framework. They may be considered to form three major chemical groups; the potassium feldspars, the sodium-calcium feldspars, and the barium feldspars. All have essentially the same structure, consisting of warped chains of four-membered rings extending in the direction of the a axis and joined together by ionic bonds through the potassium, sodium, calcium, or barium ions. The square, blocky outline of the chains, imparted by the four-membered rings, finds its outward expression in the right-angled cleavage and pseudotetragonal habit characteristic of the feldspars. The mono- or divalent cations are surrounded by ten oxygens, but these cannot be regarded as occupying the apices of a regular polyhedron.

Composition. Potassium, sodium, calcium, barium, and to a lesser extent iron, lead, rubidium, and cesium may all occupy the single type of cation site, and ionic substitution to some degree exists between all. However, the common feldspars may be considered to be solid solutions of the three components

orthoclase	$K(AlSi_3O_8)$
albite	$Na(AlSi_3O_8)$
anorthite	$Ca(Al_2Si_2O_8)$

Celsian, $BaAl_2Si_2O_8$, is of minor importance. Albite and anorthite form a continuous solid-solution series at all temperatures; anorthite and orthoclase display very limited solid solution, whereas albite and orthoclase form a continuous series at high temperatures, becoming discontinuous at lower temperatures. These relationships are expressed in the diagram of Fig. 421. Any composition on the triangle may be expressed by giving the percentages of three components, often abbreviated Ab, An, and Or; as $Ab_{95}An_2Or_3$ (a nearly pure albite), or $Ab_{20}An_2Or_{78}$ (a soda-rich orthoclase).

Exsolution, Perthite. If feldspar compositions in the vicinity of $Ab_{50}An_0Or_{50}$ are considered, it will be seen that these are homogeneous solid solutions at elevated temperature, in which potassium and sodium are distributed in a statistically random way among the cation sites. When the temperature decreases, the size requirements of the lattice become more stringent, and strong ordering forces operate to separate potassium and sodium into regions having the lattice configuration proper to each. This separation generally results in thin layers of potassium-saturated albite in a host crystal of orthoclase containing some sodium

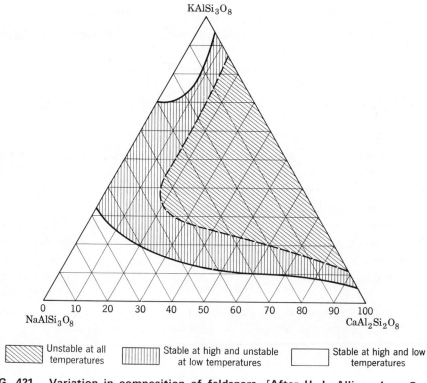

FIG. 421. Variation in composition of feldspars. [After H. L. Alling, *Jour. Geol.*, XXIX, 1921 (modified).]

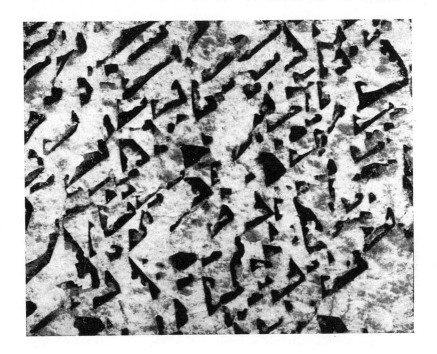

FIG. 422. Graphic granite, Hybla, Ontario. Quartz dark, feldspar light.

FIG. 423. Phenocrysts of orthoclase in lava.

still in solid solution. The layers of albite tend to form roughly parallel to {100}, perpendicular to the direction of maximum contraction of the feldspar on cooling. The growing regions of albite thus take advantage of the zones of weakness induced in the feldspar crystal by differential contraction during cooling. This process of separation is called *exsolution*, and the resulting intergrowth is called *perthite*. Perthitelike intergrowths may also arise from later replacement of one feldspar by another. When the lamellae are so finely dispersed that they can be resolved only with a microscope, the intergrowth is called *microperthite*. More rarely, the host crystal has the lattice of the sodium-calcium feldspar, and the lamellae are of orthoclase. This is called *antiperthite*.

Polymorphism. All three of the principal types of feldspar have both high- and low-temperature modifications. In the high-temperature forms the tetrahedrally coordinated aluminum is randomly distributed, whereas in the low-temperature forms aluminum and silicon have an ordered relationship. Thus, $CaAl_2Si_2O_8$ occurs not only as anorthite, in which the aluminum ions occupy definite sites but also as high-anorthite, in which aluminum and silicon are statistically distributed. Orthorhombic and hexagonal polymorphs of $CaAl_2Si_2O_8$ have been synthesized as well. Likewise, albite has a high-temperature form. Potassium feldspar comprises the ordered low-temperature *microcline*, and the partly ordered *orthoclase*, and the disordered high-temperature form, *sanidine*. Microcline is particularly characteristic of deep-seated rocks and pegmatites, orthoclase of intrusive rocks formed at intermediate temperatures and sanidine of extrusive lavas.

Although albite and anorthite differ slightly in structure, the latter having a double cell with twice the c dimension of albite, the discontinuities in the solid-solution series between them are minor, and most properties, such as specific gravity or refractive index, show linear change with chemical composition. Thus determination of a suitable property with sufficient precision permits a close approximation of the chemical composition (see Fig. 429).

ORTHOCLASE—K(AlSi₃O₈)

Crystallography. Monoclinic; $2/m$. Crystals are usually short prismatic elongated parallel to a, or elongated parallel to c and flattened on {010}. The prominent forms are b{010}, c{001}, and m{110}, often with smaller second- and fourth-order prisms, (Fig. 424). Frequently twinned according to the following laws: *Carlsbad* a penetration twin with c as twin axis (Fig. 426a); *Baveno* with {021} as twin and composition plane (Fig. 426c); *Manebach* with {001} as twin and composition plane (Fig. 426b). Commonly in crystals or in coarsely cleavable to granular masses; more rarely fine-grained, massive, and cryptocrystalline. Most abundantly in rocks as formless grains.

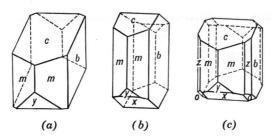

FIG. 424. Orthoclase.

Angles: $m(110) \wedge m'(1\bar{1}0) = 61° 13'$; $z(130) \wedge z'(1\bar{3}0) = 58° 48'$, $c(001) \wedge$ $y(\bar{2}01) = 80° 18'$, $c(001) \wedge x(\bar{1}01) = 50° 17'$.

$C2/m$; $a = 8.56$, $b = 13.00$, $c = 7.19$ Å; $\beta = 116° 1'$; $a:b:c = 0.658:1:0.553$. $Z = 4$. *d's:* 3.77(7), 3.46(6), 3.31(7), 3.23(10), 2.99(5).

Physical Properties. *Cleavage* {001} perfect, {010} good, {110} imperfect. **H** 6. **G** 2.57. *Luster* vitreous. *Color* colorless, white, gray, flesh-red. *Streak*

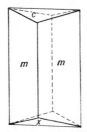

FIG. 425. Adularia.

white. *Adularia* is colorless, translucent to transparent, and is usually in pseudo-orthorhombic crystals (Fig. 425). Some adularia shows an opalescent play of colors and is called *moonstone*. *Sanidine* is glassy often transparent. *Optics:* $(-)$; $\alpha = 1.518$, $\beta = 1.524$, $\gamma = 1.526$; $2V = 10°-70°$; $Z = b$, $X \wedge a = 5°$; $r > v$.

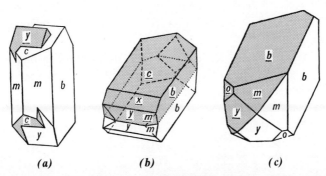

FIG. 426. Twinning in feldspar. (a) Carlsbad twin. (b) Manebach twin. (c) Baveno twin.

Composition. For $K(AlSi_3O_8)$, K_2O 16.9, Al_2O_3 18.4, SiO_2 64.7 per cent. Orthoclase and microcline (see below) are together known as *potash feldspar*. Na may replace K and in the variety *sanidine* as much as 50 per cent of the potassium is replaced. In *hyalophane*, $(K,Ba)(Al,Si)_2Si_2O_8$, barium replaces part of the potassium. Celsian, $Ba(Al_2Si_2O_8)$ is a rare barium feldspar.

Diagnostic Features. Is usually recognized by its color, hardness, and cleavage. Distinguished from the other feldspars by its right-angle cleavage and the lack of twin striations on the best cleavage surface. Difficulty fusible (5). Insoluble in acids.

Occurrence. Orthoclase is a prominent constituent of many types of intrusive igneous rocks, especially granites, syenites, and nepheline syenites; in sedimentary rocks is present in arkose and conglomerates; in metamorphic rocks in gneisses.

Sanidine is found in extrusive rocks and adularia in low-temperature hydrothermal veins.

Name. The name *orthoclase* refers to the right-angle cleavage possessed by the mineral. *Feldspar* is derived from the German word *feld*, field.

MICROCLINE—$K(AlSi_3O_8)$

Crystallography. Triclinic; $\bar{1}$. The habit, crystal forms, and interfacial angles are similar to those of orthoclase. Microcline may be twinned according to the same laws as orthoclase; Carlsbad twins are common, but Baveno and Manebach twins are rare. It is also twinned according to the *albite law* with $\{010\}$ the twin plane, and the *pericline law* with b the twin axis. These two types of twinning are usually present in microcline and on $\{001\}$ the lamellae cross at nearly $90°$ giving a characteristic grating structure. Microcline is found in cleavable masses, in crystals, and as a rock constituent in irregular grains. Microcline probably forms the largest known crystals. In a pegmatite in Karelia, U.S.S.R., a mass weighing over 2000 tons showed the continuity of a single crystal. Also in pegmatites microcline may be intimately intergrown with quartz, forming *graphic granite* (Fig. 422).

Microcline frequently has irregular and discontinuous bands crossing $\{001\}$ and $\{010\}$ that result from the exsolution of albite. The intergrowth as a whole is called *perthite*, or, if very fine, *microperthite* (see page 462).

$C\bar{1}$; $a = 8.57$, $b = 12.97$, $c = 7.22$ Å; $\alpha = 90°\ 41'$, $\beta = 115°\ 59'$, $\gamma = 87°\ 30'$; $a:b:c = 0.661:1:0.557$; $Z = 4$. *d's:* 4.21(6), 3.83(5), 3.48(5), 3.37(5), 3.24(10).

Physical Properties. *Cleavage* $\{001\}$ perfect, $\{010\}$ good at an angle of $89°\ 30'$. **H** 6. **G** 2.54–2.57. *Luster* vitreous. *Color* white to pale yellow, more rarely red or green. Green microcline is known as *Amazon stone*. Translucent to transparent. *Optics:* $(-)$; $\alpha = 1.522$, $\beta = 1.526$, $\gamma = 1.530$; $2V = 83°$; $r > v$.

Composition. Like orthoclase, $K(AlSi_3O_8)$. Sodium may replace potassium, giving rise to soda-microcline, and if sodium exceeds potassium the mineral is known as *anorthoclase*.

Diagnostic Features. Distinguished from orthoclase only by determining the presence of triclinic twinning, which can rarely be determined without the aid of the microscope. If a feldspar is a deep green it is microcline.

Occurrence. Same as for orthoclase, and often associated with it. Microcline is the common potash feldspar of pegmatites and is quarried extensively in North Carolina, South Dakota, Colorado, Virginia, Wyoming, Maine, and Connecticut. *Amazon stone* is found in the Ural Mountains and in various places in Norway and Madagascar. In the United States it is found at Pikes Peak, Colorado, and Amelia Court House, Virginia (See Fig. 427.)

Use. Feldspar is used chiefly in the manufacture of porcelain. It is ground very fine and mixed with kaolin or clay, and quartz. When heated to high temperature the feldspar fuses and acts as a cement to bind the material together. Fused feldspar also furnishes the major part of the glaze on porcelain ware. A small amount of feldspar is used in the manufacture of glass to contribute alumina to the batch. Amazon stone is polished and used as an ornamental material.

FIG. 427. Microcline and smoky quartz, Crystal Peak, Colorado.

Name. *Microcline* is derived from two Greek words meaning *little* and *inclined*, referring to the slight variation of the cleavage angle from 90°.

Plagioclase Feldspar Series

The plagioclase feldspars, also called the soda-line feldspars, form a complete solid-solution series from pure albite, $NaAlSi_3O_8$, to pure anorthite, $CaAl_2Si_2O_8$. Calcium substitutes for sodium, with a concomitant substitution of aluminum for silicon, in all proportions. The series is divided into the following six wholly arbitrary divisions, according to the relative amounts of albite and anorthite:

	Per Cent Albite	Per Cent Anorthite
Albite Na($AlSi_3O_8$)	100–90	0–10
Oligoclase	90–70	10–30
Andesine	70–50	30–50
Labradorite	50–30	50–70
Bytownite	30–10	70–90
Anorthite Ca($Al_2Si_2O_8$)	10–0	90–100

Although species names are given to the above arbitrary divisions, most of the properties vary in a uniform manner with the change in chemical composition. For this reason the series can be more easily understood if one comprehensive description is given, rather than six individual descriptions, and the dissimilarities between members indicated. The distinction between the high- and low-temperature modifications can be made only by x-ray or optical means.

ALBITE—ANORTHITE

Crystallography. Triclinic; $\bar{1}$. Crystals commonly tabular parallel to {010}; occasionally elongated on *b* (Fig. 428). In anorthite crystals may be prismatic elongated on *c*.

 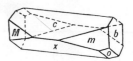

FIG. 428. Albite crystals.

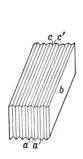

FIG. 429. Albite twinning.

Crystals are fequently twinned according to the various laws governing the twins of orthoclase, i.e., Carlsbad, Baveno, and Manebach. In addition, they are nearly always twinned according to the *albite law* and less frequently according to the pericline law. Albite twinning with twin plane {010} is commonly poly-synthetic and, since (010) ∧ (001) ≈ 86°, {001} either as a crystal face or cleavage is crossed by parallel groovings or striations (Fig. 429). Often these striations are so fine as not to be visible to the unaided eye, but on some specimens they are coarse and easily seen. The presence of the striations on the basal cleavage is one of the best proofs that a feldspar belongs to the plagioclase series. Pericline twin-ning with b as the twin axis is also polysynthetic. Striations resulting from it can be seen on {010}. Their direction on {010} is not constant but varies with the composition.

Distinct crystals are rare. Usually in twinned, cleavable masses; as irregular grains in igneous rocks.

Angles: $c(001) \wedge b(010) = 86° 24'$, albite; 85° 50′, anorthite. $m(110) \wedge M(1\bar{1}0) = 59° 14'$, albite; 59° 29′, anorthite. The corresponding angles for feldspars of intermediate composition lie between those of albite and anorthite.

$C\bar{1}$. Albite: $a = 8.14$, $b = 12.8$, $c = 7.16$ Å; $\alpha = 94° 20'$, $\beta = 116° 34'$, $\gamma = 87° 39'$. $a:b:c = 0.636:1:0.559$; $Z = 4$. Anorthite: $a = 8.18$, $b = 12.88$, $c = 14.17$ Å; $\alpha = 93° 10'$, $\beta = 115° 51'$, $\gamma = 91° 13'$; $a:b:c = 0.635:1:1.100$; $Z = 8$. *d's: ab–an* 4.02–4.03(7), 3.77–3.74(5), 3.66–3.61(6), 3.21(7)–3.20(10), 3.18(10)–3.16(7).

Physical Properties. *Cleavage* {001} perfect, {010} good. **H** 6. **G** 2.62 in albite to 2.76 in anorthite (Fig. 430). *Color* colorless, white, gray; less frequently greenish, yellowish, flesh-red. *Luster* vitreous to pearly. Transparent to translucent. A beautiful play of colors is frequently seen, especially in labradorite and andesine. *Optics.* Albite: (+); $\alpha = 1.525$, $\beta = 1.529$, $\gamma = 1.536$; 2V = 74°. Anorthite: (−); $\alpha = 1.576$, $\beta = 1.584$, $\gamma = 1.588$; 2V = 77° (see Fig. 430).

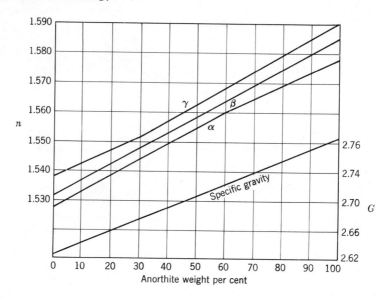

FIG. 430. Plagioclase feldspar. Variation of refractive indices and specific gravity with composition.

Composition. A complete solid-solution series extends from albite, Na(AlSi₃O₈), to anorthite, Ca(Al₂Si₂O₈). Considerable potassium may be present toward the albite end of the series.

Diagnostic Features. The plagioclase feldspars can be distinguished from other feldspars by the presence of albite twin striations on {001}. They can be placed accurately in their proper places in the series only by quantitative chemical analyses or optical tests, but they can be roughly distinguished from one another by specific gravity. Fusible at 4–4½ to a colorless glass. Albite is insoluble, but anorthite is decomposed by hydrochloric acid. Between these extremes the intermediate members show a greater solubility the greater the amount of calcium. A strong yellow sodium flame is given by members rich in sodium.

Occurrence. The plagioclase feldspars, as rock-forming minerals, are even more widely distributed and more abundant than the potash feldspars. They are found in igneous, metamorphic, and, more rarely, sedimentary rocks.

The classification of igneous rocks is based largely on the kind and amount of feldspar present (see page 488). As a rule, the greater the percentage of silica in a rock the fewer the dark minerals, the greater amount of potash feldspar, and the more sodic the plagioclase; and conversely, the lower the percentage of silica the greater the percentage of dark minerals and the more calcic the plagioclase.

Albite is included with orthoclase and microcline in what are known as the *alkali* feldspars, all of which have a similar occurrence. They are usually found together in granites, syenites, rhyolites, and trachytes. Albite is common in pegmatites where it may replace earlier microcline, also in small crystals and as the platy variety, *cleavelandite*. Some albite and oligoclase shows an opalescent play of colors and is known as *moonstone*. The name albite is derived from the Latin *albus*, meaning *white*, in allusion to the color.

Oligoclase is characteristic of granodiorites and monzonites. In some localities, notably at Tvedestrand, Norway it contains inclusions of hematite, which give the mineral a golden shimmer and sparkle. Such feldspar is called *aventurine* oligoclase, or *sunstone*. The name is derived from two Greek words meaning *little* and *fracture*, since it was believed to have less perfect cleavage than albite.

Andesine is rarely found except as grains in andesites and diorites. Named from the Andes Mountains where it is the chief feldspar in the andesite lavas.

Labradorite is the common feldspar in gabbros and basalts and in anorthosite is the only important constituent. Found on the coast of Labrador in large cleavable masses which show a fine iridescent play of colors. The name is derived from this locality.

Bytownite is rarely found except as grains in gabbros. Named from Bytown, Canada (now the city of Ottawa).

Anorthite is rarer than the more sodic plagioclase. Found in rocks rich in dark minerals and in druses of ejected volcanic blocks and in granular limestones of contact metamorphic deposits. Name derived from the Greek word meaning *oblique*, because its crystals are triclinic.

Use. Plagioclase feldspars are less widely used than potash feldspars. Albite, or *soda spar*, as it is called commercially, is used in ceramics in a manner similar to microcline. Labradorite that shows a play of colors is polished and used as an ornamental stone. Those varieties which show opalescence are cut and sold under the name of *moonstone* or *sunstone*.

Name. The name plagioclase is derived from the Greek meaning *oblique*, in allusion to the oblique angle between the cleavages. (See under "Occurrence" for names of specific species.)

Danburite—Ca($B_2Si_2O_8$)

Crystallography. Orthorhombic; $2/m2/m2/m$. Prismatic crystals, similar in habit to topaz. Commonly in crystals (see Fig. 431).

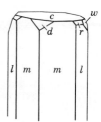

FIG. 431. Danburite.

Angles: $l(110) \wedge l'(1\bar{1}0) = 85°\ 8'$, $m(210) \wedge m'(2\bar{1}0) = 57°\ 8'$, $c(001) \wedge d(101) = 41°\ 27'$, $c(001) \wedge w(021) = 62°\ 32'$.

Pbnm; $a = 8.77$, $b = 8.03$, $c = 7.74\ Å$; $a:b:c = 1.092:1:0.964$; $Z = 4$. $d's$: 3.57(10), 3.44(7), 3.23(6), 2.96(10), 2.74(8).

Physical Properties. *Cleavage* {001} poor. **H** 7. **G** 2.97–3.02. *Luster* vitreous. *Color* colorless or pale yellow. Transparent to translucent. *Optics:* $(-)$; $\alpha = 1.630$, $\beta = 1.633$, $\gamma = 1.636$; $2V = 88°$; $X = b$, $Z = c$; $r < v$.

Composition. CaO 22.8, B_2O_3 28.4, SiO_2 48.8 per cent.

Diagnostic Features. Fusible at $3\frac{1}{2}$–4 to a colorless glass, giving a green flame. Insoluble in acids, but when previously ignited gelatinizes in hydrochloric acid. Characterized by its crystal form and high hardness. Distinguished from topaz by the test for boron.

Occurrence. Found in crystals in eastern Switzerland, Madagascar, and Japan. In the United States at Danbury, Connecticut, and Russell, New York; also found in small crystals in a salt dome in Louisiana.

Name. From the locality, Danbury, Connecticut.

Feldspathoid Group

The feldspathoids are rock-forming minerals chemically similar to the feldspars, in that they are aluminosilicates of chiefly potassium, sodium, and calcium. The chief chemical difference between feldspathoids and feldspars lies in the silica content. The feldspathoids contain about two-thirds as much silica as alkali feldspars, and hence tend to form from melts rich in alkalis (sodium and potassium) and poor in silica. Their structures are aluminosilicate frameworks in whose interstices the cations are bound and which occasionally play host to unusual anions as well. Thus, in sodalite, chlorine is an essential constituent and in cancrinite the carbonate ion, whereas noselite contains sulfate, and lazurite sulfate, sulfide, and chlorine ions. The formulas of these minerals may be thought of as three formula weights of nepheline ($NaAlSiO_4$) to one formula weight of NaCl for sodalite, one formula weight of Na_2SO_4 for noselite, etc. The structures, of course, show no such simple relation, and the strange anions are simply held in open spaces in the rather spacious aluminosilicate framework.

LEUCITE—$K(AlSi_2O_6)$

Crystallography. Isometric; $4/m\bar{3}2/m$ above 605°C. Tetragonal; $4/m$ below 605°C. Usually in trapezohedral crystals which at low temperatures are paramorphs of the high-temperature form (see Fig. 432).

Above 605°C: *Ia3d*; $a = 13.43\ Å$; $Z = 16$. Below 605°C: *I4a*; $a = 13.04$, $c = 13.85\ Å$; $a:c = 1:1.062$; $Z = 16$. $d's$: 5.39(8), 3.44(9), 2.27(10), 2.92(7), 2.84(7).

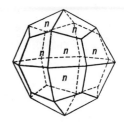

FIG. 432. Leucite.

Physical Properties. H $5\frac{1}{2}$–6. **G** 2.47. *Luster* vitreous to dull. *Color* white to gray. Translucent. *Optics:* (+); $\omega = 1.508$, $\varepsilon = 1.509$.

Composition. K_2O 21.5, Al_2O_3 23.5, SiO_2 55.0 per cent.

Diagnostic Features. Infusible. Characterized by its trapezohedral form and infusibility. Leucite is usually embedded in a fine-grained matrix, whereas analcime is usually in cavities in free-growing crystals. Moreover, analcime is fusible and yields water in the closed tube.

Occurrence. Although leucite is a rather rare mineral it is abundant in certain recent lavas; rarely observed in deepseated rocks. It is found only in silica-deficient rocks and thus never in rock containing quartz. Its most notable occurrence is as phenocrysts in the lavas of Mount Vesuvius. In the United States it is found in rocks of the Leucite Hills, Wyoming, and in certain of the rocks in the Highwood Mountains and Bear Paw Mountains, Montana. Pseudomorphs of a mixture of nepheline, orthoclase, and analcime after leucite, *pseudoleucite*, are found in syenites of Arkansas, Montana, and Brazil.

Name. From a Greek word meaning *white*.

Similar Species. *Pollucite*, $Cs_4Al_4Si_9O_{26}\cdot H_2O$, is a rare isometric mineral usually occurring in pegmatites.

NEPHELINE—(Na,K)(AlSiO₄)

Crystallography. Hexagonal; 6. Rarely in small prismatic crystals with base. Almost invariably massive, compact, and in embedded grains.

$P6_3$; $a = 10.01$, $c = 8.41$ Å; $a:c = 1:0.840$; $Z = 8$. $d's$: 4.18(7), 3.27(7), 3.00(10), 2.88(7), 2.34(6).

Physical Properties. *Cleavage* {10$\bar{1}$0} distinct. **H** $5\frac{1}{2}$–6. **G** 2.60–2.65. *Luster* vitreous in the clear crystals; greasy in the massive variety and hence called *eleolite*. *Color* colorless, white, or yellowish. In the massive variety gray, greenish, and reddish. Transparent to translucent. *Optics:* (−); $\omega = 1.529$–1.546, $\varepsilon = 1.526$–1.542.

Composition. The amount of potassium present is usually low. The percentages of oxides in the artificial compound $Na(AlSiO_4)$ are: Na_2O 21.8, Al_2O_3 35.9, SiO_2 42.3. *Kalsilite*, $K(AlSiO_4)$, forms a solid solution series with nepheline.

Diagnostic Features.   Fusible at 4 to a colorless glass giving a strong yellow flame of sodium. Readily soluble in hydrochloric acid and on evaporation yields a silica jelly. Characterized in massive varieties by its greasy luster. Distinguished from quartz by inferior hardness and from feldspar by gelatinizing in acid.

Occurrence.   Nepheline is a rock-forming mineral found in silica-deficient intrusive and extrusive rocks. Crystals are present in the lavas of Mount Vesuvius. The largest known mass of intrusive nepheline rocks is on the Kola Peninsula, U.S.S.R. where, locally, nepheline is associated with apatite. Extensive masses of nepheline rocks are found in Norway and South Africa. In the United States nepheline, both massive and in crystals, is found at Litchfield, Maine, associated with cancrinite. Found near Magnet Cove, Arkansas, and Beemerville, New Jersey. Common in the syenites of the Bancroft region of Ontario, Canada, where the associated pegmatites contain large masses of nearly pure nepheline.

Use.   Iron-free nepheline, because of its high alumina content, has been used in place of feldspar in the glass industry. Most nepheline for this purpose comes from Ontario. Nepheline produced as a byproduct of apatite mining on the Kola Peninsula is used by the Russians in several industries including ceramics, leather, textile, wood, rubber, and oil.

Name.   *Nepheline* is derived from a Greek word meaning a *cloud*, because when immersed in acid the mineral becomes cloudy. *Eleolite* is derived from the Greek word for *oil*, in allusion to its greasy luster.

Similar Species.   *Cancrinite*, $Na_6Ca_2(CO_3)(AlSiO_4)_6$, is a rare mineral similar to nepheline in occurrence and associations.

SODALITE—$Na_8(AlSiO_4)_6Cl_2$

Crystallography.   Isometric; $4/m\bar{3}2/m$. Crystals rare, usually dodecahedrons. Commonly massive, in embedded grains.

$P\bar{4}3m$; $a = 8.83$–8.91 Å; $Z = 1$. *d's:* 6.33(8), 3.64(10), 2.58(5), 2.10(2), 1.750(3).

Physical Properties.   *Cleavage* $\{011\}$ poor. **H** $5\frac{1}{2}$–6. **G** 2.15–2.3. *Luster* vitreous. *Color* usually blue, also white, gray, green, Transparent to translucent. *Optics:* refractive index 1.483.

Composition. Na_2O 25.6, Al_2O_3 31.6, SiO_2 37.2, Cl 7.3 per cent.

Diagnostic Features.   Fusible at $3\frac{1}{2}$–4, to a colorless glass, giving a strong yellow flame (sodium). Soluble in hydrochloric acid and gives gelatinous silica upon evaporation. Nitric acid solution with silver nitrate gives white precipitate of silver chloride. In a salt of phosphorus bead with copper oxide gives azure-blue copper chloride flame. Usually distinguished by its blue color, and told from lazurite by the different occurrence and absence of associated pyrite. If color is not blue, a positive test for chlorine is necessary to distinguish it from analcime, leucite, and haüynite.

Occurrence. Sodalite is a comparatively rare rock-forming mineral associated with nepheline, cancrinite, and other feldspathoids in nepheline syenites, trachytes, phonolites, etc. Found in transparent crystals in the lavas of Vesuvius. The massive blue variety is found at Litchfield, Maine; Bancroft, Ontario; and Ice River, British Columbia.

Name. Named in allusion to its sodium content.

Similar Species. Other rare feldspathoids are *haüynite*, $(Na, Ca)_{4-8}(AlSiO_4)_6$-$(SO_4)_{1-2}$ and *noselite*, $Na_8(AlSiO_4)_6SO_4$.

LAZURITE—$(Na,Ca)_8(AlSiO_4)_6(SO_4,S,Cl)_2$

Crystallography. Isometric; $4/m\bar{3}2/m$. Crystals rare, usually dodecahedral. Commonly massive, compact.

$P\bar{4}3m$; $a = 9.08$; $Z = 1$. *d's:* 3.74(10), 2.99(10), 2.53(9), 1.545(9), 1.422(7).

Physical Properties. *Cleavage* $\{011\}$ imperfect. **H** $5-5\frac{1}{2}$. **G** 2.4–2.45. *Luster* vitreous. *Color* deep azure-blue, greenish blue, Translucent. *Optics:* refractive index $1.50\pm$.

Composition. There is considerable variation in amounts of SO_4, S, and Cl. Small amounts of Rb, Cs, Sr, and Ba may substitute for Na.

Diagnostic Features. Characterized by its blue color and the presence of associated pyrite. Fusible at $3\frac{1}{2}$, giving strong yellow flame (sodium). Soluble in hydrochloric acid with slight evolution of hydrogen sulfide gas.

Occurrence. Lazurite is a rare mineral, occurring usually in crystalline limestones as a product of contact metamorphism. *Lapis lazuli* is a mixture of lazurite with small amounts of calcite, pyroxene, and other silicates, and commonly contains small disseminated particles of pyrite. The best quality of lapis lazuli comes from northeastern Afghanistan. Also found at Lake Baikal, Siberia, and in Chile.

Use. Lapis lazuli is highly prized as an ornamental stone, for carvings etc. As a powder it was formerly used as the paint pigment *ultramarine*. Now ultramarine is produced artificially.

Name. Lazurite is an obsolete synonym for azurite, and hence the mineral is named because of its color resemblance to azurite.

Petalite—$Li(AlSi_4O_{10})$

Crystallography. Monoclinic; $2/m$. Crystals rare, flattened on $\{010\}$ or elongated on $[100]$. Usually massive or in foliated cleavable masses.

$P2_1/m$; $a = 11.76, b = 5.14, c = 7.62$ Å; $\beta = 112°\ 24'$; $a:b:c = 2.288:1:1.482$; $Z = 2$. *d's:* 3.73(10), 3.67(9), 3.52(3), 2.99(1), 2.57(2).

Physical Properties. *Cleavage* $\{001\}$ perfect, $\{201\}$ good. Brittle. **H** $6-6\frac{1}{2}$. **G** 2.4. *Luster* vitreous, pearly on $\{001\}$. *Color* colorless, white, gray. Transparent to translucent. *Optics:* (+); $\alpha = 1.505$, $\beta = 1.511$, $\gamma = 1.518$; $2V = 83°$; $Z = b$, $X \wedge a = 2°-8°$. $r > v$.

Composition. LiO 4.9, Al_2O_3 16.7, SiO_2 78.4 per cent.

Diagnostic Features. Characterized by its platy habit. Distinguished from spodumene by its cleavage and lesser specific gravity. Fusible at 5 giving the red lithium flame. When gently heated it emits a blue phosphorescent light. Insoluble.

Occurrence. Petalite is found in pegmatites where it is associated with other lithium-bearing minerals such as spodumene, tourmaline, and lepidolite. Petalite was considered a rare mineral until it was found at the Varutrask pegmatite, Sweden. More recently it has been found abundantly at Bikita, Rhodesia and in South-West Africa. At these localities, associated with lepidolite and eucryptite, it is mined extensively.

Use. An important ore of lithium. (See spodumene, page 413).

Name. From the Greek, *leaf*, alluding to the cleavage.

Scapolite Series

The *scapolites* are metamorphic minerals with formulas suggestive of the feldspars and with structures consisting of endless chains of aluminosilicate framework parallel to the c axis. The crystals are tetragonal prisms elongate parallel to the chains. The structure is rather open, and accommodates large anions such as chlorine, sulfate, and carbonate in much the same way that these ions are housed in the feldspathoids. There is a solid-solution series in the scapolites between the sodium member, *marialite*, and the calcium member, *meionite*. The formula for marialite may best be recalled by considering it to consist of three formula weights of albite, $3(NaAlSi_3O_8)$ plus one formula weight of NaCl; and meionite as three formula weights of anorthite, $3(CaAl_2Si_2O_8)$ plus one formula weight of $CaCO_3$ or $CaSO_4$. There is complete ionic substitution of calcium for sodium with charge compensation effected as in the feldspars by concomitant substitution of aluminum for silicon in the chains. There is also complete substitution of carbonate, sulfate, and chlorine for each other. Therefore, most scapolite to which the name *wernerite* is given is intermediate in composition between the end members marialite and meionite.

Marialite—Meionite

Crystallography. Tetragonal; $4/m$. Crystals usually prismatic. Prominent forms are prisms {010} and {010} and dipyramid {011}; rarely shows the faces of the tetragonal dipyramid z (Fig. 433). Crystals are usually coarse, or with faint fibrous appearance.

Angles: $a(100) \wedge p(101) = 58°\ 12'$, $p(101) \wedge p'(011) = 43°\ 45'$.

$I4/m$; $a = 12.2$, $c = 7.6$ Å; $a:c = 1:0.623$; $Z = 2$. a varies slightly with composition. $d's:$ (marialite), 4.24(7), 3.78(9), 3.44(10), 3.03(10), 2.68(9).

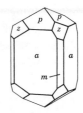

FIG. 433. Scapolite.

Physical Properties. Imperfect. *Cleavage* {100} and {110} distinct. **H** 5–6. **G** 2.55–2.74. The specific gravity and refractive indices increase with increasing Ca content. *Luster* vitreous when fresh. *Color* white, gray, pale green; more rarely bluish or reddish. Transparent to translucent. *Optics:* (−); $\omega = 1.55$–1.60, $\varepsilon = 1.54$–1.56.

Composition. Scapolite is of varied composition; the name *wernerite* is used to designate all intermediate members of a series, the end members of which are: *marialite*,

$$Na_4(AlSi_3O_8)_3(Cl,CO_3,SO_4)$$

and *meionite*,

$$Ca_4(Al_2Si_2O_8)_3(Cl,CO_3,SO_4)$$

Diagnostic Features. Fusible at 3 with intumescence to a white blebby glass and colors the flame yellow. Characterized by its crystals with square cross section and four cleavage directions at 45°. When massive resembles feldspar but has a characteristic fibrous appearance on the cleavage surfaces. It is also more readily fusible. Its fusibility with intumescence distinguishes it from pyroxene.

Occurrence. Scapolite occurs in crystalline schists, gneisses, and amphibolites, and in many cases is derived by alteration from plagioclase feldspars. It is also characteristic in crystalline limestones as a contact metamorphic mineral. Associated with diopside, amphibole, garnet, apatite, sphene, and zircon.

Crystals of gem quality with a yellow color occur in Madagascar. In the United States found in various places in Massachusetts, notably at Bolton; Orange, Lewis, and St. Lawrence counties, New York. Also found at various points in Ontario, Canada.

Name. From the Greek meaning *a shaft*, in allusion to the prismatic habit of the crystals.

Zeolite Group

The zeolites form a large group of hydrous silicates which show close similarities in composition, association and mode of occurrence. They are silicates of aluminum with sodium and calcium as the important bases. They average between $3\frac{1}{2}$ and $5\frac{1}{2}$ in hardness and between 2.0 and 2.4 in specific gravity. Many of them fuse readily with marked intumescence, hence the name *zeolite*, from two Greek words meaning *to boil* and *stone*. Traditionally, the zeolites have been thought of as well-crystallized minerals found in cavities and veins in basic igneous rocks. Recently, large deposits of zeolites have been found in western United States and in Tanzania as alterations of volcanic tuff and volcanic glass.

All the zeolites are aluminosilicates whose gross compositions somewhat resemble those of the feldspars, and, like the feldspars, are built about chains made up of 4-fold rings of SiO_4 and AlO_4 tetrahedra. The chains, bound by the interstitial cations, sodium, potassium, calcium, and barium, form an open structure with wide channelways in which water and other molecules may be readily housed. Much of the interest in zeolites derives from the presence of these spacious channels. When a zeolite is heated, the water in the channelways is given off easily and continuously as the temperature rises, leaving the structure intact. This is in sharp contrast to other hydrated compounds, as gypsum, in which the water molecules play a structural role and complete dehydration produces structural collapse. After essentially complete dehydration of a zeolite, the channels may be filled again with water or with ammonia, mercury vapor, iodine vapor, or a variety of substances. This process is selective and depends on the particular zeolite structure and the size of the molecules, and, hence, zeolites are used as "molecular sieves."

Zeolites have a further useful property that derives from their structure. Water may pass easily through the channelways, and, in the process, ions in solution may be exchanged for ions in the structure. This process is called "base exchange" or "cation exchange," and by its agency zeolites or synthetic compounds with zeolitic structure are used for water softening. The zeolite used has a composition about $Na_2Al_2Si_3O_{10} \cdot 2H_2O$ (like natrolite). "Hard" water, that is, water containing many calcium ions in solution, is passed through a tank filled with zeolite grains. The calcium ions replace the sodium ions in the zeolite, forming $CaAl_2Si_3O_{10} \cdot 2H_2O$, contributing sodium ions to the solution. Water containing sodium does not form scum and is said to be "soft." When the zeolite in the tank has become saturated with calcium, a strong NaCl brine is passed through the tank. The high concentration of sodium ion forces the reaction to go in the reverse direction, and the $Na_2Al_2Si_3O_{10} \cdot 2H_2O$ is reconstituted, calcium going into solution. By base exchange many ions, including silver, may be substituted for the alkali-metal cation in the zeolite structure.

ANALCIME—$Na(AlSi_2O_6)H_2O$

Crystallography. Isometric; $4/m\bar{3}2/m$. Usually in trapezohedrons (Fig. 434a). Cubes with trapezohedral truncations also known (Fig. 434b). Usually in crystals, also massive granular.

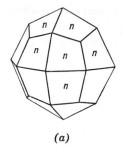

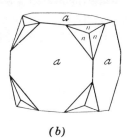

(a)   (b)

FIG. 434.   Analcime crystals.

$Ia3d$; $a = 13.71$ Å; $Z = 16$. d's: 5.61(8), 4.85(4), 3.43(10), 2.93(7), 2.51(5).

Physical Properties.   H $5–5\frac{1}{2}$. G 2.27. *Luster* vitreous. *Color* colorless or white. Transparent to translucent. *Optics:* refractive index 1.48–1.49.

Composition.   The percentages of oxides in $Na(AlSi_2O_6)H_2O$, Na_2O 14.1, Al_2O_3 23.2, SiO_2 54.5, H_2O 8.2. Some K substitutes for Na.

Diagnostic Features.   Fusible at $3\frac{1}{2}$, to a clear glass, coloring the flame yellow (sodium). Gives water in the closed tube. Usually recognized by its free-growing crystals and its vitreous luster. Crystals resemble leucite but distinguished from it by ease of fusibility, presence of water, and occurrence. (Leucite is always imbedded in rock matrix.)

Occurrence.   Analcime is characteristically found in cavities of igneous rocks associated with calcite, and other zeolites. It is also an original constituent of igneous rocks as in the analcime basalts and in places an alteration of volcanic glass. Fine crystals are found in the Cyclopean Islands near Sicily; in the Val di Fassa and on the Seiser Alpe, Trentino, Italy; in Victoria, Australia; and Kerguelen Island in the Indian Ocean. In the United States found at Bergen Hill, New Jersey; in the Lake Superior copper district; and at Table Mountain, near Golden, Colorado. Also found at Cape Blomidon, Nova Scotia.

Name.   Derived from a Greek word meaning *weak*, in allusion to its weak electric property when heated or rubbed.

NATROLITE—$Na_2(Al_2Si_3O_{10})\cdot2H_2O$

Crystallography.   Orthorhombic; $mm2$. Prismatic, often acicular with prism zone vertically striated. Usually in radiating crystal groups; also fibrous, massive, granular, or compact.

$Fdd2$; $a = 18.35$, $b = 18.70$, $c = 6.61$; $a:b:c = 0.981:1:0.353$; $Z = 8$. d's: 6.6(9), 5.89(8), 2.42(6), 3.19(9), 2.86(10).

Physical Properties. *Cleavage* {110} perfect. **H** 5–5½. **G** 2.25. *Luster* vitreous. *Color* colorless or white, rarely tinted yellow to red. Transparent to translucent. *Optics:* (+); $\alpha = 1.480$, $\beta = 1.482$, $\gamma = 1.493$; 2V = 63°; X = a, Y = b; $r < v$.

Composition. Na_2O 16.3, Al_2O_3 26.8, SiO_2 47.4, H_2O 9.5 per cent. Some K may replace Na.

Diagnostic Features. Fusible at 2½ to a clear glass, giving a yellow sodium flame. Yields water in the closed tube. Soluble in HCl and gelatinizes upon evaporation.

Occurrence. Natrolite is characteristically found lining cavities in basalt associated with other zeolites and calcite. Notable localities are Aussig and Salesel, Bohemia; Puy-de-Dôme, France; and Val di Fassa, Trentino, Italy. In the United States found at Bergen Hill, New Jersey. Also found in various places in Nova Scotia.

Name. From the Latin *natrium*, meaning *sodium*, in allusion to its composition.

Similar Species. *Scolecite* (monoclinic), $Ca(Al_2Si_3O_{10})\cdot3H_2O$, is a zeolite similar to natrolite.

CHABAZITE—$Ca_2(Al_2Si_4O_{12})\cdot6H_2O$

Crystallography. Hexagonal—R; $\bar{3}2/m$. Usually in rhombohedral crystals, {10$\bar{1}$1}, with nearly cubic angles. May show several different rhombohedrons (Fig. 435). Often in penetration twins.

Angles: $r(10\bar{1}1) \wedge r'(\bar{1}101) = 85° 14'$, $e(01\bar{1}2) \wedge r(10\bar{1}1) = 42° 37'$.

$R\bar{3}m$; $a = 13.78$, $c = 14.97$ Å; $a:c = 1:1.086$; $Z = 6$. d's: 9.46(10), 5.56(10), 5.03(10), 4.33(10), 3.87(10).

Physical Properties. *Cleavage* {10$\bar{1}$1} poor. **H** 4–5. **G** 2.05–2.15. *Luster* vitreous. *Color* white, yellow, pink, red. Transparent to translucent. *Optics:* (+); $\omega = 1.482$, $\varepsilon = 1.480$.

Composition. Usually contains Na and lesser amounts of K replacing Ca.

Diagnostic Features. Recognized usually by its rhombohedral crystals, and distinguished from calcite by its poorer cleavage and lack of effervescence in

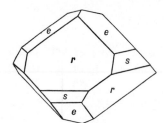

FIG. 435. Chabazite.

FIG. 436. Chabazite in trap rock cavity.

hydrochloric acid. Fuses with swelling at 3 to a blebby glass. Decomposed by hydrochloric acid with the separation of silica but without the formation of a jelly. Gives much water in the closed tube.

Occurrence. Chabazite is found, usually with other zeolites, lining cavities in basalt. Notable localities are the Faeroe Islands; the Giant's Causeway, Ireland; Aussig, Bohemia; Seiser Alpe, Trentino, Italy; and Oberstein, Germany. In the United States found at West Paterson, New Jersey, and Goble Station, Oregon. Also found in Nova Scotia and there known as *acadialite* (see Fig. 436).

Name. Chabazite is derived from a Greek word which was an ancient name for *a stone*.

HEULANDITE—Ca($Al_2Si_7O_{18}$)·6H$_2$O

Crystallography. Monoclinic; $2/m$, but crystals often simulate orthorhombic symmetry (Fig. 437); often diamond-shaped with {010} prominent.

$C2/m;$ $a = 17.71,$ $b = 17.84,$ $c = 7.46$ Å; $\beta = 116°\ 20';$ $a:b:c = 0.993:1:0.418;$ $Z = 4.$ *d's:* 8.9(7), 5.10(4), 3.97(10), 3.42(5), 2.97(7).

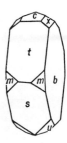

FIG. 437. Heulandite.

Physical Properties. *Cleavage* {010} perfect. **H** $3\frac{1}{2}$–4. **G** 2.18–2.2. *Luster* vitreous, pearly on {010}. *Color* colorless, white, yellow, red. Transparent to translucent. *Optics:* (+); $\alpha = 1.482$, $\beta = 1.485$, $\gamma = 1.489$; $2V = 50°$; $X = b$, $Y \wedge c = 35°$; $r > v$.

Composition. Sodium and potassium may substitute for calcium.

Diagnostic Features. Fusible at 3 with intumescence to a white glass. Yields water in the closed tube. Characterized by its crystal form and one direction of perfect cleavage with pearly luster.

Occurrence. Heulandite is usually found in cavities of basic igneous rocks associated with other zeolites and calcite. Found in notable quality in Iceland; the Faeroe Islands; Andreasberg, Harz Mountains; Tyrol, Austria; and India, near Bombay. In the United States found at West Paterson, New Jersey. Also found in Nova Scotia.

Name. In honor of the English mineral collector, H. Heuland.

STILBITE—$Ca(Al_2Si_7O_{18}) \cdot 7H_2O$

Crystallography. Monoclinic; $2/m$. Crystals usually tabular on {010} or in sheaflike aggregates (Fig. 438). They may also form cruciform penetration twins.

$C2/m$; $a = 13.63$, $b = 18.17$, $c = 11.31$ Å; $\beta = 129°\ 10'$; $a:b:c = 0.750:1:0.622$; $Z = 4$. $d's:$ 9.1(9), 4.68(7), 4.08(10), 3.41(5), 3.03(7).

Physical Properties. *Cleavage* {010} perfect. **H** $3\frac{1}{2}$–4. **G** 2.1–2.2. *Luster* vitreous; pearly on {010}. *Color* white, more rarely yellow, brown, red. Translucent. *Optics:* (−); $\alpha = 1.494$, $\beta = 1.498$, $\gamma = 1.500$; $2V = 33°$; $Y = b$, $X \wedge a = 5°$.

Composition. Na and K are usually present substituting for Ca.

Diagnostic Features. Characterized chiefly by its cleavage, pearly luster on the cleavage face, and common sheaflike groups of crystals. Fuses with intumescence at 3 to a white enamel. Decomposed by HCl with separation of silica but without the formation of a jelly. Yields water in the closed tube.

Occurrence. Stilbite is found in cavities in basalts and related rocks associated with other zeolites and calcite. Notable localities are Poonah, India; Isle of Skye,

FIG. 438. Stilbite.

FIG. 439. Stilbite, Nova Scotia.

Faeroe Islands; Kilpatrick, Scotland; and Iceland. In the United States it is found in northeastern New Jersey. Also found in Nova Scotia.

Name.   Derived from a Greek word meaning *luster*, in allusion to the pearly luster.

Similar Species.   Other zeolites of lesser importance than those described are:

Phillipsite, $KCa(Al_3Si_5O_{16})\cdot 6H_2O$, monoclinic.
Harmotome, $(Ba,K)(Al_2Si_6O_{16})\cdot 6H_2O$, monoclinic.
Thomsonite, $NaCa_2(Al_5Si_5O_{20})\cdot 6H_2O$, orthorhombic.
Gmelinite, $(Na,Ca)(Al_2Si_4O_{12})\cdot 6H_2O$, hexagonal.
Laumontite, $Ca(Al_2Si_4O_{12})\cdot 4H_2O$, monoclinic.
Scolecite, $Ca(Al_2Si_3O_{10})\cdot 3H_2O$, monoclinic.

REFERENCES

Descriptive Mineralogy

Bates, R. L., *Geology of the Industrial Rock and Minerals*. Harper and Row, New York, 1960.

Deer, W. A., R. A. Howie, and J. Zussman, *An Introduction to the Rock-forming Minerals*. John Wiley and Sons, New York, 1966.

Frondel, C., Dana's *The System of Mineralogy*, 7th ed., Vol. 3. John Wiley and Sons, New York, 1962.

Gillson, J. L., ed., *Industrial Minerals and Rocks*, 3rd ed. American Institute of Mining, Metallurgical and Petroleum Engineers, New York.

Kostov, I., *Mineralogy*. Oliver and Boyd, London, 1968.

Palache, C., H. Berman, and C. Frondel, Dana's *The System of Mineralogy*, 7th ed., Vol. 1, 1944; vol. 2, 1951. John Wiley and Sons, New York.

Ramdohr, P., and H. Strunz, Klockmann's *Lehrbuch der Mineralogie*. Ferdinand Enke Verlag, Stuttgart, 1967.

Strunz, H., *Mineralogische Tabellen*, 4th ed. Akademische Verlagsgesellschaft, Leipzig, 1966.

U.S. Bureau of Mines, *Minerals Yearbook*. Annual publication by the Government Printing Office, Washington, D.C.

6

OCCURRENCE AND ASSOCIATION OF MINERALS

Of the nearly 2000 minerals, relatively few are common. These are the rock-forming minerals that make up the bulk of the earth's crust. Which minerals form and in what proportions is largely controlled by the available chemical elements. The crust to a depth of 9 to 30 km is composed mainly of igneous rocks covered in places with a thin veneer of metamorphic and sedimentary rocks. The composition of the earth's crust is thus essentially that of the average igneous rock. Based on over 5000 rock analyses Clark and Washington (1924) calculated the weight per cent of the elements in the earth's crust. Their results, in part modified by data of more recent workers, are given in the accompanying table. It can be seen that the eight most abundant elements make up 98.59 weight per cent. Concentrations of the remaining elements have been brought about by various geologic processes. These include gravitational separation of minerals from magmas with concomitant enrichment of the rarer elements in the fluid portion. As hydrothermal solutions, the magmatic fluids are differentiated as they rise into the overlying rocks. A succession of minerals crystallize from them as they move continuously to lower pressure–lower temperature environments toward the surface. Weathering

Abundance of the Elements[a]

In Earth's Crust		In Sandstones		In Shales		In Limestones	
O	46.60	O ~	51	O ~	49	O ~	49
Si	27.72	Si	36.75	Si	27.28	Ca	3.45
Al	8.13	Al	2.53	Al	8.19	C	11.35
Fe	5.00	Ca	3.95	Fe	4.73	Mg	4.77
Ca	3.63	C	1.38	K	2.70	Si	2.42
Na	2.83	K	1.10	Ca	2.23	Al	0.43
K	2.59	F	0.99	Mg	1.48	Fe	0.40
Mg	2.09	Mg	0.71	C	1.53	P	0.11
Ti	0.44	Na	0.33	Na	0.97	S	0.11
H	0.14	S	0.28	Ti	0.43	Sr	0.049
P	0.118	Ti	0.096	S	0.26	Mn	0.039
Mn	0.100	P	0.035	P	0.074	Na	0.037
F	0.070	Rb	0.027	Cr	0.068	F	0.027
S	0.052	Cr	0.020	Mn	0.062	Cl	0.020
Sr	0.045	Ba	0.017	F	0.051	K	0.015
Ba	0.040			Ba	0.046	Ba	0.012
C	0.032			B	0.031		
Cl	0.020			Rb	0.028		
Cr	0.020			Cu	0.019		
Zr	0.016			Sr	0.017		
Rb	0.012			Zr	0.011		
V	0.110			V	0.011		

[a] In weight per cent for all elements more abundant than 0.01 per cent. For sandstones, shales, and limestones from: Jack Green, Geochemical Table.

of rocks at the surface, the life process of plants and animals, the separation of different minerals from sea water, and the sorting action of sedimentary processes all play a part in the differentiation of the chemical elements. The varying abundance of the elements in different sedimentary environments shows the results of one of these processes.

ROCKS AND ROCK-FORMING MINERALS

Since minerals occur most commonly and abundantly as rock constituents, it is appropriate that short descriptions of the most important rock types be given. Rocks are divided into three main divisions:

I. Igneous
II. Sedimentary
III. Metamorphic

I. Igneous Rocks

Igneous rocks are those that have formed by the cooling and consequent solidification of a hot fluid mass of rock material, a rock magma. A magma is a melt that, in addition to the principal elements O, Si, Al, Fe, Ca, Mg, Na, K, contains water and minor amounts of many other elements. As it cools there is, in general, a definite order of crystallization of the various mineral constituents. However, if a magma cools slowly, the mineral present in the solid rock may have formed, not by simple crystallizations, but by a series of reactions, the early-formed minerals reacting with the still fluid portion to form new minerals. This concept was developed by N. L. Bowen and is known as *Bowen's reaction principle.* Laboratory experiments and evidence from rocks themselves indicated that the major minerals of igneous rocks form in two series: the ferromagnesian minerals in a discontinuous reaction series and the feldspars in a continuous reaction series (Fig. 440).

As a magma with the composition of a basalt cools, olivine and bytownite are first to crystallize. If these minerals remain essentially in place, they will react with the melt to form pyroxene and labradorite; and the resulting rock will be a gabbro or basalt. If, however, some of the olivine and bytownite are removed, as by crystal settling, the bulk composition of the melt plus remaining crystals changes. The pyroxene reacts to form amphibole and the labradorite reacts to form a more sodic plagioclase. With continued fractionation the resulting rocks become richer in H_2O, Si, Al, Na, and K and poorer in Fe, Mg, and Ca. Thus a great variety of igneous rocks may be generated from a common parent magma.

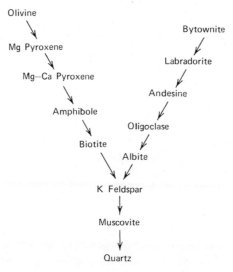

FIG. 440. The reaction series.

The mineral assemblage in any igneous rock depends on the composition of the final melt from which it crystallized. If the melt is low in silica, as a gabbro, the rock would contain silica-poor minerals and no quartz, and usually a high percentage of ferromagnesian minerals which give it a dark color. Crystallization of a melt high in silica, as a granite, results in a rock with silica-rich minerals and quartz. Because of the low percentage of dark minerals, it is usually light in color. Some rocks, as peridotites, are composed essentially of ferromagnesian minerals that, as early-formed crystals, settled out of the magma. The composition of such rocks would, therefore, not be the same as the melt from which the minerals crystallized.

In addition to the wide variation in chemical and mineral composition, igneous rocks have a great variation in the size of the constituent crystals. This is determined chiefly by the rate of cooling of the magma. A magma deeply buried in the earth's crust cools slowly and the mineral constituents crystallizing from it have time to grow to considerable size giving the rock a coarse texture. Rocks with such a deep-seated origin are called *plutonic* and the individual mineral grains can usually be differentiated by the naked eye.

On the other hand, if the magma has been extruded as lava upon the earth's surface, its subsequent cooling and solidification go on rapidly. Under these conditions the mineral particles, not already formed, have little chance to grow and the resulting rock is fine grained. In some cases the cooling has been too rapid to allow the separation of any minerals, and the resulting rock is a glass. Ordinarily the mineral constituents of fine-grained rocks can be definitely recognized only by a microscopic examination of a thin section of the rock. Such igneous rocks are known as *volcanic* or *extrusive* rocks.

Magma intruded as dikes and sills close to the earth's surface forms a group of rocks known as *hypabyssal*. The texture of these rocks is usually finer than that of the plutonic but coarser than that of the volcanic.

Porphyry and Porphyritic Texture

Some igneous rocks show distinct crystals of some minerals embedded in a much finer-grained matrix. The larger crystals are *phenocrysts*, and the finer-grained material is the *ground-mass*. (See Fig. 423, page 461.) Such rocks are known as *porphyries*. The phenocrysts may vary in size from crystals an inch or more across down to very small individuals. The groundmass may also be composed of fairly coarse-grained material, or its grains may be microscopic. It is the difference in size between the phenocrysts and the particles of the groundmass that is the distinguishing feature of a porphyry. The porphyritic texture develops when some of the crystals grow to considerable size before the main mass of the magma consolidates into the finer and uniform-grained material. Any of the types of igneous rocks described below may have a porphyritic variety, such as *granite porphyry, diorite porphyry, rhyolite porphyry.* Porphyritic varieties are more common in volcanic rocks, especially in the more siliceous types.

An igneous rock, because of the mode of its formation, consists of crystalline particles which may be said to interlock with each other, and each mineral particle is intimately and firmly imbedded in the surrounding particles. This texture will enable one ordinarily to distinguish between an igneous and a sedimentary rock; sedimentary rock being composed of grains which do not interlock and is thus not as firm and coherent as an igneous rock. Further, the texture of most igneous rocks is the same in all directions, giving rise to a fairly uniform and homogeneous mass. This characteristic will enable one to distinguish an igneous from a metamorphic rock, since the metamorphic rock commonly shows a layered structure, with a more or less definite parallel arrangement of its minerals.

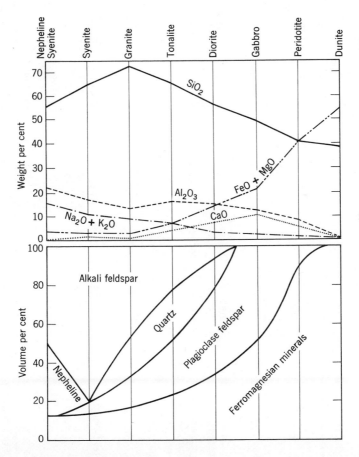

FIG. 441. Diagrams showing the relation of variations in chemical and mineral compositions of igneous rocks.

Simplified Classification of the Igneous Rocks

Feldspar	Quartz > 5 per cent		No Quartz; No Feldspathoids		Nepheline or Leucite > 5 per cent	
	Coarse	Fine	Coarse	Fine	Coarse	Fine
[a]Potash feldspar > plagioclase	Granite	Rhyolite	Syenite	Trachyte	Nepheline syenite Leucite syenite	Phonolite Leucite phonolite
Plagioclase > potash feldspar	Granodiorite	Quartz latite	Monzonite	Latite	Nepheline monzonite	
Plagioclase (oligoclase or andesine)	Tonalite	Dacite	Diorite	Andesite	Nepheline diorite	
Plagioclase (labradorite to anorthite)	Quartz gabbro		Gabbro	Basalt	Nepheline gabbro	Tephrite (−olivine) Basanite (+olivine)
No feldspar			Peridotite (olivine dominant) Pyroxenite (pyroxene dominant) Hornblendite (hornblende dominant)		Ijolite	Nephelinite (−olivine) Nepheline basalt (+olivine)

[a] Under potash feldspar are included orthoclase, microcline, anorthoclase, microperthite.

Because of the great variation in magmas both in chemistry and conditions of crystallization, igneous rocks show a wide variation in mineralogy and texture. There is a complete gradation from one rock type into another, so that the names of igneous rocks, and the boundaries between types, are largely arbitrary. (See Fig. 441.)

Classification of the Igneous Rocks

Many schemes have been proposed for the classification of igneous rocks, but the most practical for the elementary student is based on mineralogy and texture. In general, three criteria are to be considered in classifying a rock. (1) The relative amount of silica; quartz indicates an excess of silica, feldspathoids indicate a deficiency of silica. (2) The kinds of feldspar and the relative amount of each kind. (3) The texture, or size of the grains. Is the rock coarse or fine grained; that is, is it plutonic or volcanic?

It is obvious that the exact determination of the kind of feldspar or a correct estimate of the amount of each kind is impossible in the field or in the hand specimen. It is also impossible in many fine-grained rocks to recognize individual minerals. Such precise work must be left for the laboratory and carried out by the microscopic examination of thin sections of rocks. Nevertheless, it is important that the basis for the general classification be understood in order that a simplified field classification may have more meaning.

Three major divisions may be made on the basis of the silica content.[1] (1) Quartz present in amounts greater than 5 per cent. (2) Less than 5 per cent quartz or 5 per cent feldspathoids. (3) Feldspathoids in amounts greater than 5 per cent. The above divisions made on the basis of silica content are further divided according to the kind and amount (or the absence) of feldspar. Most of the rocks thus classified have a coarse- and fine-grained variety, which receive different names. The table on page 488 gives the principal rock types according to such a classification. Although these rock names are the most important, more than 600 have been proposed to indicate specific types.

The Igneous Rock-Forming Minerals

Many minerals are found in the igneous rocks, but those that can be called rock-forming minerals are comparatively few. The following list is divided into two parts: (1) the common rock-forming minerals of igneous rocks; (2) the accessory minerals of igneous rocks.

---

[1] Feldspathoids are minerals which take the place of all, or part, of the feldspar in rocks low in silica. A magma may have just the exact amount of silica to combine with the alkalis, calcium, and aluminum and form feldspar on complete crystallization of the rock. If silica is in excess of this amount, quartz will form; if there is a deficiency, feldspathoids will form.

Common Rock-Forming Minerals of Igneous Rocks	Common Accessory Minerals of Igneous Rocks
1. Quartz	1. Zircon
2. Feldspars	2. Sphene
Orthoclase	3. Magnetite
Microcline	4. Ilmenite
Plagioclase	5. Hematite
3. Nepheline	6. Apatite
4. Sodalite	7. Pyrite
5. Leucite	8. Rutile
6. Micas	9. Corundum
Muscovite	10. Garnet
Biotite	
Phlogopite	
7. Pyroxenes	
Augite	
Aegirite	
Hypersthene	
8. Amphiboles	
Hornblende	
Arfredsonite	
Riebeckite	
9. Olivine	

Plutonic Rocks

Granite–Granodiorite. Granite is a granular rock of light color and even texture consisting chiefly of feldspar and quartz. Usually both potash feldspar and oligoclase are present; potash feldspar may be flesh-colored or red, whereas oligoclase is commonly white and can be recognized by the presence of albite twinning striations. The quartz can be recognized by its greasy luster and lack of cleavage. Granites usually carry a small amount (about 10 per cent) of mica or hornblende. The mica is commonly biotite, but muscovite may also be present. The minor accessory minerals are zircon, sphene, apatite, magnetite, ilmenite.

There is a complete series grading from granite, with feldspar almost entirely the potash varieties, to granodiorite with feldspar mostly plagioclase and only slightly more than 5 per cent potash feldspar. The boundary between the two types is arbitrarily set. Granites are those rocks in which potash feldspar exceeds plagioclase; granodiorites are those in which plagioclase exceeds potash feldspar. In most instances it so happens that as the plagioclase increases in amount the percentage of dark minerals also increases, and thus, in general, granodiorites are darker than granites. However, in the field or in a hand specimen, it is usually impossible to distinguish between the two rock types with certainty.

Syenite–Monzonite. A syenite is a granular rock of light color and even texture composed essentially of potash feldspar and oligoclase, with lesser amounts of hornblende, biotite, and pyroxene. It thus resembles granite but differs from granite in that it contains less than 5 per cent quartz. Accessory minerals are apatite, sphene, zircon, and magnetite.

A series exists between syenite and monzonite, and if plagioclase feldspar exceeds potash feldspar the rock is a monzonite. Monzonites are usually darker than syenites, for an increase in dark minerals frequently accompanies an increase in plagioclase. However, without microscopic aid it is rarely possible to distinguish between the two types.

Nepheline is present in some syenites; if the amount exceeds 5 per cent, the rock is called a *nepheline syenite*. The nepheline has a greasy luster and may be mistaken for quartz, but can be distinguished by its hardness ($5\frac{1}{2}$–6). Some nepheline syenites may contain sodalite; others, corundum.

Syenites in which leucite is present in amounts greater than 5 per cent are called *leucite syenites*. The leucite can be recognized by its trapezohedral form. Such rocks are extremely rare.

Tonalite—Quartz Gabbro. A tonalite or *quartz diorite* is composed essentially of plagioclase feldspar and quartz with only minor amounts of potash feldspar (less than 5 per cent). The plagioclase is oligoclase or andesine. Dark minerals, especially biotite and hornblende, are plentiful; pyroxene is more rarely present. Apatite, sphene, magnetite are common accessory minerals. Although not essential to the classification, dark minerals are usually abundant, and thus, in general, tonalites are darker in color than granites and granodiorites.

As the plagioclase becomes richer in clacium tonalite grades into the rather uncommon rock quartz gabbro. With a lessening of the amount of quartz, it grades into *diorite*.

Diorite—Gabbro. A diorite is a granular rock characterized by plagioclase feldspar (oligoclase to andesine) but lacking quartz and potash feldspar in appreciable amounts. Hornblende is the principal dark mineral, but biotite is usually present. Pyroxenes are rare. Magnetite, ilmenite, apatite, and, less commonly, sphene and zircon are accessory minerals. Normally dark minerals are present in sufficient amount to give the rock a dark appearance.

If the plagioclase is more calcic in composition than andesine (labradorite to anorthite) the rock is a gabbro. Although the distinction is made on this criterion alone, it so happens that rocks carrying labradorite or more calcic plagioclase usually have pyroxene as the chief dark constituent, whereas the diorites with more sodic feldspar usually have amphiboles as dark minerals. Olivine is also present in most gabbros.

The name *norite* is given to a gabbro in which the pyroxene is essentially hypersthene; it is usually impossible to make this distinction without microscopic aid. A type of gabbro known as *anorthosite* is composed almost entirely of feldspar and may therefore be light in color.

If amounts of nepheline in diorites and gabbros exceed 5 per cent the rocks are called respectively *nepheline diorite* and *nepheline gabbro*. These rocks are rare and unimportant.

The term *diabase* is sometimes used to indicate a fine-grained gabbro characterized by a certain texture. This "diabasic" texture is shown microscopically to have augite filling the interstices of tabular plagioclase crystals.

Peridotite. A peridotite is a granular rock composed of dark minerals; feldspar is negligible (less than 5 per cent). The dark minerals are chiefly pyroxene and olivine in varying proportions, but hornblende may be present. If the rock is composed almost wholly of pyroxene it is called a *pyroxenite*; if it is composed almost wholly of olivine it is called a *dunite*. The name *hornblendite* is given to a rare type of rock composed almost wholly of hornblende. Magnetite, chromite, ilmenite, and garnet are frequently associated with peridotites. Platinum is associated with chromite in some peridotites, usually dunites, whereas diamond is found in a variety of altered peridotite known as *kimberlite*.

The olivine in peridotites is usually altered in whole or in part to the mineral serpentine. If the entire rock is thus altered the name *serpentine* is given to it.

Volcanic Rocks

Because of their fine-grained texture it is much more difficult to distinguish between the different types of volcanic rocks than between their plutonic equivalents. In the field only an approximate classification, depending chiefly upon whether the rock is light or dark in color, can be made. The term *felsite* is thus used to include the dense, fine-grained rocks of all colors except dark gray, dark green, or black. Felsite thus embraces the following types described below: rhyolite, trachyte, quartz latite, latite, dacite, and andesite. The experienced petrographer may be able, by the aid of a hand lens, to discern differences in texture or mineral composition which enable him to classify these rocks, but to the untrained observer they all appear much the same.

Fine-grained rocks that are a very dark green or black are called *traps*. This term is applied to dark, fine-grained rocks of indefinite mineral composition irrespective of whether they have been intruded as dikes or extruded as lava. It so happens that most rocks thus classified as traps in the field or hand specimen are basalts and satisfy the more rigorous classification based on microscopic examination.

Rhyolite is a dense fine-grained rock, the volcanic equivalent of a granite. It is thus composed essentially of alkali feldspar and quartz, but much of the silica may be present as tridymite or cristobalite. Phenocrysts of quartz, orthoclase (frequently sanidine), and oligoclase are common. Dark minerals are never abundant, but dark brown biotite is most common. Augite and hornblende are found in some rhyolites.

Rhyolites may be very uniform in appearance or may show a flow structure, giving a banded or streaked appearance to the rock. The groundmass may be partly or wholly glassy. Where the rock is completely glassy and compact, it is known as *obsidian* and is usually black. Similar glassy rocks of a brown, pitchy appearance are called *pitchstones*. *Pumice* is rhyolite glass in which expanding gas bubbles have distended the magma to form a highly vesicular material. In pumice, therefore, cavities are so numerous as to make up the bulk of the rock and give it an apparent low specific gravity.

Trachyte is the volcanic equivalent of syenite. It is thus composed chiefly of alkali feldspar with some dark minerals but lacks quartz. Small amounts of tridymite and cristobalite are often found in gas cavities. Phenocrysts of sanidine are frequently present and characteristically show Carlsbad twinning; phenocrysts of oligoclase, biotite, hornblende, and pyroxene are less common. Olivine may be present.

Banding or streaking, due to flow, is common in the trachytes. Unlike the rhyolites, glass is seldom found in the groundmass, and there are thus few glassy or vesicular types. As a result of flow the tabular feldspar frequently shows a subparallel orientation which is so common in trachytes that it is called *trachytic texture*.

Phonolite is the volcanic equivalent of nepheline syenite and is thus poorer in silica than trachyte. This is expressed mineralogically by the presence of feldspathoids. Orthoclase (sanidine) is the common feldspar; albite is rarely present. Nepheline occurs in the groundmass as minute hexagonal crystals and can be observed only by microscopic aid. Sodalite and other feldspathoids may be present, usually altered to zeolites. Lucite is present in *leucite phonolite*. It is in well-formed crystals which range from microscopic sizes to half an inch in diameter. Aegirine is the common dark mineral and normally occurs as phenocrysts, but biotite may be abundant in the leucite-rich rocks. The phonolites are completely crystalline, and there are thus no glassy varieties.

Latite and **quartz latite** are respectively the volcanic equivalents of monzonite and granodiorite. They, therefore, contain plagioclase in excess of potash feldspar. The dark minerals are chiefly biotite and hornblende. The distinction between them rests on the amount of quartz present; quartz latites would contain more than 5 per cent quartz, latites less than 5 per cent quartz. Both of these rocks are relatively unimportant.

Dacite is the dense volcanic equivalent of tonalite (quartz diorite). It contains plagioclase feldspar and quartz, both of which may occur as phenocrysts. The dark mineral is usually hornblende, but biotite is found in some varieties. Some glass may be present in the groundmass, but glassy equivalents of dacites are rare.

Andesite is the volcanic equivalent of diorite and thus is composed chiefly of oligoclase or andesine feldspar. Orthoclase and quartz are absent or present in amounts of less than 5 per cent. Hornblende, biotite, augite, or hypersthene may be present, frequently as phenocrysts. Andesites are usually named according to the dark mineral present, as *hornblende andesite, hypersthene andesite*, etc. In some andesites the groundmass is partly glassy and in rarer types completely so.

Andesites are abundant in certain localities, notably in the Andes Mountains of South America, from which locality the rock receives its name.

Basalt is a dark-colored, fine-grained rock, the volcanic equivalent of gabbro. Labradorite feldspar is the chief constituent of the groundmass, whereas more calcic plagioclase (bytownite to anorthite) may be present as phenocrysts. Augite and olivine are usually present; the augite is frequently found both as phenocrysts and in the groundmass, but olivine, as a rule, is only in phenocrysts. Brown hornblende and brown biotite are present in some basalts.

The groundmass of some basalts contains small amounts of interstitial glass and in rare instances is wholly glassy. Gas cavities near the top of basalt flows may be abundant enough to make the rock vesicular.

The presence of nepheline or leucite in basalt gives rise to the rare rock types *tephrite* and *leucite tephrite*.

Basalts are the most abundant of the volcanic rocks and form extensive lava flows in many regions; the most noted are the Columbia River flows in western United States and the Deccan "traps" of western India. Many of the great volcanos, such as form the Hawaiian Islands, are built up of basaltic material. In addition to forming extrusive rock masses, basalt is widely found forming many small dikes and other intrusives.

Fragmental Igneous Rocks

During periods of igneous activity volcanos eject much fragmental material which accumulates and forms the *fragmental igneous rocks*, or *pyroclastic rocks*. The ejectamenta vary greatly in size. Rock composed of finer particles of *volcanic ash* and *volcanic dust* is called *tuff;* that composed of coarser volcanic bombs is called *agglomerate*, or *volcanic breccia*. Such rocks are frequently waterlaid and bedded, and thus form a transition between the igneous and sedimentary rocks.

Pegmatites

Pegmatites are extremely coarse-grained igneous bodies closely related genetically and in space to large masses of plutonic rocks. They may be found as veins or dikes traversing the granular igneous rock but more commonly extend out from it into the surrounding country rock. Granites more frequently than any other rock have pegmatites genetically associated with them; consequently, unless modified by other terms, pegmatite refers to *granite pegmatite*. The minerals in most pegmatites, therefore, are the common minerals found in granite—quartz, feldspar, mica—but of extremely large size. Crystals of these minerals measuring a foot across are common, and in some localities they reach gigantic sizes. Probably the largest crystals ever found were of feldspar in pegmatites in Karelia, U.S.S.R., where material weighing thousands of tons was mined from single crystals. Quartz crystals weighing thousands of pounds and mica crystals over 10 feet across have been found. One of the characteristics of pegmatites is the simultaneous and interpenetrating crystallization of quartz and feldspar (usually microcline) to form *graphic granite*. (See Fig. 422, page 461.)

Although most pegmatites are composed entirely of the minerals found abundantly in granite, those of greatest interest contain other, and rarer, minerals. In these pegmatites there has apparently been a definite sequence in deposition. The earliest minerals are microcline and quartz, with smaller amounts of garnet and black tourmaline. These are followed, and partly replaced, by albite, lepidolite, gem tourmaline, beryl, spodumene, amblygonite, topaz, apatite, and fluorite. A host of rarer minerals such as triphylite, columbite, monazite, molybdenite, and uranium minerals may be present. In places some of the above minerals are abundant and form large crystals that are mined for their rare constituent elements. Thus spodumene crystals over 40 feet long have been found in the Black Hills of South Dakota, and beryl crystals from Albany, Maine, have measured as much as 27 feet long and 6 feet in diameter.

The formation of pegmatite dikes is believed to be directly connected with the crystallization of the larger mass of associated plutonic rock. The process of crystallization brings about a concentration of the volatile constituents in the remaining liquid portion of the magma. The presence of these volatiles (water, boron, fluorine, chlorine, and phosphorus) decreases the viscosity and thus facilitates crystallization. Such an end product of consolidation is also enriched in the rare elements originally disseminated through the magma. When this residual liquid is injected into the cooler surrounding rock, it crystallizes from the borders inward frequently giving a zonal distribution of minerals with massive quartz at the center.

Nepheline syenite pegmatites have been found in a number of localities. They are commonly rich in unusual constituents and contain numerous zirconium, titanium, and rare-earth minerals.

II. Sedimentary Rocks

The materials of which sedimentary rocks are composed have been derived from the weathering of some previously existing rock mass. They are deposited in areas of accumulation by the action of water or, less frequently, by glacial or wind action. Weathering includes both chemical decomposition and mechanical disintegration, and thus the end products consist of clay minerals, various soluble salts, and grains of inert minerals such as quartz, zircon, rutile, and magnetite. The sedimentary rocks may be divided into two classes, depending upon whether they were deposited by mechanical or chemical means. The mechanically deposited sediments such as gravel, sand, clay, or mud, known as *detrital* sediments, have been transported by streams into a body of water and deposited in layers. Sedimentary rocks of mechanical origin are composed of particles of clay minerals, or grains of minerals which have resisted chemical attack. Sedimentary rocks of chemical origin have had the materials of which they are composed dissolved by waters circulating through the rocks and brought ultimately by these waters into the sea or a lake, where, through some chemical or organic process, they are precipitated.

All sedimentary rocks are, in general, characterized by a parallel arrangement of their constituent particles forming layers or beds which are distinguished from each other by differences in thickness, size of grain, or color. In all the coarser-grained sedimentary rocks there is some material which acts as a cement and surrounds the individual mineral particles, binding them together. This cement is usually silica, calcium carbonate, or iron oxide.

Mechanical Sedimentary Rocks

Conglomerate.   Conglomerates may be considered consolidated gravels. They are composed of coarse pebbles, usually rounded by stream transportation. The individual pebbles may be composed of quartz entirely, or may be rock fragments that have not been decomposed. Fine conglomerates grade into coarse sandstones.

Sandstone.   Beds of sand that have been consolidated into rock masses are called sandstones. The constituent grains are usually rounded and waterworn but may be more or less angular. The cement which binds the sand grains together may be silica, a carbonate (usually calcite), an iron oxide (hematite or goethite), or fine-grained argillaceous material. The color of the rock depends in large measure upon the character of the cement. The rocks which have silica or calcite as their binding material are light in color, usually pale yellow, buff, white to gray; those that contain an iron oxide are red to reddish brown. It is to be noted that when a sandstone breaks it is usually the cement that is fractured, the individual grains remaining unbroken, so that the fresh surfaces of the rock have a granular appearance and feeling. The chief mineral of sandstones is quartz; if the rock contains notable amounts of feldspar it is termed *arkose*. In the finer-grained sandstones there may be considerable clayey material; such rocks grade into the shales.

Shale.   The shales are very fine-grained sedimentary rocks which have been formed by the consolidation of beds of mud, clay, or silt. They usually have a thinly laminated structure. Their color is commonly some tone of gray, although they may be white, yellow, brown, red, or green to black. They are composed chiefly of the clay minerals with quartz and mica but are too fine grained to permit the recognition of their mineral constituents by the eye alone. By the introduction of quartz and an increase in the size of grain they grade into the sandstones, and with the presence of calcite they grade into the limestones.

Chemical Sedimentary Rocks

1. Precipitation. *Limestone* composed essentially of calcite, $CaCO_3$ is the most abundant of the chemically precipitated sedimentary rocks. Although the calcite may be precipitated directly from sea water, most limestone is the result of organic precipitation. Many organisms living in the sea extract calcium carbonate from the water to build hard protective shells. On the death of the organisms the hard calcareous parts accumulate on the sea floor. When marine life is abundant, great thicknesses of shells and other hard parts may build up, which, when consolidated, become limestone. There are several varieties of limestone. *Chalk* is a porous, fine-grained limestone composed for the most part of foraminiferal shells. *Coquina*, found on the coast of Florida, consists of shells and shell fragments only partially consolidated. *Lithographic limestone* is an extremely fine-grained rock from Solenhofen, Bavaria. Although limestones are made up dominantly of calcite they may contain small amounts of other minerals. If impurities of clay are abundant, limestone grades toward shale and would be called an *argillaceous limestone*. When dolomite becomes an important constituent of a limestone, the rock grades toward the rock *dolomite*.

Oölitic Limestone. This variety of limestone, composed of small spherical concretions resembling fish roe, is believed to have been chemically precipitated. Each tiny concretion has a nucleus of a sand grain, shell fragment, or some foreign particle around which deposition has taken place. Oölitic sand which is forming at the present time on the floor of the Great Salt Lake would on consolidation, produce an oölitic limestone.

Travertine. This is a calcareous material deposited from spring waters under atmospheric conditions. If the deposit is porous it is known as *calcareous tufa*. Such deposits are prevalent in limestone regions where circulating ground water containing carbon dioxide has taken considerable calcium carbonate in solution. When the ground water reaches the surface as springs, some of the carbon dioxide is given off, resulting in the precipitation of some of the calcium carbonate as travertine.

Dolomite. The rock *dolomite* resembles limestone so closely in its appearance that it is usually impossible to distinguish between them without a chemical test. Moreover, dolomite as a rock name is not restricted to a material of the composition of the mineral dolomite but may have admixed calcite. Dolomites have formed not as original rocks but by the alteration of limestone in which part of the calcium is replaced by magnesium. This process of *dolomitization* is believed to have taken place either by the action of sea water shortly after original deposition, or by the action of circulating ground water after the rock has been consolidated and raised above sea level.

Magnesite. As a sedimentary rock *magnesite* is much more restricted in its distribution than dolomite. It has been formed by the almost complete replacement by magnesium of the calcium of an original limestone. Such magnesite rocks are found in Austria; in the Ural Mountains, U.S.S.R.; and in Stevens County, Washington.

Siliceous Sinter. In certain volcanic regions hot springs deposit an opaline material known as *siliceous sinter* or *geyserite*. The deposit apparently results from both evaporation and secretion of silica by algae.

Diatomite. Diatoms are minute one-celled organisms which live in both fresh and sea water and have the power of secreting tests of opaline material. When the organisms die, their tiny shells accumulate to build up a chalklike deposit of diatomaceous earth.

2. Evaporites. When a cut-off body of sea water or the waters of saline lakes evaporate, the dissolved salts are precipitated forming sedimentary deposits known as evaporites. Sea water contains about 3.5 per cent dissolved salts chief of which are given below.

Salts in Sea Water

Sodium chloride	2.94%	Sodium bromide	0.06%
Magnesium chloride	0.32	Potassium chloride	0.05
Magnesium sulfate	0.15	Calcium carbonate	0.01
Calcium sulfate	0.14	Iron oxide	<0.001

On evaporation the sequence of precipitation is: calcium carbonate, calcium sulfate, sodium chloride, potassium chloride, and other potassium and magnesium salts.

When the salinity of sea water has doubled through evaporation, calcium carbonate and any iron oxide present are precipitated. These are quantitatively unimportant.

On reduction of the original volume of sea water to about 20 per cent, gypsum is precipitated. With increase in salinity, anhydrite may be deposited; or if the temperature is above 34°C, anhydrite may form without prior deposition of gypsum. Under appropriate conditions extensive beds of gypsum and anhydrite have formed.

Halite, which makes up about 80 per cent of the dissolved salts, begins to precipitate when about 90 per cent of the original volume of sea water has evaporated and the density of the brine is about 1.25 g/cm^3. Thick and extensive beds of *rock salt* commonly overlie beds of gypsum and anhydrite.

Deposits of the more soluble salts such as sylvite, carnallite, and polyhalite are rare for they are not deposited until nearly complete dryness has been reached. Nevertheless, in places large deposits, particularly of sylvite, have formed and are mined extensively as the chief source of potassium.

III. Metamorphic Rocks

Metamorphic rocks are those which have undergone some chemical or physical change subsequent to their original formation. In general, metamorphic rocks are divided into two groups: (1) those formed by regional metamorphism, (2) those formed by contact metamorphism.

Regional Metamorphic Rocks

Various geologic agencies acting over wide areas bring about changes in great masses of rocks. The metamorphosis is brought about by means of high temperature and high pressure aided by the action of water and other chemical agents. The changes involve the formation of new minerals, the adding or subtracting of chemical constituents, and a physical readjustment of the mineral particles to conform to the new condition. The original rock from which a metamorphic rock has been derived may be either igneous or sedimentary, but the process of metamorphism may completely alter the original features. As rocks become involved in movements of the earth's crust, they are subjected to extreme pressures accompanied usually by high temperatures. The result will frequently be to transform the existing minerals into others, more stable under the new conditions. The physical structure of the rock will also ordinarily be changed during the process. Because of the pressure to which the rock is subjected, the mineral particles will be more or less broken and flattened or will recrystallize to form in parallel layers. This banded or laminated character given by the parallel arrangement of minerals is the most striking peculiarity of a metamorphic rock. Because of this structure a metamorphic rock can usually be distinguished from an igneous rock. Further, most metamorphic rocks have a texture of interlocking crystals which distinguishes them from sedimentary rocks. There are, of course, all gradations from a typical metamorphic rock into a sedimentary rock on the one hand and into an igneous rock on the other. The resulting metamorphic rock depends not only upon the bulk composition of the original rock but also upon the temperature and pressure at which the metamorphism took place. As the intensity of metamorphism changes with increased temperature and pressure there are corresponding mineralogical changes by means of which metamorphic rocks have been grouped into broad zones. These zones for an argillaceous rock and the principal minerals in them are listed below in order of increasing intensity.

Zone	Minerals
Chlorite	Muscovite, chlorite, quartz
Biotite	Biotite, muscovite, chlorite, quartz
Garnet	Garnet (almandite), muscovite, biotite, quartz
Staurolite	Staurolite, garnet, biotite, muscovite, quartz
Kyanite	Kyanite, garnet, biotite, muscovite, quartz
Sillimanite	Sillimanite, quartz, garnet, muscovite, biotite, oligoclase, orthoclase

The most common types of metamorphic rocks are briefly described below.

Gneiss. When the word gneiss is used alone it refers to a coarsely foliated metamorphic rock. The banding is caused by the segregation of quartz and feldspar into layers alternating with layers of dark minerals. Since the metamorphism of many igneous or sedimentary rocks may result in a gneiss, there are many varieties, with varied mineral associations. Thus such names as *granite gneiss, diorite gneiss, syenite gneiss* are used to indicate the composition; *biotite gneiss* or *hornblende gneiss* would be used to indicate rocks unusually rich in a given mineral. Granite gneisses, that is, rocks derived from the metamorphism of granites, are common, especially in regions in which rocks of Archean age are found.

Schist. Schists are metamorphic rocks which are distinguished from gneisses by the absence of coarse banding and the presence of lamination or "schistosity" along which the rock may be easily broken. There are several varieties, the most important of which is *mica schist*, composed essentially of quartz and a mica, usually either muscovite or biotite. The mica is the prominent mineral, occurring in irregular leaves and in foliated masses. The mica plates all lie with their cleavage planes parallel to each other and give to the rock a striking laminated appearance. The mica schists frequently carry characteristic accessory minerals, such as garnet, staurolite, kyanite, sillimanite, andalusite, epidote, and hornblende; thus the rock may be called *garnet schist, staurolite schist,* etc. Mica schists may have been derived from either an igneous or a sedimentary rock. Next to the gneisses the mica schists are the most common metamorphic rocks.

There are various other kinds of schistose rocks, which are chiefly derived by the metamorphism of the igneous rocks rich in ferromagnesian minerals. The most important types are *talc schist, chlorite schist,* and *amphibolite* or *hornblende schist.* They are characterized, as their names indicate, by the preponderance of some metamorphic ferromagnesian mineral.

Quartzite. As its name indicates, a quartzite is a rock composed essentially of quartz. It has been derived from a sandstone by intense metamorphism. It is a common and widely distributed rock in which solution and redeposition of silica have yielded a compact rock of interlocking quartz grains. It is distinguished from a sandstone by noting the fracture, which in a quartzite passes through the grains but in a sandstone passes around them.

Slate. Slates are exceedingly fine-grained rocks which have a remarkable property known as *slaty cleavage* that permits them to be split into thin, broad sheets. Their color is commonly gray to black but may be green, yellow, brown, and red. Slates are usually the result of the metamorphism of shales. Their characteristic slaty cleavage may or may not be parallel to the bedding planes of the original shales. Slate is rather common in occurrence.

Marble. A marble is a metamorphosed limestone. It is a crystalline rock composed of grains of calcite, or more rarely dolomite. The individual grains may be so small that they cannot be distinguished by the eye, and again they may be coarse and show clearly the characteristic calcite cleavage. Like limestone, a marble is characterized by its softness and its effervescence with acids. When pure, marble is white in color, but it may show a wide range of color, due to various impurities that it contains. It is a rock which is found in many localities and may be in thick and extensive beds. Commercially *marble* is used to indicate any lime carbonate rock capable of taking a polish and thus includes some limestones.

Serpentine. Serpentine, as a rock, is composed essentially of the mineral serpentine derived by the metamorphism of a peridotite. Such rocks are compact, of a green to greenish yellow color, and may have a slightly greasy feel. Serpentines may be a source for associated chromite and platinum, as in the Ural Mountains, or a source of nickel from the associated garnierite, as in New Caledonia.

Contact Metamorphism

When a magma is intruded into the earth's crust, it causes through the attendant heat and accompanying solutions an alteration in the surrounding rock known as *contact metamorphism.* Two kinds of contact metamorphism are recognized: (1) *thermal* or *pyrometamorphism* due to the heating of the country rock by the intrusion; (2) *hydrothermal metamorphism* or *metasomatism* in which solutions, emanating from the igneous rock, have reacted with the country rock and formed contact metamorphic minerals.

Any rock into which an igneous mass is intruded will be affected; the amount and nature of the change depends chiefly upon the size of the intruded mass and upon the chemical and physical character of the surrounding rock. Sandstones are converted to quartzites, and shales changed to "hornfels," a dense rock containing biotite, andalusite, staurolite, cordierite, garnet, and scapolite. The most striking and important contact metamorphic changes take place when the igneous rock is intruded into limestones as dolomites. When a pure limestone is subjected to thermal metamorphism, it is recrystallized and converted into a marble but without any development of new species. On the other hand, in an impure limestone or dolomite, thermal metamorphism may cause impurities such as quartz, clay, and iron oxide to react with the carbonate to form new minerals. Quartz may combine with calcite to form wollastonite and with dolomite to produce diopside. If clay is present, aluminum will enter into the reaction and such minerals as corundum, spinel, and grossularite garnet may result. If any carbonaceous materials are present, the effect of the thermal metamorphism may convert them into graphite. The common thermal contact metamorphic minerals found in impure limestone are: *graphite, spinel, corundum, wollastonite, tremolite, diopside,* and the lime garnets, *grossularite* and *andradite.*

The greatest changes are produced in carbonate rocks by *hydrothermal contact metamorphism*. In this process, solutions given off by the intruded magma react with the limestone or dolomite to produce new minerals containing elements not originally present. At the contact the country rock may be completely replaced, farther away only partly replaced, and at a distance the only evidence may be recrystallization. In this type of contact metamorphism, ore minerals are frequently present and the deposit may be of economic importance. Such deposits may grade into hydrothermal vein deposits.

The introduction of material into the hydrothermal contact metamorphic deposits gives rise to a greater abundance and variety of minerals than are formed by pure thermal metamorphism. The most important silicates are *quartz, grossularite* and *andradite* garnet, *diopside, epidote, zoisite, idocrase, wollastonite, tremolite, olivine*. Scapolite, chondrodite, axinite, topaz, tourmaline, and fluorite may be present and contain hydroxyl, fluorine, chlorine, or boron. The sulfide ore minerals include *pyrite, chalcopyrite, bornite, sphalerite, pyrrhotite, molybdenite*, and *arsenopyrite*; the oxides include *magnetite, ilmenite, hematite, spinel*, and *corundum*.

VEINS AND VEIN MINERALS

Many mineral deposits, especially those of economic importance, exist as tabular or lenticular bodies known as *veins*. The veins have been formed by the filling with mineral material of a pre-existing fracture or fissure.

The shape and general physical character of a vein depends upon the type of fissure in which its minerals have been deposited, and the type of fissure in turn depends upon the character of the rock in which it lies and the kind of force which originally caused its formation. In a firm, homogeneous rock, like a granite, a fissure will be fairly regular and clean-cut. It is likely to be comparatively narrow in respect to its horizontal and vertical extent and reasonably straight in its course. On the other hand, if a rock that is easily fractured and splintered, like a slate or a schist, is subjected to a breaking stress, a zone of narrow and interlacing fissures is more likely to have formed than one straight crack. In an easily soluble rock like a limestone, a fissure will often be extremely irregular owing to the differential solution of its walls by the waters that have flowed through it.

A typical vein consists of a mineral deposit which has filled a fissure solidly from wall to wall and shows sharply defined boundaries. There are, however, many variations from this type. Frequently irregular openings termed *vugs* may occur along the center of the vein. It is from these vugs that many well-crystallized mineral specimens are obtained. Again, the walls of a vein may not be sharply defined. The mineralizing waters that filled the fissure may have acted upon the wall rocks and partially replaced them with the vein minerals. Consequently, there may be an almost complete gradation from the unaltered rock to the pure vein filling with no sharp line of division between. Some deposits have been largely formed by the deposition of vein minerals in the wall rocks and are known as *replacement deposits*. They are more likely to be found in, but by no means confined to, the soluble rocks like limestones. There is every gradation possible, from a vein formed by filling of an open fissure with sharply defined walls to a replacement deposit with indefinite boundaries.

It is now almost universally believed that vein deposits have formed from aqueous solutions. Some deposits have formed by the method of *lateral secretion* in which ground waters circulating through the rocks have dissolved disseminated minerals and later deposited them in veins. However, the deposits of greatest economic importance are believed to have been deposited from hot ascending magmatic solutions which originated from a cooling and crystallizing magma. As such hydrothermal solutions move upward the pressure and temperature become less and deposition of dissolved material results. It has been shown that certain minerals characteristically form under given conditions of temperature and pressure. Lindgren[2] has thus divided the hydrothermal vein deposits into the following three groups, each with characteristic mineral associations.

1. **Hypothermal deposits;** deposited at great depth at high pressure and high temperature (300–500°C ±). Several types of hypothermal veins exist, each with characteristic mineral associations: (*a*) cassiterite, wolframite, and molybdenite veins; (*b*) gold-quartz veins; (*c*) copper-tourmaline veins; (*d*) lead-tourmaline veins.

The minerals of greatest importance are, therefore, cassiterite, wolframite and scheelite, molybdenite, native gold, chalcopyrite, and galena. Often associated ore minerals are pyrite, pyrrhotite, arsenopyrite, bismuthinite, and magnetite. Quartz is the predominant gangue mineral but is frequently accompanied by fluorite, tourmaline, topaz, axinite, and other minerals containing volatiles. The chief metals won from hypothermal deposits are tin, tungsten, gold, molybdenum, copper, and lead.

2. **Mesothermal deposits;** formed at intermediate depths at high pressure and temperature (200–300°C ±). The chief ore minerals are pyrite, chalcopyrite, arsenopyrite, galena, sphalerite, tetrahedrite, and native gold. Quartz is the chief gangue mineral, but carbonates such as calcite, ankerite, siderite, rhodochrosite are also common. The chief metals mined are gold, silver, copper, lead, and zinc.

3. **Epithermal deposits;** formed at slight depth under moderate pressure and temperature (50–200°C ±). Typical ore minerals of epithermal deposits are native gold, marcasite, pyrite, cinnabar, and stibnite, and the gangue minerals include quartz, opal, chalcedony, calcite, aragonite, fluorite, and barite. The chief metals found in these deposits are gold, silver, and mercury.

[2] Waldemar Lindgren. *Mineral Deposits*, McGraw-Hill Book Co., New York, 1933.

Alteration of Vein Minerals

The *primary* or *hypogene* ore minerals in many vein deposits commonly alter near the surface to *secondary* or *supergene* minerals. The primary sulfide minerals such as pyrite, chalcopyrite, galena, and sphalerite are particularly subject to alteration. Under the oxidizing influence of surface waters sphalerite alters to hemimorphite and smithsonite; galena to anglesite and cerussite; copper sulfides to malachite, azurite, cuprite, and native copper. Many other rarer oxidized minerals are found at some localities. The zone of oxidation in which these minerals form is usually relatively shallow and extends from the surface to the water table.

Pyrite is the most abundant primary vein mineral. On oxidation it forms the insoluble hydrated ferric oxide goethite (or limonite) which remains at the surface. The upper portion of a vein is thus frequently converted to a cellular rusty mass of goethite called the *gossan*. The yellow-brown, rusty appearance of the outcrop of many veins is so characteristic as to be a guide to ore.

The oxidation of pyrite and copper sulfides produces ferric sulfate and sulfuric acid which react with the copper to form soluble copper sulfate. As this copper sulfate solution moves downward into a reducing environment, it reacts with the sulfides present there, coating them with or replacing them by secondary copper sulfides, chiefly chalcocite. As oxidation penetrates deeper these minerals in turn react and there is thus a downward migration of the valuable metal in the veins. Consequently, near the water table there may be a zone of *secondary enrichment*, a concentration of copper derived from the primary minerals originally above. In some districts massive chalcocite "blankets" formed in this way have proved to be extremely rich ore bodies. Below the zone of secondary enrichment the ore grades into the original primary minerals.

REFERENCES

Rocks

Heinrich, E. W., *Microscopic Petrography*, McGraw-Hill Book Co., 1956.

Pirsson, I. V., and A. Knopf, *Rocks and Rock Minerals*, 3rd ed. John Wiley and Sons, New York, 1947.

Williams, H., F. J. Turner, and C. M. Gilbert, *Petrography*. W. H. Freeman, San Francisco, 1950.

Ore Minerals

Bateman, A. M., *The Formation of Mineral Deposits*. John Wiley and Sons, 1951.

Pegmatites

Jahns, R. H., "The Study of Pegmatites," *Economic Geology*, 50th Ann. Vol., pp. 1025–1130, 1952.

DETERMINATIVE
MINERALOGY

INTRODUCTION

The determinative tables given in this chapter are based on the physical properties of minerals that can be easily and quickly determined. They should be used with the understanding that for many minerals there is a variation in physical properties from specimen to specimen. Color for some minerals is constant but for others is extremely variable. Hardness, although more definite, may vary slightly and, by change in the state of aggregation of a mineral, may appear to vary more widely. Cleavage may also be obscured in a fine-grained aggregate of a mineral. In using the tables it is often impossible to differentiate between two or three similar species. However, by consulting the descriptions of these possible minerals in the section on Descriptive Mineralogy and the specific tests given there, a definite decision can usually be made.

In the determinative tables only the common minerals or those which, though rarer, are of economic importance have been included. The chances of having to determine a mineral that is not included in these tables are small, but it must be borne in mind that there is such a possibility. The names of the minerals have been printed in three different styles of type, as **PYRITE**, CHALCOCITE, and

Stephanite, in order to indicate their relative importance and frequency of occurrence. Whenever there is ambiguity in placing a mineral, it has been included in the two or more possible divisions.

The general scheme of classification is outlined on page 507. The proper division in which to look for a mineral can be determined by means of the tests indicated there. The tables are divided into two main sections on the basis of luster: (1) metallic and submetallic and (2) nonmetallic. The metallic minerals are opaque and give black or dark-colored streaks; whereas, nonmetallic minerals are nonopaque and give either colorless or light-colored streaks. The tables are next subdivided according to hardness. These divisions are easily determined. For nonmetallic minerals they are: $<2\frac{1}{2}$, can be scratched by the fingernail; $>2\frac{1}{2}-<3$, cannot be scratched by fingernail, can be scratched by copper cent; $>3-<5\frac{1}{2}$, cannot be scratched by copper cent, can be scratched by a knife; $>5\frac{1}{2}-<7$, cannot be scratched by a knife, can be scratched by quartz; >7, cannot be scratched by quartz. Metallic minerals have fewer divisions of hardness; they are $<2\frac{1}{2}$; $>2\frac{1}{2}-<5\frac{1}{2}$; $>5\frac{1}{2}$.

Nonmetallic minerals are further subdivided according to whether they show a *prominent* cleavage. If the cleavage is not prominent and imperfect or obscure, the mineral is included with those that have no cleavage. Minerals in which cleavage may not be observed easily, because of certain conditions in the state of aggregation, have been included in both divisions.

The minerals that fall in a given division of the tables have been arranged according to various methods. In some cases, those that possess similar cleavages have been grouped together; frequently color determines the order, etc. The column farthest to the left will indicate the method of arrangement.

The figures given in the column headed **G** are specific gravity. This is an important diagnostic physical property but less easily determined than luster, hardness, color, and cleavage. For a discussion of specific gravity and methods for its accurate determination, see page 131. However, if a specimen is pure and of sufficient size, its approximate specific gravity can be determined by simply weighing it in the hand. Below is a list of common minerals with a wide range of

Gypsum	2.32	Topaz,	3.53	Cassiterite,	6.95
Orthoclase,	2.57	Corundum,	4.02	Galena,	7.50
Quartz,	2.65	Barite,	4.45	Cinnabar,	8.10
Calcite,	2.71	Pyrite,	5.02	Copper,	8.9
Fluorite,	3.18	Arsenopyrite,	6.07	Silver,	10.5

specific gravity. By experimenting with specimens of these one can become expert in the approximate determination of specific gravity. Following the determinative tables is a list of the common minerals arranged according to increasing specific gravity.

DETERMINATIVE TABLES

General Classification of the Tables

Luster—Metallic or Submetallic

I. Hardness: $<2\frac{1}{2}$. (Will leave a mark on paper.) Page 508.

II. Hardness: $>2\frac{1}{2}$, $<5\frac{1}{2}$. (Can be scratched by knife; will not readily leave a mark on paper.) Page 509.

III. Hardness: $>5\frac{1}{2}$. (Cannot be scratched by knife.) Page 515.

Luster—Nonmetallic

I. Streak definitely colored. Page 518.

II. Streak colorless.

  A. Hardness: $<2\frac{1}{2}$. (Can be scratched by fingernail.) Page 522.

  B. Hardness: $>2\frac{1}{2}$, <3. (Cannot be scratched by fingernail; can be scratched by cent.)

    1. Cleavage prominent. Page 525.

    2. Cleavage not prominent.

      a. A small splinter is fusible in the candle flame. Page 527.

      b. Infusible in candle flame. Page 528.

  C. Hardness: >3, $<5\frac{1}{2}$. (Cannot be scratched by cent; can be scratched by knife.)

    1. Cleavage prominent. Page 529.

    2. Cleavage not prominent. Page 534.

  D. Hardness: $>5\frac{1}{2}$, <7. (Cannot be scratched by knife; can be scratched by quartz.)

    1. Cleavage prominent. Page 539.

    2. Cleavage not prominent. Page 541.

  E. Hardness: >7. (Cannot be scatched by quartz.)

    1. Cleavage prominent. Page 546.

    2. Cleavage not prominent. Page 547.

Streak	Color	G.	H.	Remarks	Name, Composition, Crystal System
Black	Iron-black	4.7	1–2	Usually splintery or in radiating fibrous aggregates.	**PYROLUSITE** MnO_2 Tetragonal
	Steel-gray to iron-black	2.3	$1-1\frac{1}{2}$	Cleavage perfect {0001}. May be in hexagonal-shaped plates. Greasy feel.	**GRAPHITE** C Hexagonal
Black to greenish black	Blue-black	4.7	$1-1\frac{1}{2}$	Cleavage perfect {0001}. May be in hexagonal-shaped leaves. Greenish streak on glazed porcelain (graphite, black). Greasy feel.	MOLYBDENITE MoS_2 Hexagonal
Gray-black	Blue-black to lead-gray	7.6	$2\frac{1}{2}$	Cleavage perfect cubic {100}. In cubic crystals. Massive granular.	**GALENA** PbS Isometric
	Blue-black	4.5	2	Cleavage perfect {010}. Bladed with cross striations. Fuses in candle flame.	**STIBNITE** Sb_2S_3 Orthorhombic
Bright red	Red to vermilion	8.1	$2-2\frac{1}{2}$	Cleavage perfect {10$\bar{1}$0}. Luster adamantine. Usually granular massive.	**CINNABAR** HgS Rhombohedral
Red-brown	Red to vermilion	5.2	1 +	Earthy. Frequently as pigment in rocks. Crystalline hematite is harder and black.	**HEMATITE** Fe_2O_3 Rhombohedral

Streak	Color	G.	H.	Remarks	Name, Composition, Crystal System
Black. May leave a slight mark on paper	Gray-black	7.3	$2-2\frac{1}{2}$	Usually massive or earthy. Easily sectile. Bright steel-gray on fresh surfaces; darkens on exposure.	Argentite Ag_2S Isometric
	Blue; may tarnish to blue-black	4.6	$1\frac{1}{2}-2$	Usually in platy masses or in thin 6-sided platy crystals. Moistened with water turns purple.	Covellite CuS Hexagonal

LUSTER: METALLIC OR SUBMETALLIC
II. Hardness: $>2\frac{1}{2}$, $<5\frac{1}{2}$
(Can be scratched by a knife; will not readily leave a mark on paper.)

Streak	Color	G.	H.	Remarks	Name, Composition, Crystal System
Black; may have brown tinge	Steel-gray. May tarnish to dead black on exposure	4.7 to 5.0	$3-4\frac{1}{2}$	Massive or in tetrahedral crystals. Often associated with silver ores.	TETRAHEDRITE $(Cu, Fe, Zn, Ag)_{12}$-Sb_4S_{13} Isometric
Gray-black		5.7	$2\frac{1}{2}-3$	Somewhat sectile. Usually compact massive. Associated with other copper minerals.	CHALCOCITE Cu_2S Orthorhombic
Black. May leave a mark on paper	Steel-gray on fresh surface. Tarnishes to dull gray	7.3	$2-2\frac{1}{2}$	Easily sectile. Usually massive or earthy. Rarely in cubic crystals.	Argentite Ag_2S Pseudoisometric

LUSTER: METALLIC OR SUBMETALLIC
II. Hardness: $>2\frac{1}{2}$, $<5\frac{1}{2}$
(Can be scratched by a knife; will not readily
leave a mark on paper.) (*Continued*)

Streak	Color	G.	H.	Remarks	Name, Composition, Crystal System
Black	Iron-black	4.7	1–2	Usually splintery or in radiating fibrous aggregates.	**PYROLUSITE** MnO_2 Tetragonal
	Steel-gray on fresh surface. Tarnishes to dull gray on exposure	6.2	2–2$\frac{1}{2}$	Fuses easily in candle flame. Usually in small irregular masses, often earthy. A rare mineral.	Stephanite Ag_5SbS_4 Orthorhombic
	Gray-black	4.4	3	Cleavage {110}. Usually in bladed masses showing cleavage. Associated with other copper minerals.	ENARGITE Cu_3AsS_4 Orthorhombic
		5.5 to 6.0	2–3	Fuses easily in candle flame. Characteristically in fibrous, featherlike masses.	Jamesonite $Pb_4FeSb_6S_{14}$ Monoclinic
		5.8 to 5.9	2$\frac{1}{2}$–3	Fuses easily in candle flame. In stout prismatic crystals; characteristically twinned with re-entrant angles.	Bournonite $PbCuSbS_3$ Orthorhombic
		6.2	2$\frac{1}{2}$	Fuses easily in candle flame. Often in thin 6-sided plates with triangular marking. Also massive and earthy.	Polybasite $Ag_{16}Sb_2S_{11}$ Monoclinic

Streak	Color	G.	H.	Remarks	Name, Composition, Crystal System
Black	Steel-gray	4.4	4	Decrepitates and fuses in candle flame. Irregular massive.	Stannite Cu_2FeSnS_4 Tetragonal
Gray-black. Will mark paper	Lead-gray	4.6	2	Cleavage perfect {010}. Fuses easily in candle flame. In bladed crystal aggregates with cross striations.	**STIBNITE** Sb_2S_3 Orthorhombic
		7.5	$2\frac{1}{2}$	Cleavage perfect {100}. In cubic crystals and granular masses. If held in the candle flame does not fuse but small globules of metallic lead collect on the surface.	**GALENA** PbS Isometric
Gray-black	Tin-white; tarnishes to dark gray	5.7	$3\frac{1}{2}$	Usually occurs in fibrous botryoidal masses. Heated in the candle flame gives off white fumes and yields a strong garlic odor.	Arsenic As Rhombohedral
	Tin-white	8 to 8.2	2	Cleavage {010}. Fuses easily in candle flame. Often as thin coatings and in lath-shaped crystals.	Sylvanite $(Au, Ag)Te_2$ Monoclinic
	Tin-white to brass yellow	9.4	$2\frac{1}{2}$	Fuses easily in candle flame. In irregular masses or in thin deeply striated lath-shaped crystals. Told from sylvanite by its lack of cleavage.	Calaverite $AuTe_2$ Monoclinic
Black	Pale copper-red. May be silver-white, pinkish	7.8	$5-5\frac{1}{2}$	Usually massive. May be coated with green nickel bloom. Associated with cobalt and nickel minerals.	NICCOLITE NiAs Hexagonal

LUSTER: METALLIC OR SUBMETALLIC
II. Hardness: $> 2\frac{1}{2}$, $< 5\frac{1}{2}$
(Can be scratched by a knife; will not readily
leave a mark on paper.)   (*Continued*)

Streak	Color	G.	H.	Remarks	Name, Composition, Crystal System
Black	Fresh surface brownish bronz; purple tarnish	5.1	3	Usually massive. Associated with other copper minerals, chiefly chalcocite and chalcopyrite.	**BORNITE** Cu_5FeS_4 Isometric
	Brown-ish bronze	4.6	4	Small fragments magnetic. Usually massive. Often associated with chalcopyrite and pyrite.	**PYRRHOTITE** $Fe_{1-x}S$ Hexagonal
		4.6 to 5.0	$3\frac{1}{2}$–4	Octahedral parting. Resembles pyrrhotite with which it usually is associated but nonmagnetic	Pentlandite $(Fe,Ni)_9S_8$ Isometric
	Brass-yellow	4.1 to 4.3	$3\frac{1}{2}$–4	Usually massive, but may be in crystals resembling tetrahedrons. Associated with other copper minerals and pyrite.	**CHALCOPYRITE** $CuFeS_2$ Tetragonal
	Brass-yellow. Slender crystals almost greenish	5.5	3–$3\frac{1}{2}$	Cleavage $\{10\bar{1}1\}$, rarely seen. Usually in radiating groups of hairlike crystals.	MILLERITE NiS Rhombohedral
Dark brown to black	Steel-gray to iron-black	4.3	4	In radiating fibrous or crystalline masses. Distinct prismatic crystals often grouped in bundles. Associated with pyrolusite.	MANGANITE $MnO(OH)$ Orthorhombic

Streak	Color	G.	H.	Remarks	Name, Composition, Crystal System
Dark brown to black	Iron-black to brown-ish black	4.6	$5\frac{1}{2}$	Luster pitchy. May be accompanied by yellow or green oxidation products. Usually in masses in peridotites.	**CHROMITE** $FeCr_2O_4$ Isometric
	Brown to black	7.0 to 7.5	$5-5\frac{1}{2}$	Cleavage perfect {010}. With greater amounts of Mn streak and color are darker.	WOLFRAMITE $(Fe,Mn)WO_4$ Monoclinic
Black	Blue; may tarnish to blue black	4.6	$1\frac{1}{2}-2$	Usually in platy masses or in thin 6-sided platy crystals. Moistened with water turns purple.	Covellite CuS Hexagonal
Usually black. May be brownish	Black	3.7 to 4.7	$5-6$	Massive botryoidal and stalactitic. Usually associated with pyrolusite.	PSILOMELANE $(Ba,H_2O)_2Mn_5O_{10}$ Appears amorphous
Light to dark brown	Dark brown to coal-black. More rarely yellow or red	3.9 to 4.1	$3\frac{1}{2}-4$	Cleavage perfect {110} (6 directions). Usually cleavable granular; may be in tetrahedral crystals. The streak is always of a lighter color than the specimen.	**SPHALERITE** ZnS Isometric
Red brown to Indian red	Dark brown to steel-gray to black	4.8 to 5.3	$5\frac{1}{2}-6\frac{1}{2}$	Usually harder than knife. Massive, radiating, reniform, micaceous.	**HEMATITE** Fe_2O_3 Rhombohedral
	Deep red to black	5.85	$2\frac{1}{2}$	Cleavage {10$\bar{1}$1}. Fusible in candle flame. Shows dark ruby-red color in thin splinters. Associated with other silver minerals.	Pyrargyrite Ag_3SbS_3 Rhombohedral

LUSTER: METALLIC OR SUBMETALLIC
II. Hardness: $>2\frac{1}{2}$, $<5\frac{1}{2}$
(Can be scratched by a knife; will not readily leave a mark on paper.) *(Continued)*

Streak	Color	G.	H.	Remarks	Name, Composition, Crystal System
Red brown to Indian red	Red-brown to deep red. Ruby-red if transparent	6.0	$3\frac{1}{2}$–4	Massive or in cubes or octahedrons. May be in very slender crystals. Associated with malachite, azurite, native copper.	**CUPRITE** Cu_2O Isometric
Bright red	Ruby-red	5.55	2–$2\frac{1}{2}$	Cleavage $\{10\bar{1}1\}$. Fusible in candle flame. Associated with pyrargyrite.	Proustite Ag_3AsS_3 Rhombohedral
Yellow brown. Yellow ocher	Dark brown to black	4.37	5–$5\frac{1}{2}$	Cleavage $\{010\}$. In radiating fibers, mammillary and stalactitic forms. Rarely in crystals.	**GOETHITE** $HFeO_2$ Orthorhombic
Dark red. (Some varieties mark paper.)	Dark red to vermilion	8.10	$2\frac{1}{2}$	Cleavage $\{10\bar{1}1\}$. Usually granular or earthy. Commonly impure and dark red or brown. When pure, translucent or transparent and bright red.	**CINNABAR** HgS Rhombohedral
Copper-red, shiny	Copper-red on fresh surface; black tarnish	8.9	$2\frac{1}{2}$–3	Malleable. Usually in irregular grains. May be in branching crystal group or in rude isometric crystals.	COPPER Cu Isometric
Silver-white, shiny	Silver-white on fresh surface. Black tarnish	10.5	$2\frac{1}{2}$–3	Malleable. Usually in irregular grains. May be in wire, plates, branching crystal groups.	SILVER Ag Isometric

Streak	Color	G.	H.	Remarks	Name, Composition, Crystal System
Gray, shiny	White or steel-gray	14 to 19	$4-4\frac{1}{2}$	Malleable. Irregular grains or nuggets. Unusually hard for a metal. Rare.	Platinum Pt Isometric
Silver-white, shiny	Silver-white with reddish tone	9.8	$2-2\frac{1}{2}$	Cleavage perfect {0001}. Sectile. Fusible in candle flame. When hammered, at first malleable but soon breaks into small pieces.	Bismuth Bi Rhombohedral
Gold-yellow, shiny	Gold-yellow	15.0 to 19.3	$2\frac{1}{2}-3$	Malleable. Irregular grains, nuggets, leaves. Very heavy; specific gravity varies with silver content.	GOLD Au Isometric

LUSTER: METALLIC OR SUBMETALLIC
III. Hardness: $>5\frac{1}{2}$
(Cannot be scratched by a knife.)

Streak	Color	G.	H.	Remarks	Name, Composition, Crystal System
Black	Silver- or tin-white	6.0 to 6.2	$5\frac{1}{2}-6$	Usually massive. Crystals pseudorthorhombic.	**ARSENOPYRITE** FeAsS Monoclinic
		6.1 to 6.9	$5\frac{1}{2}-6$	Usually massive. Crystals pyritohedral. May be coated with pink nickel bloom.	Skutterudite-Nickel skutterudite $(Co,Ni,Fe)As_3$-$(Ni,Co,Fe)As_3$ Isometric
		6.33	$5\frac{1}{2}$	Commonly in pyritohedral crystals with pinkish cast. Also massive.	Cobaltite-Gersdorffite CoAsS—NiAsS Isometric

Streak	Color	G.	H.	Remarks	Name, Composition, Crystal System
Black	Copper-red to silver-white with pink tone	7.5	$5-5\frac{1}{2}$	Usually massive. May be coated with green nickel bloom.	NICCOLITE NiAs Hexagonal
	Pale brass-yellow	5.0	$6-6\frac{1}{2}$	Often in pyritohedrons or striated cubes. Massive granular. Most common sulfide.	**PYRITE** FeS_2 Isometric
	Pale yellow to almost white	4.9	$6-6\frac{1}{2}$	Frequently in "cock's comb" crystal groups and radiating fibrous masses.	**MARCASITE** FeS_2 Orthorhombic
	Black	5.18	6	Strongly magnetic. Crystals octahedral. May show octahedral parting.	**MAGNETITE** Fe_3O_4 Isometric
Dark brown to black	Black	9.0 to 9.7	$5\frac{1}{2}$	Luster pitchy. Massive granular, botryoidal crystals.	Uraninite UO_2 Isometric
		4.7	$5\frac{1}{2}-6$	May be slightly magnetic. Often associated with magnetite. Massive granular; platy crystals; as sand.	**ILMENITE** $FeTiO_3$ Rhombohedral
Dark brown to black	Black	3.7 to 4.7	$5-6$	Compact massive, stalactitic, botryoidal. Associated with other manganese minerals and told by greater hardness.	PSILOMELANE $(Ba,H_2O)_2Mn_5O_{10}$ Monoclinic

Streak	Color	G.	H.	Remarks	Name, Composition, Crystal System
Dark brown to black		5.3 to 7.3	6	Luster black and shiny on fresh surface. May have slight bluish tarnish. Granular or in stout prismatic crystals.	Columbite- Tantalite $(Fe,Mn)(Nb,Ta)_2O_6$ Orthorhombic
Dark brown	Iron-brown to brown-ish black	7.0 to 7.5	5–5½	Cleavage perfect {010}. With greater amounts of Mn streak and color are darker.	WOLFRAMITE $(Fe, Mn)WO_4$ Monoclinic
		4.6	5½	Luster pitchy. Frequently accompanied by green oxidation products. Usually in granular masses in peridotites.	**CHROMITE** $FeCr_2O_4$ Isometric
		5.15	6	Slightly magnetic. Granular or in octahedral crystals. Common only at Franklin, N. J., associated with zincite and willemite.	FRANKLINITE (Fe,Zn,Mn)- $(Fe,Mn)_2O_4$ Isometric
Red-brown, Indian red	Dark brown to steel-gray to black	4.8 to 5.3	5½–6½	Radiating, reniform, massive, micaceous. Rarely in steel-black rhombohedral crystals. Some varieties softer.	**HEMATITE** Fe_2O_3 Rhombohedral
Pale brown	Brown to black	4.18 to 4.25	6–6½	In prismatic crystals vertically striated; often slender acicular. Crystals frequently twinned. Found in black sands.	**RUTILE** TiO_2 Tetragonal
Yellow-brown to yellow-ocher	Dark brown to black	4.37	5–5½	Cleavage {010}. Radiating, colloform, stalactitic.	**GOETHITE** $HFeO_2$ Orthorhombic

Streak	Color	G.	H.	Remarks	Name, Composition, Crystal System
Dark red	Dark red to vermil-ion	8.10	$2\frac{1}{2}$	Cleavage $\{10\bar{1}0\}$. Usually granular or earthy. Commonly impure and dark red or brown. When pure, translucent or transparent and bright red.	**CINNABAR** HgS Rhombohedral
	Red-brown. Ruby-red when trans-parent	6.0	$3\frac{1}{2}$–4	Massive or in cubes or octahedrons. May be in very slender crystals. Associated with malachite, azurite, native copper.	CUPRITE Cu_2O Isometric
Red-brown, Indian brown	Dark brown to steel-gray to black	4.8 to 5.3	$5\frac{1}{2}$–$6\frac{1}{2}$	Radiating, reniform, massive, micaceous. Rarely in steel-black rhombohedral crystals. Some varieties softer.	**HEMATITE** Fe_2O_3 Rhombohedral
	Deep red to black	5.8	$2\frac{1}{2}$	Cleavage $\{10\bar{1}1\}$. Fusible in candle flame. Shows dark ruby-red color in thin splinters. Associated with other silver minerals.	Pyrargyrite Ag_3SbS_3 Rhombohedral
Bright red	Ruby-red	5.55	2–$2\frac{1}{2}$	Cleavage $\{10\bar{1}1\}$. Fusible in candle flame. Light "ruby silver." Associated with pyrargyrite.	Proustite Ag_3AsS_3 Rhombohedral
Pink	Red to pink	2.95	$1\frac{1}{2}$–$2\frac{1}{2}$	Cleavage perfect $\{010\}$. Usually reniform or as pulverulent or earthy crusts. Found as coatings on cobalt minerals.	Erythrite (cobalt bloom) $Co_3As_2O_8\cdot8H_2O$ Monoclinic

Streak	Color	G.	H.	Remarks	Name, Composition, Crystal System
Yellow-brown to yellow-ocher	Dark brown to black	4.4	$5-5\frac{1}{2}$	Cleavage {010}. In radiating fibers, mammillary and stalactitic forms. Rarely in crystals.	**GOETHITE** $HFeO_2$ Orthorhombic
Brown	Dark brown	7.0 to 7.5	$5-5\frac{1}{2}$	Cleavage perfect {010}. With greater amounts of manganese the streak and color are darker.	WOLFRAMITE $(Fe, Mn)WO_4$ Monoclinic
	Light to dark brown	3.83 to 3.88	$3\frac{1}{2}-4$	In cleavable masses or in small curved rhombohedral crystals. Becomes magnetic after heating in candle flame.	**SIDERITE** $FeCO_3$ Rhombohedral
Light brown	Light to dark brown	3.9 to 4.1	$3\frac{1}{2}-4$	Cleavage perfect {011} (6 directions). Usually cleavable granular; may be in tetrahedral crystals. The streak is always of a lighter color than the specimen.	**SPHALERITE** ZnS Isometric
	Brown to black	6.8 to 7.1	$6-7$	Occurs in twinned crystals. Fibrous, reniform and irregular masses; in rolled grains.	**CASSITERITE** SnO_2 Tetragonal
	Reddish brown to black	4.18 to 4.25	$6-6\frac{1}{2}$	Crystals vertically striated; often acicular. Twinning common.	**RUTILE** TiO_2 Tetragonal
Orange-yellow	Deep red to orange-yellow	5.68	$4-4\frac{1}{2}$	Cleavage {0001}. Found only at Franklin, N. J., associated with franklinite and willemite.	ZINCITE ZnO Hexagonal

Streak	Color	G.	H.	Remarks	Name, Composition, Crystal System
Orange-yellow	Bright red	5.9 to 6.1	$2\frac{1}{2}$–3	Luster adamantine. In long slender crystals, often interlacing groups. Decrepitates in candle flame.	Crocoite $PbCrO_4$ Monoclinic
	Deep red	3.48	$1\frac{1}{2}$–2	Frequently earthy. Associated with orpiment. Fusible in candle flame.	REALGAR AsS Monoclinic
Pale yellow	Lemon-yellow	3.49	$1\frac{1}{2}$–2	Cleavage {010}. Luster resinous. Associated with realgar. Fusible in candle flame.	ORPIMENT As_2S_3 Monoclinic
	Pale yellow	2.05 to 2.09	$1\frac{1}{2}$–$2\frac{1}{2}$	Burns with blue flame, giving odor of SO_2. Crystallized, granular, earthy.	**SULFUR** S Orthorhombic
Light green	Dark emerald-green	3.75 to 3.77	3–$3\frac{1}{2}$	One perfect cleavage {010}. In granular cleavable masses or small prismatic crystals.	Atacamite $Cu_2Cl(OH)_3$ Orthorhombic
		3.9±	$3\frac{1}{2}$–4	One good cleavage {010}. In small prismatic crystals or granular masses.	Antlerite $Cu_3(SO_4)(OH)_4$ Orthorhombic
	Bright green	3.9 to 4.03	$3\frac{1}{2}$–4	Radiating fibrous, mammillary. Associated with azurite and may alter to it. Effervesces in cold acid.	**MALACHITE** $Cu_2CO_3(OH)_2$ Monoclinic
Light blue	Intense azure-blue	3.77	$3\frac{1}{2}$–4	In small crystals, often in groups. Radiating fibrous, usually as alteration from malachite. Effervesces in cold acid.	**AZURITE** $Cu_3(CO_3)_2(OH)_2$ Monoclinic

Streak	Color	G.	H.	Remarks	Name, Composition, Crystal System
Light blue	Blue	2.12 to 2.30	$2\frac{1}{2}$	Soluble in water. Metallic taste. In crystals, massive, stalactitic.	Chalcanthite $CuSO_4 \cdot 5H_2O$ Triclinic
Very light blue	Light green to turquoise blue	2.0 to 2.4	2–4	Massive compact. Associated with oxidized copper minerals.	CHRYSOCOLLA Amorphous
Grayish blue	Very dark blue. Bluish green	2.58 to 2.68	$1\frac{1}{2}$–2	One perfect cleavage {010}. Usually in prismatic crystals.	Vivianite $Fe_3(PO_4)_2 \cdot 8H_2O$ Monoclinic

LUSTER: NONMETALLIC

II. Streak colorless

A. Hardness: $< 2\frac{1}{2}$.   (Can be scratched by fingernail.)

Cleavage, Fracture	Color	G.	H.	Remarks	Name, Composition, Crystal System
Perfect cleavage in one direction — The micas or related micaceous minerals, which possess such a perfect cleavage that they can be split into exceedingly thin sheets. They may occur as aggregates of minute scales, when the micaceous structure may not be readily apparent.	Pale brown, green, yellow, white	2.76 to 3.0	$2-2\frac{1}{2}$	In foliated masses and scales. Crystals tabular with hexagonal or diamond-shaped outline. Cleavage flakes elastic.	**MUSCOVITE** $KAl_2(AlSi_3O_{10})$-$(OH)_2$ Monoclinic
	Dark brown, green to black; may be may be yellow	2.95 to 3	$2\frac{1}{2}-3$	Usually in irregular foliated masses. Crystals have hexagonal outline, but rare. Cleavage flakes elastic.	**BIOTITE** $K(Mg, Fe)_3$ $(AlSi_3O_{10})(OH)_2$ Monoclinic
	Yellowish brown, green, white	2.86	$2\frac{1}{2}-3$	Often in 6-sided tabular crystals; in irregular foliated masses. May show copperlike reflection from cleavage.	PHLOGOPITE $KMg_3(AlSi_3O_{10})$-$(OH)_2$ Monoclinic
	Green of various shades	2.6 to 2.9	$2-2\frac{1}{2}$	Usually in irregular foliated masses. May be in compact masses of minute scales. Thin sheets flexible but not elastic.	**CHLORITE** Monoclinic
	White, apple-green, gray. When impure, as in soapstone,	2.7 to 2.8	1	Greasy feel. Frequently distinctly foliated or micaceous. Cannot be positively identified by physical tests.	**TALC** $Mg_3(Si_4O_{10})(OH)_2$ Monoclinic
	dark gray, dark green to almost black	2.8 to 2.9	1–2		Pyrophyllite $Al_2(Si_4O_{10})(OH)_2$ Monoclinic
	White, gray, green	2.39	$2\frac{1}{2}$	Pearly luster on cleavage face, elsewhere vitreous. Sectile. Commonly foliated massive, may be in broad tabular crystals. Thin sheets flexible but not elastic.	BRUCITE $Mg(OH)_2$ Rhombohedral

Cleavage, Fracture	Color	G.	H.	Remarks	Name, Composition, Crystal System
Pinacoidal {010}	Blue, bluish green to colorless	2.58 to 2.68	$1\frac{1}{2}$–2	Streak grayish blue. Usually in prismatic crystals.	Vivianite $Fe_3(PO_4)_2 \cdot 8H_2O$ Monoclinic
	Colorless to white	1.75	2–$2\frac{1}{2}$	Occurs in crusts and capillary fibers. Soluble in water.	Epsomite $MgSO_4 \cdot 7H_2O$ Orthorhombic
Cubic {100}	Colorless or white	1.99	2	Soluble in water, bitter taste. Resembles halite but softer. In granular cleavable masses or cubic crystals.	Sylvite KCl Isometric
{010} perfect, {100} {011} good	Colorless, white, gray. May be colored by impurities	2.32	2	Occurs in crystals, broad cleavage flakes. May be compact massive without cleavage, or fibrous with silky luster.	**GYPSUM** $CaSO_4 \cdot 2H_2O$ Monoclinic
Rhomb. {10$\bar{1}$1} poor	Colorless or white	2.29	1–2	Occurs in saline crusts. Readily soluble in water; cooling and salty taste. Fusible in candle flame.	SODA NITER $NaNO_3$ Rhombohedral
Prismatic {110} seldom seen. F. Conchoidal		2.09 to 2.14	2	Usually in crusts, silky tufts and delicate acicular crystals. Readily soluble in water; cooling and salty taste. Fusible in the candle flame.	Niter KNO_3 Orthorhombic
{001} perfect; seldom seen. F. earthy	White; may be darker	2.6 to 2.63	2–$2\frac{1}{2}$	Compact, earthy. Breathed upon gives argillaceous odor. Will adhere to the dry tongue.	**KAOLINITE** $Al_4(Si_4O_{10})(OH)_8$ Monoclinic

Cleavage, Fracture	Color	G.	H.	Remarks	Name, Composition, Crystal System
Fract. uneven	Pearl-gray or color-less. Turns to pale brown on exposure to light	5.5 ±	2–3	Perfectly sectile. Translucent in thin plates. In irregular masses, rarely in crystals. Distinguished from other silver halides only by chemical tests.	Cerargyrite AgCl Isometric
	Pale yellow	2.05 to 2.09	$1\frac{1}{2}$–$2\frac{1}{2}$	Burns with blue flame, giving odor of SO_2. Crystallized, granular, earthy.	**SULFUR** S Orthorhombic
	Yellow, brown, gray, white	2.0 to 2.55	1–3	In rounded grains, often earthy and clay-like. Usually harder than $2\frac{1}{2}$.	**BAUXITE** A mixture of Al hydroxides
Cleavage seldom seen	White	1.95	1	Usually in rounded masses of fine fibers and acicular crystals.	ULEXITE $NaCaB_5O_9 \cdot 8H_2O$ Triclinic

II. Streak colorless

B. Hardness: $>2\frac{1}{2}$, <3. (Cannot be scratched by fingernail; can be scratched by cent.)

1. Cleavage prominent

Cleavage, Fracture	Color	G.	H.	Remarks	Name, Composition, Crystal System
{001}	Lilac, gray-ish white	2.8 to 3.0	$2\frac{1}{2}$–4	Crystals 6-sided prismatic. Usually in small irregular sheets and scales. A pegmatite mineral.	LEPIDOLITE $K_2Li_3Al_3(AlSi_3O_{10})_2$- $(O, OH, F)_4$ Monoclinic
{001}	Pink, gray, white	3.0 to 3.1	$3\frac{1}{2}$–5	Usually in irregular foliated masses; folia brittle. Associated with emery.	MARGARITE $CaAl_2(Al_2Si_2O_{10})$- $(OH)_2$ Monoclinic
{010}	Bluish green to colorless	2.58 to 2.68	$1\frac{1}{2}$–2	Prismatic crystals often in stellate groups. Also fibrous, earthy. Streak grayish blue. A rare mineral.	Vivianite $Fe_3(PO_4)_2 \cdot 8H_2O$ Monoclinic
{010}	Color-less or white	4.3	$3\frac{1}{2}$	Usually massive, with radiating habit. Effervesces in cold acid.	**WITHERITE** $BaCO_3$ Orthorhombic
{001}	Pale yellow or gray	2.7 to 2.85	$2\frac{1}{2}$–3	Crystals tabular. Found associated with other salts in salt lake deposits.	Glauberite $Na_2Ca(SO_4)_2$ Monoclinic
Cleavage in 2 directions {001} {100}	Color-less or white to gray	1.95	3	Occurs in cleavable crystalline aggregates.	KERNITE $Na_2B_4O_7 \cdot 4H_2O$ Monoclinic

Left margin: Perfect cleavage in one direction which may be harder than the fingernail. See also the minerals of the mica group, p. 522

LUSTER: NONMETALLIC

II. Streak colorless

B. Hardness: $>2\frac{1}{2}$, <3. (Cannot be scratched by fingernail; can be scratched by cent.)

1. Cleavage prominent. (*Continued*)

Cleavage, Fracture		Color	G.	H.	Remarks	Name, Composition, Crystal System
Cleavage in 3 directions at right angles	Cu-bic {100}	Color-less, white, red, blue	2.1 to 2.3	$2\frac{1}{2}$	Common salt. Soluble in water, taste salty, fusible in candle flame. In granular masses or in cubic crystals.	**HALITE** NaCl Isometric
	Cu-bic {100}	Color-less or white	1.99	2	Resembles halite, but distinguished from it by more bitter taste and lesser hardness.	Sylvite KCl Isometric
	{001} {010} {100}	Color-less, white, blue, gray, red	2.89 to 2.98	$3–3\frac{1}{2}$	Commonly in massive aggregates, not show-ing cleavage; then distinguished only by chemical tests.	**ANHYDRITE** $CaSO_4$ Orthorhombic
Cleavage in 3 directions not at right angles. Rhombo-hedral {10$\bar{1}$1}		Color-less, white, and vari-ously tinted	2.71	3	Effervesces in cold acid. Crystals show many forms. Occurs as limestone and marble. Clear varieties show strong double refraction.	**CALCITE** $CaCO_3$ Rhombohedral
		Color-less white, pink	2.85	$3\frac{1}{2}–4$	Usually harder than 3. Often in curved rhom-bohedral crystals with pearly luster; as dolo-mitic limestone and marble. Powdered mineral will effervesce in cold acid.	**DOLOMITE** $CaMg(CO_3)_2$ Rhombohedral
Cleavage in 3 directions. Basal{001} at right angles to prismatic {210}		Color-less, white, blue, yellow red	4.5	$3–3\frac{1}{2}$	Frequently in aggre-gates of platy crystals. Pearly luster on basal cleavage. Character-ized by high specific gravity and thus dis-tinguished from celes-tite.	**BARITE** $BaSO_4$ Orthorhombic

Cleavage, Fracture	Color	G.	H.	Remarks	Name, Composition, Crystal System
Cleavage in 3 directions. Basal {001} at right angles to prismatic {210}	Colorless, white, blue, red	3.95 to 3.97	$3-3\frac{1}{2}$	Similar to barite but lower specific gravity. A crimson strontium flame will distinguish it.	**CELESTITE** $SrSO_4$ Orthorhombic
	Colorless or white. Gray, brown when impure	6.2 to 6.4	3	Adamantine luster. Usually massive but may be in small tabular crystals. Alteration of galena. When massive may need test for SO_4 to distinguish from cerrusite ($PbCO_3$).	**ANGLESITE** $PbSO_4$ Orthorhombic

2. Cleavage not prominent
a. A small splinter is fusible in the candle flame

Color	G.	H.	Remarks	Name, Composition, Crystal System
Colorless or white	$1.7\pm$	$2-2\frac{1}{2}$	Soluble in water. One good cleavage, seldom seen. In crusts and prismatic crystals. In candle flame swells and then fuses. Sweetish alkaline taste.	**BORAX** $Na_2B_4O_7 \cdot 10H_2O$ Monoclinic
	2.95 to 3.0	$2\frac{1}{2}$	Massive with peculiar translucent appearance. Fine powder becomes nearly invisible when placed in water. Ivigtut, Greenland, only important locality. Pseudocubic parting.	CRYOLITE Na_3AlF_6 Monoclinic
	6.55	$3-3\frac{1}{2}$	Luster adamantine. In granular masses and platy crystals, usually associated with galena. Effervesces in cold nitric acid. Reduced in candle flame, producing small globules of lead.	**CERUSSITE** $PbCO_3$ Orthorhombic
Colorless, white, red	1.6	1	Bitter taste. Commonly massive granular. In candle flame swells, then fuses. Soluble in water.	Carnallite $KMgCl_3 \cdot 6H_2O$ Orthorhombic
Colorless, yellow, orange, brown	7.0 to 7.2	$3\frac{1}{2}$	Luster resinous. In small prismatic crystals. Prism faces may be curved, giving barrel shapes. In granular masses.	Mimetite $Pb_5Cl(AsO_4)_3$ Hexagonal

II. Streak colorless

B. Hardness: $>2\frac{1}{2}$, <3. (Cannot be scratched by fingernail; can be scratched by cent.)

2. Cleavage not prominent   b. Infusible in candle flame.

Color	G.	H.	Remarks	Name, Composition, Crystal System
Colorless or white	4.3	$3\frac{1}{2}$	Often in radiating masses; granular; rarely in pseudohexagonal crystals. Effervesces in cold acid.	**WITHERITE** $BaCO_3$ Orthorhombic
	2.6 to 2.63	$2-2\frac{1}{2}$	Usually compact, earthy. When breathed upon gives an argilaceous odor. The basis of most clays.	**KAOLINITE** $Al_4(Si_4O_{10})(OH)_8$ Monoclinic
Colorless, white, blue, gray, red	2.89 to 2.98	$3-3\frac{1}{2}$	Commonly in massive fine aggregates, not showing cleavage, and can be distinguished only by chemical tests.	**ANHYDRITE** $CaSO_4$ Orthorhombic
Honey-, citron-, or orange-yellow	4.9	$3-3\frac{1}{2}$	Usually found as a coating of fine powder on sphalerite. Rarely in crystals. A rare mineral.	Greenockite CdS Hexagonal
Yellow, brown, gray, white	2.0 to 2.55	$1-3$	Usually pisolitic; in rounded grains and earthy masses. Often impure.	**BAUXITE** A mixture of aluminum hydroxides
Ruby-red, brown, yellow	6.7 to 7.1	3	Luster resinous. In slender prismatic and cavernous crystals; in barrel-shaped forms.	Vanadinite $Pb_5Cl(VO_4)_3$ Hexagonal
Yellow, green, white, brown	2.33	$3\frac{1}{2}-4$	Characteristically in radiating hemispherical globular aggregates. Cleavage seldom seen.	Wavellite $Al_3(OH)_3(PO_4)_2\cdot 5H_2O$ Orthorhombic
Olive- to blackish-green, yellow-green, white	2.2	$2-5$	Massive. Fibrous in the asbestos variety, chrysotile. Frequently mottled green in the massive variety.	**SERPENTINE** $Mg_6Si_4O_{10}(OH)_8$ Monoclinic

C. Hardness: >3, $<5\frac{1}{2}$. (Cannot be scratched by cent; can be scratched by knife.)

1. Cleavage prominent

Cleavage	Color	G.	H.	Remarks	Name, Composition, Crystal System
{100}	Blue, usually darker at center.	3.56 to 3.66	5–7	In bladed aggregates with cleavage parallel to length. Can be scratched by knife parallel to the length of the crystal but not in a direction at right angles to this.	**KYANITE** Al_2SiO_5 Triclinic
{010}	White, yellow, brown, red	2.1 to 2.2	$3\frac{1}{2}$–4	Characteristically in sheaflike crystal aggregates. May be in flat tabular crystals. Luster pearly on cleavage face.	**STILBITE** $Ca(Al_2Si_7O_{18})\cdot 7H_2O$ Monoclinic
{001}	Colorless, white, pale green, yellow, rose	2.3 to 2.4	$4\frac{1}{2}$–5	In prismatic crystals vertically striated. Crystals often resemble cubes truncated by the octahedron. Luster pearly on base, elsewhere vitreous.	APOPHYLLITE $Ca_4K(Si_4O_{10})_2F\cdot 8H_2O$ Tetragonal
{010}	White, yellow, red	2.18 to 2.20	$3\frac{1}{2}$–4	Luster pearly on cleavage, elsewhere vitreous. Crystals often tabular parallel to cleavage plane. Found in cavities in igneous rocks.	HEULANDITE $Ca(Al_2Si_7O_{18})\cdot 6H_2O$ Monoclinic
{010}	Colorless, white	2.42	4–$4\frac{1}{2}$	In crystals and in cleavable aggregates. Decrepitates violently in the candle flame.	COLEMANITE $Ca_2B_6O_{11}\cdot 5H_2O$ Monoclinic
{010}	Colorless, white	4.3	$3\frac{1}{2}$	Often in radiating crystal aggregates; granular. Rarely in pseudohexagonal crystals.	WITHERITE $BaCO_3$ Orthorhombic
{010} {110} poor	Colorless, white	2.95	$3\frac{1}{2}$–4	Effervesces in cold acid. Frequently in radiating groups of acicular crystals; in pseudohexagonal twins.	**ARAGONITE** $CaCO_3$ Orthorhombic

One cleavage direction

II. Streak colorless

C. Hardness: >3, $<5\frac{1}{2}$. (Cannot be scratched by cent; can be scratched by knife.)

1. Cleavage prominent. *(Continued)*

Two cleavage directions

Cleavage	Color	G.	H.	Remarks	Name, Composition, Crystal System
{001} {010}	Blueish gray, salmon to clove-brown	3.42 to 3.56	$4\frac{1}{2}$–5	Commonly cleavable massive. Found in pegmatites with other lithium minerals.	Triphylite-lithiophilite $Li(Fe, Mn)PO_4$ Orthorhombic
{001} {100}	Color-less, white, gray	2.8 to 2.9	5–$5\frac{1}{2}$	Usually cleavable massive to fibrous. Also compact. Associated with crystalline limestone.	WOLLASTONITE $Ca(SiO_3)$ Triclinic
{001} {100}	Color-less, white, gray	2.7 to 2.8	5	In radiating aggregates of sharp acicular crystals. Associated with zeolites in cavities in igneous rocks.	Pectolite $Ca_2NaH(SiO_3)_3$ Triclinic
{110}	Color-less, white	2.25	5–$5\frac{1}{2}$	Slender prismatic crystals, prism faces vertically striated. Often in radiating groups. Found lining cavities in igneous rocks.	**NATROLITE** $Na_2Al_2Si_3O_{10}\cdot2H_2O$ Orthorhombic
{110}	Color-less, white	3.7	$3\frac{1}{2}$–4	In prismatic crystals and pseudohexagonal twins. Also fibrous and massive. Effervesces in cold acid.	**STRONTIANITE** $SrCO_3$ Orthorhombic
{110}	White, pale green, blue	3.4 to 3.5	$4\frac{1}{2}$–5	Often in radiating crystal groups. Also stalactitic, mammillary. Prismatic cleavage seldom seen.	**HEMIMORPHITE** $Zn_4(Si_2O_7)(OH)_2\cdot H_2O$ Orthorhombic

Cleavage		Color	G.	H.	Remarks	Name, Composition, Crystal System
Two cleavage directions	prismatic at angles of 55° and 125°	White, green, black	3.0 to 3.3	5–6	Crystals usually slender, fibrous, asbestiform. Tremolite (white, gray, violet), actinolite (green) common in metamorphic rocks. Hornblende and arfvedsonite (dark green to black) common in igneous and metamorphic rocks.	**AMPHIBOLE GROUP** Essentially calcium magnesium silicates Monoclinic
		Gray, clovebrown, green	2.85 to 3.2	$5\frac{1}{2}$–6	An amphibole. Distinct crystals rare. Commonly in aggregates and fibrous massive.	Anthophyllite $(Mg,Fe)-(Si_8O_{22})-(OH)_2$ Orthorhombic
	Prismatic at nearly 90° angles	White, green, black	3.1 to 3.5	5–6	In stout prisms with rectangular cross section. Often in granular crystalline masses. Diopside (colorless, white, green), aegirite (brown, green), augite (dark green to black).	**PYROXENE GROUP** Essentially calcium magnesium silicates Monoclinic
		Rosered, pink, brown	3.58 to 3.70	$5\frac{1}{2}$–6	Color diagnostic. Usually massive, cleavable to compact, in imbedded grains; in large rough crystals, with rounded edges.	RHODONITE $Mn(SiO_3)$ Triclinic

C. Hardness: >3, $<5\frac{1}{2}$. (Cannot be scratched by cent; can be scratched by knife.)

1. Cleavage prominent. (*Continued*)

Cleavage	Color	G.	H.	Remarks	Name, Composition, Crystal System
Two cleavage directions {110}	Brown, gray, green, yellow	3.4 to 3.55	$5-5\frac{1}{2}$	Luster adamantine to resinous. In thin wedge-shaped crystals with sharp edges. Prismatic cleavage seldom seen.	**SPHENE** $CaTiSiO_5$ Monoclinic
Three cleavage directions · Three directions not at right angles. Rhombohedral {10$\bar{1}$1}	Colorless, white, and variously tinted	2.72	3	Effervesces in cold acid. Crystals show many forms. Occurs as limestone and marble. Clear varieties show strong double double refraction.	**CALCITE** $CaCO_3$ Rhombóhedral
	Colorless, white, pink	2.85	$3\frac{1}{2}-4$	Often in curved rhombohedral crystals with pearly luster. In coarse masses as dolomitic limestone and marble. Powdered mineral will effervesce in cold acid.	**DOLOMITE** $CaMg(CO_3)_2$ Rhombohedral
	White, yellow, gray, brown	3.0 to 3.2	$3\frac{1}{2}-5$	Commonly in dense compact masses; also in fine to coarse cleavable masses. Effervesces in hot hydrochloric acid.	**MAGNESITE** $MgCO_3$ Rhombohedral
	Light to dark brown	3.83 to 3.88	$3\frac{1}{2}-4$	In cleavable masses or in small curved rhombohedral crystals. Becomes magnetic after heating.	**SIDERITE** $FeCO_3$ Rhombohedral
	Pink, rose-red, brown	3.45 to 3.6	$3\frac{1}{2}-4\frac{1}{2}$	In cleavable masses or in small rhombohedral crystals. Chacterized by its color.	RHODOCROSITE $MnCO_3$ Rhombohedral

Cleavage		Color	G.	H.	Remarks	Name, Composition, Crystal System
Three cleavage directions	Rhombohedral $\{10\bar{1}1\}$	Brown, green, blue, pink, white	4.35 to 4.40	5	In rounded botryoidal aggregates and honeycombed masses. Effervesces in cold hydrochloric acid. Cleavage rarely seen.	SMITHSONITE $ZnCO_3$ Rhombohedral
		White, yellow, flesh-red	2.05 to 2.15	4–5	In small rhombohedral crystals with nearly cubic angles. Found lining cavities in igneous rocks.	CHABAZITE $(Ca,Na)_2(Al_2Si_4O_{12}) \cdot 6H_2O$ Rhombohedral
	$\{001\}$ $\{010\}$ $\{100\}$	Colorless, white, blue, gray, red	2.89 to 2.98	$3–3\frac{1}{2}$	Commonly in massive fine aggregates, not showing cleavage, and can be distinguished only by chemical tests.	**ANHYDRITE** $CaSO_4$ Orthorhombic
	$\{001\}$ at rt. angles to $\{110\}$	Colorless, white, blue, yellow, red	4.5	$3–3\frac{1}{2}$	Frequently in aggregates of platy crystals. Pearly luster on basal cleavage. Characterized by high specific gravity and thus distinguished from celestite.	**BARITE** $BaSO_4$ Orthorhombic
		Colorless, white, blue, red	3.95 to 3.97	$3–3\frac{1}{2}$	Similar to barite but lower specific gravity. A crimson strontium flame will distinguish it.	**CELESTITE** $SrSO_4$ Orthorhombic
Four cleavage directions	$\{111\}$ Oct.	Colorless, violet, green, yellow, pink.	3.18	4	In cubic crystals often in penetration twins. Characterized by cleavage.	**FLUORITE** CaF_2 Isometric

II. Streak colorless

C. Hardness: >3, $<5\frac{1}{2}$. (Cannot be scratched by cent; can be scratched by knife.)

1. Cleavage prominent. (*Continued*)

Cleavage	Color	G.	H.	Remarks	Name, Composition, Crystal System
Four cleavage directions {100} {110}	White, pink, gray, green, brown	2.65 to 2.74	5–6	In prismatic crystals, granular or massive. Commonly altered. Prismatic cleavage obscure.	SCAPOLITE Essentially Na,Ca aluminum silicate Tetragonal
Six cleavage directions — Dodecahedral {011}	Yellow, brown, white	3.9 to 4.1	$3\frac{1}{2}$–4	Luster resinous. Small tetrahedral crystals rare. Usually in cleavable masses. If massive, difficult to determine.	**SPHALERITE** ZnS Isometric
	Blue, white, gray, green	2.15 to 2.3	$5\frac{1}{2}$–6	Massive or in imbedded grains; rarely in crystals. A feldspathoid associated with nepheline, never with quartz.	SODALITE $Na_8(AlSiO_4)_6Cl_2$ Isometric

2. Cleavage not prominent

Color	G.	H.	Remarks	Name, Composition, Crystal System
Colorless, pale green, yellow	2.8 to 3.0	5–$5\frac{1}{2}$	Usually in crystals with many brilliant faces. Occurs with zeolites lining cavities in igneous rocks.	DATOLITE $CaB(SiO_4)(OH)$ Monoclinic
White, pale green, blue	3.4 to 3.5	$4\frac{1}{2}$–5	Often in radiating crystal groups. Also stalactitic, mammillary. Prismatic cleavage seldom seen.	HEMIMORPHITE $Zn_4(Si_2O_7)(OH)_2 \cdot H_2O$ Orthorhombic

LUSTER: NONMETALLIC
II. Streak colorless
C. Hardness: >3, $<5\frac{1}{2}$.   (Cannot be scratched by cent; can be scratched by knife.)
2. Cleavage not prominent.   (*Continued*)

Color	G.	H.	Remarks	Name, Composition, Crystal System
White, pink, gray, green, brown	2.65 to 2.74	5–6	In prismatic crystals, granular or massive. Commonly altered. Prismatic cleavage obscure.	SCAPOLITE Essentially Na,Ca aluminum silicate Tetragonal
Colorless, white	2.95	$3\frac{1}{2}$–4	Effervesces in cold acid. Falls to powder in the candle flame. Frequently in radiating groups of acicular crystals; in pseudohexagonal twins. Cleavage indistinct.	**ARAGONITE** $CaCO_3$ Orthorhombic
	2.27	5–$5\frac{1}{2}$	Usually in trapezohedrons with vitreous luster. A zeolite, found lining cavities in igneous rocks.	ANALCIME $NaAlSi_2O_6 \cdot H_2O$ Isometric
	3.7	$3\frac{1}{2}$–4	Occurs in prismatic crystals and pseudohexagonal twins. Also fibrous and massive. Effervesces in cold acid.	**STRONTIANITE** $SrCO_3$ Orthorhombic
	3.0 to 3.2	$3\frac{1}{2}$–5	Commonly in dense compact masses showing no cleavage. Effervesces in hot hydrochloric acid.	**MAGNESITE** $MgCO_3$ Rhombohedral
	4.3	$3\frac{1}{2}$	Often in radiating masses; granular; rarely in pseudohexagonal crystals. Effervesces in cold hydrochloric acid.	**WITHERITE** $BaCO_3$ Orthorhombic
	2.7 to 2.8	5	Commonly fibrous in radiating aggregates of sharp acicular crystals. Associated with zeolites in cavities in igneous rocks.	Pectolite $Ca_2NaH(SiO_3)_3$ Triclinic

LUSTER: NONMETALLIC
II. Streak colorless
C. Hardness: >3, $<5\frac{1}{2}$.   (Cannot be scratched by cent; can be scratched by knife.)
2. Cleavage not prominent.   (*Continued*)

Color	G.	H.	Remarks	Name, Composition, Crystal System
Colorless, white	2.25	5–5½	In slender prismatic crystals, prism faces vertically striated. Often in radiating groups. Found lining cavities in igneous rocks. Poor prismatic cleavage.	NATROLITE $Na_2(Al_2Si_3O_{10})\cdot 2H_2O$ Orthorhombic
White, grayish, red	2.6 to 2.8	4	May be in rhombohedral crystals. Usually massive granular. Definitely determined only by blow-pipe tests. Cleavage {0001}, poor.	ALUNITE $KAl_3(OH)_6(SO_4)_2$ Rhombohedral
Colorless, white, yellow, red, brown,	1.9 to 2.2	5–6	Conchoidal fracture. Precious opal shows internal play of colors. Specific gravity and hardness less than fine-grained quartz.	**OPAL** $SiO_2\cdot nH_2O$ Amorphous
Brown, green, blue, pink, white	4.35 to 4.40	5	Usually in rounded botryoidal aggregates and in honeycombed masses. Effervesces in cold hydrochloric acid. Cleavage rarely seen.	**SMITHSONITE** $ZnCO_3$ Rhombohedral
Brown, gray, green, yellow	3.4 to 3.55	5–5½	Adamantine to resinous luster. In thin wedge-shaped crystals with sharp edges. Prismatic cleavage seldom seen.	SPHENE $CaTiSiO_5$ Monoclinic
Colorless, white, yellow, red, brown	2.72	3	May be fibrous or fine granular, banded in Mexican onyx variety. Effervesces in cold hydrochloric acid.	**CALCITE** $CaCO_3$ Rhombohedral
Yellowish to reddish brown	5.0 to 5.3	5–5½	In small crystals or as rolled grains. Found in pegmatites.	Monazite $(Ce,La,Y,Th)PO_4$ Monoclinic

Color	G.	H.	Remarks	Name, Composition, Crystal System
Light to dark brown	3.83 to 3.88	$3\frac{1}{2}$–4	Usually cleavable but may be in compact concretions in clay or shale—clay ironstone variety. Becomes magnetic on heating.	**SIDERITE** $FeCO_3$ Rhombohedral
White, yellow, green, brown	5.9 to 6.1	$4\frac{1}{2}$–5	Luster vitreous to adamantine. Massive and in octahedral-like crystals. Frequently associated with quartz. Will fluoresce.	SCHEELITE $CaWO_4$ Tetragonal
Yellow, orange, red, gray, green	6.8±	3	Luster adamantine. Usually in square tabular crystals. Also granular massive. Characterized by color and high specific gravity.	WULFENITE $PbMoO_4$ Tetragonal
Colorless, yellow, orange, brown	7.0 to 7.2	$3\frac{1}{2}$	Resinous luster. In small hexagonal prisms. Faces often curved, giving barrel shapes. In granular masses. Fuses slowly in candle flame.	Mimetite $Pb_5Cl(AsO_4)_3$ Hexagonal
White, yellow, brown, gray	2.6 to 2.9	3–5	Occurs massive as rock phosphate. Difficult to identify without chemical tests.	**APATITE** $Ca_5(F, Cl, OH)-(PO_4)_3$ Appears amorphous
Yellow, brown, gray, white	2.0 to 2.55	1–3	Usually pisolitic; in rounded grains and earthy masses. Often impure.	**BAUXITE** A mixture of aluminum hydroxides
Green, blue, violet, brown, colorless	3.15 to 3.20	5	Usually in hexagonal prisms with pyramid. Also massive. Poor basal cleavage.	**APATITE** $Ca_5(F,Cl,OH)-(PO_4)_3$ Hexagonal
Green, brown, yellow, gray	6.5 to 7.1	$3\frac{1}{2}$–4	In small hexagonal crystals, often curved and barrel-shaped. Crystals may be cavernous. Often globular and botryoidal.	Pyromorphite $Pb_5Cl(PO_4)_3$ Hexagonal

C. Hardness: >3, $<5\frac{1}{2}$. (Cannot be scratched by cent; can be scratched by knife.)

2. Cleavage not prominent. (*Continued*)

Color	G.	H.	Remarks	Name, Composition, Crystal System
Yellow, green, white, brown	2.33	$3\frac{1}{2}$–4	Characteristically in radiating hemispherical globular aggregates. Cleavage seldom seen.	Wavellite $Al_3(OH)_3(PO_4)_2 \cdot 5H_2O$ Orthorhombic
Olive- to blackish green, yellow-green, white	2.2	2–5	Massive. Fibrous in the asbestos variety, chrysotile. Frequently mottled green in the massive variety.	**SERPENTINE** $Mg_6(Si_4O_{10})(OH)_8$ Monoclinic
Yellow-green, white, blue, gray, brown	3.9 to 4.2	$5\frac{1}{2}$	Massive and in disseminated grains. Rarely in hexagonal prisms. Associated with red zincite and black franklinite at Franklin, N.J. will fluoresce.	WILLEMITE $Zn_2(SiO_4)$ Rhombohedral
White, gray, blue, green	2.15 to 2.3	$5\frac{1}{2}$–6	Massive or in imbedded grains; rarely in crystals. A feldspathoid associated with nepheline, never with quartz. Dodecahedral cleavage poor.	SODALITE $Na_8(AlSiO_4)_6Cl_2$ Isometric
Deep azure-blue, greenish blue	2.4 to 2.45	5–$5\frac{1}{2}$	Usually massive. Associated with feldspathoids and pyrite. Poor dodecahedral cleavage.	Lazurite $(Na,Ca)_8(AlSiO_4)_6$-$(SO_4,S,Cl)_2$ Isometric

1. Cleavage prominent

	Cleavage	Color	G	H	Remarks	Name, Composition, Crystal System
One cleavage direction	{010}	White, gray, pale lavender yellow-green	3.35 to 3.45	$6\frac{1}{2}$–7	In thin tabular crystals. Luster pearly on cleavage face. Associated with emery, margarite, chlorite.	Diaspore $HAlO_2$ Orthorhombic
	{010} perfect	Hair-brown, grayish green	3.23	6–7	In long, prismatic crystals. May be in parallel groups— columnar or fibrous. Found in schists.	SILLIMANITE Al_2SiO_5 Orthorhombic
	{001}	Yellowish to blackish green	3.35 to 3.45	6–7	In prismatic crystals striated parallel to length. Found in metamorphic rocks and crystalline limestones.	**EPIDOTE** Monoclinic
	{100}	Blue. May be gray, or green	3.56 to 3.66	5–7	In bladed aggregates with cleavage parallel to length. H5 parallel to length of crystal;	**KYANITE** Al_2SiO_5
Two cleavage directions	{100} good {110} poor	White, pale green, or blue	3.0 to 3.1	6	Usually cleavable, resembling feldspar. Found in pegmatites associated with other lithium minerals.	AMBLYGONITE $LiAlFPO_4$ Triclinic
	{100} good {001}	Colorless, white, gray	2.8 to 2.9	5–$5\frac{1}{2}$	Usually cleavable massive to fibrous. Also compact. Associated with crystalline limestone.	WOLLASTONITE $CaSiO_3$ Triclinic
	{001} {100}	Grayish white, green, pink	3.25 to 3.37	6–$6\frac{1}{2}$	In prismatic crystals deeply striated. Also massive, columnar, compact. Luster pearly on cleavage, elsewhere vitreous.	Clinozoisite $Ca_2Al_3O(SiO_4)$-$(Si_2O_7)(OH)$ Monoclinic
	{110}	Colorless, white	2.25	5–$5\frac{1}{2}$	In slender prismatic crystals, prism faces vertically striated Often in radiating groups lining cavities in igneous	**NATROLITE** $Na_2(Al_2Si_3O_{10})\cdot$ $2H_2O$ Orthorhombic

D. Hardness: $> 5\frac{1}{2}$, < 7.   (Cannot be scratched by knife; can be scratched by quartz.)

1. Cleavage prominent.   (*Continued*)

Cleavage	Color	G.	H.	Remarks	Name, Composition, Crystal System
{001} {010}	Colorless, white, gray, cream, red, green	2.54 to 2.56	6	In cleavable masses or in irregular grains as rock constituents. May be in crystals in pegmatite. Distin- with certainty only with the microscope. Green amazonstone is microcline.	**ORTHOCLASE** (Monoclinic) **MICROCLINE** (Triclinic) $K(AlSi_3O_8)$
{001} {010}	Colorless, white, gray, bluish. Often shows play of colors	2.62 (al-bite) to 2.76 (an-or-thite)	6	In cleavable masses or in irregular grains as a rock constituent. On the better cleavage can be seen a series of fine parallel striations due to albite twinning; these distinguish it from orthoclase.	**PLAGIOCLASE** Various proportions of albite, $Na(AlSi_3O_8)$ and anorthite $Ca(Al_2Si_2O_8)$ Triclinic
{110}	White, gray, pink, green	3.15 to 3.20	$6\frac{1}{2}$–7	In flattened prismatic crystals, vertically striated. Also massive cleavable. Found in pegmatites. Frequently shows good {100} parting.	SPODUMENE $LiAl(Si_2O_6)$ Monoclinic
{110}	White, green, black	3.1 to 3.5	5–6	In stout prisms with rectangular cross sec-tion. Often in granular crystalline masses. Di-opside (colorless, white, green), aegirite (brown, green), augite (dark green to black). Characterized by cleavage.	**PYROXENE GROUP** Essentially calcium magnesium silicates Monoclinic
{110}	Gray-brown, green, bronze-brown, black	3.2 to 3.5	$5\frac{1}{2}$	Crystals usually pris-matic but rare. Commonly massive, fibrous, lamellar. Fe may replace Mg and mineral is darker.	ENSTATITE $Mg_2(Si_2O_6)$ Orthorhombic

Two cleavage directions at or nearly 90° angles

Cleavage	Color	G.	H.	Remarks	Name, Composition, Crystal System
Two directions at nearly 90° {110}	Rose-red, pink, brown	3.58 to 3.70	$5\frac{1}{2}$–6	Color diagnostic. Usually massive; cleavable to compact; in imbedded grains; in large rough crystals with rounded edges.	RHODONITE $Mn(SiO_3)$ Triclinic
Two cleavage directions at 55° and 125° angles {110}	White, green, black	3.0 to 3.3	5–6	Crystals usually slender fibrous asbestiform. Tremolite (white, gray, violet) and actinolite (green) are common in metamorphic rocks. Hornblende and arfvedsonite (dark green to black) are common in igneous rocks. Characterized by cleavage angle.	**AMPHIBOLE GROUP** Essentially calcium magnesium silicates Monoclinic Monoclinic
{110}	Gray, clove-brown, green	2.85 to 3.2	$5\frac{1}{2}$–6	An amphibole. Distinct crystals rare. Commonly in aggregates and fibrous massive.	Anthophyllite $(Mg,Fe)_7(Si_8O_{22})$-$(OH)_2$ Orthorhombic
{110}	Brown, gray, green, yellow	3.4 to 3.55	5–$5\frac{1}{2}$	Luster adamantine to resinous. In thin wedge-shaped crystals with sharp edges. Prismatic cleavage seldom seen.	SPHENE $CaTiO(SO_4)$ Monoclinic
Six cleavage directions {110}	Blue, gray, white, green	2.15 to 2.3	$5\frac{1}{2}$–6	Massive or in imbedded grains; rarely in crystals. A feldspathoid associated with nepheline, never with quartz.	SODALITE $Na_4Al_3Si_3O_{12}Cl$ Isometric

2. Cleavage not prominent

Color	G.	H.	Remarks	Name, Composition, Crystal System
Colorless	2.26	7	Occurs as small crystals in cavities in volcanic rocks. Difficult to determine without optical aid.	Tridymite SiO_2 Pseudohexagonal

LUSTER: NONMETALLIC

II. Streak colorless

D. Hardness: $>5\frac{1}{2}$, <7.   (Cannot be scratched by knife; can be scratched by quartz.)

2. Cleavage not prominent.   (*Continued*)

Color	G.	H.	Remarks	Name, Composition, Crystal System
Colorless or white	2.27	$5-5\frac{1}{2}$	Usually in trapezohedrons with vitreous luster. A zeolite, found lining cavities in igneous rocks.	ANALCIME $Na(AlSi_2O_6)H_2O$ Isometric
	2.32	7	Occurs in spherical aggregations in volcanic rocks. Difficult to determine without optical aid.	Cristobalite SiO_2 Pseudoisometric
Colorless, yellow, red, brown, green, gray, blue	1.9 to 2.2	5–6	Conchoidal fracture. Precious opal shows internal play of colors. Gravity and hardness less than fine-grained quartz.	**OPAL** $SiO_2 \cdot nH_2O$ Amorphous
Gray, white, colorless	2.45 to 2.50	$5\frac{1}{2}-6$	In trapezohedral crystals embedded in dark igneous rock. Does not line cavities as analcime does.	LEUCITE $K(AlSi_2O_6)$ Pseudoisometric
Colorless, pale green, yellow	2.8 to 3.0	$5-5\frac{1}{2}$	Usually in crystals with many brilliant faces. Occurs with zeolites lining cavities in igneous rocks.	DATOLITE $CaB(SiO_4)(OH)$ Monoclinic
Colorless, white, pale yellow	2.97 to 3.02	7	In prismatic crystals resembling topaz but distinguished by poor cleavage. Also in irregular masses and indistinct crystals. A rare mineral.	Danburite $CaB_2Si_2O_8$ Orthorhombic
White, gray, light to dark green, brown	2.65 to 2.74	5–6	In prismatic crystals, granular or massive. Commonly altered and prismatic cleavage obscure.	SCAPOLITE Essentially Na-Ca aluminum Silicate Tetragonal

Color	G.	H.	Remarks	Name, Composition, Crystal System
Colorless, white, smoky, amethyst. Variously colored.	2.65	7	Crystals usually show horizontally striated prism with rhombohedral terminations.	**QUARTZ** SiO_2 Rhombohedral
Colorless, gray, greenish, reddish	2.55 to 2.65	$5\frac{1}{2}$–6	Greasy luster. A rock constituent, usually massive; rarely in hexagonal prisms. Poor prismatic cleavage. A feldspathoid.	NEPHELINE $(Na, K)AlSiO_4$ Hexagonal
Light yellow, brown, orange	3.1 to 3.2	6–$6\frac{1}{2}$	Occurs in disseminated crystals and grains. Commonly in crystalline limestones.	Chondrodite $Mg_5(SiO_4)_2(F, OH)_2$ Monoclinic
Light brown, yellow, red, green	2.65	7	Luster waxy to dull. Commonly colloform. May be banded or lining cavities.	**CHALCEDONY** SiO_2 Cryptocrystalline quartz
Blue, bluish green, green	2.6 to 2.8	6	Usually appears amorphous in reniform and stalactitic masses.	TURQUOISE $CuAl_6(PO_4)_4(OH)_8 \cdot 4H_2O$ Triclinic
Apple-green, gray, white	2.8 to 2.95	6–$6\frac{1}{2}$	Reniform and stalactitic with crystalline surface. In subparallel groups of tabular crystals.	PREHNITE $Ca_2Al_2Si_3O_{10}(OH)_2$ Orthorhombic
Yellow-green, white, blue, gray, brown	3.9 to 4.2	$5\frac{1}{2}$	Massive and in disseminated grains. Rarely in hexagonal prisms. Associated with red zincite and black franklinite at Franklin, N. J. Will fluoresce.	WILLEMITE Zn_2SiO_4 Rhombohedral
Olive to grayish green, brown	3.27 to 3.37	$6\frac{1}{2}$–7	Usually in disseminated grains in basic igneous rocks. May be massive granular.	**OLIVINE** $(Mg, Fe)_2(SiO_4)$ Orthorhombic

LUSTER: NONMETALLIC
II. Streak colorless
D. Hardness: $>5\frac{1}{2}$, <7.   (Cannot be scratched by knife; can be scratched by quartz.)
2. Cleavage not prominent. (*Continued*)

Color	G.	H.	Remarks	Name, Composition, Crystal System
Black, green, brown, blue, red, pink, white	3.0 to 3.25	$7-7\frac{1}{2}$	In slender prismatic crystals with triangular cross section. Crystals may be in radiating groups. Found usually in pegmatites. Black most common, other colors associated with lithium minerals.	**TOURMALINE** Rhombohedral
Green, brown, yellow, blue, red	3.35 to 4.45	$6\frac{1}{2}$	In square prismatic crystals vertically striated. Often columnar and granular massive. Found in crystalline limestones.	IDOCRASE Tetragonal
Clove-brown, gray, green, yellow	3.27 to 3.35	$6\frac{1}{2}-7$	In wedge-shaped crystals with sharp edges. Also lamellar.	AXINITE $Ca_2(FeMn)$-$Al_2(BO_3)(Si_4O_{12})$-(OH) Triclinic
Red-brown to brownish black	3.65 to 3.75	$7-7\frac{1}{2}$	In prismatic crystals; commonly in cruciform penetration twins. Frequently altered on the surface and then soft. Found in schists.	STAUROLITE $Fe_2Al_9O_6(SiO_4)_4$-$(O, OH)_2$ Pseudoorthorhombic
Reddish brown, flesh-red, olive-green	3.16 to 3.20	$7\frac{1}{2}$	Prismatic crystals with nearly square cross section. Cross section may show black cross (chiastolite). May be altered to mica and then soft. Found in schists.	**ANDALUSITE** Al_2SiO_5 Orthorhombic
Brown, gray, green, yellow	3.4 to 3.55	$5-5\frac{1}{2}$	Luster adamantine to resinous. In thin wedge-shaped crystals with sharp edges. Prismatic cleavage seldom seen.	**SPHENE** $CaTiO(SiO_4)$ Monoclinic

Color	G.	H.	Remarks	Name, Composition, Crystal System
Yellowish to reddish brown	5.0 to 5.3	$5-5\frac{1}{2}$	In isolated crystals, granular. Commonly found in pegmatites.	Monazite $(Ce,La,Y,Th)PO_4$ Monoclinic
Brown to black	6.8 to 7.1	$6-7$	Rarely in prismatic crystals, twinned. Fibrous, giving reniform surface. Rolled grains. Usually gives light brown streak.	**CASSITERITE** SnO_2 Tetragonal
Reddish brown to black	4.18 to 4.25	$6-6\frac{1}{2}$	In prismatic crystals vertically striated; often slender acicular. Crystals frequently twinned. A constituent of black sands.	**RUTILE** TiO_2 Tetragonal
Brown to pitch-black	3.5 to 4.2	$5\frac{1}{2}-6$	Crystals often tabular. Massive and in imbedded grains. An accessory mineral in igneous rocks.	Allanite Monoclinic
Blue, rarely colorless	2.60 to 2.66	$7-7\frac{1}{2}$	In imbedded grains and massive, resembling quartz. Commonly altered and foliated; then softer than a knife.	Cordierite $(Mg, Fe)Al_3-$ $(AlSi_5O_{18})$ Orthorhombic (Pseudohexagonal)
Deep azure-blue, greenish blue	2.4 to 2.45	$5-5\frac{1}{2}$	Usually massive. Associated with feldspathoids and pyrite. Poor dodecahedral cleavage.	LAZURITE $(Na,Ca)_8(AlSiO_4)_6-$ $(SO_4,S,Cl)_2$ Isometric
Azure-blue	3.0 to 3.1	$5-5\frac{1}{2}$	Usually in pyramidal crystals, which distinguishes it from massive lazurite. A rare mineral.	LAZULITE $(Mg,Fe)Al_2(OH)_2-$ $(PO_4)_2$ Monoclinic
Blue, green, white, gray	2.15 to 2.3	$5\frac{1}{2}-6$	Massive or in imbedded grains; rarely in crystals. A feldspathoid associated with nepheline, never with quartz. Poor dodecahedral cleavage.	SODALITE $Na_8(AlSiO_4)_6Cl_2$ Isometric

	Cleavage	Color	G.	H.	Remarks	Name, Composition, Crystal System
One cleavage direction	{001}	Color-less, yellow, pink, bluish, greenish	3.4 to 3.6	8	Usually in crystals, also coarse to fine granular. Found in pegmatites.	**TOPAZ** $Al_2(SiO_4)(F,OH)_2$ Orthorhombic
One cleavage direction	{010}	Brown, gray, green-ish gray	3.23	6–7	Commonly in long slender prismatic crys-tals. May be in parallel groups, columnar or fibrous. Found in schistose rocks.	SILLIMANITE Al_2SiO_5 Orthorhombic
Two cleavage directions	{110}	White, gray, pink, green	3.15 to 3.20	$6\frac{1}{2}$–7	In flattened prismatic crystals, vertically striated. Also massive cleavable. Pink va-riety, kunzite; green, hiddenite. Found in pegmatites. Frequent-ly shows good {100} parting.	SPODUMENE $LiAl(Si_2O_6)$ Monoclinic
Three cleav-age directions	{010} {110}	Color-less, pale blue, gray	3.09	8	Commonly in tabular or prismatic crystals in schists.	Lawsonite $CaAl_2(Si_2O_7)(OH)_2$ H_2O Orthorhombic
Four cleav-age directions	{111}	Color-less, yellow red, blue, black	3.5	10	Adamantine luster. In octahedral crystals, frequently twinned. Faces may be curved.	Diamond C Isometric
	No cleav-age. Rhom-bohedral and basal parting.	Color-less, gray, blue, red, yellow, brown, green	3.95 to 4.1	9	Luster adamantine to vitreous. Parting frag-ments may appear nearly cubic. In rude barrel-shaped crystals.	**CORUNDUM** Al_2O_3 Rhombohedral

2. Cleavage not prominent

Color	G.	H.	Remarks	Name, Composition, Crystal System
Colorless, white, smoky, Variously colored	2.65	7	Crystals usually show horizontally striated prism with rhombohedral terminations.	**QUARTZ** SiO_2 Rhombohedral
Colorless, white, pale yellow	2.97 to 3.02	7	In prismatic crystals resembling topaz but distinguished by lack of good cleavage. Also in irregular masses and indistinct crystals. A rare mineral.	Danburite $Ca(B_2Si_2O_8)$ Orthorhombic
White, colorless	2.97 to 3.0	$7\frac{1}{2}$–8	In small rhombohedral crystals. A rare mineral.	Phenacite $Be_2(SiO_4)$ Rhombohedral
White and almost any color	3.95 to 4.1	9	Luster adamantine to vitreous. Parting fragments may appear nearly cubic. In rude barrel-shaped crystals.	**CORUNDUM** Al_2O_3 Rhombohedral
Red, black, blue, green, brown	3.6 to 4.0	8	In octahedrons; twinning common. Associated with crystalline limestones.	**SPINEL** $MgAl_2O_4$ Isometric
Bluish green, yellow, pink, colorless	2.75 to 2.8	$7\frac{1}{2}$–8	Commonly in hexagonal prisms terminated by the base; pyramid faces are rare. Crystals large in places. Poor basal cleavage.	**BERYL** $Be_3Al_2(Si_6O_{18})$ Hexagonal
Yellowish to emerald-green	3.65 to 3.8	$8\frac{1}{2}$	In tabular crystals frequently in pseudohexagonal twins. Found in pegmatites.	CHRYSOBERYL $BeAl_2O_4$ Orthorhombic
Green, brown, blue, red, pink, black	3.0 to 3.25	7–$7\frac{1}{2}$	In slender prismatic crystals with triangular cross section. Found usually in pegmatites. Black most common, other colors associated with lithium minerals.	**TOURMALINE** Rhombohedral

LUSTER: NONMETALLIC

II. Streak colorless
E. Hardness: >7.   (Cannot be scratched by quartz.)
1. Cleavage not prominent.   (*Continued*)

Color	G.	H.	Remarks	Name, Composition, Crystal System
Green, gray, white	3.3 to 3.5	$6\frac{1}{2}$–7	Massive, closely compact. Poor prismatic cleavage at nearly 90° angles. A pyroxene.	Jadeite $NaAl(Si_2O_6)$ Monoclinic
Olive to grayish green, brown	3.27 to 3.37	$6\frac{1}{2}$–7	Usually in disseminated grains in basic igneous rocks. May be massive granular.	**OLIVINE** $(Mg,Fe)_2(SiO_4)$ Orthorhombic
Green, brown, yellow, blue, red	3.35 to 3.45	$6\frac{1}{2}$	In square prismatic crystals vertically striated. Often columnar and granular massive. Found in crystalline limestones.	IDOCRASE Tetragonal
Dark green	4.55	$7\frac{1}{2}$–8	Usually in octahedrons characteristically striated. A zinc spinel.	Gahnite $ZnAl_2O_4$ Isometric
Reddish brown to black	6.8 to 7.1	6–7	Rarely in prismatic crystals twinned. Fibrous, giving reniform surface. Rolled grains. Usually gives light brown streak.	**CASSITERITE** SnO_2 Tetragonal
Reddish brown, flesh-red, olive-green	3.16 to 3.20	$7\frac{1}{2}$	Prismatic crystals with nearly square cross section. Cross section may show black cross (chiastolite). May be altered to mica and then soft. Found in schists.	**ANDALUSITE** Al_2SiO_5 Orthorhombic
Clove-brown, green, yellow, gray	3.27 to 3.35	$6\frac{1}{2}$–7	In wedge-shaped crystals with sharp-edges. Also lamellar.	AXINITE $Ca_2(Fe,Mn)Al_2(BO_3)(Si_4O_{12})(OH)$ Triclinic

Color	G.	H.	Remarks	Name, Composition, Crystal System
Red-brown to brownish black	3.65 to 3.75	$7-7\frac{1}{2}$	In prismatic crystals; commonly in cruciform penetration twins. Frequently altered on the surface and then soft. Found in schists.	STAUROLITE $Fe_2Al_9O_6(SiO_4)_4-$ (O,OH) Pseudo-orthorhombic
Brown, red, gray, green, colorless	4.68	$7\frac{1}{2}$	Usually in small prisms truncated by the pyramid. An accessory mineral in igneous rocks. Found as rolled grains in sand.	**ZIRCON** $ZrSiO_4$ Tetragonal
Usually brown to red. Also yellow, green, pink	3.5 to 4.3	$6\frac{1}{2}-7\frac{1}{2}$	Usually in dodecahedrons or trapezohedrons or in combinations of the two. An accessory mineral in igneous rocks and pegmatites. Commonly in metamorphic rocks. As sand.	**GARNET** $A_3B_2(SiO_4)_3$ Isometric

MINERALS ARRANGED ACCORDING TO INCREASING
SPECIFIC GRAVITY

G.	Name	G.	Name	G.	Name
		2.6–2.79		3.0–3.1	**Lazulite**
1.6	Carnallite			3.0–3.2	**Magnesite**
1.7	**Borax**	2.55–2.65	**Nepheline**	3.0–3.1	Margarite
1.75	Epsomite	2.6–2.63	**Kaolinite**	3.0–3.25	**Tourmaline**
1.95	**Kernite**	2.62	**Albite**	3.0–3.3	**Tremolite**
1.96	**Ulexite**	2.60–2.66	Cordierite	3.09	Lawsonite
1.99	Sylvite	2.58–2.68	Vivianite	3.1–3.2	Autunite
		2.65	**Oligoclase**	3.1–3.2	Chondrodite
2.0–2.19		2.65	**Quartz**	3.15–3.20	**Apatite**
		2.69	**Andesine**	3.15–3.20	**Spodumene**
2.0–2.55	**Bauxite**	2.6–2.8	Alunite	3.16–3.20	**Andalusite**
2.0–2.4	**Chrysocolla**	2.6–2.8	Turquois	3.18	**Fluorite**
		2.71	**Labradorite**		
2.05–2.09	**Sulfur**	2.65–2.74	**Scapolite**	**3.2–3.39**	
2.05–2.15	**Chabazite**	2.72	**Calcite**		
1.9–2.2	**Opal**	2.6–2.9	**Chlorite**	3.1–3.3	Scorodite
2.09–2.14	Niter			3.2	**Hornblende**
2.1–2.2	**Stilbite**	2.62–2.76	**Plagioclase**	3.23	**Sillimanite**
2.16	**Halite**	2.6–2.9	**Collophane**	3.2–3.3	**Diopside**
2.12–2.30	Chalcanthite	2.74	**Bytownite**	3.2–3.4	**Augite**
2.18–2.20	**Heulandite**	2.7–2.8	Pectolite	3.25–3.37	Clinozoisite
		2.7–2.8	**Talc**	3.26–3.36	Dumortierite
2.2–2.39		2.70–2.85	Glauberite	3.27–3.35	Axinite
		2.76	**Anorthite**	3.27–4.37	**Olivine**
2.12–2.30	Chalcanthite	2.75–2.8	**Beryl**	3.2–3.5	**Enstatite**
2.0–2.4	**Chrysocolla**				
2.2–2.65	**Serpentine**			**3.4–3.59**	
2.25	**Natrolite**	**2.8–2.99**			
2.26	Tridymite			3.27–4.27	Olivine
2.27	**Analcime**	2.6–2.9	**Collophane**	3.3–3.5	Jadeite
2.29	Soda niter	2.8–2.9	Pyrophyllite	3.35–3.45	**Diaspore**
2.30	Cristobalite	2.8–2.9	**Wollastonite**	3.35–3.45	**Epidote**
2.30	**Sodalite**	2.85	**Dolomite**	3.35–3.45	**Idocrase**
2.3	**Graphite**	2.86	Phlogopite	3.4–3.5	**Hemimorphite**
2.32	**Gypsum**	2.76–3.1	**Muscovite**	3.45	Arfvedsonite
2.33	Wavellite	2.8–2.95	**Prehnite**	3.40–3.55	Aegirine
2.3–2.4	**Apophyllite**	2.8–3.0	**Datolite**	3.4–3.55	**Sphene**
2.39	**Brucite**	2.8–3.0	**Lepidolite**	3.48	**Realgar**
		2.89–2.98	**Anhydrite**	3.42–3.56	Triphylite
2.4–2.59		2.9–3.0	Boracite	3.49	**Orpiment**
		2.95	**Aragonite**	3.4–3.6	**Topaz**
		2.95	Erythrite	3.5	**Diamond**
2.0–2.55	**Bauxite**	2.8–3.2	**Biotite**	3.45–3.60	**Rhodochrosite**
2.2–2.65	Serpentine	2.95–3.0	**Cryolite**	3.5–4.3	Garnet
2.42	**Colemanite**	2.97–3.00	Phenacite		
2.42	Petalite			**3.6–3.79**	
2.4–2.45	**Lazurite**	**3.0–3.19**			
2.45–2.50	**Leucite**			3.27–4.37	Olivine
2.2–2.8	**Garnierite**	2.97–3.02	Danburite	3.5–4.2	Allanite
2.54–2.57	**Microcline**	2.85–3.2	Anthophyllite	3.5–4.3	**Garnet**
2.57	**Orthoclase**	3.0–3.1	**Amblygonite**	3.6–4.0	**Spinel**

G.	Name	G.	Name	G.	Name
3.56–3.66	**Kyanite**	4.6	**Chromite**	5.85	**Pyrargyrite**
3.58–3.70	**Rhodonite**	4.58–4.65	**Pyrrhotite**		
3.65–3.75	**Staurolite**	4.7	**Ilmenite**	**6.0–6.49**	
3.7	**Strontianite**	4.75	**Pyrolusite**		
3.65–3.8	**Chrysoberyl**	4.6–4.76	Covellite	5.9–6.1	Crocoite
3.75–3.77	Atacamite	4.62–4.73	**Molybdenite**	5.9–6.1	**Scheelite**
3.77	**Azurite**	4.68	**Zircon**	6.0	**Cuprite**
				6.07	**Arsenopyrite**
3.8–3.99		**4.8–4.99**		6.0–6.2	Polybasite
				6.2–6.3	Stephanite
3.7–4.7	**Psilomelane**			6.2–6.4	**Anglesite**
3.6–4.0	**Spinel**	4.6–5.0	Pentlandite	5.3–7.3	Columbite
3.6–4.0	Limonite	4.6–5.1	**Tetrahedrite-**	6.33	**Cobaltite**
3.83–3.88	**Siderite**		**Tennantite**		
3.5–4.2	Allanite	4.89	**Marcasite**	**6.5–6.99**	
3.5–4.3	**Garnet**	4.9	Greenockite		
3.9	Antlerite			6.5	**Skutterudite**
3.9–4.03	**Malachite**	**5.0–5.19**		6.55	**Cerussite**
3.95–3.97	**Celestite**			6.78	Bismuthinite
		5.02	**Pyrite**	6.5–7.1	Pyromorphite
4.0–4.19		4.8–5.3	**Hematite**	6.8	**Wulfenite**
		5.06–5.08	**Bornite**	6.7–7.1	Vanadinite
3.9–4.1	**Sphalerite**	5.15	**Franklinite**	6.8–7.1	**Cassiterite**
4.02	**Corundum**	5.0–5.3	Monazite		
3.9–4.2	**Willemite**	5.18	**Magnetite**	**7.0–7.49**	
				7.0–7.2	Mimetite
4.2–4.39		**5.2–5.39**		7.0–7.5	**Wolframite**
				7.3	**Argentite**
4.1–4.3	**Chalcopyrite**	**5.4–5.59**			
3.7–4.7	**Psilomelane**			**7.5–7.99**	
4.18–4.25	**Rutile**	5.5	Millerite		
4.3	**Manganite**	5.5±	**Cerargyrite**	7.4–7.6	**Galena**
4.3	**Witherite**	5.55	**Proustite**	7.3–7.9	Iron
4.37	**Goethite**			7.78	**Niccolite**
4.35–4.40	**Smithsonite**	**5.6–5.79**			
				>8.0	
4.4–4.59		5.5–5.8	**Chalcocite**		
		5.68	**Zincite**	8.0–8.2	Sylvanite
4.4	Stannite	5.7	Arsenic	8.10	**Cinnabar**
4.43–4.45	**Enargite**	5.5–6.0	Jamesonite	8.9	**Copper**
4.5	**Barite**	5.3–7.3	Columbite	9.0–9.7	Uraninite
4.55	Gahnite			9.35	Calaverite
4.52–4.62	**Stibnite**			9.8	Bismuth
				10.5	**Silver**
4.6–4.79		**5.8–5.99**		15.0–19.3	**Gold**
				14–19	**Platinum**
3.7–4.7	**Psilomelane**	5.8–5.9	Bournonite		

INDEX

MINERAL INDEX

In this index the mineral name is followed by the commonly sought information: composition, crystal system (Xl Sys.), specific gravity (**G**), hardness (**H**), and index of refraction (n). For uniaxial crystals $n = \omega$, for biaxial crystals $n = \beta$. The refractive index is given as a single entry when the range is usually no greater than ± 0.01.

Name, page	Composition	Xl. Sys.	G.	H.	n	Remarks
Acadialite, 479						See chabazite
Acanthite, 244	Ag_2S	Mon	7.2–7.3	$2-2\frac{1}{2}$	—	Low temp. Ag_2S
Achroite, 403						Colorless tourmaline
Acmite, 415	$NaFe'''(Si_2O_6)$	Mon	3.5	$6-6\frac{1}{2}$	1.82	A pyroxene
Actinolite, 420	$Ca_2(Mg,Fe)_5(Si_8O_{22})(OH)_2$	Mon	3.0–3.2	5–6	1.64	An amphibole
Adularia, 463						See orthoclase
Aegirine, 415	$NaFe(Si_2O_6)$	Mon	3.40–3.55	$6-6\frac{1}{2}$	1.82	A pyroxene
Aenigmatite, 390	$(Na,Ca)_4(Fe'',Fe''',$ $Mn,Ti,Al)_{13}(Si_2O_7)_6$	Tric	3.75	$5\frac{1}{2}$	1.80	Cl {110}
Agate, 454						See quartz
Alabandite, 249, 289	MnS	Iso	4.0	$3\frac{1}{2}-4$	—	Black
Alabaster, 348						See gypsum
Albite, 466	$Na(AlSi_3O_8)—Ab_{90}An_{10}$	Tric	2.62	6	1.53	A feldspar
Alexandrite, 299						Gem chrysoberyl
Allanite, 395	$X_2Y_3O(SiO_4)(Si_2O_7)(OH)$	Mon	3.5–4.2	$5\frac{1}{2}-6$	1.70–1.81	Brown-black
Allemontite, 233	$AsSb$	Hex	5.8–6.2	3–4	—	One cleavage
Almandite, 379	$Fe_3Al_2(SiO_4)_3$	Iso	4.25	7	1.83	A garnet
Altaite, 249, 269	$PbTe$	Iso	8.16	3	—	Tin-white
Alumstone, 352						See alunite
Alunite, 351	$KAl_3(OH)_6(SO_4)_2$	Rho	2.6–2.8	4	1.57	Usually massive
Amalgam, 227						See silver
Amazonstone, 464						Green microcline
Amblygonite, 364		Tric	3.0–3.1	6	1.60	Fusible at 2
Amethyst, 452						Purple quartz
Amphibole, 418						A mineral group
Analcime, 476	$Na(AlSi_2O_6)·H_2O$	Iso	2.27	$5-5\frac{1}{2}$	1.49	A feldspathoid
Anatase, 287	TiO_2	Tet	3.9	$5\frac{1}{2}-6$	2.6	Adamantine luster
Anauxite, 436		Mon	2.6	2	1.56	Si-rich kaolinite
Andalusite, 383	Al_2SiO_5	Orth	3.16–3.20	$7\frac{1}{2}$	1.64	Infusible
Andesine, 469	$Ab_{70}An_{30}—Ab_{50}An_{50}$	Tric	2.69	6	1.55	Plagioclase feldspar
Andradite, 379	$Ca_3Fe_2(SiO_4)_3$	Iso	3.75	7	1.89	A garnet
Anglesite, 344	$PbSO_4$	Orth	6.2–6.4	3	1.88	Cl {001}{110}
Anhydrite, 345	$CaSO_4$	Orth	2.89–2.98	$3-3\frac{1}{2}$	1.58	Cl {100}{010}{001}

Mineral	Formula	System	G	H	n	Remarks
Ankerite, 326	$Ca(Fe,Mg)(CO_3)_2$	Rho	2.95–3	$3\frac{1}{2}$	1.70–1.75	Cl $\{10\bar{1}1\}$
Annabergite, 363	$Ni_3(AsO_4)_2 \cdot 8H_2O$	Mon	3.0	$2\frac{1}{2}$–3	1.68	Nickel bloom. Green
Anorthite, 466	$Ab_{10}An_{90}$—$CaAl_2Si_2O_8$	Tric	2.76	6	1.58	Plagioclase feldspar
Anorthoclase, 465	$K(AlSi_3O_8)$ — $Na(AlSi_3O_8)$	Tric	2.58	6	1.53	A feldspar
Anthophyllite, 419	$(Mg,Fe)_7(Si_8O_{22})(OH)_2$	Orth	2.85–3.2	$5\frac{1}{2}$–6	1.61–1.71	An amphibole
Antigorite, 437					1.57	See serpentine
Antimony, 233	Sb	Rho	6.7	3	—	Cl $\{0001\}$
Antlerite, 350	$Cu_3(OH)_4SO_4$	Orth	3.9±	$3\frac{1}{2}$–4	1.74	Green
Apatite, 359	$Ca_5(F,Cl,OH)(PO_4)_3$	Hex	3.15–3.20	5	1.63	Cl $\{0001\}$ poor
Apophyllite, 434	$Ca_4K(Si_4O_{10})_2F \cdot 8H_2O$	Tet	2.3–2.4	$4\frac{1}{2}$–5	1.54	Cl $\{001\}$
Aquamarine, 401						See beryl
Aragonite, 328	$CaCO_3$	Orth	2.95	$3\frac{1}{2}$–4	1.69	Cl $\{010\}\{110\}$
Arfvedsonite, 422	$Na_3Mg_4Al(Si_8O_{22})(OH,F)_2$	Mon	3.45	6	1.69	An amphibole
Argentite, 244	Ag_2S	Iso	7.3	2–$2\frac{1}{2}$	—	Sectile
Arsenic, 233	As	Rho	5.7	$3\frac{1}{2}$	—	Cl $\{0001\}$
Arsenopyrite, 266	$FeAsS$	Mon	6.07 ± 0.15	$5\frac{1}{2}$–6	—	Pseudo-orth.
Asbestos, 438						See amphibole and serpentine
Astrophyllite, 390	$(Na,Ca)_5(Fe'',Al,Ti)_{15}-$ $(Si_2O_7)_6(F,OH)_8$	Tri	3.35	3	1.71	Micaceous cl.
Atacamite, 314	$Cu_2Cl(OH)_3$	Orth	3.75–3.77	3–$3\frac{1}{2}$	1.86	Cl $\{010\}$
Augite, 411	$(Ca,Na)(Mg,Fe,Al)(Si,Al)_2O_6$	Mon	3.2–3.4	5–6	1.67–1.73	A pyroxene
Aurichalcite, 335	$2(Zn,Cu)CO_3 \cdot 3(Zn,Cu)(OH)_2$	Mon	3.64	2	1.74	Green to blue
Autunite, 366	$Ca(UO_2)_2(PO_4)_2 \cdot 10-12H_2O$	Tet	3.1–3.2	2–$2\frac{1}{2}$	1.58	Yellow-green
Aventurine, 453						Oligoclase or qtz.
Axinite, 399	$Ca_2(Fe,Mn)Al_2(BO_3)-$ $(Si_4O_{12})(OH)$	Tric	3.27–3.35	$6\frac{1}{2}$–7	1.69	Crystal angles acute
Azurite, 334	$Cu_3(CO_3)_2(OH)_2$	Mon	3.77	$3\frac{1}{2}$–4	1.76	Always blue
Balas ruby, 294						Red gem spinel
Barite, 342	$BaSO_4$	Orth	4.5	3–$3\frac{1}{2}$	1.64	Cl $\{001\}\{110\}$
Bauxite, 305	A mixture of aluminum hydroxides	Amor	2.0–2.55	1–3	—	An earthy rock
Beidellite, 436	$Al_8(Si_4O_{10})_3(OH)_{12} \cdot 12H_2O$	Orth?	2.6	$1\frac{1}{2}$	1.55	See kaolinite
Benitoite, 390	$BaTiSi_3O_9$	Hex	3.6	$6\frac{1}{2}$	1.76	Blue
Bentonite, 436						Montmorillonite

Name, page	Composition	XI. Sys.	G.	H.	n	Remarks
Beryl, 400	$Be_3Al_2(Si_6O_{18})$	Hex	2.75–2.8	$7\frac{1}{2}$–8	1.56–1.61	Usually green
Biotite, 443	$K(Mg,Fe)_3(AlSi_3O_{10})(OH)_2$	Mon	2.8–3.2	$2\frac{1}{2}$–3	1.61–1.70	Black mica
Bismuth, 233	Bi	Rho	9.8	2–$2\frac{1}{2}$	—	Cl {0001}
Bismuthinite, 262	Bi_2S_3	Orth	6.78 ± 0.03	2	—	Cl {010}
Black jack, 250						See sphalerite
Bloodstone, 454						Green and red chalcedony
Blue vitriol, 349						See chalcanthite
Boehmite, 304	$AlO(OH)$	Orth	3.01–3.06	—	1.65	In bauxite
Bog-iron ore, 304						See limonite
Boracite, 340	$Mg_3B_7O_{13}Cl$	Orth	2.9–3.0	7	1.66	Pseudo-isometric
Borax, 338	$Na_2B_4O_7 \cdot 10H_2O$	Mon	$1.7\pm$	2–$2\frac{1}{2}$	1.47	Cl {100}
Bornite, 246	Cu_5FeS_4	Iso	5.06–5.08	3	—	Purple-blue tarnish
Boulangerite, 275	$Pb_5Sb_4S_{11}$	Orth	$6.0\pm$	$2\frac{1}{2}$–3	—	Cl {001}{010}
Bournonite, 274	$PbCuSbS_3$	Orth	5.8–5.9	$2\frac{1}{2}$–3	—	Fusible at 1
Bravoite, 264	$(Ni,Fe)S_2$	Iso	4.66	$5\frac{1}{2}$–6	—	Steel gray
Breithauptite, 255	$NiSb$	Hex	8.23	$5\frac{1}{2}$	—	Copper red
Brittle mica, 445						See margarite
Brochantite, 351	$Cu_4(OH)_6SO_4$	Mon	3.9	$3\frac{1}{2}$–4	1.78	Cl {010}. Green
Bromyrite, 311	$AgBr$	Iso	5.9	1–$1\frac{1}{2}$	2.25	Sectile
Bronzite, 410	$(Mg,Fe)_2(Si_2O_6)$	Orth	$3.3\pm$	$5\frac{1}{2}$	1.68	See enstatite
Brookite, 287	TiO_2	Orth	3.9–4.1	$5\frac{1}{2}$–6	2.6	Adamantine luster
Brucite, 301	$Mg(OH)_2$	Rho	2.39	$2\frac{1}{2}$	1.57	Cl {0001}
Bytownite, 469	$Ab_{30}An_{70}$—$Ab_{10}An_{90}$	Tric	2.74	6	1.57	Plag. feldspar
Cairngorm stone, 452						See quartz
Calamine, 392						See hemimorphite
Calaverite, 268	$AuTe_2$	Mon	9.35	$2\frac{1}{2}$	—	Fusible at 1
Calcite, 318	$CaCO_3$	Rho	2.71	3	1.66	Cl {10$\bar{1}$1}
Californite, 396						See idocrase
Cancrinite, 472	$(Na,K)_{6-8}Al_6Si_6O_{24}$-$(CO_3)_{1-2} \cdot 2$-$3H_2O$	Hex	2.45	5–6	1.52	A feldspathoid
Capillary pyrites, 255						See millerite

Name, page	Composition	Xl. Sys.	G.	H.	n	Remarks
Clay ironstone, 323						See siderite
Cleavelandite, 469						White, platy albite
Cliachite, 306	$Al(OH)_3$	Amor	$2.5\pm$	$1-3$	—	See bauxite
Clinochlore, 445					1.58	See chlorite
Clinoenstatite, 411	$Mg_2(Si_2O_6)$	Mon	3.19	6	1.66	Prismatic cl.
Clinohumite, 388	$Mg_9(SiO_4)_4(F, OH)_2$	Mon	$3.1-3.2$	6	1.64	See chondrodite
Clinohypersthene, 411	$(Mg, Fe)_2(Si_2O_6)$	Mon	$3.4-3.5$	$5-6$	$1.68-1.72$	A pyroxene
Clinozoisite, 393	$Ca_2Al_3O(SiO_4)(Si_2O_7)(OH)$	Mon	$3.25-3.37$	$6-6\frac{1}{2}$	$1.67-1.72$	Crystals striated
Cobaltite, 264	$(Co,Fe)AsS$	Iso	6.33	$5\frac{1}{2}$	—	In pyritohedrons
Coseite 449	SiO_2	Mon	2.93	7	1.59	High pressure SiO_2
Colemanite, 340	$Ca_2B_6O_{11}\cdot5H_2O$	Mon	2.42	$4-4\frac{1}{2}$	1.59	Cl {010} perfect
Collophane, 359						See apatite
Columbite, 299	$(Fe, Mn)Nb_2O_6$	Orth	$5.3-7.3$	6	—	Luster submetallic
Common salt, 308						See halite
Copper, 228	Cu	Iso	8.9	$2\frac{1}{2}-3$	—	Malleable
Copper glance, 244						See chalcocite
Copper pyrites, 251						See chalcopyrite
Cordierite, 402	$(Mg, Fe)_2Al_3(AlSi_5O_{18})$	Orth	$2.60-2.66$	$7-7\frac{1}{2}$	$1.53-1.57$	Blue
Corundum, 278	Al_2O_3	Rho	4.02	9	1.77	Rhomb. parting
Cotton-balls, 339						See ulexite
Covellite, 257	CuS	Hex	$4.6-4.76$	$1\frac{1}{2}-2$	—	Blue
Cristobalite, 456	SiO_2	Tet?	2.30	7	1.49	In volcanic rocks
Crocidolite, 422	$Na_2Fe_3{}^2Fe_2{}^3(Si_8O_{22})(OH)_2$	Mon	$3.2-3.3$	4	1.70	Blue asbestos
Crocoite, 346	$PbCrO_4$	Mon	$5.9-6.1$	$2\frac{1}{2}-3$	—	Orange-red
Cryolite, 311	Na_3AlF_6	Mon	$2.95-3.0$	$2\frac{1}{2}$	1.34	White
Cummingtonite, 419	$(Mg,Fe)_7(Si_8O_{22})(OH)_2$	Mon	$2.85-3.2$	6	1.66	An amphibole
Cuprite, 276	Cu_2O	Iso	6.0	$3\frac{1}{2}-4$	—	In red crystals
Cymophane, 299						See chrysoberyl
Danburite, 469	$Ca(B_2Si_2O_8)$	Orth	$2.97-3.02$	7	1.63	In crystals
Datolite, 388	$CaB(SiO_4)(OH)$	Mon	$2.8-3.0$	$5-5\frac{1}{2}$	1.65	Usually in crystals
Demantoid, 379						Green andradite

Mineral	Formula	System	G	H	n	Remarks
Diallage, 412						See diopside
Diamond, 238	C	Iso	3.5	10	2.42	Adamantine luster
Diaspore, 303	AlO(OH)	Orth	3.35–3.45	$6\frac{1}{2}$–7	1.72	Cl {010} perfect
Diatomaceous earth, 458						See opal
Diatomite, 458						See opal
Dichroite, 402						See cordierite
Dickite, 436	$Al_4(Si_4O_{10})_3(OH)_{12}\cdot 3H_2O$	Mon	2.6	2–$2\frac{1}{2}$	1.56	Clay mineral
Digenite, 245	Cu_9S_5	Iso	5.6	$2\frac{1}{2}$–3	—	Like chalcocite
Diopside, 411	$CaMg(Si_2O_6)$	Mon	3.2–3.3	5–6	1.68	A pyroxene
Dioptase, 405	$Cu_6(Si_6O_{18})\cdot 6H_2O$	Rho	3.3	5	1.65	Green
Dolomite, 326	$CaMg(CO_3)_2$	Rho	2.85	$3\frac{1}{2}$–4	1.68	Cl {10$\bar{1}$1}
Dry-bone ore, 324						See smithsonite
Dumortierite, 385	$(Al,Fe)_7O_3(BO_3)(SiO_4)_3$	Orth	3.26–3.36	7	1.69	Radiating
Edenite, 422	$Ca_2NaMg_5(AlSi_7O_{22})(OH,F)_2$	Mon	3.0	6	1.63	See hornblende
Electrum, 225						See gold
Eleolite, 472						See nepheline
Embolite, 311	$Ag(Cl,Br)$	Iso	5.3–5.4	1–$1\frac{1}{2}$	—	Sectile
Emerald, 400						See beryl
Emery, 279						Corundum with magnetite
Enargite, 273	Cu_3AsS_4	Orth	4.43–4.45	3	—	Cl {110}
Endlichite, 362						See vanadinite
Enstatite, 410	$Mg_2(Si_2O_6)$	Orth	3.2–3.5	$5\frac{1}{2}$	1.66	A pyroxene
Epidote, 393	$Ca_2(Al,Fe)Al_2O(SiO_4)\text{-}(Si_2O_7)(OH)$	Mon	3.35–3.45	6–7	1.73–1.78	Cl {001}
Epsomite, 349	$MgSO_4\cdot 7H_2O$	Orth	1.75	2–$2\frac{1}{2}$	1.46	Bitter taste
Epsom salt, 350						See epsomite
Erythrite, 363	$Co_3(AsO_4)_2\cdot 8H_2O$	Mon	2.95	$1\frac{1}{2}$–$2\frac{1}{2}$	1.66	Pink. Cobalt bloom
Essonite, 379						See grossularite
Euclase, 402	$B_2Al_2(SiO_4)_2(OH)_2$	Mon	3.1	$7\frac{1}{2}$	1.66	Cl {010}
Eucryptite, 414	$Li(Al,Si)_2O_4$	Hex	2.67	—	1.55	Spodumene alter.
Fahlore, 272						See tetrahedrite
Famatinite, 274	Cu_3SbS_4	Tet	4.52	$3\frac{1}{2}$	—	Gray
Fayalite, 376	Fe_2SiO_4	Orth	4.14	$6\frac{1}{2}$	1.86	See olivine

Name, page	Composition	Xl. Sys.	G.	H.	n	Remarks
Feather ore, 275	....					See jamesonite
Feldspar, 458	....					A mineral group
Feldspathoid, 470	....					A mineral group
Ferberite, 353	$FeWO_4$	Mon	7.0–7.5	5	—	See wolframite
Fergusonite, 301	$R'''(Nb,Ta)O_4$	Tet	5.8	$5\frac{1}{2}$–6	—	Brown-black
Fersmannite, 390	$Ca_4Na_2Ti_4Si_3O_{18}F_2$	Mon	3.44	$5\frac{1}{2}$	1.93	Brown
Fibrolite, 384	....					See sillimanite
Flint, 455	SiO_2	—	2.65	7	1.54	Cryptocryst. qtz.
Flos ferri, 329	....					See aragonite
Fluorite, 312	CaF_2	Iso	3.18	4	1.43	Cl octahedral
Forsterite, 376	Mg_2SiO_4	Orth	3.2	$6\frac{1}{2}$	1.63	See olivine
Fowlerite, 418	....					Zinc-bearing rhodonite
Franklinite, 297	$(Fe,Zn,Mn)(Fe,Mn)_2O_4$	Iso	5.15	6	—	At Franklin, N. J.
Freibergite, 272	....					Argentiferous tetrahedrite
Gadolinite, 402	$Y_2Fe''Be_2(SiO_4)_2O_2$	Mon	4.0–4.5	$6\frac{1}{2}$–7	1.79	Black
Gahnite, 295	$ZnAl_2O_4$	Iso	4.55	$7\frac{1}{2}$–8	1.80	Green octahedrons
Galaxite, 295	$MnAl_2O_4$	Iso	4.03	$7\frac{1}{2}$–8	1.92	Mn spinel
Galena, 246	PbS	Iso	7.4–7.6	$2\frac{1}{2}$	—	Cl cubic
Garnet, 377	$A_3''B_2'''(SiO_4)_3$	Iso	3.5–4.3	$6\frac{1}{2}$–$7\frac{1}{2}$	1.71–1.88	In crystals
Garnierite, 438	$(Ni,Mg)SiO_3·nH_2O$	Amor	2.2–2.8	2–3	1.59	Green
Gaylussite, 335	$Na_2Ca(CO_3)_2·5H_2O$	Mon	1.99	2–3	1.52	Fusible at 1
Gedrite, 419	....					See anthophyllite
Geikielite, 284	$(Mg,Fe)TiO_3$	Rho	4.05	$5\frac{1}{2}$–6	—	Cl $\{10\overline{1}1\}$
Geocronite, 275	$Pb_5(Sb,As)_2S_8$	Orth	6.4±	$2\frac{1}{2}$	—	
Gersdorffite, 265	$NiAsS$	Iso	5.9	$5\frac{1}{2}$	—	See cobaltite
Geyserite, 458	....					See opal
Gibbsite, 304	$Al(OH)_3$	Mon	2.3–2.4	$2\frac{1}{2}$–$3\frac{1}{2}$	1.57	Basal cl.
Glaucodot, 267	$(Co,Fe)AsS$	Orth?	6.04	5	—	Tin-white
Glauconite, 444	$K_2(Mg,Fe)_2Al_6(Si_4O_{10})_3(OH)_{12}$	Mon	2.3±	2	1.62	In green sands
Glaucophane, 422	$Na_2Mg_3Al_2(Si_8O_{22})(OH)_2$	Mon	3.0–3.2	6–$6\frac{1}{2}$	1.62–1.67	An amphibole

Mineral	Formula	System	G	H	n	Remarks
Gmelinite, 481	$(Na,Ca)_6Al_6(Al,Si)$-$Si_{13}O_{40}\cdot20H_2O$	Rho	2.1 ±	$4\frac{1}{2}$	1.49	A zeolite
Goethite, 304	$HFeO_2$	Orth	4.37	$5-5\frac{1}{2}$	—	Cl{010}
Gold, 224	Au	Iso	15.0–19.3	$2\frac{1}{2}-3$	—	Yellow. Soft
Graphite, 242	C	Hex	2.3	1–2	—	Black. Platy
Gray copper, 272						See tetrahedrite
Greenockite, 253	CdS	Hex	4.9	$3-3\frac{1}{2}$	—	Yellow-orange
Grossularite, 379	$Ca_3Al_2(SiO_4)_3$	Iso	3.53	$6\frac{1}{2}$	1.73	A garnet
Gypsum, 347	$CaSO_4\cdot2H_2O$	Mon	2.32	2	1.52	Cl {010}{100}{011}
Halite, 308	$NaCl$	Iso	2.16	$2\frac{1}{2}$	1.54	Cl cubic. Salty
Halloysite, 436	$Al_4(Si_4O_{10})(OH)_8$	Amor	2.0–2.2	1–2	1.54	A clay mineral
Harmotome, 481	$Ba(Al_2Si_6O_{16})\cdot6H_2O$	Mon	2.45	$4\frac{1}{2}$	1.51	A zeolite
Hastingsite, 422	$Ca_2NaMg_4Al_3Si_6O_{22}(OH,F)_2$	Mon	3.2	6	1.66	See hornblende
Hauynite, 473	$(Na,Ca)_{6-8}Al_6Si_6O_{24}\cdot(SO_4)_{1-2}$	Iso	2.4–2.5	$5\frac{1}{2}-6$	1.50	A feldspathoid
Heavy spar, 342						See barite
Hectorite, 436	$(Mg,Li)_6Si_8O_{20}(OH)_4$	Mon	2.5	$1-1\frac{1}{2}$	1.52	Li montmorillonite
Hedenbergite, 411	$CaFe(Si_2O_6)$	Mon	3.55	5–6	1.71	A pyroxene
Heliotrope, 454						Green and red chalcedony
Hematite, 281	Fe_2O_3	Rho	5.26	$5\frac{1}{2}-6\frac{1}{2}$	—	Red streak
Hemimorphite, 391	$Zn_4(Si_2O_7)(OH)_2\cdot H_2O$	Orth	3.4–3.5	$4\frac{1}{2}-5$	1.62	Cl {110}
Hercynite, 295	$FeAl_2O_4$	Iso	4.39	$7\frac{1}{2}-8$	1.80	Iron spinel
Hessite, 269	Ag_2Te	Iso	8.4	$2\frac{1}{2}-3$	—	
Heulandite, 479	$Ca(Al_2Si_7O_{18})\cdot6H_2O$	Mon	2.18–2.20	$3\frac{1}{2}-4$	1.49	Cl {010} perfect
Hiddenite, 413						Green spodumene
Holmquistite, 422						Lithium-bearing glaucophane
Hornblende, 421	$Ca_2Na(Mg,Fe^2)_4(Al,Fe^3,Ti)_3$-$Si_8O_{22}(O,OH)_2$	Mon	3.2	5–6	1.62–1.71	An amphibole
Horn silver, 311						See cerargyrite
Huebnerite, 353	$MnWO_4$	Mon	7.0	5	—	See wolframite
Humite, 388	$Mg_7(SiO_4)_3(F,OH)_2$	Orth	3.1–3.2	6	1.64	See chondrodite
Hyacinth, 381						See zircon
Hyalite, 457						Globular, colorless opal

Name, page	Composition	Xl. Sys.	G.	H.	n	Remarks
Malachite, 333	$Cu_2CO_3(OH)_2$	Mon	3.9–4.03	$3\frac{1}{2}$–4	1.88	Green
Manganite, 302	$MnO(OH)$	Orth	4.3	4	—	Prismatic crystals
Manganotantalite, 300	$MnO(Ta,Cb)_2O_5$	Orth	6.6±	$4\frac{1}{2}$	—	See columbite
Marcasite, 265	FeS_2	Orth	4.89	6–$6\frac{1}{2}$	—	White iron pyrites
Margarite, 445	$CaAl_2(Al_2Si_2O_{10})(OH)_2$	Mon	3.0–3.1	$3\frac{1}{2}$–5	1.65	A brittle mica
Marialite, 475	$(Na,Ca)_4Al_3(Al,Si)_3Si_6O_{24}(Cl,CO_3,SO_4)$	Tet	2.60±	$5\frac{1}{2}$–6	1.55	See scapolite
Martite, 281						See hematite
Meerschaum, 438						See sepiolite
Meionite, 474	$(Ca,Na)_4Al_3(Al,Si)_3Si_6O_{24}(Cl,CO_3,SO_4)$	Tet	2.69	$5\frac{1}{2}$–6	1.59	See scapolite
Melaconite, 277						See tenorite
Melanite, 378	Black andradite	Iso	3.7	7	1.94	A garnet
Melanterite, 350	$FeSO_4 \cdot 7H_2O$	Mon	1.90	2	1.48	Green-blue
Menaccanite,						See ilmenite
Meneghinite, 275	$Pb_{13}Sb_7S_{23}$	Orth	6.36	$2\frac{1}{2}$	—	
Mercury, 222	Hg	—	13.6	0	—	Fluid. Quicksilver
Metacinnabar, 258	HgS	Iso	7.65	3	—	Grayish black
Mica, 440						A mineral group
Microcline, 464	$K(AlSi_3O_8)$	Tric	2.54–2.57	6	1.53	A feldspar
Microlite, 301	$Ca_2Ta_2O_7$	Iso	5.48–5.56	$5\frac{1}{2}$	1.92–1.99	Ore of tantalum
Microperthite, 464						Microcline and albite
Millerite, 255	NiS	Rho	5.5 ± 0.2	3–$3\frac{1}{2}$	—	Capillary crystals
Mimetite, 361	$Pb_5Cl(AsO_4)_3$	Hex	7.0–7.2	$3\frac{1}{2}$	—	Like pyromorphite
Molybdenite, 268	MoS_2	Hex	4.62–4.73	1–$1\frac{1}{2}$	—	Black. Platy
Monazite, 358	$(Ce,La,Y,Th)PO_4$	Mon	5.0–5.3	5–$5\frac{1}{2}$	1.79	Parting {001}
Monticellite, 377	$CaMgSiO_4$	Orth	3.2	5	1.65	See olivine
Montmorillonite, 436	$(Al,Mg)_8(Si_4O_{10})_3(OH)_{10} \cdot 12H_2O$	Mon	2.5	1–$1\frac{1}{2}$	1.50–1.64	A clay mineral
Moonstone, 463						See albite and orthoclase
Morganite, 401						See beryl

Mineral	Formula	System	G	H	n	Remarks
Mullite, 385	$Al_6Si_2O_{13}$	Orth	3.23	6–7	1.65	Cl {100}
Muscovite, 441	$KAl_2(AlSi_3O_{10})(OH)_2$	Mon	2.76–3.1	2–2½	1.60	Cl {001} perfect
Nacrite, 436	$Al_4(Si_4O_{10})(OH)_8$	Mon	2.6	2–2½	1.56	See kaolinite
Nagyagite, 267	$Pb_5Au(Te,Sb)_4S_{5-8}$	Mon?	7.4	1–1½	—	
Natroalunite, 351						Soda alunite
Natrolite, 477	$Na_2(Al_2Si_3O_{10})\cdot 2H_2O$	Mon	2.25	5–5½	1.48	Cl {100} perfect
Nepheline, 471	$(Na,K)(AlSiO_4)$	Hex	2.55–2.65	5½–6	1.54	Greasy luster
Nephrite, 420						See tremolite
Neptunite, 390	$(Na,K)(Fe'',Mn,Ti)Si_2O_6$	Mon	3.23	5–6	1.70	Black
Niccolite, 254	NiAs	Hex	7.78	5–5½	—	Copper-red
Nickel bloom, 363						See annabergite
Nickel iron, 231	Ni, Fe	Iso	7.8–8.2	5	—	In meteorites
Nickel skutterudite, 270	$(Ni,Co,Fe)As_3$	Iso	6.5 ± 0.4	5½–6	—	Tin white
Niter, 336	KNO_3	Orth	2.09–2.14	2	1.50	Saltpeter
Nontronite, 436	$Fe_4(AlSi_8)O_{20}(OH)_4$	Mon	2.5	1–1½	1.60	Clay mineral
Norbergite, 388	$Mg_3(SiO_4)_1(F,OH)_2$	Orth	3.1–3.2	6	1.57	See chondrodite
Noselite, 473		Iso	2.3±	6	1.50	A feldspathoid
Octahedrite, 287						See anatase
Oligoclase, 469	$Ab_{90}An_{10}$—$Ab_{70}An_{30}$	Tric	2.65	6	1.54	Plag. feldspar
Olivine, 375	$(Mg,Fe)_2SiO_4$	Orth	3.27–4.37	6½–7	1.69	Green rock mineral
Onyx, 454						Layered chalcedony
Opal, 457	$SiO_2\cdot nH_2O$	Amor	1.9–2.2	5–6	1.44	Conch. fracture
Orpiment, 259	As_2S_3	Mon	3.49	1½–2	—	Cl {010}. Yellow
Orthite, 395						See allanite
Orthoclase, 462	$K(AlSi_3O_8)$	Mon	2.57	6	1.52	A feldspar
Orthoferrosilite, 410	$Fe_2(Si_2O_6)$	Orth	3.9	6	1.79	A pyroxene
Ottrelite, 445	$(Fe,Mn)(Al,Fe)(Al_2Si_2O_{10})(OH)_2$	Mon	3.5	6–7	1.73	A brittle mica
Palladium, 231	Pd	Iso	11.9	4½–5	—	See platinum
Paragonite, 442	$NaAl_2(AlSi_3O_{10})(OH)_2$	Mon	2.85	2	1.60	Like muscovite
Pargasite, 422	$Ca_4Na_2Mg_9Al_4Si_{13}O_{44}(OH,F)_4$	Mon	3–3.5	5½	1.62	See hornblende
Patronite, 362						An ore of vanadium
Pectolite, 416	$Ca_2NaH(SiO_3)_3$	Tric	2.7–2.8	5	1.61	Crystals acicular

Name, page	Composition	Xl. Sys.	G.	H.	n	Remarks
Penninite, 445					1.58	See chlorite
Pentlandite, 256	$(Fe,Ni)_9S_8$	Iso	4.6–5.0	$3\frac{1}{2}$–4	—	In pyrrhotite
Peridot, 376						Gem olivine
Perovskite, 287	$CaTiO_3$	Iso	4.03	$5\frac{1}{2}$	2.38	Yellow
Perthite, 464						Microcline and albite
Petalite, 473	$Li(AlSi_4O_{10})$	Mon	2.4	6–$6\frac{1}{2}$	1.51	Good cleavage
Petzite, 269	$(Ag,Au)_2Te$	Iso?	8.7–9.0	$2\frac{1}{2}$–3	—	—
Phenacite, 374	$Be_2(SiO_4)$	Rho	2.97–3.00	$7\frac{1}{2}$–8		In pegmatites
Phillipsite, 481	$KCa(Al_3Si_5O_{16})\cdot6H_2O$	Mon	2.2	$4\frac{1}{2}$–5	1.65	A zeolite
Phlogopite, 443	$KMg_3(AlSi_3O_{10})(OH)_2$	Mon	2.86	$2\frac{1}{2}$–3	1.50	Brown mica
Phosgenite, 333	$Pb_2Cl_2CO_3$	Tet	6.0–6.3	3	1.56–1.64	Fusible at 1
Phosphorite, 360					2.12	Phosphate rock
Picotite, 294						See spinel
Piedmontite, 394	Mn'' epidote	Mon	3.4	$6\frac{1}{2}$	1.75–1.81	Reddish brown
Pinite, 442						See muscovite
Pitchblend, 291						See uraninite
Plagioclase, 466	$Ab_{100}An_0$—Ab_0An_{100}	Tric	2.62–2.76	6	1.53–1.59	A feldspar series
Plagionite, 275	$Pb_5Sb_8S_{17}$	Mon	5.56	$2\frac{1}{2}$		—
Platinum, 230	Pt	Iso	14–19	4–$4\frac{1}{2}$		As grains in placers
Pleonaste, 295						See spinel
Polianite, 287	MnO_2	Tet	5.0	6–$6\frac{1}{2}$		Steel-gray
Pollucite, 471	$Cs_4Al_4Si_9O_{26}\cdot H_2O$	Iso	2.9	$6\frac{1}{2}$	1.52	Colorless
Polybasite, 272	$Ag_{16}Ab_2S_{11}$	Mon	6.0–6.2	2–3		Pseudorhombohedral
Polyhalite, 310	$K_2Ca_2Mg(SO_4)_4\cdot2H_2O$	Tric	2.78	$2\frac{1}{2}$–3	1.56	Bitter taste
Potash feldspar, 464						See orthoclase
Powellite, 355	$CaMoO_4$	Tet	4.23	$3\frac{1}{2}$–4	—	Fluoresces
Prase, 455						See quartz
Prehnite, 397	$Ca_2Al_2(Si_3O_{10})(OH)_2$	Orth	2.8–2.95	6–$6\frac{1}{2}$	1.63	Tabular crystals
Prochlorite, 445					1.60	See chlorite
Proustite, 271	Ag_3AsS_3	Rho	5.55	2–$2\frac{1}{2}$		Light ruby silver
Pseudoleucite, 471						See leucite
Pseudowollastonite, 416						See wollastonite
Psilomelane, 303	$(Ba,H_2O)_2Mn_5O_{10}$	Orth	3.7–4.7	5–6	—	Botryoidal

Mineral	Formula	System	G	H	R.I.	Remarks
Pyrargyrite, 271	Ag_3SbS_3	Rho	5.85	$2\frac{1}{2}$	—	Dark ruby silver
Pyrite, 262	FeS_2	Iso	5.02	$6-6\frac{1}{2}$	—	Crystals striated
Pyrochlore, 301	$(Na,Ca)_2(Nb,Ti)(O,F)_7$?	Iso	$4.3\pm$	5	—	Infusible
Pyrolusite, 287	MnO_2	Tet	4.75	1-2	—	Sooty
Pyromorphite, 360	$Pb_5Cl(PO_4)_3$	Hex	6.5-7.1	$3\frac{1}{2}-4$	2.06	Adamantine luster
Pyrope, 379	$Mg_3Al_2(SiO_4)_3$	Iso	3.51	7	1.71	A garnet
Pyrophyllite, 439	$Al_2(Si_4O_{10})(OH)_2$	Mon	2.8-2.9	1-2	1.59	Smooth feel
Pyroxene, 408						A mineral group
Pyrrhotite, 254	$Fe_{1-x}S$	Hex	4.58-4.65	4	—	Magnetic
Quartz, 451	SiO_2	Rho	2.65	7	1.54	No cleavage
Ramsayite, 390	$Na_2Ti_2Si_2O_9$	Orth	3.43	6	2.01	From Kola
Realgar, 258	AsS	Mon	3.48	$1\frac{1}{2}-2$	—	Cl {010}. Red
Red ocher, 282						See hematite
Rhodochrosite, 324	$MnCO_3$	Rho	3.45-3.6	$3\frac{1}{2}-4\frac{1}{2}$	1.82	Cl {1011}. Pink
Rhodolite, 379	$3(Mg,Fe)O\cdot Al_2O_3\cdot 3SiO_2$	Iso	3.84	7	1.76	See garnet
Rhodonite, 417	$Mn(SiO_3)$	Tric	3.58-3.70	$5\frac{1}{2}-6$	1.73	Pink
Riebeckite, 422	$Na_3Fe_3{''}Fe_2{'''}(Si_8O_{22})(OH)_2$	Mon	3.44	4	1.66-1.71	An amphibole
Rock crystal, 452						See quartz
Rock salt, 308						See halite
Roscoelite, 362	$K_2V_4Al_2Si_6O_{20}(OH)_4$	Mon	2.97	$2\frac{1}{2}$	1.69	Vanadium mica
Rubellite, 404						See tourmaline
Ruby, 279						Red gem corundum
Ruby copper, 276						See cuprite
Ruby silver, 271						See pyrargyrite and proustite
Rutile, 285	TiO_2	Tet	4.18-4.25	$6-6\frac{1}{2}$	2.61	Adamantine luster
Saltpeter, 336						See niter
Sanidine, 463						See orthoclase
Saponite, 436	$(Mg,Al)_6(Si,Al)_8O_{20}(OH)_4$	Mon	2.5	$1-1\frac{1}{2}$	1.52	Clay mineral
Sapphire, 279						Blue gem corundum
Sard, 454						Brown chalcedony
Sardonyx, 454						Onyx with sard
Satin spar, 348						Fibrous gypsum

Name, page	Composition	XI. Sys.	G.	H.	n	Remarks
Scapolite, 474	Various	Tet	2.65–2.74	5–6	1.55–1.60	Cl {010}{110}
Scheelite, 354	$CaWO_4$	Tet	5.9–6.1	$4\frac{1}{2}$–5	1.92	Cl {011}
Schorlite, 403						See tourmaline
Scolecite, 478, 481	$Ca(Al_2Si_3O_{10})\cdot 3H_2O$	Mon	2.2±	5–$5\frac{1}{2}$	1.52	A zeolite
Scorzalite, 364	$(Fe,Mg)Al_2(OH)_2(PO_4)_2$	Mon	3.35	$5\frac{1}{2}$–6	1.67	—
Selenite, 348						See gypsum
Semseyite, 275	$Pb_9Sb_8S_4$	Mon	5.8	$2\frac{1}{2}$	—	—
Sepiolite, 438	$Mg_4(Si_6O_{15})(OH)_2\cdot 6H_2O$	Mon?	2.0	2–$2\frac{1}{2}$	1.52	Meerschaum
Sericite, 442						Fine-grained muscovite
Serpentine, 437	$Mg_6(Si_4O_{10})(OH)_8$	Mon	2.2–2.6	2–5	1.55	Green to yellow
Siderite, 323	$FeCO_3$	Rho	3.83–3.88	$3\frac{1}{2}$–4	1.88	Cl {10Ī1}
Sillimanite, 384	Al_2SiO_5	Orth	3.23	6–7	1.66	Cl {010} perfect
Silver, 227	Ag	Iso	10.5	$2\frac{1}{2}$–3	—	White, malleable
Silver glance, 244						See argentite
Skutterudite, 270	$(Co,Ni,Fe)As_3$	Iso	6.5 ± 0.4	5	—	Tin white
Smaltite, 270						See skutterudite
Smithsonite, 324	$ZnCO_3$	Rho	4.35–4.40	5	1.85	Reniform
Soapstone, 439						See talc
Sodalite, 472	$Na_8(AlSiO_4)_6Cl_2$	Iso	2.15–2.3	$5\frac{1}{2}$–6	1.48	Usually blue
Sodamicrocline, 465						See microcline
Soda niter, 336	$NaNO_3$	Rho	2.29	1–2	1.59	Cooling taste
Specular iron, 281						See hematite
Sperrylite, 231	$PtAs_2$	Iso	10.50	6–7	—	See platinum
Spessartite, 379	$Mn_3Al_2(SiO_4)_3$	Iso	4.18	7	1.80	A garnet
Sphalerite, 249	ZnS	Iso	3.9–4.1	$3\frac{1}{2}$–4.1	—	Cl {110} 6 directions
Sphene, 389	$CaTiO(SiO_4)$	Mon	3.40–3.55	5–$5\frac{1}{2}$	1.91	Wedge-shaped xls
Spinel, 293	$MgAl_2O_4$	Iso	3.6–4.0	8	1.72	In octahedrons
Spodumene, 413	$LiAl(Si_2O_6)$	Mon	3.15–3.20	$6\frac{1}{2}$–7	1.67	Cl {110}. Part {100}
Staurolite, 386	$Fe_2Al_9O_6(SiO_4)_4(O,OH)_2$	Orth	3.65–3.75	7–$7\frac{1}{2}$	1.75	In cruciform twins
Steatite, 439						See talc
Stephanite, 272	Ag_5SbS_4	Orth	6.2–6.3	2–$2\frac{1}{2}$	—	Pseudohexagonal